PHYSICS RESEARCH AND TECHNOLOGY

SCALAR STRONG INTERACTION HADRON THEORY

PHYSICS RESEARCH AND TECHNOLOGY

Additional books in this series can be found on Nova's website
under the Series tab.

Additional E-books in this series can be found on Nova's website
under the E-books tab.

PHYSICS RESEARCH AND TECHNOLOGY

SCALAR STRONG INTERACTION HADRON THEORY

FANG CHAO HOH

nova publishers
New York

For permission to use material from this book please contact us:
Telephone 631-231-7269; Fax 631-231-8175
Web Site: http://www.novapublishers.com

NOTICE TO THE READER

LIBRARY OF CONGRESS CATALOGING-IN-PUBLICATION DATA

Hoh, Fang Chao.
Scalar strong interaction hadron theory / Fang Chao Hoh.
p. cm.
Includes index.
ISBN 978-1-61122-039-1 (hardcover)
1. Hadron interactions. 2. Hadrons--Decay. 3. Wave equation. 4. Scalar field theory. I. Title.
QC793.5.H328H64 2010
539.7'216--dc22
2010036162

Published by Nova Science Publishers, Inc. ✝ New York

IN MEMORY OF MY FATHER

HOH CHIH-HSIANG(郝 志 翔)

AND MY MOTHER

HWANG CHIA TSENG (黄 家 珍)

THEY PROVIDED THE INSPIRATIONS UNDERLYING
THE PRESENT THEORY

CONTENTS

PREFACE

This is not a finished research monograph, but gives an account of the present state of a theory supposedly still at its early stage of development. The chapters herein have been revised and updated many times over the years. This work will hopefully facilitate researchers interested in entering into this field and serve as a basis for future development of this theory.

The scalar strong interaction hadron theory is a first principles' (Lorentz and Gauge invariant) and nonlocal theory at quantum mechanical level that provides an alternative to low energy QCD and Higgs related part of the standard model. The quark-quark interaction is scalar rather than color-vectorial. A set of equations of motion for mesons and another set for baryons have been constructed. These play the roles like those played by the Schroedinger-Dirac equations in atomic physics. Work consists of solving these equations of motion to account for data, analogous to solving the Schroedinger equation for atomic problems. The theory has been rather successful in accounting for many basic low energy hadronic data(spectra, decays) not derivable from QCD. Physics takes place in the relative space between the quarks and the relative time among the quarks plays essential roles. The theory gives rise to new concepts such as confinement, gauge boson mass generation without Higgs, unified electro-strong coupling hypothesis, nonconservation of angular momentum in free neutron decay, etc.

The present equations of motion were first proposed in 1993-4. Since then, they has been rather successfully anchored on a large number of hadronic data points in a dozen or so papers. There are however "infant diseases" in these papers in form of insufficient explanations, gaps in mathematical derivations, errors, inconsistent formalisms and notations, etc. The chronological development has not been systematic.

The purpose of this book is to address to these problems and present the scalar strong interaction hadron theory in a more coherent and systematic manner. Considerable amount of new material that successfully enlarge the contact surface with data has been incorporated. The background of this theory and how the theory has been received by the particle physics community are given in Appendices G and H.

Inasmuch as this book has not been read by any colleague, it will contain errors. If any reader finds such or have questions or comments, I'd appreciate it if he will let me know.

My wife Birgitta has put up a great deal throughout the decades during which the ground work underlying the present theory was laid down and has also typed the text.

July 2010, Uppsala, Sweden. Fang Chao Hoh(郝方照)

Chapter 1

INTRODUCTION

1.1. BACKGROUND

A detailed background of this Chapter is given in Appendix G.

The background for proposing and developing the scalar strong interaction hadron theory is the fact that Quantum Chromodynamics (QCD) [P1, P2], the current main stream theory for strong interactions, does not work at low energies. At high energies, QCD is perturbative (pQCD) and appears to work well, even if some important experimental verifications refer to form factors and structural functions, not deducible from first principles.

At low energies, QCD is nonperturbative and, without further modelling, has been unable to provide predictions on almost all of the data provided by the Particle Data Group [P1], including the fundamental property confinement. Also, the Higgs boson needed in the associated standard model has not been found after over four decades of search, including the latest experiments in the new LHC at CERN.

There is no conclusive evidence of gluon predicted by QCD; the three jet experiments can be account for by the presence of diquark instead of gluon. The predicted glueball has not been found. The evidence behind and reasoning for this standpoint have been given in [H1, H2], reproduced in Appendices D to G. Some complementary remarks are given in Part I of [H18].

That pQCD works well at higher energies does not mean that it is unique; other theories that also possess this property can emerge. At the end of Sec. 12.8, it is noted that degrees of freedom corresponding to color are present in the theory to be developed below. The quarks are also "asymtotically free" at high energies, although no wrok has been done yet in this energy region.

pQCD may be compared to classical mechanics(CM) in that both hold at higher energies. At low energies, CM breaks down and quantum mechanics(QM) steps in. Similarly, pQCD must be replaced by a new low energy theory containig ingredients not present in QCD. Just like that QM goes over CM at high energies or momenta, one must first have this new low energy theory, which goes over to some form equivalent to pQCD at high energies. Just like that one cannot start from CM and go down in energy to recover QM, one cannot start from pQCD and go down in energy to recover this new low energy theory.

Therefore, a first principles' hadron theory to provide an alternative to or eventually a replacement of QCD, initially at low energies, is looked for.

1.2. On Construction of New Physical Theories

Construction of modern physical theories follows the teachings of logical positivism, propounded by R. Carnap and his associates of the "Vienna circle" in the 1920's. This branch of epistemology downplays the significance of physical concepts and principles and is rather pragmatic. In practice, any self-consistent mathematical construction that leads to predictions which systematically account for data with sufficient coverage is a good theory. This approach is to considerable degree guided by mathematics.

In all known basic physics disciplines, classical mechanics, classical electrodynamics, general relativity, quantum mechanics, quantum electrodynamics, fluid mechanics, etc, the starting point is a set of partial differential equations which mostly can be converted into action integrals. The low energy, zeroth order or ground state solutions for simpler configurations are easily obtained and are simple. Complications arise when complicated geometries or higer order effects are considered.

QCD however disregards this historical lesson and starts from an action integral that cannot be converted into differential equations to be solved. Hence, the above-mentioned zeroth order solutions cannot be found. This is the root cause for all the difficulties encountered by nonperturbative QCD. From this point of view, QCD should be replaced by a theory that starts with a set of partial differential equations.

1.3. Reformulation of Old Theories and Generalization

The history of the development of physical theories has shown that an older theory must be cast into a suitable form for it to be generalized to a new theory, valid in a new domain. The original forms of the laws of Coulomb, Ampère and Faraday are in such a crude form that they cannot be generalized to Maxwell's equations. Only after reformulating them into suitable differential forms could Maxwell recognize their inconsistency and propose his displacement current and his set of equations [J2]. The original equations of classical mechanics written down by Newton are also not suitable for generalization to quantum mechanics. First after recasting these equations into the Hamilton-Jacobi form, which is analogous to that of the geometrical optics limit of a wave motion [G1], could Schrödinger [S2] generalize it into his equation. The Schrödinger equation was not generalized to Dirac's equation directly. Only after its generalization to the Klein-Gordon equation could Dirac's equation be constructed [D1].

After having casted the older theories into suitable forms, they are to be generalized. The generalization is basically heuristic and intuitive. It consists of one or several leaps, equivalent to postulates, hypotheses or assumptions, from some suitable starting platform anchored in the associated old theory to be generalized. In the case of Maxwell's equations, the leap is the introduction of the displacement current. For the Schrödinger equation, the leaps are the introduction of the Planck constant and the replacement of geometrical optics form by wave optics form. In special relativity, the leap is postulating the constancy of the speed of light. In Einstein's equations, the main leap is the acceptance of the equivalence principle.

These leaps are not derivable from any older theory, just like that statistical mechanics cannot be derived from thermodynamics. There may, however, exist different sets of such leaps that lead to the same final mathematical construction, here a set of partial differential equations. Therefore, these leaps are generally not unique. After having found the new set of differential equations, this set by itself can be taken as the basic postulate, to be justified only by its ability to account for data. Differential equations were arrived at first, from which integral forms or Lagrangians are subsequently obtained.

New mathematical entities arising from the new set of differential equations are interpreted as or associated with some new physical concepts (words) later. This is in line with the Kantian thesis that "the thing-itself is unknowable". The consequences of this new set of equations together with the new physical concepts coming from it lead to new physical insights, in agreement with what V.F. Weisskopf once wrote: "physics comes out of mathematics". Thus, the concept that electromagnetic waves can propagate without any medium comes out of Maxwell's theory. Time dilitation emerges from special relativity. Curved space-time is a result from general relativity. Quantum mechanical tunneling arises in the Schrödinger theory. In Sec. 4.5 and 10.2 below, it is shown that confinement is a consequence of the theory developed below.

These concepts could hardly be imagined before the emergences of the associated theories. They may be compared to the simple concepts of time and angle in Kepler's time and temperature and pressure in the time of Boyle. These verify Einstein's observation that the distance between experimentally measured results and physical concepts capable to account for them grows with time and will continue to grow in the future.

The epistemological and historical implications for elementary particle theories are extensively considered in Appendix G.

1.4. Higher Spin Wave Equations for Point Particle

Dirac looked for a wave equation linear in the first derivative operator and obtained for a free, spin ½ particle (A3) and (A5a) or

$$\left(-i\gamma^{\mu}\partial_{X\mu}+m\right)\psi(X)=0 \tag{1.4.1}$$

The great success of (1.4.1) has apparently greatly influenced physicists who looked for wave equations representing particles with spin > ½. In the three decades following Dirac's proposal, most of the proposed equations are linear in the first derivative and are of or can be rewritten in the form (1.4.1) with more components in the wave function ψ. Strictly speaking, only one equation for a spin 1 particle, the Kemmer equation [K1, R1], has been found. It has the form of (1.4.1) with ψ containing 10 components and the coresponding γ's are 10×10 matrices. The algebra of these matrices is fairly complicated and leads to rather cumbersome handling. For spin >1, further complications arise [H3]. Therefore, this line of generalization is not aesthetically appealing.

The original form of Dirac's equation is suitable for going to the nonrelativistic limit, in which ψ_3 and ψ_4 are the small and ψ_1 and ψ_2 are the large components representing spins up and down, respectively. This form may therefore not be convenient for possible application to quarks which are relativistic. In (1.4.1), $\partial_{X\mu}$ is a Lorentz vector and m is a Lorentz scalar but the wave function ψ does not transform as any basis vector generating an irreducible representation of the Lorentz group. Therefore, (1.4.1) is not manifestly covariant with respect to Lorentz transformations. This is in contrast to the manifestly covariant forms of classical mechanics and Maxwell's equations, exemplified by (B22). Lorentz invariance automatically accompanies such forms.

Such a manifestly covariant form of Dirac's equation has been given by van der Wareden [W1] in terms of the so-called spinors. These spinors are right and left handed and hence are suitable for representing the generally relativistic quarks. This form of Dirac's equation is that of (C10). It has been generalized by Weinberg [W3, J4, S12] to represent a free particle having spin s ≥ ½;

$$\begin{aligned}&\partial_X^{a(1)\dot{b}(1)}.....\partial_X^{a(2s)\dot{b}(2s)}\chi_{\dot{b}(1).....\dot{b}(2s)}(X)=(im)^{2s}\psi^{a(1).....a(2s)}(X)\\&\partial_{X\dot{c}(1)a(1)}.....\partial_{X\dot{c}(2s)a(2s)}\psi^{a(1).....a(2s)}(X)=(im)^{2s}\chi_{\dot{c}(1).....\dot{c}(2s)}(X)\end{aligned} \tag{1.4.2}$$

Unlike the Kemmer equation, Weinberg's equations are of higher order in the derivatives, more elegant and not limited to spin 1. The Kemmer and Weinberg equations have however not found any application. The only massive point particles observed so far have spin ½ (leptons…) or spin 1 (gauge bosons). Wave functions of the gauge bosons do not have 10 components in general and are therefore not describable by the Kemmer equation. They turn out to possess one dotted index and one undotted index (see §7.1.1, §7.1.3 and §7.4.2 below) and can hence not be represented by those in (1.4.2).

1.5. Higher Spin Wave Equations for Composite Particle, the BS Equation

Apart from point particles of spin ½ and the gauge field bosons, all other observed particles are composite consisting of two or more point particles. Consider the two fermion case, e.g. the positronium or a meson. The Bethe-Salpeter (BS) equation governing the former can be put in the form [S3]

$$\left(i\gamma_I^{\mu}\partial_{I\,X\mu}-m_I\right)\left(i\gamma_{II}^{\nu}\partial_{II\,X\nu}-m_{II}\right)\psi_{BS}(x_I,x_{II})=\text{interaction terms} \tag{1.5.1}$$

where I and II differentiate the both fermions. The left side can be construed as the product of the left sides of two Dirac equations of the original form (1.4.1) with the product wave functions in x_I and x_{II} generalized to a nonseparable $\psi_{BS}(x_I,x_{II})$. The right side is of the order g^2, the coupling constant of the interaction. This equation has produced highly successful numerical results in the nonrelativistic and perturbational or low energy and small g region.

In spite of this success, the BS equation is beset by a number of unresolved and basic theoretical issues, as were raised by Wick [W4] shortly after the proposal of this equation. Some of these are i) the physical role of the relative time and the boundary conditions of the

BS wave functions at large relative times, ii) the lack of a positive definite norm for the wave functions and iii) the presence of strong singularities in the interaction kernel.

Prescriptions for calculations in the BS frame work turn out to be very complicated [H4]. For relativistic constituents and strong coupling or large g, as in the case for hadrons, only numerical solutions are possible. To date, such results are insufficient and unreliable in accounting for data.

The BS formalism does not contain a mechanism for quark confinement. It is also basically incompatible with QCD. Under these circumstances, QCD oriented, phenomenological potentials including a confining part have for instance [M1] been introduced into simplified BS equations obtained by various approximations. In this manner, predictions on hadron properties have been made and compare favorably with data. Nevertheless, these results are phenomenological and not obtainable from a first principles' theory.

Although the BS equation works well for positronium, it does not have to, indeed, should not be applicable to hadrons. This is because that the constituents in the positronium, the electron and the positron, can be separated off and each of them is observable and describable by Dirac's equation. The quarks in the hadrons can however not be separated from each other and be observed individually. This difference is so fundamental that these two systems are to be described differently [H1, H2].

1.6. Approach to Equations of Motion for Hadron

The above discussions show that the original form of the Dirac equation (1.4.1) is not suitable for generalization to equations of motion or wave equations, which will be used synonymously, for higher spin point particles (Sec. 1.4) as well as for composite particles such as hadrons (Sec. 1.5). Therefore, it is natural to look into the possibility of generalizing Dirac's equation in the manifestly Lorentz invariant spinor form (C10) into wave equations for hadrons. This aproach is supported by the aesthetically pleasing Weinberg equations (1.4.2) for a hypothetical point particle with spin s, which may be regarded as generalization of the spin ½ equations (C10) putting V and q there to zero.

Chapter 2

CONSTRUCTION OF EQUATIONS OF MOTION FOR MESON

The construction of a set of wave equations describing two quark mesons is guided by the principles of Sec. 1.5 and initially follows the basic steps of the related parts in [H5]. There are three major steps.

2.1. QUARK WAVE AND POTENTIAL EQUATIONS

The standard text [P2] for QCD is based upon the assumption that quarks are femions and obeys Pauli's principle. However, after the invention of QCD, it was found that quarks cannot exist individually, unlike fermions. Therefore, quarks are not fermions and hence need not obey Pauli's principle. Color is no longer necessary and quarks can interact via scalar force analogous to nuclear force.

Based upon these considerations, the *first step* is to assume that the quark and antiquark in a two quark meson interact via a *massless scalar* or *pseudoscalar* interaction. Vector forces among quarks are reserved for the U(1) electromagnetic and the SU(2) electro-weak interactions. Let the subscripts A and B refer to the quark at x_I and the antiquark at x_{II}, respectively. For the quark, (C10) and (C12), dropping the pseudoscalar interaction V_{PB} and vector potential A, can be put in the form

$$\partial_I^{a\dot{b}} \chi_{A\dot{b}}(x_I) - iV_{SB}(x_I)\psi_A^a(x_I) = im_A \psi_A^a(x_I) \tag{2.1.1a}$$

$$\partial_{I\dot{b}c} \psi_A^c(x_I) - iV_{SB}(x_I)\chi_{A\dot{b}}(x_I) = im_A \chi_{A\dot{b}}(x_I) \tag{2.1.1b}$$

$$\Box_I V_{SB}(x_I) = \tfrac{1}{2} g_s^2 \left(\psi_B^b(x_I)\chi_{Bb}(x_I) + \psi_B^{\dot{b}}(x_I)\chi_{B\dot{b}}(x_I)\right) \tag{2.1. 2}$$

Here, ∂_I and ∂_{II} refer to differentiations with respect to x_I and x_{II}, respectively. g_s^2 is the scalar strong coupling constant for quark-antiquark or quark-quark interaction. The corresponding equations for the antiquark is basically found by taking the complex conjugate of (2.1.1, 2) and letting $I\to II$ and $A\leftrightarrow B$. With (B8) and (B7), they become

$$\partial_{II\dot{e}f}\chi_B^f(x_{II})-iV_{SA}(x_{II})\psi_{B\dot{e}}(x_{II})=im_B\psi_{B\dot{e}}(x_{II}) \tag{2.1.3a}$$

$$\partial_{II}^{d\dot{e}}\psi_{B\dot{e}}(x_{II})-iV_{SA}(x_{II})\chi_B^d(x_{II})=im_B\chi_B^d(x_{II}) \tag{2.1.3b}$$

$$\Box_{II}V_{SA}(x_{II})=\tfrac{1}{2}g_s^2\left(\psi_A^b(x_{II})\chi_{Ab}(x_{II})+\psi_A^{\dot{b}}(x_{II})\chi_{A\dot{b}}(x_{II})\right) \tag{2.1.4}$$

The terms in (2.1.1, 3) are grouped such that the left sides only contain operators in space time. Their right sides contain the quark masses which will be generalized to quark mass operators pertaining to a different, internal or flavor space in §2.3.3 below. The quark masses m_A and m_B will turn out to be separation constants between functions in space time and functions in internal space, as is illustrated in (2.3.15) below.

Performing the interchange $\psi\leftrightarrow\chi$ in (2.1.3) leads to the form (2.1.1) and vice versa. This corresponds to the charge conjugation transformation $\psi\to\psi_c=-i\gamma^2\psi^*$ of the Dirac equation (A3) together with (A5), excluding the electromagnetic interaction (A5d). Invariance of this equation under such a transformation implies that a quark and an antiquark not interacting electromagnetically cannot be distinguished from each other by their space time wave functions alone; they are equivalent in space time. From this viewpoint, the purely space time strong interaction between two quarks or two antiquarks or a quark and an antiquark should be the same. The scalar interaction (A5c) has been assumed. Its replacement by the pseudoscalar interaction (A5b) will lead to no different result here, as is mentioned at the end of Sec. 2.2 below.

2.2. Spinor Meson Wave Equations in Space Time

The construction of meson wave equations from the two-spinor quark and antiquark equations (2.1.1-4) has been discussed in Sec. 3 and 4 of [H5]. The BS equation (1.5.1) has been regarded to have been built up from two Dirac equations of the form (A3, A5). Analogously, the left and right sides of the quark wave equations (2.1.1) are multiplied into the left and right sides, respectively, of the antiquark wave equations (2.1.3). Similarly, the left and right sides of the potential equation (2.1.2) are consistently multiplied into the corresponding sides of (2.1.4).

In the five equations that result after the multiplications, place the operators to the left of the wave functions. The *second step* is to generalize the separable product wave functions for the quark A at x_I and the antiquark B at x_{II} into two-quark wave functions not separable in x_I and x_{II};

$$\chi_{A\dot{b}}(x_I)\chi_B^f(x_{II})\to\chi_{\dot{b}}^f(x_I,x_{II}),\qquad \psi_A^c(x_I)\psi_{B\dot{e}}(x_{II})\to\psi_{\dot{e}}^c(x_I,x_{II}) \tag{2.2.1.a}$$

$$\chi_b^{*\dot{f}}(x_I, x_{II}) = \left(\chi_{\dot{b}}^{f}(x_I, x_{II})\right)^*, \qquad \chi \rightarrow \psi \tag{2.2.1.b}$$

$$\chi_{A\dot{b}}(x_I)\psi_{B\dot{e}}(x_{II}) \rightarrow \chi_{\dot{b}\dot{e}}(x_I, x_{II}) \tag{2.2.2a}$$

$$\psi_A^c(x_I)\chi_B^f(x_{II}) \rightarrow \psi^{cf}(x_I, x_{II}) \tag{2.2.2b}$$

$$V_{SA}(x_{II})V_{SB}(x_I) \rightarrow \Phi_m(x_I, x_{II}) \tag{2.2.3}$$

The complex conjugates of the mixed spinors of second rank (2.2.1b) are still mixed spinors and behave in the same way under SL(2,C) transformations. Analogous to $X^{a\dot{b}}$ of (C1a), $\chi_{\dot{b}}^{f}$ has also four components decomposable into a singlet and triplet, which can be associated with the pseudoscalar and vector mesons, respectively. It will therefore be assigned to represent meson wave functions, together with $\psi_{\dot{e}}^{c}$.

On the other hand, the complex conjugate of the dotted spinor of second rank (2.2.2a) no longer transforms under SL(2,C) in the same way but transforms as an undotted spinor of second rank like (2.2.2b). Further, χ and ψ on the left of (2.2.2) can be transposed and this leads to that the right members of (2.2.2) are symmetric spinors of second rank each having three components only. They can therefore not represent the pseudoscalar meson. In view of the indistinguishability of a quark and an antiquark in the context mentioned at the end of Sec. 2.1, ψ^{cf} and $\chi_{\dot{b}\dot{e}}$ of (2.2.2) may be associated with a diquark or an antidiquark from the space time and transformation point of view. No diquark has, however, been observed. Further, this assignment will be incompatible with internal or flavor assignments to quarks in Sec. 2.3 below, where quark and antiquark are distinguishable, as in (2.3.11) ff below. Therefore, ψ^{cf} and $\chi_{\dot{b}\dot{e}}$ are put to zero here.

The quark wave functions ψ and χ multiplied by V_S in the product equations are not paired off according to (2.2.1) and are therefore also put to zero. The justification is that quarks are not observed and any quark wave function cannot exist in a set of meson wave equations. If quark wave functions were to be present, they can be used to obtain expectation values of an observable operator, in contradiction to the nonobservation of quarks.

With these interpretations, the five product equations reduce to three coupled meson wave equations,

$$\partial_I^{a\dot{b}}\partial_{II\dot{e}f}\chi_{\dot{b}}^{f}(x_I, x_{II}) + (m_A m_B - \Phi_m(x_I, x_{II}))\psi_{\dot{e}}^{a}(x_I, x_{II}) = 0 \tag{2.2.4a}$$

$$\partial_{I\dot{b}c}\partial_{II}^{d\dot{e}}\psi_{\dot{e}}^{c}(x_I, x_{II}) + (m_A m_B - \Phi_m(x_I, x_{II}))\chi_{\dot{b}}^{d}(x_I, x_{II}) = 0 \tag{2.2.4b}$$

$$\Box_I \Box_{II}\Phi_m(x_I, x_{II}) = -\frac{g_s^4}{4}\left(\psi^{b\dot{a}}(x_I, x_{II})\chi_{\dot{a}b}^{*}(x_I, x_{II}) + \psi^{*a\dot{b}}(x_I, x_{II})\chi_{\dot{b}a}(x_I, x_{II})\right) \tag{2.2.5}$$

Multiplication of (2.1.1) with (2.1.4) and of (2.1.2) with (2.1.3) give no further contribution for the same reason given above (2.2.4); the product terms conatin an odd power of the quark wave functions ψ or χ and one of them cannot be paired off according to (2.2.1).

Return to the first step of Sec. 2.1 and replace the scalar interaction V_S by the pseudoscalar interaction V_P in (C12a). Repeating the steps from Sec. 2.1 and on, one finds that (2.2.4) and (2.2.5) remain unaltered if V_S in (2.2.3) is replaced by V_P . As far as the meson wave equations are concerned, it is immaterial where the $\bar{q}q$ interaction is scalar or pseudoscalar; such interactions cannot be directly observed anyway (see also the discussion above Sec. 2.2).

The quark wave functions and equations (2.1.1-4) play in a sense the role of a scaffold, or a mold which form the meson wave equations (2.2.4) and (2.2.5), which is discarded afterwards. Equation (2.2.4) contains eight meson wave function components, half of 16, the number of the equivalent Bethe-Salpeter wave function components. This is due to that the diquarks represented by (2.2.2) have been discarded above in this section. The equations of motion for meson (2.2.4) are therefore far simpler than the full BS equation for mesons.

2.3. Mass Operators and Internal Space

2.3.1. Mass Operators

After the discovery of the Schrödinger and Klein-Gordon (KG) equations, it became obvious that these equations may be obtained formally by replacing the energy and momentum in the classical energy equation by operators operating on some wave function. Thus,

$$E \rightarrow i\hbar\partial / \partial X^0, \qquad \underline{p} \rightarrow -i\hbar\partial / \partial \underline{X} \tag{2.3.1}$$

are substituted into

$$E^2 - p^2 - m^2 = 0 \tag{2.3.2}$$

which is then multiplied by a wave function $\psi_{KG}(X^0, \underline{X})$ from the right to obtain the KG equation

$$\left(\partial^2 / \partial (X^0)^2 - \partial^2 / \partial (\underline{X})^2 + m^2\right)\psi_{KG}\left(X^0, \underline{X}\right) = 0 \tag{2.3.3}$$

Here, h slash $=h/2\pi$ is unity according to the top of Appendix A. The mass m of the particle however remains a constant input parameter. With the proposal of the broken global SU(3) internal symmetry of the early 1960's [G2, N1] for classification of the many then-known hadron masses, masses of hadrons were taken to be expectation values of some mass operator [O1]. The entities upon which this operator operates need not be explicitly specified as long as they transform as an SU(3) octet or decuplet. These multiplets can in their turn be constructed from triplets formed by the *u, d* and *s* quarks [G3, Z1].

In a more complete theory for hadrons, therefore, the constant mass, as m in the KG equation (2.3.3), should be replaced by a mass operator operating on some internal function ξ, just like the energy-momentum operators (2.3.1) operate on the wave function $\psi_{KG}(X^0, \underline{X})$ in (2.3.3). In this way, the mass term is put on equal footing with the energy and momentum terms in the classical relation (2.3.2).

The broken SU(3) scheme, also successful in relating the baryon masses, depends upon the relatively small differences of the *u, d* and *s* quark masses. The breaking of SU(3) symmetry is small and can be treated as perturbation. In the 1970's, the *c* and *b* quarks were discovered and in the 1990's, the *t* quark was identified. These three quarks are however much heavier than the *u, d* and *s* quarks. An extension of the broken SU(3) above to broken SU(4) or SU(5) to incorporate the *c* or *c* and *b* quarks turned out to be not useful in relating hadron masses, as expected.

To obtain a theory that treats all these quarks on the same footing, we have to abandon the successful, conventional broken SU(3) type of approach in which hadron masses were replaced by hadron mass operator. Instead, quark masses are to be replaced by quark mass operator. In this way, quark-quark scalar interaction Φ_m in (2.2.4) will contribute to the meson mass on equal footing as does the quark mass term $m_A m_B$ in (2.2.4). This potential bonding Φ_m in physical space is naturally the basis for strong forces among nucleons and is not obtainable if only meson or baryon mass operator, but not quark mass operator, is present in the formalism.

Since quark charges are, like quark masses, an internal property of quarks, they must also consistently be elevated to quark charge operator.

2.3.2. Internal or Flavor Space

In the SU(3) case, hadron mass operators operate on octets and decuplets, but a quark mass operator operates correspondingly on a quark triplet. Since there are six quarks, the triplet is enlarged to a sextet. Historically, the first internal multiplet is the isospin doublet, the proton and the neutron, proposed by Heisenberg in the 1930's. After the quark hypothesis of 1963, this doublet was also associated with the *u* and *d* quarks.

The concept of isospin is carried over from the conventional spin which in general is associated with two complex variables, such as ψ_b in (B1). The corresponding complex variables in the isospin case are chosen to be z^b [B1] here. Other authors have used other symbols [K2, S4]. A main difference is that the SL(2,C) group represented in (B1) for ψ_b is further restricted by a unitarity condition to become the SU(2) group when taken over to the z^b space. z^b provides a point field for implementing the SU(2) isospin transformations.

Since there are six quarks, z^b is enlarged to z^p, p=1,2,…6, associated with *u, d, s, c, b* and *t* quarks, respectively. The indices $p,q,\ldots t$ run from 1 to 6 and are called flavor indices. The complex conjugate of z^p,

$$z_p = \left(z^p\right)^* \tag{2.3.4}$$

is associated with antiquarks. These z's have been called internal coordinates [H5] and the space spanned by z^p and z_p can be called internal space or flavor space, which will be used synonymously. The internal or intrinsic properties of a quark, such as its mass and charge, are determined by its flavor index.

By analogy to z^b, global SU(6) transformations can be carried out in this enlarged flavor space. Although the SU(6) symmetry is badly broken in the internal space due to the large quark mass differences, it will turn out that rotations in the internal space do not affect the interquark wave functions in space time, as will be shown in Sec. 4.4 below. It is therefore of interest in classifying hadrons and predicting the possible existence of still unobserved states. Such a classification has been rather successfully carried out for SU(4) [L2 Ch. 10] despite that the associated symmetry breaks down completely by the large c quark mass.

The "length" R_f in the scalar product

$$z_p z^p = R_f^2 \to 1 \tag{2.3.5}$$

is an invariant quantity under U(6) tranformations. Repeated flavor indices $p,q...t,$ even if there are three of the same, are summed over from 1 to 6. Since this z space has no direct physical meaning, R_f can be conveniently set to 1 [B1, R2]. z^p can be regarded as a basis vector generating the first fundamental representation of SU(6). Similarly, z_p constitutes a basis vector generating the equivalent second fundamental representation (see [L2 §6.3]) of SU(6). The different quark flavors transform into each other under SU(6) transformations.

z^p can be parametrized in spherical coordinates. For the subgroups SU(2) and SU(3), these are [B1]

$$z^1 = \cos\vartheta_2 \exp(i\varphi_1), \qquad z^2 = \sin\vartheta_2 \exp(i\varphi_2) \tag{2.3.6a}$$

$$\begin{aligned} &z^1 = \sin\vartheta_3 \cos\vartheta_2 \exp(i\varphi_1), \qquad z^2 = \sin\vartheta_3 \sin\vartheta_2 \exp(i\varphi_2) \\ &z^3 = \cos\vartheta_3 \exp(i\varphi_3) \end{aligned} \tag{2.3.6b}$$

For U(q) [R2],

$$z^r = \sin\vartheta_q z^{q-1}, \qquad r = 1,2,....q-1 \tag{2.3.7}$$

$$z^q = \cos\vartheta_q \exp(i\varphi_q), \qquad 0 \le \vartheta_p \le \pi/2, \qquad 0 \le \varphi_p \le 2\pi \tag{2.3.8}$$

The normalized volume element in the z^q, z_q space is

$$dv_{zq} = \prod_{s=1}^{q} dz_s dz^s = N_q \prod_{r=2}^{q} \sin^{2r-3}\vartheta_r \cos\vartheta_r d\vartheta_r \prod_{t=1}^{q} d\varphi_t \tag{2.3.9}$$

where N_q is a normalization constant. The orthogonality condition is

$$\int dv_{zq} z_r z^p \Big/ \int dv_{zq} = \delta_r{}^p / q \tag{2.3.10}$$

2.3.3. Quark Mass and Charge Operators and Flavor Functions

The generalization mentioned at the end of §2.3.1 can now be carried out. The quark wave equation (C10) is multiplied by a quark internal function $\xi^p(z)$, the quark mass and charge are replaced by a quark mass operator $m_{1op}(z)$ and a quark charge operator $q_{op}(z)$, respectively. These operators operate on $\xi^p(z)$, which transforms under SU(6) as the basis vector z^p. If (C10) represents an antiquark, $\xi^p(z)$ is replaced by $\xi_p(z)$ transforming like z_p. The generalized (C10) for a quark with flavor p reads

$$\begin{aligned}
&\left(\partial_X^{a\dot{b}} + iq_{op}(z)A^{a\dot{b}}(X)\right)\chi_{\dot{b}}(X)\xi^p(z) = i\left(m_{1op}(z) + V_{SB}(X) - V_{PB}(X)\right)\psi^a(X)\xi^p(z) \\
&\left(\partial_{X\dot{b}c} + iq_{op}(z)A_{\dot{b}c}(X)\right)\psi^c(X)\xi^p(z) = i\left(m_{1op}(z) + V_{SB}(X) + V_{PB}(X)\right)\chi_{\dot{b}}(X)\xi^p(z)
\end{aligned} \tag{2.3.11}$$

where the gauge field $A(X)$ will be further considered in §6.3.1 below. The total quark wave functions are now $\chi_{\dot{b}}(X)\xi^p(z)$ and $\psi^a(X)\xi^p(z)$. For antiquarks, these are replaced by $\chi_b(X)\xi_p(z)$ and $\psi^{\dot{a}}(X)\xi_p(z)$, respectively. Although a quark and an antiquark are not distinguishable by their space time wave functions, as was pointed out at the end of Sec. 2.1, they can now be distinguished by the flavor functions $\xi^p(z)$ and $\xi_p(z)$.

The forms of the internal quantities $\xi^p(z)$, $m_{1op}(z)$ and $q_{op}(z)$ cannot be derived from some known starting point and must be suitably chosen. $m_{1op}(z)$ needs even not be specified at the quark level, since quarks are not observed. It is possible to leave the specification to the meson level (§2.3.5 below), built up from (2.3.11) and a corresponding set of equations for an antiquark.

The choice is guided by Einstein's epistemological thesis §A3-Ed) of Appendix G which was applied to general relativity in §3.2-7) there. It is essentially the simplest possible, nontrivial choice and reads

$$\xi^p(z) = z^p, \qquad\qquad \xi_p(z) = z_p \tag{2.3.12}$$

In order to reproduce (C10) from (2.3.11) or its complex conjugate, z^p and z_p must be eigenfunctions of $m_{1op}(z)$ and $q_{op}(z)$ which in their simplest forms are

$$m_{1op}(z) = m_q\left(z^q \partial/\partial z^q + z_q \partial/\partial z_q\right) \tag{2.3.13}$$

$$q_{op}(z) = q_r\left(z^r \partial/\partial z^r - z_r \partial/\partial z_r\right) = -q^*_{op}(z) \tag{2.3.14a}$$

$$q_1 = q_4 = q_6 = 2e/3, \qquad\qquad q_2 = q_3 = q_5 = -e/3 \tag{2.3.14b}$$

where $-e$ is the electron charge and m_q the mass of the quark with flavor q.

To be consistent, the space time quark wave functions on the right of (C12) for the potentials V in (2.3.11) should also be multiplied by the flavor functions (2.3.12). This generalization however does not alter (C12) due to (2.3.5).

2.3.4. Meson Wave Equations Including Internal Functions

Like the generalization of (C10) to (2.3.11) and its complex conjugate, (2.1.1, 3) are now analogously generalized. Multiply (2.1.1) by $\xi_A{}^P(z_I)$ and (2.1.3) by $\xi_{Br}(z_{II})$, where I and II are associated with quark A and antiquark B, respectively. Also, let the quark masses m_A and m_B therein be replaced by the quark mass operators $m_{1op}(z_I)$ and $m_{1op}(z_{II})$, respectively, in accordance with (2.3.13). These generalized quark equations are separable in x and z. For example, the so-generalized (2.1.1a) can be put in the form

$$-i\frac{\partial_I^{a\dot b}\chi_{A\dot b}(x_I)}{\psi_A^a(x_I)}-V_{SB}(x_I)=\frac{m_{1op}(z_I)\xi_A^P(z_I)}{\xi_A^P(z_I)}=m_A \tag{2.3.15}$$

where the quark mass m_A is the separation constant mentioned below (2.1.4) and (2.3.12) has been used. Note that this separation is removed by the gauge field term $q_{op}(z)A(x)$, which is of order e, in (2.3.11). This term couples e, a quantity pertaining with the internal space, to space time functions.

The so-generalized (2.1.1) and (2.1.3) are again multiplied together, and the second step in Sec. 2.2 is repeated. Introduce the simplified notations,

$$z = z_I, \qquad u = z_{II}, \qquad \partial z/\partial u = 0 \tag{2.3.16}$$

$$\partial_{zs} = \partial/\partial z^s,\ \ \partial_{us} = \partial/\partial u^s,\ \ \partial_z^{\ s} = \partial/\partial z_s = (\partial_{zs})^*,\ \ \partial_u^{\ s} = \partial/\partial u_s = (\partial_{us})^* \tag{2.3.17}$$

Generalizations in space time and in internal space are now put on equal footing. Generalizations of the products of internal expressions corresponding to (2.2.1) and (2.2.3) are

$$\xi_A^P(z)\xi_{Br}(u) = z^P u_r \leftrightarrow \xi_r^P(z,u) \tag{2.3.18a}$$

$$m_{1op}(z)m_{1op}(u) \rightarrow m_{2op}(z,u) \tag{2.3.18b}$$

where (2.3.18a) denotes the internal or flavor function of the meson. The internal product on the left sides of (2.3.18b) separable in z and u becomes now generally nonseparable quantities corresponding to the nonseparable $\Phi_m(x_I,x_{II})$ in (2.2.3). These generalizations to include internal coordinates including (2.3.18) make up the *third step*, which leads to the generalized form of (2.2.4);

$$\partial_I^{a\dot{b}} \partial_{II\dot{e}f} \chi_{\dot{b}}^f (x_I, x_{II}) \xi_r^p (z,u) + (m_{2op}(z,u) - \Phi_m(x_I, x_{II})) \psi_{\dot{e}}^a (x_I, x_{II}) \xi_r^p (z,u) = 0$$

(2.3.19a)

$$\partial_{I\dot{b}c} \partial_{II}^{d\dot{e}} \psi_{\dot{e}}^c (x_I, x_{II}) \xi_r^p (z,u) + (m_{2op}(z,u) - \Phi_m(x_I, x_{II})) \chi_{\dot{b}}^d (x_I, x_{II}) \xi_r^p (z,u) = 0$$

(2.3.19b)

The generalization to include internal function $\xi(z)$ in the form of (2.3.12) also does not affect (2.1.2) and (2.1.4), just like that (C12) is not altered by such an inclusion mentioned below (2.3.14). More generally, the potentials V_{SA} and V_{SB} appear on the left of (2.1.1, 3) and depend only upon the space time coordinates, as was shown in (2.3.15). They cannot depend upon any internal coordinate z or u. In this sense, the generalizations to include internal functions mentioned above (2.3.11) and of the type in (8.1.1, 2, 7, 8) below do not apply to the source functions of V_{SA} and V_{SB}. For these reasons, the potential equation (2.2.5) is analogously not affected by the above generalizations to include internal functions.

Equations (2.3.19) and (2.2.5) are the set of wave equations or equations of motion for two quark mesons, from which all results relating to two quark mesons in the remainder of this book are derived. This set of equations of motion for meson may themselves be considered as a basic postulate, independent of the three steps of assumptions given above in this chapter, of the *scalar strong interaction hadron theory* for two quark mesons. The sole justification of these equations lies in their ability to account for data [P1].

This theory is so named because the strong interaction among the quarks is scalar. It is built up from van der Waerden's two-spinors or elementary spinors each having two components, treated in Appendix B. These differ from the conventional Dirac wave function ψ of Appendix A, which has four components and has been called a four-spinor or bispinor. This Dirac ψ does not generate the fundamental representation of the SL(2,C) group.

Invariance of (2.3.19) and (2.2.5) under proper Lorentz transformations $L_p{}^{\mu}{}_{\nu}$ and space inversion are manifest due to their spinor form. This follows from such invariances of (2.1.1-4) by virtue of the statement below (C12), that x_I and x_{II} transform under the same $L_p{}^{\mu}{}_{\nu}$ and that the both sides of (2.2.1-3) transform in the same way. Application of (B31) to (2.2.1) leads to

$$\chi_{\dot{b}}^f (x_I^0, \underline{x}_I, x_{II}^0, \underline{x}_{II}) \rightarrow \psi_{\dot{f}}^b (x_I^0, -\underline{x}_I, x_{II}^0, -\underline{x}_{II}), \qquad \chi \leftrightarrow \psi \qquad (2.3.20)$$

under space inversion. The z's and u's in this set of meson wave equations are independent of x_I and x_{II} and are not affected by $L^{\mu}{}_{\nu}$.

Equation (2.3.19) is separable in the space time functions and internal functions if the meson internal function $\xi^p{}_r(z,u)$ is an eigenfunction to m_{2op};

$$m_{2op}(z,u) \xi_r^p (z,u) = M_m^2 \xi_r^p (z,u) \qquad (2.3.21)$$

$$\partial_I^{a\dot{b}} \partial_{II}^{f\dot{e}} \chi_{\dot{b}f} (x_I, x_{II}) - (M_m^2 - \Phi_m(x_I, x_{II})) \psi^{a\dot{e}} (x_I, x_{II}) = 0 \qquad (2.3.22a)$$

$$\partial_{I\dot{b}c}\partial_{II\dot{e}d}\psi^{c\dot{e}}(x_I,x_{II})-\left(M_m^2-\Phi_m(x_I,x_{II})\right)\chi_{\dot{b}d}(x_I,x_{II})=0 \tag{2.3.22b}$$

where some indices have been raised and lowered with the aid of (B7, B8, B16, B17). The eigenvalue M_m^2 is the separation constant, analogous to m_A in (2.3.15) at the quark level, and Φ_m is determined from (2.2.5);

$$\Box_I\Box_{II}\Phi_m(x_I,x_{II})=-\tfrac{1}{2}g_s^4\,\mathrm{Re}\,\psi^{b\dot{a}}(x_I,x_{II})\chi^*_{\dot{a}b}(x_I,x_{II}) \tag{2.3.23}$$

If there exists no eigenvalue, (2.3.22) can still be obtained. Multiplying (2.3.21) by the complex conjugate of the meson internal function and integrating over the z and u spaces leads to

$$M_m^2=\frac{\int dv_{zq}dv_{uq}\xi_p^r(z,u)m_{2op}(z,u)\xi_r^p(z,u)}{\int dv_{zq}dv_{uq}\xi_p^r(z,u)\xi_r^p(z,u)},\qquad \xi_p^r=\left(\xi_r^p\right)^* \tag{2.3.24}$$

when dv_{uq} is the same as dv_{zq} of (2.3.9) with u replacing z. This M_m^2 is a quasi-separation constant and not a unique, genuine separation constant like M_m^2 in (2.3.21) but both represent the quark mass contribution to the meson mass.

2.3.5. Internal or Flavor Functions and Mass Opersator

According to the remark above (2.3.12), the simplest form of (2.3.18b) is

$$m_{2op}(z,u)=m_{1op}^2(z,u) \tag{2.3.25}$$

where

$$m_{1op}(z,u)=\frac{1}{2}m_q\left(z^q\partial_{zq}+z_q\partial_z^{\ q}+u^q\partial_{uq}+u_q\partial_u^{\ q}\right) \tag{2.3.26}$$

is a simple generalization of (2.3.13) and is called quark mass summing operator in which the quark flavors q are summed over. The mass operator (2.3.26) is not separable in z and u, analogous to that $\Phi_m(x_I,x_{II})$ in (2.2.3) is not separable in x_I and x_{II}. The factor ½ has been introduced to uphold the conventional relation between the quark and meson masses (see (4.4.1) below).

The internal or flavor function $\xi^p{}_r(z,u)$ of (2.3.18a) is an eigenfunction of the mass operator (2.3.25) with the eigenvalue

$$M_m^2=\tfrac{1}{4}\left(m_p+m_r\right)^2 \tag{2.3.27}$$

according to (2.3.21). The forms (2.3.25) and (2.3.27) hold for two flavor mesons, such as kaons, but not for mesons like π^0 and η, which have closed flavor and contain four and six quark flavors, respectively. For such mesons, the eigenvalue form (2.3.21) no longer holds and may be replaced by (2.3.24), which also holds for nonseparable $\xi^p{}_r(z,u)$. However, this may lead to an approximate result only because (2.3.24), unlike (2.3.21), is not a unique solution to the original (2.3.19). Here, (2.3.18a) is decomposed into some irreducible SU(n≤6) multiplet characterized by e. g. hypercharge Y, isospin I, etc, as will be considered in §2.4.3 below.

The complex internal spaces z and u are considered as unobservable "hidden variables" on par with the unobservable relative space x^μ introduced in (3.1.3a) below.

2.4. Global SU(3) Transformations and Meson Wave Functions

2.4.1. Labelling the Meson Wave Functions

Considering the generalizations in the paragraph containing (2.3.18), it is seen that the subscript A in the quark wave functions in (2.1.1-4) can suitably be replaced by p to show the association with the flavor index of the quark. Similarly, the corresponding index B there is suitably replaced by r to show the association with the flavor index of the antiquark. With these replacements, the quantities on the right sides of (2.2.1) will be modified to include a subscript pr to indicate that the meson wave functions refer to a quark of flavor p and an antiquark with antiflavor $\bar{r}$. Similarly,

$$M_m^2 \rightarrow M_{mpr}^2 = \tfrac{1}{4}\left(m_p + m_r\right)^2 \qquad (2.4.1)$$

in (2.3.21, 22, 27). Using (2.3.18a) and rearranging some indices in the same way that led to (2.3.22), (2.3.19) becomes

$$\partial_I^{a\dot{b}}\partial_{II}^{f\dot{e}}\chi_{pr\dot{b}f}\left(x_I, x_{II}\right)z^p u_r - \left(m_{2op}(z,u) - \Phi_m\left(x_I, x_{II}\right)\right)\psi_{pr}^{a\dot{e}}\left(x_I, x_{II}\right)z^p u_r = 0 \qquad (2.4.2a)$$

$$\partial_{I\dot{b}c}\partial_{II\dot{e}d}\psi_{pr}^{c\dot{e}}\left(x_I, x_{II}\right)z^p u_r - \left(m_{2op}(z,u) - \Phi_m\left(x_I, x_{II}\right)\right)\chi_{pr\dot{b}d}\left(x_I, x_{II}\right)z^p u_r = 0 \qquad (2.4.2b)$$

Similarly, (2.3.22) becomes

$$\partial_I^{a\dot{b}}\partial_{II}^{f\dot{e}}\chi_{(pr)\dot{b}f}\left(x_I, x_{II}\right) - \left(M_{mpr}^2 - \Phi_m\left(x_I, x_{II}\right)\right)\psi_{(pr)}^{a\dot{e}}\left(x_I, x_{II}\right) = 0 \qquad (2.4.3a)$$

$$\partial_{I\dot{b}c}\partial_{II\dot{e}d}\psi_{(pr)}^{c\dot{e}}\left(x_I, x_{II}\right) - \left(M_{mpr}^2 - \Phi_m\left(x_I, x_{II}\right)\right)\chi_{(pr)\dot{b}d}\left(x_I, x_{II}\right) = 0 \qquad (2.4.3b)$$

The parentheses around pr indicate that the internal properties carried by $z^p u_r$ are now transferred to the wave functions ψ and χ themselves. The flavor indices p and r run from 1 to 6, as is in (2.3.4) (The repeated flavor indices are not summed over here).

The total meson wave functions $\psi^{a\dot{e}}_{pr}(x_I, x_{II})z^p u_r$ together with $\psi \to \chi$ in (2.4.2) completely describe a free meson consisting of a quark of flavor p and antiquark of antiflavor $\bar{r}$. Thus, p=1 and r=2 refers to π^+ or ρ^+, p=3 and r=1 to K^-or K^{*-}, p=2 and r=5 to B^0 or B^{*0}, p=r=3 to ϕ, p=r=4 to J/ψ, etc. The total meson wave functions as they stand in (2.4.2) involve only two flavors and can formally not describe π^0, η, η', η_c and ω which contain four or more flavors, as was pointed out beneath (2.3.27).

2.4.2. Broken Global SU(6) Symmetry and Octet Internal Functions

Consider the linear combination $a_{pr}z^p u_r$, where a_{pr} are constants, basically associated with SU(6) Clebsch-Gordan coefficients. Under global SU(6) transformations, it transforms like

$$a_{pr}z^p u_r \to a'_{pr}z'^p u'_r = U_6(z)a_{pr}z^p u_r U_6^{-1}(u) \tag{2.4.4a}$$

$$U_6(z) = \exp(iJ_l(z)\vartheta_l), \qquad U_6^{-1}(u) = \exp(-iJ_l(u)\vartheta_l) \tag{2.4.4b}$$

J_l are infinitesimal generators of the SU(6) group in a representation like that given for the SU(3) case in (2.4.5) below. ϑ_l corresponds to angles in a complex six dimensional space z^p or u_p and l runs from 1 to $6^2-1=35$.

If all quark masses were the same, $m_1=m_2 \ldots =m_6$, the eigenvalues (2.4.1) of m_{2op} in (2.4.2) degenerate. ψ_{pr} will therefore be the same for all p,r =1,2,...6 according to (2.4.3) so that only $z^p u_r$ differentiates the different mesons. Application of the SU(6) transformations (2.4.4) to (2.4.2) will therefore only affect $z^p u_r$ and, with suitable choices of ϑ_l, will take it into $z^q u_s$, which is special form of (2.4.4a). In the equal quark mass limit, (2.4.2) can be considered as form invariant under the global SU(6) transformations (2.4.4).

As the SU(6) symmetry is badly broken by the large differences between the quark masses according to Table 5.2 below; M^2_{mpr} in (2.4.1) and hence ψ_{pr} in (2.4.2) differ for different p, r values. The table however shows that m_1 is very close to m_2 which in its turn is not too different from m_3. Using these m_p values, (2.4.1) shows that M^2_{m11} and M^2_{m22} differ only by 0.65 % and M^2_{m12} differs from M^2_{m23} by about 12%. Thus, SU(2) symmetry is a very good one and even SU(3) symmetry holds approximately. Deviations from these symmetries are regarded as first order perturbations. This point will be further discussed at the end of Sec. 7.2 below.

In the limit of SU(3) symmetry, (2.4.4) holds with p,r=1,2,3 and l runs from 1 to 8. The generators are [K2, B1]

$$J_1(z) = \sqrt{6}(E_{+1}(z) + E_{-1}(z)), \qquad J_2(z) = -i\sqrt{6}(E_{+1}(z) - E_{-1}(z))$$
$$J_4(z) = \sqrt{6}(E_{+2}(z) + E_{-2}(z)), \qquad J_5(z) = -i\sqrt{6}(E_{+2}(z) - E_{-2}(z))$$

$$
\begin{aligned}
&J_6(z)=\sqrt{6}\left(E_{+3}(z)+E_{-3}(z)\right), && J_7(z)=-i\sqrt{6}\left(E_{+3}(z)-E_{-3}(z)\right)\\
&J_3(z)=2\sqrt{3}H_1(z), && J_8(z)=2\sqrt{3}H_2(z)\\
&J_l(u)=J_l(z\to u)
\end{aligned}
\tag{2.4.5}
$$

$$
\begin{aligned}
&E_{+1}(z)=\left(z^1\partial_{z2}-z_2\partial_z^{\,1}\right)/\sqrt{6}, && E_{+2}(z)=\left(z^1\partial_{z3}-z_3\partial_z^{\,1}\right)/\sqrt{6}\\
&E_{+3}(z)=\left(z^2\partial_{z3}-z_3\partial_z^{\,2}\right)/\sqrt{6}, && E_{-1}(z)=-E_{+1}^*(z)\\
&E_{-2}(z)=-E_{+2}^*(z), && E_{-3}(z)=-E_{+3}^*(z)\\
&H_1(z)=\left(z^1\partial_{z1}-z^2\partial_{z2}-c.c.\right)/2\sqrt{3}\\
&H_1(z)=\left(z^1\partial_{z1}+z^2\partial_{z2}-2z^3\partial_{z3}-c.c.\right)/6
\end{aligned}
\tag{2.4.6}
$$

where the notations of (2.3.17) have been used.

The commutation relations

$$
\left[J_k,\ J_l\right]=2if_{klm}J_m \tag{2.4.7}
$$

hold and f_{klm} are the structural constants for the SU(3) group [C3 §2.2, L2 §6.3]. In the limit of this symmetry, the internal functions $z^p u_r$ in (2.4.2) are conventionally decomposed into a U(3) singlet and an SU(3) octet by means of SU(3) Clebsch-Gordan coefficients,

$$
\begin{aligned}
&\pi^+:\ z^1u_2\to\left(z^1u_2+u^1z_2\right)/\sqrt{2}, \qquad \pi^-:\ z^2u_1\to\left(z^2u_1+u^2z_1\right)/\sqrt{2}\\
&\pi^0:\ \left(z^1u_1-z^2u_2\right)/\sqrt{2}\to\left(z^1u_1-z^2u_2+u^1z_1-u^2z_2\right)/2\\
&K^+:\ z^1u_3\to\left(z^1u_3+u^1z_3\right)/\sqrt{2}, \qquad K^-:\ z^3u_1\to\left(z^3u_1+u^3z_1\right)/\sqrt{2}\\
&K^0:\ z^2u_3\to\left(z^2u_3+u^2z_3\right)/\sqrt{2}, \qquad \overline{K}^0:\ z^3u_2\to\left(z^3u_2+u^3z_2\right)/\sqrt{2}\\
&\eta:\ \left(z^1u_1+z^2u_2-2z^3u_3\right)/\sqrt{6}\to\left(z^1u_1+z^2u_2-2z^3u_3+u^1z_1+u^2z_2-2u^3z_3\right)/\sqrt{12}
\end{aligned}
\tag{2.4.8}
$$

where the conventional pseudoscalar meson symbols precede the internal functions. Under suitable SU(3) transformations, these states transform into each other. The first three members of (2.4.8) transform into each other under the isospin subgroup $SU(2)_I$ of the SU(3) group of transformations. Note that linear combinations of $z^p u_p$ are required to represent π^0 and η. These are not separable in z and u and hence cannot be represented by (2.3.18a), unlike the other mesons in (2.4.8). The expressions after the arrows indicate that z and u are interchangeable because they cannot be detected [H5 (9.1a)]. In this way, for example, π^+ and π^- in (2.4.8) are complex conjugate of each other.

2.4.3. Wave Functions For π^0 And η

Since the SU(3) symmetry is broken to first order, ψ_{pr} in (2.4.2) is no longer the same for different p,r values. For $p{\neq}r$, (2.3.25-26) and (2.4.1-2) leads to that (2.4.3) holds for these open flavor mesons.

For π^0 and η, $p{=}r$ and the internal functions in (2.4.8) are not of the form (2.3.18a). The total meson wave functions in (2.4.2) are to be replaced by

$$\pi^0 : \left(\psi_{11}^{a\dot{e}}(x_I, x_{II})z^1u_1 - \psi_{22}^{a\dot{e}}(x_I, x_{II})z^2u_2\right)/\sqrt{2}, \qquad \psi \to \chi \tag{2.4.9a}$$

$$\eta: \left(\psi_{11}^{a\dot{e}}(x_I, x_{II})z^1u_1 + \psi_{22}^{a\dot{e}}(x_I, x_{II})z^2u_2 - 2\psi_{33}^{a\dot{e}}(x_I, x_{II})z^3u_3\right)/\sqrt{6}, \quad \psi \to \chi \tag{2.4.9b}$$

according to the linear combinations in (2.4.8). Equations (2.4.9) are not separable into a space time part and an internal part. The internal part cannot be cancelled out in the so-modified (2.4.2) and (2.4.3) does not hold for these two mesons. These mesons have to be treated approximately.

The rest frame solutions for (2.3.19, 21) and (2.2.5) to be given in Chapter 3 differ only in their mass E_0, defined in (3.1.1, 3, 5, 6) below. For different p,r pairs, M^2_{mpr} are different and only the factor

$$\exp\left(-iE_0X^0\right) = \exp\left(-iE_{0pr}X^0\right) \tag{2.4.10}$$

entering ψ_{pr} and χ_{pr} is different. According to the E_{0pp} values for pseudoscalar mesons given in the legend to Table 5.5 below,

$$E_{011} \approx< E_{022} << E_{033} \tag{2.4.11}$$

Thus, one may approximately put

$$\psi_{11}(x_I, x_{II}) \approx \psi_{22}(x_I, x_{II}) \propto \psi_{\pi 0}(x_I, x_{II}) \tag{2.4.12}$$

The expression (2.4.9a) can now be separated into a space time part $\psi_{\pi o}$ and an internal part given by the π^0 member of (2.4.8). This separated form can now be substituted into (2.4.2). Multiply the resulting equations by the complex conjugate of the π^0 internal function in (2.4.8) and integrate over the z and u spaces. These operations leads to (2.4.3) with

$$\psi_{(pr)} \to \psi_{(\pi 0)}, \qquad \chi_{(pr)} \to \chi_{(\pi 0)}, \qquad M^2_{mpr} \to M^2_{m\pi 0} = \tfrac{1}{2}\left(m_1^2 + m_2^2\right) \tag{2.4.13}$$

where (2.3.24) with $\xi^p{}_r$ replaced by the π^0 member in (2.4.8) and (2.3.26) have been employed.

The right inequality of (2.4.11) prevents an extension of (2.4.12) to include ψ_{33}. Note that this inequality does not invalidate the approximate SU(3) symmetry mentioned above (2.4.5) (It is the result of the particular values of the constants d_m and d_{m0} entering Φ_m in (2.4.2) shown in (3.2.8a), (5.2.3) and Table 5.2 below). Unlike the π^0 case, the η wave function of (2.4.9b) is thus not separable and purely space time wave equation of the type (2.4.3) cannot be obtained for η. To obtain some estimate, the condition

$$\psi_{11} = \psi_{22} = \psi_{33} \propto \psi_\eta(x_I, x_{II}) \qquad (2.4.14)$$

albeit inappropriate, is nevertheless imposed. Repeating the steps intervening (2.4.12) and (2.4.13) leads to (2.4.3) with

$$\psi_{(pr)} \to \psi_{(\eta)}, \qquad \chi_{(pr)} \to \chi_{(\eta)}, \quad M^2_{mpr} \to M^2_{m\eta} = \tfrac{1}{6}\left(m_1^2 + m_2^2 + 4m_3^2\right) \qquad (2.4.15)$$

This treatment may analogously be extended to η' with the internal function

$$\eta': \quad \left(z^1u_1 + z^2u_2 + z^3u_3\right)/\sqrt{3} \to \left(z^1u_1 + z^2u_2 + z^3u_3 + u^1z_1 + u^2z_2 + u^3z_3\right)/\sqrt{6} \qquad (2.4.16)$$

In this case, $\eta \to \eta'$ in (2.4.14) and (2.4.15) is replaced by

$$\psi_{(pr)} \to \psi_{(\eta')}, \qquad \chi_{(pr)} \to \chi_{(\eta')}, \qquad M^2_{mpr} \to M^2_{m\eta'} = \tfrac{1}{3}\left(m_1^2 + m_2^2 + m_3^2\right) \qquad (2.4.17)$$

2.4.4. Octet Meson Wave Functions in Space Time

As was mentioned beneath (2.4.3), the internal properties of a meson carried by the internal functions $z^p u_r$ in (2.4.2) have been transferred to the subscripts (pr) in the meson equations in space time (2.4.3). The internal properties of the meson can now suitably be characterized by arranging $\psi_{(pr)}$ into a matrix. With (2.4.8, 9, 12-15), one can write

$$\psi_{(pr)} \to \begin{pmatrix} \psi_{(11)} & \psi_{(12)} & \psi_{(13)} \\ \psi_{(21)} & \psi_{(22)} & \psi_{(23)} \\ \psi_{(31)} & \psi_{(32)} & \psi_{(33)} \end{pmatrix} - \frac{1}{3}\left(\psi_{(11)} + \psi_{(22)} + \psi_{(33)}\right)$$

$$= \begin{pmatrix} \frac{1}{\sqrt{2}}\psi_{(\pi 0)} + \frac{1}{\sqrt{6}}\psi_{(\eta)} & \psi_{(\pi+)} & \psi_{(K+)} \\ \psi_{(\pi-)} & -\frac{1}{\sqrt{2}}\psi_{(\pi 0)} + \frac{1}{\sqrt{6}}\psi_{(\eta)} & \psi_{(K0)} \\ \psi_{(K-)} & \psi_{(\overline{K}0)} & -\frac{2}{\sqrt{6}}\psi_{(\eta)} \end{pmatrix}, \qquad \psi \to \chi \qquad (2.4.18)$$

$\psi_{(pr)}$ transforms like

$$\psi_{(pr)} \rightarrow \psi'_{(pr)} = U_{3ps}\psi_{(sq)}U^{-1}_{3qr} \tag{2.4.19a}$$

$$U_{3ps} = \exp\left(i(\lambda_l)_{ps}\vartheta_l\right) \tag{2.4.19b}$$

where λ_l are the eight Gell-Mann matrices satisfying the same commutation relations as J_l does in (2.4.7). The first subscript denotes the row number and the second the column number in the 3×3 matrices. In the limit of $m_1=m_2=m_3$, $M^2{}_{mpr}$ in (2.4.3) degenerates and (2.4.3) with (2.4.18) is invariant under the SU(3) transformations (2.4.19). This is entirely equivalent to the invariance of (2.4.2) mentioned in §2.4.2 using the generators J_l of (2.4.5) instead of λ_l here.

As was indicated in §2.4.2, the SU(3) symmetry of (2.4.3) is actually broken to first order. The isospin SU(2) symmetry is however very weakly broken so that (2.4.12) approximately holds. This is equivalent to that (2.4.3) holds for $\psi_{(\pi o)}$ using (2.4.12). The pion triplet represented by $\psi_{(12)}$, $\psi_{(\pi o)}$ and $\psi_{(21)}$ in (2.4.18) approximately transform into each other under the isospin SU(2) group.

Chapter 3

MESON WAVE EQUATIONS IN RELATIVE SPACE

The equatons of motion for meson (2.3.22, 23) will be the starting point in this chapter which initially follows [H5].

3.1. LABORATORY AND RELATIVE COORDINATES AND MESON EQUATIONS

The meson wave functions in (2.3.22, 23) are decomposed into a singlet and a triplet analogous to (C1a);

$$\chi_{\dot{b}a}(x_I, x_{II}) = \delta_{\dot{b}a}\chi_0(x_I, x_{II}) + \underline{\sigma}_{\dot{b}a}\,\underline{\chi}(x_I, x_{II}) \tag{3.1.1a}$$

$$\psi^{b\dot{e}}(x_I, x_{II}) = \delta^{b\dot{e}}\psi_0(x_I, x_{II}) - \underline{\sigma}^{b\dot{e}}\,\underline{\psi}(x_I, x_{II}) \tag{3.1.1b}$$

Application of (B5, B16) to (3.1.1) yields

$$\begin{aligned}\psi_0 &= \tfrac{1}{2}\left(\psi^{1\dot{1}} + \psi^{2\dot{2}}\right) = \tfrac{1}{2}\left(\psi^{1}_{\dot{2}} - \psi^{2}_{\dot{1}}\right)\\ \underline{\psi} &= \tfrac{1}{2}\left(\psi^{2}_{\dot{2}} - \psi^{1}_{\dot{1}},\ i\left(\psi^{2}_{\dot{2}} + \psi^{1}_{\dot{1}}\right),\ \psi^{1}_{\dot{2}} + \psi^{2}_{\dot{1}}\right)\end{aligned} \tag{3.1.2a}$$

$$\psi \rightarrow \chi \tag{3.1.2b}$$

The singlet ψ_0 is antisymmetric and the triplet $\underline{\psi}$ is symmetric with respect to the interchange of upper and lower indices.

Introduce the laboratory meson coordinates X^μ and the relative coordinates x^μ according to

$$x^\mu = x^\mu_{II} - x^\mu_I, \qquad X^\mu = (1 - a_m)x^\mu_I + a_m x^\mu_{II} \tag{3.1.3a}$$

where a_m is a real constant. x^μ is considered as a unobservable "hidden variable", like the complex internal variables z and u at the end of Sec. 2.3. If x^μ were obsevable, then (3.1.3a) shows that the both quark coordinates x_I and x_{II} are also observables, contrary to experience. Nevertheless, the essential and new physics will turn to reside in such "hidden spaces".

Equation (3.1.3a) leads to

$$\begin{aligned}
&\partial_{I\mu} = \partial/\partial x_I^\mu = (1-a_m)\partial_{X\mu} - \partial_\mu, \qquad \partial_{II\mu} = \partial/\partial x_{II}^\mu = a_m\,\partial_{X\mu} + \partial_\mu \\
&\partial_{X\mu} = \partial/\partial X^\mu = \left(\partial/\partial X^0,\ \partial/\partial \underline{X}\right) = \left(\partial_{X0},\ \partial_{\underline{X}}\right) \\
&\partial_\mu = \partial/\partial x^\mu = \left(\partial/\partial x^0,\ \partial/\partial \underline{x}\right) = \left(\partial_0,\ \underline{\partial}\right)
\end{aligned} \tag{3.1.3b}$$

With these relations, we find that

$$\int d^4x_I d^4x_{II} = \int d^4X d^4x \tag{3.1.3c}$$

because the Jacobian relating the both sides turns out to be unity. Combining (3.1.3a, 3b) with (C3) leads to

$$\begin{aligned}
&\partial_I^{a\dot{b}} = (1-a_m)\partial_X^{a\dot{b}} - \partial^{a\dot{b}} = (1-a_m)\left(-\delta^{a\dot{b}}\partial_{X0} - \underline{\sigma}^{a\dot{b}}\partial_{\underline{X}}\right) + \delta^{a\dot{b}}\partial_0 + \underline{\sigma}^{a\dot{b}}\underline{\partial} \\
&\partial_{I\dot{b}a} = (1-a_m)\partial_{X\dot{b}a} - \partial_{\dot{b}a} = (1-a_m)\left(-\delta_{\dot{b}a}\partial_{X0} + \underline{\sigma}_{\dot{b}a}\partial_{\underline{X}}\right) + \delta_{\dot{b}a}\partial_0 - \underline{\sigma}_{\dot{b}a}\underline{\partial} \\
&\partial_{II\dot{e}f} = a_m\partial_{X\dot{e}f} + \partial_{\dot{e}f} = a_m\left(-\delta_{\dot{e}f}\partial_{X0} + \underline{\sigma}_{\dot{e}f}\partial_{\underline{X}}\right) - \delta_{\dot{e}f}\partial_0 + \underline{\sigma}_{\dot{e}f}\underline{\partial} \\
&\partial_{II}^{f\dot{e}} = a_m\partial_X^{f\dot{e}} + \partial^{f\dot{e}} = a_m\left(-\delta^{f\dot{e}}\partial_{X0} - \underline{\sigma}^{f\dot{e}}\partial_{\underline{X}}\right) - \delta^{f\dot{e}}\partial_0 - \underline{\sigma}^{f\dot{e}}\underline{\partial}
\end{aligned} \tag{3.1.4}$$

Here, $\partial_X^{a\dot{b}}$ and $\partial_{X\dot{b}a}$ go over to $\partial^{a\dot{b}}$ and $\partial_{\dot{b}a}$, respectively, when $X \to x$. a_m is real and can be suitably chosen. In classical mechanics, a_m will depend upon the quark masses if X is the center mass coordinate of the meson. This concept does not apply here because quarks are not observable.

In this and the next two chapters, we shall limit ourselves largely to a two quark meson at rest or moving slowly and freely with momentum $\underline{K}$ in the laboratory space. The following ansatz (3.1.5), that the meson wave functions are separable in X and x and can be expanded into plane wave components, is made. It will turn out in Sec. 4.6 and Chapter 5 below that this ansatz holds only for ground state mesons and not for excited ones.

$$\begin{aligned}
&\psi_0(x_I, x_{II}) = \sum_{\underline{K}} \exp\left(-iK_\mu X^\mu\right)\psi_{0K}(x) \\
&\underline{\psi}(x_I, x_{II}) = \sum_{\underline{K}} \exp\left(-iK_\mu X^\mu\right)\underline{\psi}_K(x)
\end{aligned} \tag{3.1.5a}$$

$$\psi \to \chi \tag{3.1.5b}$$

$$K_\mu = (E_K, -\underline{K}) \tag{3.1.6}$$

E_K is the total energy of the meson. In the following, we shall only consider one term in the sums (3.1.5) at a time and can therefore suppress the index K according to

$$\psi_{0K}(x),\ \underline{\psi}_K(x) \rightarrow \psi_0(x),\ \underline{\psi}(x), \qquad\qquad \psi \rightarrow \chi \tag{3.1.7a}$$

for simplified notation. Here, $x=x^\mu=(x^0, \underline{x})$ and any of the following component forms can be used;

$$\underline{x} = (x^1,\ x^2,\ x^3) = (x,\ y,\ z) = r(\hat{x},\ \hat{y},\ \hat{z}) = r\hat{r} = r(\sin\vartheta\sin\phi, \sin\vartheta\cos\phi, \cos\vartheta) \tag{3.1.7b}$$

Inserting (3.1.1, 4-7) into (2.3.22) leads to

$$\begin{aligned}
&\left[a_m(1-a_m)\left(E_K^2+\underline{K}^2\right)+\partial_0^2+\Delta+i(1-2a_m)\left(E_K\partial_0-\underline{K}\,\underline{\partial}\right)\right]\chi_0(x)\\
&+\left[2\partial_0\underline{\partial}+i(1-2a_m)\left(E_K\underline{\partial}-\underline{K}\partial_0\right)-2a_m(1-a_m)E_K\underline{K}+\underline{K}\times\underline{\partial}\right]\underline{\sigma}\chi_0(x)\\
&+\left[a_m(1-a_m)\left(E_K^2-\underline{K}^2\right)+\partial_0^2-\Delta+i(1-2a_m)\left(E_K\partial_0+\underline{K}\,\underline{\partial}\right)\right]\underline{\sigma}\,\underline{\chi}(x)\\
&+E_K\underline{\sigma}\left(\underline{\partial}\times\underline{\chi}(x)\right)+\left[2\underline{\sigma}\,\underline{\partial}+2\partial_0+i(1-2a_m)\left(E_K-\underline{\sigma}\,\underline{K}\right)\right]\underline{\partial}\,\underline{\chi}(x)\\
&-\left[i(1-2a_m)\left(\partial_0+\underline{\sigma}\,\underline{\partial}\right)+2a_m(1-a_m)\left(E_K-\underline{\sigma}\,\underline{K}\right)\right]\underline{K}\,\underline{\chi}(x)\\
&+\left(\underline{\partial}+\partial_0\underline{\sigma}\right)\left(\underline{K}\times\underline{\chi}(x)\right)=\left(\Phi_m-M_m^2\right)\left(\psi_0(x)-\underline{\sigma}\,\vec{\underline{\psi}}(x)\right), \qquad \Delta=\underline{\partial}\,\underline{\partial}
\end{aligned} \tag{3.1.8a}$$

$$\chi \leftrightarrow \psi \text{ and cross products change sign} \,(\times \rightarrow -\times) \tag{3.1.8b}$$

where $\underline{\sigma}=(\sigma^1, \sigma^2, \sigma^3)$ are the Pauli matrices in (A4). We shall further limit ourselves to the simple cases in which the relative time dependence is similar to that of the laboratory time;

$$\varphi(x) = \varphi(\underline{x})\exp(i\omega_K x^0), \qquad\qquad \varphi(x) = \chi_0(x),\ \underline{\chi}(x),\ \psi_0(x),\ \underline{\psi}(x) \tag{3.1.9}$$

where ω_K is the unknown relative energy among the quarks. The choice

$$a_m = \frac{1}{2} + \frac{\omega_K}{E_K} \tag{3.1.10a}$$

is now made which together with (3.1.9) causes the unknown relative energy ω_K to cancel out in (3.1.8) when $\underline{K}\rightarrow 0$. It also allows for the all important decoupling of the singlet $\chi_0(x)$ from the triplet $\underline{\chi}(x)$ in the general (3.1.8) so that they obey separate equations (3.2.1) and (3.2.2) below.

The constant ½ in (3.1.10a) is so chosen that the rest frame meson mass E_0 in (3.2.3b, 4a) below becomes the sum of the quark masses via (2.3.27) in the absence of the interquark potential Φ_m and of quark motion represented by Δ there. With (3.1.10a), the time dependence of the meson wave functions (3.1.5) takes the form

$$\exp\left(-iE_0\left(x_I^0 + x_{II}^0\right)/2\right) \tag{3.1.10b}$$

Inserting (3.1.1, 5, 7a) into (2.3.23) leads to

$$\Delta\Delta\Phi_m(\underline{x}) = g_s^4 \,\mathrm{Re}\left(\underline{\psi}(\underline{x})\underline{\chi}^*(\underline{x}) - \psi_0(\underline{x})\chi_0^*(\underline{x})\right) \tag{3.1.11}$$

This equation together with (3.1.8) are three coupled higher order partial differential equations and cannot be solved generally. Assuming that Φ_m and M_m^2 are known, (3.1.8) by itself still consists of two coupled second order partial differential equations and can in general not be solved analytically.

The relative coordinates x^μ are not expected to be directly observable. Since the laboratory coordinates X^μ are observable, the measurability of x^μ would by (3.1.3a) imply that the quark coordinates x_I and x_{II} are also observable, contrary to experience. On the other hand, x^μ are required to transform as an ordinary four vector so that the meson wave equations (2.3.22, 23) together with (3.1.3) are Lorentz covariant.

In conventional local theories of hadrons, effects of dependencies upon relative coordinates are put in the form of new parameters or functions, such as form factors and structure functions, which must be assumed or determined from other data. These assumptions are not needed in the present nonlocal theory, in which the structure of the hadron are given by the dependence of the hadron wave function upon $\underline{x}$. Furthermore, the relative time x^o will turn out to be of crucial importance in the normalization of meson wave functions in Sec. 4.2 and in providing mass to the gauge bosons in (6.2.14) below.

Therefore, the relative coordinates x^μ may be of the nature of "hidden variable" occassionally mentioned in the literature.

3.2. Rest Frame Meson Equations and Confining Potential

Assuming that $\Phi_m(\underline{x})$ is known, the two coupled second order equations (3.1.8) have four solutions. These equations are greatly simplified in the rest frame of the meson $\underline{K}$=0. Using (3.1.9, 10), the singlet and triplet parts in (3.1.8) separate and (3.1.8) decomposes into

$$\left(E_0^2/4 + \Delta\right)\chi_0(\underline{x}) = \left(\Phi_m(\underline{x}) - M_m^2\right)\psi_0(\underline{x}) \tag{3.2.1a}$$

$$\left(E_0^2/4 + \Delta\right)\psi_0(\underline{x}) = \left(\Phi_m(\underline{x}) - M_m^2\right)\chi_0(\underline{x}) \tag{3.2.1b}$$

$$\left(-E_0^2/4+\Delta\right)\underline{\chi}(\underline{x})-2\underline{\partial}\left(\underline{\partial}\underline{\chi}(\underline{x})\right)-E_0\underline{\partial}\times\underline{\chi}(\underline{x})=\left(\Phi_m(\underline{x})-M_m^2\right)\underline{\psi}(\underline{x}) \tag{3.2.2a}$$

$$\left(-E_0^2/4+\Delta\right)\underline{\psi}(\underline{x})-2\underline{\partial}\left(\underline{\partial}\underline{\psi}(\underline{x})\right)+E_0\underline{\partial}\times\underline{\psi}(\underline{x})=\left(\Phi_m(\underline{x})-M_m^2\right)\underline{\chi}(\underline{x}) \tag{3.2.2b}$$

The fourth order (3.2.1) can be further reduced to two decoupled second order equations

$$\left(\Delta+E_0^2/4-\Phi_m(\underline{x})+M_m^2\right)\chi_0(\underline{x})=0, \qquad \psi_0(\underline{x})=\chi_0(\underline{x}) \tag{3.2.3a}$$

$$\left(\Delta+E_0^2/4+\Phi_m(\underline{x})-M_m^2\right)\chi_0(\underline{x})=0, \qquad \psi_0(\underline{x})=-\chi_0(\underline{x}) \tag{3.2.3b}$$

Similarly, (3.2.2) can also be reduced to two decoupled second order equations

$$\left(\Delta-E_0^2/4-\Phi_m(\underline{x})+M_m^2\right)\underline{\chi}(\underline{x})-2\underline{\partial}\left(\underline{\partial}\underline{\chi}(\underline{x})\right)=0, \qquad \underline{\psi}(\underline{x})=\underline{\chi}(\underline{x}) \tag{3.2.4a}$$

$$\left(\Delta-E_0^2/4+\Phi_m(\underline{x})-M_m^2\right)\underline{\chi}(\underline{x})-2\underline{\partial}\left(\underline{\partial}\underline{\chi}(\underline{x})\right)=0, \qquad \underline{\psi}(\underline{x})=-\underline{\chi}(\underline{x}) \tag{3.2.4b}$$

with the subsidiary conditions

$$\underline{\partial}\times\underline{\psi}(\underline{x})=0, \qquad \underline{\partial}\times\underline{\chi}(\underline{x})=0 \tag{3.2.5a}$$

These are fullfiled for rest frame ground state vector mesons (3.2.13b) below if

$$\underline{\psi}(\underline{x})=\hat{r}\psi_1(r), \qquad \underline{\chi}(\underline{x})=\hat{r}\chi_1(r), \qquad r=|\underline{x}|, \qquad \hat{r}=\underline{x}/r \tag{3.2.5b}$$

where r is the interquark distance.

Green's function for (3.1.11) satisfies

$$\Delta\Delta G_m(\underline{x},\underline{x}')=\delta(\underline{x}-\underline{x}') \tag{3.2.6a}$$

$$G_m(\underline{x},\underline{x}')=-\frac{1}{8\pi}|\underline{x}-\underline{x}'| \tag{3.2.6b}$$

This result can be seen as follows. Noting that $|\underline{x}-\underline{x}'|$ is a regular function so that

$$\Delta|\underline{x}-\underline{x}'|=\frac{2}{|\underline{x}-\underline{x}'|} \tag{3.2.7a}$$

and (3.2.6) becomes the well-known

$$\Delta\frac{1}{|\underline{x}-\underline{x}'|}=-4\pi\delta|\underline{x}-\underline{x}'| \tag{3.2.7b}$$

Combining (3.1.11) and (3.2.6) yields the intrameson potential

$$\Phi_m(\underline{x})=-\Phi_c(\underline{x})+\frac{d_m}{r}+d_{m0}+d_{m2}r^2 \tag{3.2.8a}$$

$$\Phi_c(\underline{x})=\frac{g_s^4}{8\pi}\int d^3\underline{x}'|\underline{x}-\underline{x}'|\mathrm{Re}\left(\underline{\psi}(\underline{x}')\underline{\chi}^*(\underline{x}')-\psi_0(\underline{x}')\chi_0^*(\underline{x}')\right) \tag{3.2.8b}$$

For large r, we have

$$\Phi_c(\underline{x}\to\infty)=\frac{g_s^4}{8\pi}r\int d^3\underline{x}'\mathrm{Re}\left(\underline{\psi}(\underline{x}')\underline{\chi}^*(\underline{x}')-\psi_0(\underline{x}')\chi_0^*(\underline{x}')\right)=\beta_{m1l}r \tag{3.2.8c}$$

The last three terms in (3.2.8a) are homogenous solutions and the d_m's three integration constants to the fourth order potential equation (3.1.11). This can be seen by writing out the $\Delta\Delta$ operator in (3.2.6a) in spherical coordinates. The angular part of this operator gives no contribution when operating on the d_m terms. The radial part

$$\Delta\Delta=\frac{1}{r^2}\frac{\partial}{\partial r}r^2\frac{\partial}{\partial r}\frac{1}{r^2}\frac{\partial}{\partial r}r^2\frac{\partial}{\partial r} \tag{3.2.9a}$$

also gives no contribution to the d_m terms exept possibly for the d_m/r term near the irregular r=0. Using (3.2.7b) with $\underline{x}'$=0, (3.2.9a) yields

$$\Delta\Delta\frac{1}{r}=-4\pi\frac{1}{r^2}\frac{\partial}{\partial r}r^2\frac{\partial}{\partial r}\delta(r) \tag{3.2.9b}$$

Integrating this expression over a small sphere with radius δr and letting $\delta r\to 0$ yields

$$\int d^3\underline{x}\Delta\Delta\frac{1}{r}=-16\pi^2\left[r^2\frac{\partial}{\partial r}\delta(r)\right]_0^{\delta r}=0 \tag{3.2.9c}$$

In the $r\to 0$ limit. Therefore, (3.2.8) holds for all r.

Substituting (3.2.8) into (3.2.3, 4) leads to

$$\left(\Delta+\Phi_{c0}(\underline{x})-\frac{d_m}{r}-d_{m2}r^2-d_{m0}+\frac{E_{00}^2}{4}+M_m^2\right)\chi_0(\underline{x})=0,\quad \psi_0(\underline{x})=\chi_0(\underline{x})$$

(3.2.10a)

$$\left(\Delta - \Phi_{c0}(\underline{x}) + \frac{d_m}{r} + d_{m2}r^2 + d_{m0} + \frac{E_{00}^2}{4} - M_m^2\right)\chi_0(\underline{x}) = 0, \quad \psi_0(\underline{x}) = -\chi_0(\underline{x})$$

(3.2.10b)

$$\left(\Delta + \Phi_{c1}(\underline{x}) - \frac{d_m}{r} - d_{m2}r^2 - d_{m0} - \frac{E_{10}^2}{4} + M_m^2\right)\underline{\chi}(\underline{x}) - 2\underline{\partial}(\underline{\partial}\underline{\chi}(\underline{x})) = 0, \quad \underline{\psi}(\underline{x}) = \underline{\chi}(\underline{x})$$

(3.2.11a)

$$\left(\Delta - \Phi_{c1}(\underline{x}) + \frac{d_m}{r} + d_{m2}r^2 + d_{m0} - \frac{E_{10}^2}{4} - M_m^2\right)\underline{\chi}(\underline{x}) - 2\underline{\partial}(\underline{\partial}\underline{\chi}(\underline{x})) = 0, \quad \underline{\psi}(\underline{x}) = -\underline{\chi}(\underline{x})$$

(3.2.11b)

$$\Phi_{cJ}(\underline{x}) = \frac{g_s^4}{8\pi}\int d^3\underline{x}'|\underline{x} - \underline{x}'||\psi_J(\underline{x}')|^2 \qquad (3.2.12)$$

subjected to the subsidiary conditions (3.2.5). E_0 in (3.2.3, 4) has been replaced by E_{J0} here to reflect the spin J=0 or singlet and spin J=1 or triplet mesons.

For meson wave functions depending only upon the interquark distance r,

$$\chi_0(\underline{x}) = \chi_0(|\underline{x}|) = \chi_0(r), \qquad \chi \to \psi \qquad (3.2.13a)$$

$$\underline{\chi}(\underline{x}) = \underline{\chi}(|\underline{x}|) = \underline{\chi}(r), \qquad \chi \to \psi \qquad (3.2.13b)$$

as is the case for ground state mesons considered below, (3.2.12) can be further reduced to radial integrals. With the last of (3.1.7b), (3.2.12) becomes

$$\Phi_{cJ}(\underline{x}) = \frac{g_s^4}{8\pi}\int_0^{2\pi} d\phi' \int_0^{\pi} d\vartheta' \sin\vartheta' \int_0^{\infty} dr' r'^2 |\psi_J(r')|^2 \sqrt{(\underline{x} - \underline{x}')^2} \qquad (3.2.14)$$

The geometry is shown in §Figure 3.1. The integrand represents the contribution of a volume element on a thin sphere with thickness dr' to a field point $\underline{x}$, which can be inside or outside the sphere. Here,

$$R = \sqrt{(\underline{x} - \underline{x}')^2} = \sqrt{r'^2 + r^2 + 2r'r\cos\vartheta'} \qquad (3.2.15)$$

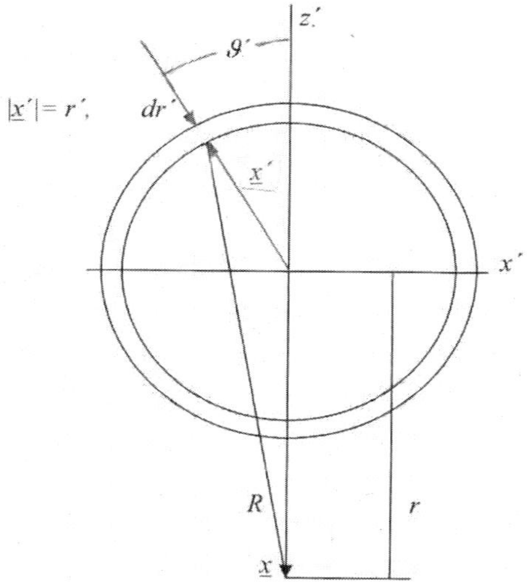

Figure 3.1. Illustration of geometrical relationships of quantities entering the angular integration in (3.2.14).

Putting $cos\,\vartheta'=y$ and integrating over ϕ' converts (3.2.14) to

$$\Phi_{cJ}(r)=\frac{g_s^4}{8\pi}\int_0^\infty dr' r'^2 \left|\psi_J(r')\right|^2 I_y(r,r') \tag{3.2.16a}$$

$$I_y(r,r')=\int_{-1}^{1} dy\sqrt{2rr'y+r^2+r'^2}=\begin{pmatrix} 2\left(3r^2+r'^2\right)/3r & \text{for} & r>r' \\ 2\left(3r'^2+r^2\right)/3r' & \text{for} & r<r' \end{pmatrix} \tag{3.2.16b}$$

Whence

$$\Phi_{cJ}(r)=\frac{g_s^4}{6}\left[\int_0^r dr' r'^2 \left|\psi_J(r')\right|^2\left(3r+\frac{r'^2}{r}\right)+\int_r^\infty dr' r' \left|\psi_J(r')\right|^2\left(3r'^2+r^2\right)\right] \tag{3.2.17}$$

In the $r\to 0$ and ∞ limits,

$$\Phi_{cJ}(r\to 0)=\frac{g_s^4}{2}\int_0^\infty dr r^3 \left|\psi_J(r)\right|^2 \tag{3.2.18}$$

$$\Phi_{cJ}(r\to \infty)=\frac{g_s^4}{2}r\int_0^\infty dr' r'^2 \left|\psi_J(r')\right|^2=\beta_{mJ}r \tag{3.2.19}$$

Φ_{cJ} is the confining potential for the ground state singlet and triplet mesons, since they rise linearly with the interquark distance r. This rersult is in agreement with observation which calls for a linear plus Coulomb type of potential [L3, L4 §14.3.2], the latter being represented by the homogenous solution term d_m/r in (3.2.8a). On this ground the homogenous solution $d_{m2}r^2$ in (3.2.8a), corresponding to harmonic type of confining potential not supported by the rather successful potential models [L3], will be dropped;

$$d_{m2} = 0 \qquad (3.2.20)$$

so that the leading confining potential at large r is the nonlinear (3.2.8c). Possible significance of this term is left to future investigations.

For large r, the dominant term in (3.2.10, 11) is now the confinement term $\Phi_{cJ}(\underline{x})$, which according to (3.2.12) is independent of the angles, assuming that $\psi_J(\underline{x})$ is appreciable for finite $|\underline{x}|$ only. Therefore, (3.2.19) holds for general $\psi_J(\underline{x})$ in (3.2.12). The radial asymtotic solutions to (3.2.10, 11) with (3.2.5b) become

$$\psi_J(r \to \infty) = d_{\infty J}\sqrt{\beta_{mJ}^{1/3} r}\, K_{1/3}\left(\tfrac{2}{3}\left(\beta_{mJ}^{1/3} r\right)^{3/2}\right) \qquad (3.2.21)$$

where K denotes the modified Bessel function of the second kind and $d_{\infty J}$ is a constant related to the volume Ω_c, occupied by the meson in laboratory space, in the normalization conditions (4.2.5, 8) below. $\psi_J(r \to \infty)$ is confined in relative space and decreases exponentially in a scale inversely proportional to the 2/3 power of some average of the amplitude of the meson wave function $\psi_J(\underline{x})$ itself.

3.3. Parity and Nonexistence of Scalar and Axial Vector Mesons

Consider the parities of the wave functions of ground state mesons at rest, $\underline{K}=0$. Under space inversion, (3.1.5, 6) shows that only those parts of the wave functions dependent upon the relative space $\underline{x}$ are affected. Applying (2.3.20) with (3.1.3a) to (3.1.2) and suppressing the common time factor (3.1.10b) leads to

$$\chi_0(\underline{x}) = \tfrac{1}{2}\left(\chi_2^1(\underline{x}) - \chi_1^2(\underline{x})\right) \to \tfrac{1}{2}\left(\psi_1^2(-\underline{x}) - \psi_2^1(-\underline{x})\right) = -\psi_0(-\underline{x}) \qquad (3.3.1a)$$

$$\begin{aligned}&\underline{\chi}(\underline{x}) = \tfrac{1}{2}\left(\chi_2^2(\underline{x}) - \chi_1^1(\underline{x}),\ i\left(\chi_2^2(\underline{x}) + \chi_1^1(\underline{x})\right),\ \chi_2^1(\underline{x}) + \chi_1^2(\underline{x})\right) \to \\ &\tfrac{1}{2}\left(\psi_2^2(-\underline{x}) - \psi_1^1(-\underline{x}),\ i\left(\psi_2^2(-\underline{x}) + \psi_1^1(-\underline{x})\right),\ \psi_1^2(-\underline{x}) + \psi_2^1(-\underline{x})\right) = \underline{\psi}(-\underline{x})\end{aligned} \qquad (3.3.1b)$$

The wave functions of ground state singlet mesons cannot depend upon the angles and (3.2.13a) holds. The wave functions of ground state triplet mesons satisfy (3.2.5b) and hence also (3.2.13b). Therefore, (3.2.17) holds. Because Φ_{cJ} now only depends upon r, (3.2.13a) applies to (3.2.10) and (3.2.13b) or (3.2.5b) applies to (3.2.11). Put (3.2.13) in the form

$$\psi_0(\underline{x})=\psi_0(r)=\psi_0(-\underline{x}) \tag{3.3.2a}$$

$$\underline{\psi}(\underline{x})=\underline{\psi}(r)=\underline{\psi}(-\underline{x}) \tag{3.3.2b}$$

Inserting (3.3.2) into (3.3.1) shows that they are compatible with (3.2.10b) and (3.2.11a) but not with (3.2.10a) and (3.2.11b). Thus, space inversion considerations exclude the unphysical (3.2.10a) and (3.2.11b), in which the total meson energy E_{J0} decreases with increasing quark masses in M_m and may need to become imaginary for heavy mesons.

Therefore, the ground state solutions of (3.2.10b) and (3.2.11a) are assigned to the ground state pseudoscalar ($J^P=0^-$) and vector ($J^P=1^-$) mesons, respectively. J denotes the total angular momentum and P the parity of the meson. The remaining (3.2.10a) and (3.2.11b) can only be assigned to the ground state scalar (0^+) and axial vector (1^+) mesons, respectively. Thus, *scalar and axial vector mesons cannot exist*. These results agree with data and explain the long and futile efforts to find or identify ground state scalar mesons.

The adjectives "pseudoscalar" and "scalar" are misleading here because (3.2.10) refers to the time component of a four vector. In current litterature, a ground state pseudoscalar meson is implicity a point particle with one wave function component that changes sign under inversion in laboratory space $\underline{X}$. Similarly, a ground state vector meson is implicity a point particle with three wave function components that transform like the laboratory coordinates $\underline{X}$. These descriptions holds whether the mesons are in motion or not. The Lagrangians describing them are generally local, i.e., dependent upon X^μ only. The effect of the finite size of the mesons are assumed to be represented by form factors.

On the other hand, the basic meson wave equations (2.3.19) are nonlocal and depend upon both the laboratory and relative coordinates X^μ and x^μ. No form factor considerations enter.

Further, the ground state of $\psi_0(\underline{x})$ in (3.2.10b), assigned to a rest frame 0^- meson, is actually the time component of a four vector. In motions, $\underline{K}\neq 0$, the eight wave function components $\psi_0(\underline{x})$, $\underline{\psi}(\underline{x})$, $\chi_0(\underline{x})$, and $\underline{\chi}(\underline{x})$ in (3.1.8) cannot be separated and the full wave function assigned 0^- and 1^- has each eight components. In this case, as can be seen in Sec. 3.5, $\psi_0(\underline{x})$ and $\chi_0(\underline{x})$ and will correspond to the "large components" and $\underline{\psi}(\underline{x})$ and $\underline{\chi}(\underline{x})$ to the "small components" of the wave functions assigned to 0^-. The "large components" correspond to u_1 and u_2 and the "small components" to u_3 and u_4 in the Dirac wave function (A14). These "small components" are to first order in $\underline{K}$ (see §3.5.1 below) and have therefore negative parity in $\underline{X}$ space. Thus, (3.1.5a) and (3.2.10b), when perturbed by a small $\underline{K}\neq 0$ part by means of (3.1.8), will be associated with negative parity "small components" and hence compatible with that they are assigned to 0^-.

Analogously, $\underline{\psi}(\underline{x})$ in (3.1.11a), assigned to a rest frame 1^- meson, consists actually of the space components of a four vector. In motion, the eight wave function components of the preceeding paragraph are again coupled. $\underline{\psi}(\underline{x})$ and $\underline{\chi}(\underline{x})$ then correspond to the "large components" and $\psi_0(\underline{x})$ and $\chi_0(\underline{x})$ to the "small components".

3.4. Classification of Mesons in Relative Space

The rest frame meson equations obtained in Sec. 3.2 allow for a classification of mesons according to the types of wave equations describing them. Denoting the orbital quantum number of the meson associated with the polar coordinate ϑ mentioned above (3.2.14) by l, there are four basic types of mesons.

3.4.1. Singlet, $l = 0$ Mesons

In the absence of angular excitations, (3.2.13a) holds and (3.2.10b) with (3.2.20) becomes

$$\left(\frac{1}{r^2}\frac{\partial}{\partial r}r^2\frac{\partial}{\partial r} - \Phi_{c0}(r) + \frac{d_m}{r} + d_{m0} + \frac{E_{00}^2}{4} - M_m^2\right)\psi_0(r) = 0, \quad \psi_0(r) = -\chi_0(r) \tag{3.4.1}$$

which together with (3.2.17) forms a set of nonlinear integro-differential equations that hold for the ground state 0^- mesons as well as their radially excited states.

3.4.2. Singlet, $l > 0$ Mesons

The basic governing equations are (3.2.10b) and (3.2.12). They are however not separable in r and ϑ if the meson wave function $\psi_0(\underline{x})$ depends upon both of them. This renders the concept angular momentum l strictly not applicable, because this concept is based upon the separability of the wave function under consideration.

Near the origin r=0, the d_m/r term dominates over $\Phi_{co}(\underline{x})$. For large r, $\Phi_{co}(\underline{x}) \to \Phi_{co}(r)$ (see end of Sec. 3.2). In these both limits, separation is again possible;

$$\psi_0(\underline{x}) = \psi_{0l}(r) Y_{lm}(\vartheta, \phi) \tag{3.4.2a}$$

where Y_{lm} is the usual spherical harmonics and m is the azimuthal quantum number. For $r \to 0$, (3.2.10b) and (3.2.20) become

$$\left(\frac{1}{r^2}\frac{\partial}{\partial r}r^2\frac{\partial}{\partial r} - \frac{l(l+1)}{r^2} - \Phi_{c0}(r \to 0) + \frac{d_m}{r} + d_{m0} + \frac{E_{00}^2}{4} - M_m^2\right)\psi_{0l}(r) = 0$$
$$\psi_{0l}(r) = -\chi_{0l}(r) \tag{3.4.2b}$$

The asymtotic wave function has been given by (3.2.21) with J=0 and is, unlike (3.4.2a), independent of the angles.

3.4.3. Triplet, $l = 0$ Mesons

In the absence of angular excitations, (3.2.13b) holds. Together with (3.2.5b, 20), (3.2.11a) bcomes

$$\left(\frac{1}{r^2}\frac{\partial}{\partial r}r^2\frac{\partial}{\partial r}-\frac{2}{r^2}-\Phi_{cl}(r)+\frac{d_m}{r}+d_{m0}+\frac{E_{10}^2}{4}-M_m^2\right)\psi_1(r)=0,\quad \psi_1(r)=\chi_1(r) \tag{3.4.3}$$

This equation and (3.2.17) constitute a set of nonlinear integro-differential equations that hold for ground state 1^- mesons as well as their radially excited states.

3.4.4. Triplet, $l > 0$ Mesons

For this case, (3.2.5) no longer holds. Therefore, (3.2.4, 11) cannot be obtained and we have to return to (3.2.2) and (3.2.8), putting $\psi_0(\underline{x})$ and $\chi_0(\underline{x})$ to zero. These both equations are again not separable in r and ϑ. As in §3.4.2, separation is again possible in the $r\to 0$ and ∞ limits for similar reasons. The meson wave functions can then be expanded into vector spherical harmonics [B2]. For a given harmonic, we have

$$\underline{\psi}(\underline{x})=(\psi^1,\psi^2,\psi^3),\qquad \psi\to\chi \tag{3.4.4a}$$

$$(\psi^1\mp\psi^2)+(\chi^1\mp\chi^2)=\mp\sqrt{\frac{(l\mp m+1)(l\mp m+2)}{l(2l+3)}}Y_{l+1m\mp1}f(r)\pm\sqrt{\frac{(l\pm m-1)(l\pm m)}{l(2l-1)}}Y_{l-1m\mp1}g(r)$$

$$\psi^3+\chi^3=\sqrt{\frac{(l-m+1)(l+m+1)}{l(2l+3)}}Y_{l+1m}f(r)+\sqrt{\frac{(l-m)(l+m)}{l(2l-1)}}Y_{l-1m}g(r) \tag{3.4.4b}$$

$$(\psi^1\mp\psi^2)-(\chi^1\mp\chi^2)=i\sqrt{\frac{(l\pm m)(l\mp m+1)}{l(l+1)}}Y_{lm\mp1}h(r)$$

$$\psi^3-\chi^3=-i\frac{m}{\sqrt{l(l+1)}}Y_{lm}h(r) \tag{3.4.4c}$$

$\underline{\psi}(\underline{x})$ and $\underline{\chi}(\underline{x})$ are eigenfunctions of $\underline{J}^2$, J_3 and $\underline{S}^2$ with eigenvalues $j(j+1)$, m and 2, respectively. Here, $\underline{J}$ denotes the total angular momentum operator, J_3 its third component and $\underline{S}$ the spin operator. $j=l-1,\ l,\ l+1$ follows from the general theory of angular momentum. For $l=0$, only one component $Y_1f(r)$ exists which corresponds to (3.2.5b). For $l>0$, (3.4.4b, 4c) and (3.2.2) can be separated into Y_{l+1}, Y_l and Y_{l-1} components associated, respectively, with the following radial equations

$$\left(-\frac{E_{j0}^2}{4}+M_m^2-\Phi_m(\underline{x}\to 0)-\frac{1}{2l+1}\Delta_{l+1}\right)f(r)-\frac{2l}{2l+1}\Delta_{(l-1)-}g(r)-E_{j0}\frac{l}{2l+1}\left(\frac{\partial}{\partial r}-\frac{l}{r}\right)h(r)=0$$

$$\left(\frac{E_{j0}^2}{4}+M_m^2-\Phi_m(\underline{x}\to 0)-\Delta_l\right)h(r)-E_{j0}\left(\frac{\partial}{\partial r}+\frac{l+2}{r}\right)f(r)+E_{j0}\left(\frac{\partial}{\partial r}-\frac{l-1}{r}\right)g(r)=0$$

$$\left(-\frac{E_{j0}^2}{4}+M_m^2-\Phi_m(\underline{x}\to 0)+\frac{1}{2l+1}\Delta_{l-1}\right)g(r)-\frac{2l+2}{2l+1}\Delta_{(l+1)+}f(r)$$

$$+E_{j0}\frac{l+1}{2l+1}\left(\frac{\partial}{\partial r}+\frac{l+1}{r}\right)h(r)=0 \qquad (3.4.5a)$$

$$\Delta_l=\frac{\partial^2}{\partial r^2}+\frac{2}{r}\frac{\partial}{\partial r}-\frac{l(l+1)}{r^2}=\left(\frac{\partial}{\partial r}+\frac{l+2}{r}\right)\left(\frac{\partial}{\partial r}-\frac{l}{r}\right)$$

$$\Delta_{l+}=\left(\frac{\partial}{\partial r}+\frac{l}{r}\right)\left(\frac{\partial}{\partial r}+\frac{l+1}{r}\right),\qquad \Delta_{l-}=\left(\frac{\partial}{\partial r}-\frac{l+1}{r}\right)\left(\frac{\partial}{\partial r}-\frac{l}{r}\right) \qquad (3.4.5b)$$

Equation (3.4.5) holds in the small r region where the d_m/r term dominates in (3.2.8a) so that $\Phi_m(\underline{x})$ depends only upon r and the separated wave functions (3.4.4) can be applied to (3.2.2).

These radial equations are sixfold degenerate, i.e., the six component equations of (3.2.2) yield the same (3.4.5). Near r=0, the radial wave functions are expanded as

$$f(r)=r^{\lambda}(f_0+f_1r+..),\ g(r)=r^{\lambda}(g_0+g_1r+..),\ h(r)=r^{\lambda}(h_0+h_1r+..) \qquad (3.4.6)$$

where λ is determined by the indicial equation for (3.4.5) and takes on the values,

$$\lambda=l-1,\quad l,\quad l+1,\quad -l(f_0,\ g_0),\quad -l-1(h_0),\quad -l-2(f_0) \qquad (3.4.7)$$

associated with the six solutions of (3.4.5). Solutions for the last three λ values, with the associated wave function components given in the parentheses, generally diverge at r=0 and are discarded. For the first three λ values in (3.4.7), one finds

λ	f_0	h_0	g_0	
$l-1$	0	0	g_0	(3.4.8a)
l	0	h_0	0	(3.4.8b)
$l+1$	f_0	0	$f_0(l+1)(2l+3)$	(3.4.8c)

The three solutions associated with (3.4.8) provide the framework for the classification of the triplet, l>0 mesons in §5.7.2 below.

In the $r\to\infty$ limit, (3.2.2) and (3.2.8) yield

$$\chi_r(\underline{x} \to \infty) = \psi_r(\underline{x} \to \infty) = \text{constant} \times \sqrt{\beta_{m1l}^{1/3} r} K_{1/3}\left(\tfrac{2}{3}\left(\beta_{m1l}^{1/3} r\right)^{3/2}\right)$$
$$\chi_\vartheta(\underline{x} \to \infty),\ \psi_\vartheta(\underline{x} \to \infty),\ \chi_\phi(\underline{x} \to \infty),\ \psi_\phi(\underline{x} \to \infty) \to 0 \tag{3.4.9}$$

where ψ_0, χ_0=0 in (3.2.8). The indices r, ϑ and ϕ refer to the radial and angular components of $\underline{\psi}$ and $\underline{\chi}$. As in (3.2.21), the asymtotic solutions (3.4.9) are also independent of the angles.

For an eventual solution of (3.4.5), it is of interest to convert it into a standard form of a first order system of equations [C4 Ch 4, 5]. Among the many ways to cary out such a conversion, only the following one was found not to strengthen the sigularities in (3.4.5);

$$w_{t1} = f, \qquad w_{t2} = w_{t1}' + \frac{l+2}{r} w_{t1}, \qquad w_{t3} = h, \qquad w_{t4} = w_{t3}' - \frac{l}{r} w_{t3}$$
$$w_{t5} = g, \qquad w_{t6} = w_{t5}' - \frac{l-1}{r} w_{t5} \tag{3.4.10}$$

For small r, (3.2.8a) yields

$$\Phi_m(x \to 0) = \frac{d_m}{r} \tag{3.4.11}$$

Substituting (3.4.10, 11) into (3.4.5) gives

$$w_{tk}'(r) = \left(\frac{A_{tkn}}{r} + B_{tkn}\right) w_{tn}(r) \tag{3.4.12}$$

$$A_{tkn} = \begin{pmatrix} -l-2 & & & & & \\ -\frac{d_m}{2l+1} & -\frac{l(2l+3)}{2l+1} & \frac{2l(l+1)}{2l+1}E_{j0} & & -\frac{2l}{2l+1}d_m & \frac{2l}{2l+1} \\ & & l & & & \\ & & -d_m & -l-2 & & \\ & & & & l-1 & \\ -\frac{2l+2}{2l+1}d_m & \frac{2l+2}{2l+1} & -\frac{l+1}{2l+1}E_{j0} & & \frac{d_m}{2l+1} & \frac{(2l-1)(l+1)}{2l+1} \end{pmatrix} \tag{3.4.13a}$$

$$B_{tkn} = \begin{pmatrix} & 1 & & & & \\ \frac{1}{2l+1}M_{-E} & & & \frac{2l}{2l+1}E_{j0} & \frac{2l}{2l+1}M_{-E} & \\ & & & 1 & & \\ & -E_{j0} & M_{+E} & & & E_{j0} \\ & & & & & 1 \\ \frac{2l+2}{2l+1}M_{-E} & & & -\frac{2l+2}{2l+1}E_{j0} & -\frac{1}{2l+1}M_{-E} & \end{pmatrix}$$

$$M_{\pm E} = M_m^2 \pm E_{j0}^2/4 \tag{3.4.13b}$$

where k, n=1, 2,…6. The indicial equation

$$\det\left|\lambda_t \delta_{kn} - A_{tkn}\right| = 0 \tag{3.4.14a}$$

yields

$$\lambda_t = -l-2(f), \qquad -l-2, \qquad -l-1, \qquad l-1(g), \qquad l, \qquad l(h) \tag{3.4.14b}$$

where the associated wave function components, shown in the parentheses, are evident from the first, third and fifth lines of (3.4.13a) and agree with the last of (3.4.7) and (3.4.8a, 8b).

3.5. Mesons in Motion

In the last two sections, mesons at rest have been treated. The singlet and triplet parts of the meson wave functions in (3.1.8) can then be decoupled to yield (3.2.1) and (3.2.2). As was mentioned at the end of Sec. 3.3, such a decoupling is not possible if the meson is moving with momentum $\underline{K}$ in the laboratory frame. In this case, the full (3.1.8) must be used which is however not amenable to analytical solutions. In this section, (3.1.8) will be treated in the small and large $\underline{K}$ limits, where analytical expressions and estimates can be obtained. The material comes largely from [H10 Appendix B].

3.5.1. Nonrelativistic Pseudoscalar Mesons

Because vector mesons have too short a lifetime to be seen in motion, only slowly moving, ground state pseudoscalar mesons are treated. Introducing the small expansion parameter

$$\varepsilon_0 = \left|\underline{K}\right|/E_{00} << 1 \tag{3.5.1}$$

Expand E_K in (3.1.6) and the wave functions in (3.1.7a), suppressing the relative time factor in (3.1.9), in the form

$$\psi_0(\underline{x}) = \sum_i \psi_{0i}(\underline{x}), \qquad \underline{\psi}(\underline{x}) = \sum_i \underline{\psi}_{0i}(\underline{x}), \qquad \psi \to \chi, \qquad E_K = \sum_i E_{0i} \tag{3.5.2}$$

where i denotes the *ith* order in ε_0. Further, only freely moving mesons are considered so that Φ_c in (3.2.8a) drops out, as will be seen in §4.3.1 below. Therefore, Φ_m of (3.2.8a) is independent of the meson wave functions (3.5.2).

To zeroth order in ε_0, (3.1.8) goes over to (3.2.10b) and

$$\chi_{00}(r) = -\psi_{00}(r), \qquad \underline{\chi}_{00}(r) = \underline{\psi}_{00}(r) = 0, \qquad E_K = E_0 = E_{00} \tag{3.5.3}$$

To first order in ε_0, (3.1.8) and (3.2.8) yield

$$\chi_{01} = 0, \qquad \psi_{01} = 0, \qquad E_{01} = 0 \tag{3.5.4a}$$

$$\begin{aligned}&\left(E_{00}^2/4 - \Delta\right)\underline{\chi}_{01} + 2\underline{\partial}\left(\underline{\partial}\,\underline{\chi}_{01}\right) + E_{00}\underline{\partial} \times \underline{\chi}_{01} + \left(d_m/r + d_{m0} - M_m^2\right)\underline{\psi}_{01} \\ &= \tfrac{1}{2}E_{00}\underline{K}\chi_{00} - \underline{K} \times \underline{\partial}\chi_{00}\end{aligned} \tag{3.5.4b}$$

$$\begin{aligned}&\left(E_{00}^2/4 - \Delta\right)\underline{\psi}_{01} + 2\underline{\partial}\left(\underline{\partial}\,\underline{\psi}_{01}\right) - E_{00}\underline{\partial} \times \underline{\psi}_{01} + \left(d_m/r + d_{m0} - M_m^2\right)\underline{\chi}_{01} \\ &= \tfrac{1}{2}E_{00}\underline{K}\psi_{00} + \underline{K} \times \underline{\partial}\psi_{00}\end{aligned} \tag{3.5.4c}$$

To second order in ε_0, the singlet part of (3.1.8) becomes

$$\begin{aligned}&\left(E_{00}^2/4 + \Delta\right)\chi_{02} - \left(d_m/r + d_{m0} - M_m^2\right)\psi_{02} \\ &= -\underline{\partial}\left(\underline{K} \times \underline{\chi}_{01}\right) - \left(\tfrac{1}{2}E_{00}E_{02} + \tfrac{1}{4}\underline{K}^2\right)\chi_{00} + \tfrac{1}{2}E_{00}\underline{K}\,\underline{\chi}_{01} - i2\omega_K \underline{K}\,\underline{\partial}\chi_{00}/E_{00}\end{aligned} \tag{3.5.5a}$$

$$\begin{aligned}&\left(E_{00}^2/4 + \Delta\right)\psi_{02} - \left(d_m/r + d_{m0} - M_m^2\right)\chi_{02} \\ &= \underline{\partial}\left(\underline{K} \times \underline{\psi}_{01}\right) - \left(\tfrac{1}{2}E_{00}E_{02} + \tfrac{1}{4}\underline{K}^2\right)\psi_{00} + \tfrac{1}{2}E_{00}\underline{K}\,\underline{\psi}_{01} - i2\omega_K \underline{K}\,\underline{\partial}\psi_{00}/E_{00}\end{aligned} \tag{3.5.5b}$$

3.5.2. Null Relative Energy, $\omega_K = 0$

The null relative energy condition

$$\omega_K = 0 \tag{3.5.6}$$

is assumed. In this way, (3.5.5) becomes real. If this not done, the perturbed wave functions in (3.5.5) become complex and cannot be determined definitely because ω_K is unknown. This in its turn will invalidate the classical energy-momentum relation (3.5.32) for small K and jeopardize the solution (3.5.28) for large K below. In addition, $\omega_K \neq 0$ would lead to an inconsistency in the dependence on the relative time x^0 between (6.4.14) and (6.4.18) below so that the radiative decay amplitude of a vector meson can in principle not be evaluated.

In addition to these practical considerations, the null relative energy assumption (3.5.6) may further be justified as follows. A meson consists of quark A with flavor p at x_I, associated with the total quark wave function $\psi_A{}^a(x_I)\xi_A{}^p(z_I)$, and antiquark B with antiflavor $\bar{r}$ at x_{II}, associated with $\psi_{B\dot{a}}(x_{II})\xi_{Br}(z_{II})$, as is implied in §2.3.4. Since the quarks are not observable, the relabelling achieved by interchanging x_I and x_{II} should not affect the observable laboratory coordinate X^μ of the meson. This condition yields $a_m=1-a_m$ in (3.1.3a) and (3.5.6) follows from the choice (3.1.10a).

Further, $M_m{}^2$ in the basic meson wave equations (2.3.22) is symmetric in the both quark masses according to (2.3.27). It is an internal property not affected by manipulations in the relative space, such as $x_I \leftrightarrow x_{II}$. This is related to the flavor independence in relative space considered in Sec. 4.4. These results are independent of whether the meson in motion or not. Thus ω_0 vanishes, even if this is not necessary in (3.1.10a) to arrive at the rest frame meson equations (3.2.1-2). The physical role of ω_K is not understood presently, although the assocoiated relative time x^0 is of central importance in the present theory, as will be shown in Sec. 4.2 and §6.2.2 below.

With (3.5.6), (3.1.10a) becomes

$$a_m = 1/2 \tag{3.5.7}$$

This relation together with (3.1.9) simplifies (3.1.8) to

$$\left\{\tfrac{1}{4}\left(E_K^2 + \underline{K}^2\right)\chi_0 - \tfrac{1}{2}E_K \underline{K}\,\underline{\chi}\right\} + \left[\Delta\chi_0 + \underline{\partial}\left(\underline{K}\times\underline{\chi}\right)\right] = \left(d_m/r + d_{m0} - M_m^2\right)\psi_0 \tag{3.5.8a}$$

$$\begin{aligned}&\left\{-\tfrac{1}{4}\left(E_K^2 - \underline{K}^2\right)\underline{\chi} - \tfrac{1}{2}\underline{K}\left(\underline{K}\,\underline{\chi}\right) + \tfrac{1}{2}E_K \underline{K}\chi_0\right\}\\ &-\left[2\underline{\partial}\left(\underline{\partial}\,\underline{\chi}\right) - \Delta\underline{\chi} + E_K\underline{\partial}\times\underline{\chi} + \underline{K}\times\underline{\partial}\chi_0\right] = \left(d_m/r + d_{m0} - M_m^2\right)\underline{\psi}\end{aligned} \tag{3.5.8b}$$

$$\text{(3.5.8a) with } \chi \leftrightarrow \psi \text{ and cross products change sign } (\times \to -\times) \tag{3.5.8c}$$

$$\text{(3.5.8b) with } \chi \leftrightarrow \psi \text{ and cross products change sign } (\times \to -\times) \tag{3.5.8d}$$

Here, the singlet and triplet parts of (3.1.8) have been separated.

3.5.3. Dimensional Approximation of Heavy Meson Wave Functions

Even (3.5.4, 5, 6) is not simply solved because the spherical symmetry present in the ε_0=0 limit is broken by the momentum $\underline{K}$ so that separation of variables in the relative space $\underline{x}$ cannot be carried out. Therefore, (3.5.4, 5, 6) is treated approximately by the following "dimensional" approximation, in which the scalar operators operating on meson wave functions are replaced by constants.

In the remainder of this chapter, the form of of the zeroth order wave function ψ_{00} of (3.5.3) for a free pseudoscalar meson at rest given by (4.3.2) below will be employed. For heavy mesons,

$$E_{00} >> d_m \tag{3.5.9a}$$

and the first term dominates over the last term on the right sides of the first order equations (3.5.4b, 4c). Here, the magnitude of $\underline{\partial}$ is $d_m/2$ according to (4.3.2) and d_m is given by (5.2.3) below. Further, the choice

$$\underline{K} = (0,\ 0,\ K) \tag{3.5.9b}$$

will be made. In this limit, subtraction and addition of (3.5.4b) and (3.5.4c) leads to the ansatz

$$\underline{\chi}_{01}(\underline{x}) = -\underline{\psi}_{01}(\underline{x}) \tag{3.5.10a}$$

$$\left(E_{00}^2/4 - \Delta\right)\underline{\psi}_{01} + 2\underline{\partial}\left(\underline{\partial}\underline{\psi}_{01}\right) - \left(d_m/r + d_{m0} - M_m^2\right)\underline{\psi}_{01} = \tfrac{1}{2}E_{00}\underline{K}\psi_{00} \tag{3.5.10b}$$

$$-E_{00}\underline{\partial}\times\underline{\psi}_{01} = \underline{K}\times\underline{\partial}\psi_{00} \tag{3.5.10c}$$

An exact solution of (3.5.10c) with (3.5.9b) is

$$\underline{\psi}_{01}(\underline{x}) = \frac{\underline{K}}{E_{00}}\psi_{00}(r) = (0,\ 0,\ \psi_{01z0}(\underline{x})) \tag{3.5.11}$$

In the dimensional approximation, the left operator of (3.5.10b) is replaced by a scalar constant c_{E1}. Consistency with (3.5.11) requires that

$$c_{E1} = c_{E10} = E_{00}^2/2 \tag{3.5.12}$$

This assumption is however not self-consistent because the $2\underline{\partial}(\underline{\partial}\ )$ operator is not a scalar operator, like the other operators on the left side of (3.5.10b), and can therefore not be replaced by a scalar constant. It is readily seen that the $2\underline{\partial}(\underline{\partial}\underline{\psi}_{01})$ term mixes the three components of $\underline{\psi}_{01}$ = (ψ_{01x}, ψ_{01y}, ψ_{01z}). The effect of this term is estimated by a perturbational approach. The $2\underline{\partial}(\underline{\partial}\underline{\psi}_{01})$ term is neglected at first to obtain (3.5.12) and is subsequently reintroduced into (3.5.10b).

Write (3.5.10b) in the form

$$O_{E1} = E_{00}^2/4 - \Delta - \left(d_m/r + d_{m0} - M_m^2\right) \tag{3.5.13a}$$

$$O_{E1}\underline{\psi}_{01} = \tfrac{1}{2}E_{00}\underline{K}\psi_{00} - 2\underline{\partial}\left(\underline{\partial}\underline{\psi}_{01}\right) \tag{3.5.13b}$$

Inserting (3.5.11) into the right side of (3.5.13b), making use of (4.3.2) below and the first of (3.5.2) and putting $O_{E1} = c_{E1}$ according to (3.5.12) leads to

$$\psi_{01x} = -2\frac{d_m^2}{E_{00}^2}\left(\frac{1}{2} + \frac{1}{d_m r}\right)\hat{x}\hat{z}\psi_{01z0}(r) \approx -2\frac{d_m^2}{E_{00}^2}\hat{x}\hat{z}\psi_{01z0}(r), \quad x \to y \tag{3.5.14a}$$

$$\psi_{01z} = \psi_{01z0}(r) + \frac{d_m^2}{E_{00}^2}\left(1 - 2\hat{z}^2\right)\psi_{01z0}(r) \approx \psi_{01z0}(r) + \frac{d_m^2}{3E_{00}^2}\psi_{01z0}(r) \tag{3.5.14b}$$

where the results on the right sides are obtained by putting

$$\frac{1}{r} = \frac{d_m}{2} \qquad\qquad \hat{z}^2 = \frac{1}{3} \tag{3.5.15}$$

(see (3.1.7b) for notation). The first relation gives some average value for $1/r$ according to (4.3.2) below, from which

$$\int d^3\underline{x}\frac{1}{r}\psi_{00}^2(r) = \frac{2}{d_m}\int d^3\underline{x}\psi_{00}^2(r)$$

follows. The last relation in (3.5.15) can be seen from (3.5.7b). The perturbational effect represented by the last term on the right of (3.5.14b) is small compared to the zeroth order solution (3.5.11) due to (3.5.9a). The terms on the right sides in (3.5.14a) are analogously small. In fact, they vanish altogether after the integration of (3.5.13b) over the angles. Therefore, treating $2\underline{\partial}(\underline{\partial}\underline{\psi}_{01})$ as perturbation is largely justified under these circumstances.

The second order equations (3.5.5) are treated analogously. The third term dominates over the first term on the right of (3.5.5) due to (3.5.9.a). Subtraction and addition of (3.5.5a) and (3.5.5b) leads to the ansatz

$$\chi_{02}(\underline{x}) = -\psi_{02}(\underline{x}) \tag{3.5.16a}$$

$$\left(E_{00}^2/4 + \Delta + d_m/r + d_{m0} - M_m^2\right)\psi_{02} = -\left(\tfrac{1}{2}E_{00}E_{02} + \tfrac{1}{4}\underline{K}^2\right)\psi_{00} + \tfrac{1}{2}E_{00}\underline{K}\underline{\psi}_{01} \tag{3.5.16b}$$

$$\underline{\partial}(\underline{K}\times\underline{\psi}_{01})=0 \tag{3.5.16c}$$

The last equation is satisfied for $\underline{\psi}_{01}$ given by (3.5.11) as well as by (3.5.14). Comparison of (3.5.16a, 16b) with (3.2.10b), putting $\Phi_{c0}=0$ and using (3.2.20) there, shows that they are compatible if $\psi_{02}(\underline{x})$ has the same form as $\psi_0(\underline{x})$, the wave function of a pseudoscalar meson at rest according to Sec. 3.3, and

$$\tfrac{1}{2}E_{00}\underline{K}\underline{\psi}_{01}=\left(\tfrac{1}{2}E_{00}E_{02}+\tfrac{1}{4}\underline{K}^2\right)\psi_{00} \tag{3.5.17}$$

Inserting (3.5.11) into this equation yields

$$E_{02}=K^2/2E_{00} \tag{3.5.18}$$

Observing the first of (3.5.2), one can put

$$\psi_{02}(\underline{x})=c_{E2}\psi_{00}(r) \tag{3.5.19}$$

where c_{E2} is a constant of second order which can however not be determined from (3.5.16c). Thus,

$$c_{E2}\approx\varepsilon_0^2=K^2/E_{00}^2 \tag{3.5.20}$$

is tentatively set here.

3.5.4. Dimensional Approximation of Light Meson Wave Functions

In the limit opposite to (3.5.9a),

$$E_{00}<<d_m \tag{3.5.21}$$

and the last term dominates over the first term on the right sides of (3.5.4b, 4c). In this limit, (3.5.10) is replaced by

$$\underline{\chi}_{01}(\underline{x})=\underline{\psi}_{01}(\underline{x}) \tag{3.5.22a}$$

$$\left(E_{00}^2/4-\Delta\right)\underline{\psi}_{01}+2\underline{\partial}(\underline{\partial}\underline{\psi}_{01})+\left(d_m/r+d_{m0}-M_m^2\right)\underline{\psi}_{01}=\underline{K}\times\underline{\partial}\psi_{00} \tag{3.5.22b}$$

$$\underline{\partial}\times\underline{\psi}_{01}=-\tfrac{1}{2}\underline{K}\psi_{00} \tag{3.5.22c}$$

Replacing the left operator in (3.5.22b) by a constant $c_{D1,}$ one finds

$$\underline{\psi}_{01} = \frac{1}{c_{D1}} \underline{K} \times \underline{\partial} \psi_{00} = \frac{K d_m}{2 c_{D1}} (\hat{y},\ -\hat{x},\ 0) \psi_{00} \tag{3.5.23}$$

where (4.3.2) below and the first of (3.5.2) have again been consulted. Inserting (3.5.23) into (3.5.22b) and making use of (4.3.2), it is seen that the $2\underline{\partial}(\underline{\partial}\,\underline{\psi}_{01})$ terms vanishes so that the left operator in (3.5.22b) is a scalar operator and is consistently approximated by c_{D1}. Inserting (3.5.23) into (3.5.22c), multiplying it by ψ_{00} and integrating over $\underline{x}$ yields

$$c_{D1} = d_m^2/3 \tag{3.5.24}$$

Turning to the second order equations (3.5.5, 6), the first term on the right side dominates over the third term. Equation (3.5.16) is here replaced by

$$\chi_{02}(\underline{x}) = \psi_{02}(\underline{x}) \tag{3.5.25a}$$

$$\left(E_{00}^2/4 + \Delta - d_m/r - d_{m0} + M_m^2\right)\psi_{02} = \tfrac{1}{2} E_{00} \underline{K}\,\underline{\psi}_{01} = 0 \tag{3.5.25b}$$

$$\underline{\partial}\left(\underline{K} \times \underline{\psi}_{01}\right) = \left(\tfrac{1}{4} K^2 + \tfrac{1}{2} E_{02} E_{00}\right)\psi_{00} \tag{3.5.25c}$$

where (3.5.23) has been employed for the right member of (3.5.25b). Substituting the left member of (3.5.23) into (3.5.25c) and (3.5.22c) and combining the former equation with the z component of the latter equation leads to

$$E_{02} = K^2/2E_{00} \tag{3.5.18=3.5.25d}$$

The value of c_{D1} given by (3.5.24) has not been used but c_{D1} in (3.5.23) has been formally regarded as an operator and eliminated here.

Since the left operator of (3.5.25b) differs from that of (3.5.16b) and hence is not of the form (3.2.10b), solution of the form (3.5.19) does not apply. Therefore, (3.5.25a, 25b) yields

$$\chi_{02}(\underline{x}) = \psi_{02}(\underline{x}) = 0 \tag{3.5.26}$$

3.5.5. Higher Order Meson Wave Functions

From (3.5.8) it can be seen that

$$\underline{\chi}_{02},\ \underline{\psi}_{02},\ \chi_{03},\ \psi_{03} = 0 \tag{3.5.27}$$

because there are no source terms of order K^2 and K^3 for these quantities. For $\underline{\psi}_{03}$, $\underline{\chi}_{03}$, ψ_{04}, and χ_{04}, the governing equations obtained from (3.5.8) are more complicated than (3.5.4, 5, 6) so that even the dimensional approximations of §3.5.3 and §3.5.4 cannot be simply carried out in a self-consistent and definite manner. Guesses or assumptions need be made for these higher order wave functions.

3.5.6. Extremely Relativistic Mesons

The following dimensional estimates are given. In limit of $\underline{K}$=(0, 0, $K\to\infty$), the right sides of (3.1.8) as well as the $\underline{\partial}$ operators can be dropped. With (3.1.9, 10a), the singlet and triplet parts of (3.1.8) give a consistent solution only if the null relative energy condition (3.5.6) is fulfilled. The solution is

$$\omega_K = 0, \qquad \psi^3 = \psi_0, \qquad \chi^3 = \chi_0, \qquad E_K^2 = K^2 \tag{3.5.28}$$

Since (3.5.8) follows from (3.1.8) and (3.5.6, 7), it can be used equivalently. Terms in the braces of (3.5.8) are of order K^2 and dominate and yield (3.5.28). If the rest terms in these braces are of order K, they may balance off the terms in the brackets in (3.5.8) which are of order K and lower. If however the rest terms in these braces are of order lower than K, then the terms in the brackets in (3.5.8) have to balance off among themselves. In this case, it is required that

$$\left|\underline{\partial}\right| \approx K \tag{3.5.29}$$

which in its turn implies that

$$\chi_0,\ \underline{\chi},\ \psi_0,\ \underline{\psi} = (\text{finite or nondiverging power series in } \underline{x}) \times \exp(-Kr) \tag{3.5.30}$$

where Kr indicates a prototype dependence on $\underline{x}$ and can actually be more complicated. The amplitudes in (3.5.30) are determined from the same normalization condition (4.2.6) below and are thus of the same order as in (4.3.2) below with $d_m\to 2K$.

Should solutions of the type (3.5.30) exist, they imply a decrease of the size of a meson at rest, $r_{0h}\approx 1$ fm in (4.7.3) below, to a size of the order of $1/K<<2/d_m$ at high meson momenta. This reduction is in the relative space and parallels the corresponding Lorentz contraction in laboratory space $\underline{X}$. Because of this reduction in size, mesons may be used to probe the hadronic structure of nucleons, similar to that the electrons are used to probe their electromagnetic structure. For this purpose, meson energies $K >> d_m/2 \approx 0.432$ Gev according to (5.2.3) below as well as >> the inverse of the nucleon size are required.

3.5.7. Classical Energy-Momentum Relation

The classical relation

$$E_K^2 = E_{00}^2 + K^2 \tag{3.5.31}$$

holds in the $K \to \infty$ limit by (3.5.28) and for K=0 by definition. To first order in ε_0, (3.5.31) also holds according to the last of (3.5.4a). To order ε_0^2, (3.5.31) can be written as

$$E_K = \sqrt{E_{00}^2 + K^2} \approx E_{00} + E_{02}, \qquad E_{02} \approx K^2/2E_{00} \tag{3.5.32}$$

which has been shown for heavy and light pseudoscalar mesons in (3.5.18) and (3.5.25d), respectively. To higher orders in ε_0, (3.5.31) has not been established presently.

Since (3.5.18, 25d) are based upon (3.5.6) so that (3.5.5) becomes real, the agreement of (3.5.18, 25d) with (3.5.32) apparently requires that the the null relative energy condition (3.5.6, 7) holds.

Chapter 4

NORMALIZATION, SUPERPOSITION PRINCIPLE, FLAVOR-INDEPENDENCE, AND CONFINEMENT

4.1. REMARKS ON EARLIER NORMALIZATIONS

Consider a Schrödinger wave function $\psi_{Sch}(X^{\mu})$. It is conventionally normalized according to

$$\int \left|\psi_{Sch}\left(X^{\mu}\right)\right| d^{3}\underline{X} = N_{Sch} = 1 \tag{4.1.1}$$

This normalization is based upon the probability density interpretation of the integrand. The existence of a positive definite norm $N_{Sch}=1$ is of great interpretational importance but there is no physical constraint underlying (4.1.1); N_{Sch} can be any positive constant. For a plane wave, (4.1.1) leads to

$$\psi_{Sch}\left(X^{\mu}\right) = \frac{1}{\sqrt{\Omega}} e^{-iEX^{0}+i\underline{K}\,\underline{X}} \tag{4.1.2}$$

which becomes very small for large normalization volume Ω.

Consider next the normalization of the free particle Klein-Gordon (KG) wave function Ψ_{KG} satisfying (2.3.3) or

$$\left(\Box + m_{KG}^{2}\right)\Psi_{KG}\left(X^{\mu}\right) = 0 \tag{4.1.3}$$

which has the plane wave solution

$$\Psi_{KG} = A_{KG} e^{-iEX^{0}+i\underline{K}\,\underline{X}}, \qquad E^{2} = K^{2} + m_{KG}^{2} \tag{4.1.4}$$

Multiplying (4.1.3) by Ψ_{KG}^{*} and substract it from its complex conjugate yields

$$\frac{\partial}{\partial X^{\mu}}\left(\Psi^{*}_{KG}\frac{\partial}{\partial X_{\mu}}\Psi_{KG}-c.c.\right)=0 \tag{4.1.5}$$

Inserting (4.1.4) into (4.1.5) and integrating over X yields

$$\frac{\partial}{\partial X^{0}}N_{KG}=0, \qquad N_{KG}=\Omega A^{2}_{KG}2iE, \qquad \Omega=\int d^{3}\underline{X} \tag{4.1.6}$$

If the first of (4.1.6) is further integrated over X^0, it will have the same dimension as the action of the KG particle, a dimensionless Lorentz scalar. Integrating the Lorentz invariant (4.1.5) over X^μ will lead to the same conclusion. Therefore, N_{KG} is also a Lorentz scalar and can for instance not depend linearly upon E, which transforms as the time component of a four vector. Analogous to (4.1.1), N_{KG} is conventionaly chosen as

$$N_{KG}=i \tag{4.1.7}$$

so that

$$A_{KG}=1/\sqrt{2E\Omega} \tag{4.1.8}$$

which is the form used throughout the literature. Unlike the positive norm (4.1.1), N_{KG} in (4.1.6) can be a positive or negative imaginary number. The lack of a positive definite norm in the KG case has led to quite some debate during the early development of quantum mechanics and remains a basic, unsettled but ignored issue to the present time. There is also the question, why is the amplitude of the wave function of a KG particle energy-dependent, as in (4.1.8), while its Schrödinger counterpart (4.1.2) is not?

The answer to these two fundamental points is related to the following. The KG equation (4.1.3) describes a point particle of spin 0, which however does not exist. Now the scalar Ψ_{KG} in (4.1.3) can be replaced by a vector KG wave function $\Psi_{KG}{}^{\mu}$ in order to describe a spin 1 point particle. However, such a massive point particle of spin 1 transforming as a U(1) isosinglet also does not exist according to 6.2.2-3 below. These nonexsistences are consistent with the fact that the Schrödinger equation cannot be derived from these KG equations in some nonrelativistic limit.

The observed massless spin 1 point particle, the photon, cannot be described nonrelativistically. The observed massive point particles of spin 1 (gauge bosons) transforming according to SU(2)×U(1) in Sec. 7.4 below have as their source lepton wave functions so that their wave functions do not have the energy dependence of (4.1.8) (see (7.4.10, 7) below).

On the other hand, the Schrödinger equation is the nonrelativistic limit of the Dirac equation (A3) which describes an observable particle of spin ½, the lowest possible spin for a point particle.

On these grounds, application of the energy-dependent KG amplitude (4.1.8) in the literature hitherto may have led to erroneous conclusions.

4.2. Normalization of Meson Wave Functions

Following the method employed in the normalization of the KG amplitude, multiply (2.3.22a) by

$$\chi^{*}_{\dot{e}a} = \left(\chi_{\dot{a}e}\right)^{*} \tag{4.2.1}$$

and the complex canjugate of (2.3.22b) by $\psi^{b\dot{d}}$. Subtraction of the resulting equations from each other, after some rearrangements using (B5, B7, B16), gives the analog of (4.1.5)

$$\begin{aligned} C_m &= \partial_{I\dot{b}a}\chi^{*a\dot{e}}\partial_{II\dot{e}f}\chi^{f\dot{b}} - \partial^{a\dot{b}}_{II}\psi_{\dot{b}f}\partial^{f\dot{e}}_{I}\psi^{*}_{\dot{e}a} \\ = D_m &= \left(\partial_{I\dot{b}a}\chi^{*a\dot{e}}\right)\left(\partial_{II\dot{e}f}\chi^{f\dot{b}}\right) - \left(\partial^{a\dot{b}}_{II}\psi_{\dot{b}f}\right)\left(\partial^{f\dot{e}}_{I}\psi^{*}_{\dot{e}a}\right) \end{aligned} \tag{4.2.2.}$$

If the subtraction operation here is replaced by an additon operation, (4.2.2) would take a form similar to the first two terms in the Lagrangian density L_m of (6.1.1b) below and has the same dimension. The associated action (6.1.1a) is a real dimensionless Lorentz scalar. C_m and D_m, like L_m, are by virtue of their spinor form manifestly Lorentz invariants. Therefore,

$$\begin{aligned} &\int d^4x_I d^4x_{II}\left(C_m - C^{*}_m\right) = \int d^4x_I d^4x_{II}\left(D_m - D^{*}_m\right) \\ &= \text{imaginary, dimensionless Lorentz scalar} \end{aligned} \tag{4.2.3}$$

We now specialize to the rest frame mesons characterized by the second of (3.2.10b) and of (3.2.11a) together with (3.2.5). With the aid of (3.1.1, 3, 4, 5, 9) and (3.5.6, 7), D_m can be simply shown to vanish. Noting that the meson wave functions must vanish at large r, (4.2.3) with (3.1.3c) becomes [H13 (3.6)]

$$\frac{1}{2}\int d^4x_I d^4x_{II}\left(C_m - C^{*}_m\right) = \int dX^0 \int \frac{\partial}{\partial X^0}\int dx^0 N_{mJ} = 0 \tag{4.2.4}$$

$$N_{mJ} = -i\frac{E_{J0}}{2}\Omega\int d^3\underline{x}\psi^2_J(r), \qquad \Omega = \int d^3\underline{X}, \qquad J = 1, 0 \tag{4.2.5}$$

Equations (4.2.3) and (4.2.4) require that N_{mJ} be a conserved quantity, at least in the limit of motionless mesons, of dimension energy. For a ground state meson at rest, the only observable quantity of this type is its mass E_{J0}. Therefore, the identification [H13 (3.7a)]

$$N_{mJ} = -iE_{J0}\,\Omega/2\Omega_c \tag{4.2.6}$$

is made. Ω/Ω_c is a Lorentz scalar and Ω_c is interpreted as some finite volume in laboratory space $\underline{X}$ occupied by meson wave functions which acquire finite amplitudes due to interaction with other hadrons (see Sec. 4.5 below). For mesons far away from other hadrons, i.e., free mesons, these amplitudes vanish as [H13 (3.7b)]

$$\Omega_c \to \Omega \to \infty \tag{4.2.7}$$

and the meson wave function is spread over the entire laboratory space.

The normalization assignment (4.2.6) is the consequence of the appearance of the *relative time* x^0 in the Lorentz invariant formulation (see (4.2.4)). It largely resolves simultaneously the two seemingly unrelated questions raised by Wick [W4] on the Bethe-Salpeter amplitude [S3], namely, the lack of a positive definite norm and the unknown role of the relative time x^0.

From (4.2.5, 6), one finds the normalization condition

$$\int d^3\underline{x}\psi_J^2(r) = 1/\Omega_c, \qquad \psi_J \propto 1/\sqrt{\Omega_c} \tag{4.2.8}$$

which is independent of the meson energy E_{J0} and thus differs in this respect from the KG case of (4.1.4, 8). This fundamental difference stems from the presence of relative time x^0 here absent in the point particle Klein-Gordon case. On the other hand, (4.2.8) with (4.2.7) is in harmony with the amplitudes of the Schrödinger wave function in (4.1.2) and the Dirac wave function in (A13, A14).

The choice of i in (4.1.7) is not aestheticaaly appealing and is formally at odds with the normalization (4.1.1). On the other hand, the factor i in (4.2.6) is necessitated by (4.2.5) and is naturally associated with E_{J0} to form the time component of a four vector required by (4.2.3, 4). The arbitrariness of (4.1.1, 7), hence also of the amplitudes in (4.1.2, 4), is absent here and the amplitudes of the meson wave functions in (4.2.8) are no longer scalable and are now fixed (see (4.3.2, 3) below).

4.3. Superposition Principle and Meson Wave Functions

4.3.1. Superposition Principle and Linear Meson Wave Equations

The linear superposition principle is a basic property of quantum mechanics. It is based upon the linearity of the Dirac equation. Different Fourier components of the wave functions can be added together to form wave packets, such as those present in a hydrogen atom. If such a wave packet is free, i.e., not affected by any external potential, and behaves like a Schrödinger particle in laboratory space, it will spread out in time accordance with the uncertainty principle [S5 §12].

The π^- meson can be captured by a proton to form a mesic atom. Comparison of this atom to the hydrogen atom above requires that this π^- meson must obey linear wave equations so that wave packets can be formed, at least when it is free or is captured by the Coulomb potential of the proton. In the latter case, the π^- meson is still very far away from the proton so that it is not in strong interaction with it and therefore can also be regarded as free. This linearity requirement is achieved by the free meson condition (4.2.7); whence the wave function amplitude (4.2.8) as well as the nonlinear confining potential Φ_{co} in (3.2.12) vanish and the meson wave equation (3.4.1) become linear. Although this linearity result has been obtained for rest frame π^- mesons, it obviously holds for the $\underline{K}\neq 0$ case as well except that (3.4.1) is to be replaced by the more general (3.1.8) with the integral in (3.2.8) put to zero.

Such a state of the π^- meson is also obtainable by applying the energy minimization principle to (3.4.1). The meson energy E_0 there is minimized if the confining potential Φ_{co} vanishes. This is achieved by letting the π^- meson wave packet spread out over a large volume $\Omega\to\infty$, whereby the amplitude of its wave function vanishes according to (4.2.7, 8) and $\Phi_{co}\to 0$ follows from (3.2.12). In the more general $\underline{K}\neq 0$ case, (3.1.8) shows that the confining potential Φ_c of (3.2.8) enters only as $\Phi_c+M_m^2$, which may be regarded as an effective average quark mass squared. Since the total meson energy E_K must increase with the last mentioned quantity, it is minimized by $\Phi_c\to 0$, just as the $\underline{K}=0$ case above.

4.3.2. Solutions to Linear Meson Wave Equations

The linearized (3.4.1) is special case of the linearized (3.4.2b) which has the same form as the radial equation for the hydrogen atom [S5 §16] and, together with the normalization conditions (4.2.5-7), has the solution

$$\psi_J(\rho) = A_{Nl} L_{N+l}^{2l+1}(\rho)\rho^l \exp(-\rho/2) \tag{4.3.1a}$$

$$\rho = r\sqrt{4M_m^2 - 4d_{m0} - E_{J0}^2} \tag{4.3.1b}$$

$$A_{Nl} = -\sqrt{\frac{(N-l-1)!}{2N(N(N+l)!)^3}\frac{d_m^3}{4\pi\Omega}} \tag{4.3.1c}$$

$$N = n_r + l + 1 = d_m \Big/ \sqrt{4M_m^2 - 4d_{m0} - E_{J0}^2}\,, \qquad n_r, l = 0,1,2... \tag{4.3.1d}$$

where n_r is the radial quantum number, N the total quantum number and L the associatied Laguerre polynomial. For π^-, $l=n_r=0$, (4.3.1) reduces to the relative space solution [H13 (2.2a), (2.3a)],

$$\psi_0(r) = \sqrt{\frac{d_m^3}{8\pi\Omega}} \exp(-d_m r/2) \tag{4.3.2}$$

which holds for all two quark pseudoscalar mesons according to Sec. 4.4 below.

Similarly, $\Phi_{c1}(r)\to 0$ will linearize (3.4.3) which is also a special case of the linearized (3.4.2b). The solution (4.3.1) restricted to $l=1$ applies to $\psi_1(r)$ of (3.2.5b), as will be justified in Sec. 4.4. Thus, for two quark vector mesons, $n_r=0$ and $l=1$, and (4.3.1) leads to the solution in relative space [H13 (2.2b), (2.3b)]

$$\psi_1(r)=\sqrt{\frac{d_m^5}{3072\pi\Omega}}\, r\exp(-d_m r/4) \tag{4.3.3}$$

4.4. Flavor Independence in Relative Space

The meson wave function in relative space (4.3.2) was derived for π^- but shows no dependence on flavor. Therefore, it holds for all rest frame two quark pseudoscalar mesons. This is in contrast to the KG wave function (4.1.4, 8) which depends upon the particle energy E and consequently the flavor content of the particle.

If the pseudoscalar meson moves slowly, the singlet wave function (4.3.2) will be augmented by a "small component" triplet part determined from (3.1.8) (see Sec. 3.5). In this case, $\Phi_{c0}(r)\to 0$ of §4.3.1 is replaced by putting $\Phi_c(\underline{x})$ of (3.2.8b) to zero so that (3.1.8) becomes linear. This linearized (3.1.8) also leads in the $\underline{K}=0$ limit, as in Sec. 3.3, to (3.4.1, 3) with $\Phi_{cJ}\to 0$. Equation (4.3.2) is the ground state solution of (3.4.1) and has been assigned to the pseudoscalar mesons. Similarly, (4.3.3) is the ground state solution of (3.4.3) and isanalogously assigned to the corresponding vector mesons. Since the relative space solution (4.3.3) is also flavor independent, it holds for all rest frame two quark vector mesons.

Putting $N=1$ and 2, (4.3.1d) gives the ground state pseudoscalar meson mass E_{00} and vector meson mass E_{10}, respectively. With (2.3.27), (4.3.1d) can be written as

$$E_{J0}^2-(m_p+m_r)^2=-4d_{m0}-\left(d_m/(J+1)\right)^2 \tag{4.4.1}$$

The first term has its origin in the laboratory space time X^μ and the second term, the quark mass term, in the internal space z of Sec. 2.3. The right side of (4.4.1) is associated with the relative space $\underline{x}$ and depends only upon the integration constants d_m and d_{m0} in the quark-quark strong interaction potential (3.2.8a) and does not depend on the flavor of the quarks. It corresponds to the "kinetic energy" of relative quark motion in Schrödinger type of description. The flavor independence of (4.4.1) has already exhibited itself in the flavor independence of the two quark ground state meson wave functions in relative space, i.e., (4.3.2, 3). It can be considered as a symmetry under exchange of quark flavors, $p,r=1,2,\ldots 6$ in (4.4.1), and be called *flavor symmetry in relative space.* The corresponding conserved quantity is the left side of (4.4.1) with a value given by the integration constants in relative space on the right side of (4.4.1).

It is this flavor symmetry that renders the by itself badly broken SU(6) symmetry in internal space of §2.3.2 useful in classifying ground state mesons.

4.5. Confinement of Ground State Mesons

The hydrogen atom, describable by (4.3.1a), can disintegrate when interacting with another particle because there is no confining mechanism for the electron. This is no longer the case for ground state hadrons; the confinement term $\Phi_{cJ}(\underline{x})$ of (3.2.12) can be called into action to prevent their disintegration. Such confinement in rest frame has already been demonstrated in (3.2.21). The actual problem of a meson interacting with another hadron or lepton has not been treated in the framework of the scalar strong interaction hadron theory. Therefore, only a qualitative account for this process will be given.

Consider a free, ground state pseudoscalar meson at rest. It is described by (3.1.5a, 7a) together with (4.3.2), which is uniformly distributed in the laboratory space $\underline{X}$ and has vanishing amplitudes in the $\Omega\rightarrow\infty$ limit. This has led to that the confining potential Φ_{c0} vanishes in (3.4.1). When another hadron approaches this free meson, the quarks in the meson starts to experience the massless scalar forces from the quarks in the approaching hadron according to equations of the form of (2.1.3, 4). These forces produce a perturbation in the meson wave functions in (3.1.5) so that they acquire gradients in $\underline{X}$ and hence also nonzero amplitudes in the interaction region. This in turn makes the right side of (2.3.23) nonzero and hence also the confining potential corresonding to Φ_{cJ} of (3.2.12).

This tendency is reinforced as the hadron moves closer to the meson. But then the confining potential will no longer be of the simple form (3.2.17) or even (3.2.8b) but will depend upon $\underline{X}$ and requires the application of the more general (2.3.23). The free meson wave functions (3.1.5a, 7a) together with (4.3.2), originally spread out over a large volume $\Omega\rightarrow\infty$, are "pushed back" by the approaching hadron so that Ω now depends upon $\underline{X}$ and approaches some effective finite volume Ω_c, partially indicated by the reverse of (4.2.7). This tends to make (4.3.2) finite which in its turn causes the confining potentials corresonding to (3.2.17) or (3.2.12) to become nonzero so that it starts to confining the 0^- meson wave functions indicatd by (3.2.21).

4.6. Excited Mesons and Confinement

The ground state singlet and triplet l=0 mesons in §3.4.1 and §3.4.3 have been treated in §4.3.2 and Sec. 4.5. All other mesons in the classification scheme §3.4.1 to §3.4.4 are excited mesons. For the excited mesons classified according to §3.4.1 to §3.4.3, the masses predicted from (4.3.1d), obtained by neglecting the confinement term Φ_{cJ}, turn out to be much too low(see Tables 5.6, 5.7 and 5.9 below). Actually, it is the spectra of both the ground state and excited mesons that led to the linear plus Coulomb type of potentials in the potential models of the late 1970's [L3]. These models are supported by the meson equations of Sec. 3.4, noting the linear confinement (3.2.19) at large interquark distances.

To account for the masses of the excited mesons, therefore, the confining potential Φ_c (3.2.8b) or rather Φ_m of (2.3.23) needs be included. This inclusion renders that the meson wave equations are no longer linear and the superposition principle of §4.3.1 breaks down. Thus, it is not possible to form wave packets for the excited mesons. Practically, this implies that they cannot be captured by a proton to form an atom, of the type of the mesic atom of §4.3.1, and that they cannot be observed as freely moving particles, as the pions or kaons can. Further, the wave functions are no longer spread out in a large volume, as in the free meson case (4.2.7), but are confined in some limited volume Ω_c in laboratory space. This in its turn leads via (2.3.23) to that Φ_m depends now upon both $\underline{X}$ and $\underline{x}$ so that the separable ansatz (3.1.5) no longer applies to the wave functions in (2.3.22).

The excited mesons therefore differ basically farom the ground state mesons but there is no contradiction to data. These excited states are more resonances than particles and have, with a few exceptions, all very large widths of the order of 100 Mev. By the uncertainty principle, this width corresponds to a time interval of $.66\times10^{-23}$ s. Even with the speed of light, such an excited meson can only travel 2 fm prior to its disintegration. In such short times and distances, there is no experimental technique to determine the wave packet nature of such mesons. A couple of excited $\bar{b}b$ states have width of tens of kev, but they have also not been seen as freely moving wave packets.

4.7. Sizes of Mesons in Laboratory and Relative Space

Combining (4.2.8) and (3.2.19), we find that, except for triplet, $l>0$ mesons in §3.4.4,

$$\frac{1}{\Omega_c} = \int d^3\underline{x}\left|\psi_J(r)\right|^2 = \frac{8\pi}{g_s^4}\beta_{mJ} = \left[\frac{8\pi}{g_s^4 r}\Phi_{cJ}(r)\right]_{r\to\infty} \tag{4.7.1}$$

4.7.1. Sizes of Ground State Mesons

According to (4.3.2, 3), the size of all free pseudoscalar mesons and vector mesons in laboratory space $\underline{X}$ is

$$\sqrt[3]{\Omega} \to \infty \tag{4.7.2}$$

In this limit, ψ_J as well as the confinement potential Φ_{cJ} vanish by (4.7.1) with (4.2.7). In relative space, the half width r_{0h} and the scale r_0 of $\psi_0^2(r)$, the square of the pseudoscalar wave function (4.3.2), are

$$r_{0h} = \ln 2/d_m \approx 1 \text{ fm}, \qquad r_0 = 1/d_m \tag{4.7.3}$$

where (5.2.3) below has been used. The maximum of (4.3.3) characterizes the size of all free vector mesons

$$r_1 = 4/d_m \approx 5.7 \text{ fm} \tag{4.7.4}$$

4.7.2. Sizes of Excited Mesons

In Sec. 4.6, it has been pointed out that the wave functions of excited mesons are no longer spread out over a large volume, as in (4.7.2), but is limited to some volume Ω_c in laboratory space. The sizes of these wave functions in relative space are determined by both the Coulomb type of potential d_m/r and the confinment potential Φ_{cJ} in (3.4.1-3) or more correctly by Φ_m of (2.3.23). In the limit of large Ω_c, $\Phi_{cJ}\to 0$ and the sizes of the wave functions in relative space can be obtained from the linear results (4.3.1). These are typically >15 fm and depend upon the radial and orbital quantum numbers.

However, the small Ω_c limit is much closer to the actual cases. As was mentioned in Sec. 4.6, the wave functions are no longer evenly distributed in the laboratory space $\underline{X}$ and separable in X and x according to the ansatz (3.1.5). The above results derived from this ansatz therefore do not apply to the excited mesons. But an adequate treatment of the nonseparable and nonlinear problem (2.3.22, 23) is beyond the scope of this book. To proceed, it will be *assumed* that the wave functions are independent of $\underline{X}$ inside a sphere of volume Ω_c and zero outside it,

$$\begin{aligned} \psi_{exc}(X,x) &= \psi_{exc}(x)\exp\left(-iE_{exc}X^0\right) && \text{for} \quad \left|\underline{X}\right|^3 < 3\Omega_c/4\pi \\ \psi_{exc}(X,x) &= 0 && \text{elsewhere} \end{aligned} \tag{4.7.5a}$$

where the subscripts *exc* refers to excited mesons, and that results derived from (3.1.5) can be used. The effect of the discontinuity in (4.7.5a) is neglected. Obviously, these results may serve to provide some indication or some gross features but cannot be fully trusted.

For small Ω_c, Φ_{cJ} by (4.7.1) is appreciable relative to the $1/d_m$ term in (3.4.1-3). The asymtotic scale $r_{exc.asymt}$ of the square of the excited meson wave functions in relative space is then estimated from (3.2.21) and is related to their size in laboratory space $\Omega_c^{1/3}$ via (4.7.1),

$$r_{exc.asymt} = \left(\frac{3}{4}\right)^{2/3} \beta_{mJ}^{-1/3} = \left(\frac{9\pi}{2g_s^4}\right)^{1/3} \sqrt[3]{\Omega_c} \tag{4.7.5b}$$

If the strong coupling $g_s^2/4\pi$ is related to the electromagnetic coupling according to the conjecture (8.3.11) below, the proportional constant on the right side of (4.7.5b) is 1.02.

4.8. Fundamental Mesonic Length $1/D_M$

§4.7.1 shows that free, ground state mesons have a size of the magnitude of $1/d_m$ in relative space. This is due to that these mesons are confined by d_m/r term in (3.4.1, 3), with the confining term Φ_{cJ} put to zero. These equations, as was pointed out in §4.3.2, have the same form as the radial wave equation of the hydrogen atom. Therefore, d_m/r plays the same role as the Coulomb potential e^2/r plays in the hydrogen atom. This identification lends itself to the suggestion that r_0 of (4.7.3) is a fundamental mesonic length scale, just like that e is the fundamental charge constant in electromagnetic interactions. r_0 provides a reference length scale for basic mesonic phenomena, analogous to that e, together with the electronic mass, controls the hydrogen atom radius, which provides a reference length scale for atomic phenomena. d_m itself then provides a fundamental mesonic energy scale for mesonic phenomena.

Chapter 5

MESON SPECTRA AND CLASSIFICATION

In this chapter, the theoretical results of the last chapters will be applied to the mesons in the suggested quark model assignments for most of the known mesons in the Particle Data Group data [P1]. The development follows closely that of [H6].

5.1. ELECTROMAGNETIC MASSES OF MESONS

Equation (4.4.1) can be written as

$$E_{J0} = \pm\sqrt{(m_p + m_r)^2 - 4d_{m0} - (d_m/(J+1))^2} \tag{5.1.1}$$

The upper sign is used, unless stated otherwise. As was pointed out below (4.4.1), this meson mass is due to the quark masses of flavors p and r and their mutual strong interaction only. However, there is also a small electromagnetic contribution E_{emJ} due to the meson charge Q_m.

The electromagnetic radii r_{em0} of π^- and K^- are about 0.7 and 0.6 fm, respectively [D2, B3]. These are of the same magnitude as the strong interaction radius r_{0h} of (4.7.3), which holds for all pseudoscalar mesons due to the flavor independence of Sec. 4.4. The strong interaction and electromagnetic radii are known to be of the same magnitude. This is supported by the conjecture of (8.3.11) below that the electromagnetic and strong interaction couplings are directly related to each other. Therefore, flavor independence is also extended to r_{em0} so that it is the same for all pseudoscalar mesons having the same absolute charge. Take r_{em0}=0.7 fm, the corresponding electromagnetic mass is

$$Q_m^2 E_{em0} = Q_m^2\, e^2 / r_{em0} = Q_m^2 \times 2.1 \text{ Mev} \tag{5.1.2}$$

The entire consideration for pseudoscalar mesons can be taken over to apply for vector mesons. Due to their much larger radius (4.7.4), we can put

$$E_{em1} \approx 0 \tag{5.1.3}$$

relative to the much larger E_{10}. The total meson mass to be compared to data is

$$E_{mJ} = E_{J0} + Q_m^2 E_{emJ} \tag{5.1.4}$$

5.2. Value of Fundamental Mesonic Energy Scale D_M

Subtract (4.4.1) with J=0 and 1 from each other and make use of (5.1.4). One obtains the flavor independent relation

$$E_{m1}^2 - \left(E_{m0} - Q_m^2 E_{em0}\right)^2 = 3/4d_m^2 \tag{5.2.1}$$

This prediction, that the difference of the squares of the vector and pseudoscalar mesons is constant, has long ago been approximately deduced from data [F1, L3]. The masses of two quark ground state mesons [P1], (5.2.1) and the so-determined integration constant d_m are given in Table 5.1.

Table 5.1. Masses in units of Gev of two quark ground state meson [P1], the differences of the squares of the vector and pseudoscalar meson masses and the fundamental mesonic energy scale d_m according to (5.2.1)

	$\pi^\pm$	$K^\pm$	K^0	$D^\pm$	D^0	$D_s^{\ +}$	$B^\pm$	B^0	$B_s^{\ 0}$	η_c
E_{m0}	.13957	.493677	.497648	1.8693	1.8645	1.9682	5.279	5.2794	5.3675	2.9804
	$\rho^\pm$	$K^{*\pm}$	K^{*0}	$D^{*\pm}$	D^{*0}	$D_s^{\ *+}$	B^*	B^*	$B_s^{\ *0}$	J/ψ
E_{m1}	.7755	.89166	.8961	2.01	2.0067	2.1124	5.325	5.325	5.4166	3.09687
(5.2.1)	.574	.55341	.55532	.55366	.5501	.5955	.50994	.4846	.50692	.71145
d_m	.8813	.859	.8605	.8592	.85644	.89107	.82457	.8038	.82213	.974

One sees that the flavor independent (5.2.1) holds well despite the large mass differences between the π and the B_s mesons. The average value of d_m, excluding the last column, is 0.85 Gev. A value closer to the light meson end, which is of greater basic importance, will be chosen to represent the average of the third row in Table 5.1;

$$E_{m1}^2 - \left(E_{m0} - Q_m^2 E_{em0}\right)^2 \approx 0.56 \ \text{Gev}^2 \tag{5.2.2}$$

This yields the fundamental mesonic energy scale

$$d_m = 0.864 \ \text{Gev} \tag{5.2.3}$$

which is not too far from the average value above and will be used in this book.

The deviations of the values in the third row of Table 5.1 from a constant value, as in (5.2.2), show that the flavor symmetry of (5.2.1) and Sec. 4.4 is broken and also give the magnitude of this *flavor symmetry breaking*, which can be up to 10%. Here, it may be noted that D_s is the singlet member of a broken internal triplet, the pions and the ρ's belong to an isotriplet, while the remaining mesons belong to isodublets, except for B_s. However, η_c is a mixed meson containing $u\bar{u}$ and is therefore not a two quark meson; (2.3.27) and hence (5.1.1) do not strictly hold here (see (2.3.27) ff and Sec. 5.4 below). Also, such flavorless mesons may involve some conjectured coupling between the internal space (z_I, z_{II}) and relative space $\underline{x}$, both "hidden", mentioned in Sec. 5.4 of Appendix G. In this case, the total separable wave function in (2.3.19) using (3.1.5b) is generalized;

$$\chi^f_{\dot{b}}(x_I, x_{II})\xi^p_r(z_I, z_{II})$$
$$\to \exp(-iK_\mu X^\mu)\left(\chi^f_{\dot{b}}(\underline{x})\xi^p_r(z_I, z_{II}) + \chi^{f\ \ p}_{1\xi\dot{b}_r}(\underline{x}, \text{real combinations of } z_I z_{II})\right) \tag{5.2.4}$$

where the last term χ_I is small but not separable into $\underline{x}$ and z parts. This conjecture is aimed at a possible account of the fact that the last column in Table 5.1 yields a stronger breaking of the flavor independence symmetry. Such a scenario is also supported by deviations from flavor symmetry in the baryon case mentioned at the end of Sec. 11.1 and Sec. 11.4 below.

5.3. Quark Masses and Ground State Meson Spectra

Next, the five quark masses m_1,... m_5 and the integration constant d_{m0} in (5.1.1) are determined. Inserting the masses of π^+, K^+, K^0, D^0 and B^0 in Table 5.1 into (5.1.1-4), the five quark masses are expressed as functions of d_{m0}, which in its turn is fixed by the D_s^+ mass. The result is given in Table 5.2.

Table 5.2. Quark masses and d_{m0} obtained from (5.1.1-4) using the masses π^+, K^+, K^0, D^0, D_s^+ and B^0 in Table 5.1 and quark contents of [P1 Table 13.2]

m_1(Gev)	m_2-m_1	m_3	m_4	m_5	d_{m0}(Gev^2)
0.6592	0.00215	0.7431	1.6215	4.7786	0.24455

It may be noted that the u and d quark masses difference 2.15 Mev is about the same as the electromagnetic mass E_{emo} of (5.1.2).

With these numbers, the remaining masses of two quark pseudoscalar and vector mesons are predicted from (5.1.1-4). The results are compared to data in Tables 5.3 and 5.4.

Table 5.3. Predicted and measured [P1, except where stated otherwise] masses of flavored pseudoscalar mesons in Mev. The particle symbols stand for their masses.

	Prediction	Data
B_s^0	5363.3	5367.5±1.8 5368.6±5.6 [A1]
B_c^+	6266	6286±5
$D^+–D^0$	4.73	4.8±0.8
$B^0–B^+$	0.11	0.4±1

Table 5.4. Predicted and measured [P1] masses of flavored vector mesons in Mev. The particle symbols stand for their masses.

	Prediction	Data
ρ^+	760.9	775.5±0.4
K^{*+}	895.35	891.66±0.26
$K^{*0}-K^{*+}$	3.4	4.34±0.51
D^{*0}	2009.2	2006.7±0.5
$D^{*+}–D^{*0}$	2.4	3.3±0.8
D_s^{*+}	2104	2112±0.6
B^{*0}	5331.8	5325±0.6
$B^{*0}–B^{*+}$	2.2	-
B_s^{*0}	5415.2	5411.6±1.6
$B_s^*-B_s$	52	46.1±1.5
B_c^*	6308.4	-

The predictions of Table 5.3 are within the error limits of data. The predictions of Table 5.4 are also in relatively good agreement with data, in view of the very large width of ρ and K^* and that the effect of the breaking of the flavor symmetry mentioned below (5.2.3) has not been considered. Note that one of the predictions in Table 5.4 must be removed because it has been used in (5.2.2) to determine d_m. Further, if the vector meson radius r_l in (4.7.4) were of the same magnitude as the pseudoscalar meson radius r_0 so that the electromagnetic radius of the vector meson is about equal to the r_{em0} in (5.1.2), (5.1.3) would be replaced by $E_{eml}=E_{em0}$. This would alter the predicted $K^{*0}–K^{*+}$=3.4 to 1.3 and $D^{*+}–D^{*0}$=2.4 to 4.5 and deteriorate the rough agreements with data in Table 5.4. This lends support to that r_l>>r_0 in (4.7.3, 4).

For pions, the quark mass term in (4.4.1) nearly cancels the "pseudo-kinetic energy" terms on the right side of (4.4.1). This results in a very small mass E_{00} for pions. In conventional terms, this corresponds to a highly relativistic case. For heavy mesons, such as the *B* mesons, the quark mass term in (4.4.1) is much greater than the same "pseudo-kinetic energy" term. The *B* meson mass is nearly the same as the sum of the both quark masses. This corresponds conventionally to quarks moving nonrelativistically.

5.4. Ground State Mesons with Closed Flavor and GMO Formula

The mesons considered in the last section have open flavor and hence contains at most two quarks. Mesons with closed flavor can contain four or six flavors, as are exemplified by

π^0 and η. The present results for two quark mesons is strictly not applicable (see §2.4.3) and additional approximations need to be introduced to obtain estimations of their masses. One such approximation a) is the replacement of (2.3.27) by the third of (2.4.13), (2.4.15) and (2.4.17) for π^0, η and η', respectively. As was pointed out in §2.4.3, this approximation is good for π^0 but not for η and η'. Another approximation b) is to regard the meson masses as some average of the masses E_{011}, E_{022} and E_{033} (see (2.4.11)) of the fictitious mesons consisting of $u\bar{u}$, $d\bar{d}$ and $s\bar{s}$, respectively, properly weighed by the internal wave functions in (2.4.8, 16).

In Sec. 6.2 below, it will be shown that pseudoscalar isosinglets with quark contents $u\bar{u}+ d\bar{d}$, $s\bar{s}$, $c\bar{c}$, $b\bar{b}$ or $\bar{t}\,t$, which include the fictitious states mentioned in the last paragraph, cannot exist on the grounds of U(1) gauge invariance and the masslessness of the photon. Therefore, a pseudoscalar meson with closed flavor does not have to have a counterpart in a vector meson with the same quark content. This differs from the open flavor cases in Table 5.1 where each pseudoscalar meson has a vector meson counterpart with the same quark content. The spectra of the closed flavor mesons are obtained from (5.1.1-4), using the approximations a) and b) above in the applicable cases, and are compared to data in Table 5.5.

Table 5.5. Estimated and measured [P1] masses in Mev for ground state mesons with closed flavor. *I* denotes the isospin of the multiplet the meson belongs to, *J* the spin and *P* the parity. The superscripts a) and b) refer to te approximations mentioned above in this section. Under b), the masses of the fictitious pseudoscalar mesons $u\bar{u}$, $d\bar{d}$ and $s\bar{s}$ are obtained from (5.1.1-4) and are in the notation of (2.4.11) E_{011}=115.018, E_{022}=156.767 and E_{033}=695.8, respectively. For vector mesons, they are E_{111}=757.1 for $u\bar{u}$ and E_{122}=764.6 for $d\bar{d}$.

IJ(*P*)	Meson	Quark Content	Prediction	Data
10(–)	π^0	$u\bar{u}- d\bar{d}$	137.486[a)] 135.893[b)]	134.9766
00(–)	η	$u\bar{u}+ d\bar{d}-2\, s\bar{s}$	573.65[a)] 509.164[b)]	547.51
00(–)	η'	$u\bar{u}+ d\bar{d}+ s\bar{s}$	417.1[a)] 322.528[b)]	957.78
00(–)	η_c	$c\bar{c}+\ldots$	2965.2	2980.4
01(–)	ω	$u\bar{u}+ d\bar{d}$	760.8[b)]	782.65
01(–)	ϕ	$s\bar{s}$	1021.84	1019.46
01(–)	J/ψ	$c\bar{c}+\ldots$	3058.7	3096.916
01(–)	$\Upsilon(1S)$	$b\bar{b}$	9496	9460.3

Differences between the predicted and measured masses in Table 5.5 are on the average much greater than the corresponding ones for mesons with open flavor in Tables 5.3 and 5.4. These are presently attributed to the approximations a) and b) mentioned above in this section and to the uncertain quark content due to mixing, in addition to the conjecture of coupling of space-time functions to internal functions indicated in (5.2.4).

The predicted π^0 mass using approximation b) is rather close to data. In fact, it can be adjusted to fit data exactly if the electromagnetic radius r_{em0} in Sec. 5.1 is taken to be 0.55 fm instead of 0.7 fm so that E_{em0} is 2.67 Mev instead of 2.1 Mev in (5.1.2). The η mass lies between the two approximatively predicted values.

The predicted η' masses are much too low. However, the conventional quark content of η' is an SU(3) singlet (2.4.16) and falls into the category of nonexisting pseudoscalar isosinglets $\bar{u}u+\bar{d}d$, $\bar{s}s$, $\bar{c}c$... mentioned above in this section. Such an η' can therefore not exist from this point of view. It has been suggested [H7] that η' has also some $\bar{c}c$ content so that it is no longer an isosinglet. Such a $\bar{c}c$ content can enhance its mass to the observed mass level.

η_c is known to be mixed, as was mentioned at the end of Sec. 5.2. The predicted ϕ mass agrees rather well with data.

The mass squared operator and its eigenvalue in (2.3.21) are now compared to the Gell-Mann-Okubo(GMO) formula which also operates in the same kind of internal space. For mesons, (9.46) with the lower member of (9.49) in [L2] give the GMO squared mass

$$M_{GMO}^2 = a_{8m} + c_{8m}\left(I(I+1) - Y^2/4\right) \tag{5.4.1}$$

Here, I is the isospin, Y the hypercharge, a_{8m} and b_{8m} constants, and the small isospin symmetry breaking is neglected. M_{GMO} has been assigned to be the obseeved meson masses. It however does not include contributions from quark motion and their interactions as is indicated by the d_{m0} and d_m constants in (5.1.1). It is more appropriate to compare M_{GMO} to the eigenvalue $2M_m$ in (2.4.1) representing the sum of the bare quark masses. Equating (5.4.1) to (2.4.1) with $m_1=m_2$ for pion and kaon gives

$$a_{8m} = \frac{4}{3}(2m_1 + m_3)m_3, \qquad c_{8m} = \frac{2}{3}(3m_1 + m_3)(m_1 - m_3) \tag{5.4.2}$$

Application of (5.4.1-2) and (2.4.15) to η leads to

$$M_{m\eta}^2 - M_{GMO\eta}^2 = \frac{4}{3}(m_3 - m_1)^2 \tag{5.4.3}$$

With Table 5.2, this discrepancy turns out to be very small, 0.45% only. The square of the mass summation operator (2.3.26) and (2.3.25) is compatible with the GMO formula here despite their formal difference.

Furthermore, the quark masses in Table 5.2 obtained from meson spectra are also used for baryons in (9.3.19) below. The coefficients in the GMO formula for mesons (5.4.1) are different from those for baryons in (9.3.21) below. Thus, use of GMO formula instead of the simpler but not exact quark mass summation operator in (2.3.26) and (9.3.14) below will lead to three more parameters to be fixed by data. Apart from much reduced predictive power, such a use would greatly complicate the computer calculations in Sec. 11.xc1 below.

5.5. RADIALLY EXCITED MESONS

The ground state meson spectra have been treated in the last two sections. They belong to the categories of singlet and triplet, l=0 mesons of §3.4.1 and §3.4.3, respectively. The remaining mesons in these categories are radially excited states of the same ground state mesons. According to Sec. 4.6, the confinement potential Φ_{cJ} in (3.4.1, 3) contributes substantially to the masses of these excited mesons. As was pointed out in §4.7.2, the assumption (4.7.5a) ff had to be introduced and the results below are indicative only. In this simplified case, the masses of excited mesons are determined by the set of nonlinear integro-differential equations (3.4.1) and (3.4.3) with Φ_{cJ} given by (3.2.17).

The solutions to these equations at small r are ψ_0=constant and $\psi_1 \propto r$, respectively, according to (4.3.1a). At large r, they are given by (3.2.21) together with (3.2.19). A full solution would require numerical treatment, as will be considered in the next section. Even so, the amplitude of the radially excited meson wave function, related to the volume Ω_c of the meson in laboratory space via (4.7.1), remains an unknown parameter. Thus, Φ_{cJ} and the mass of the excited mesons E_{NlJ} for l=0 are determined up to an unknown parameter.

Multiplying the first of (3.4.1) and (3.4.3) by $\psi_0(r)$ and $\psi_1(r)$, respectively, from the left and integrating over $\underline{x}$ leads to

$$E_{N0J}^2 - \left(m_p + m_r\right)^2 = 4\Lambda_{N0J} \tag{5.5.1}$$

$$\Lambda_{N0J} = -\frac{\int dr r^2 \psi_{JN}(r)\left(\frac{1}{r^2}\frac{\partial}{\partial r}r^2\frac{\partial}{\partial r} - \frac{J(J+1)}{r^2} + \frac{d_m}{r} + d_{m0} - \Phi_{cJN}(r)\right)\psi_{JN}(r)}{\int dr r^2 \psi_{JN}^2(r)} \tag{5.5.2}$$

where the total quantum number N has been added as an extra subscript. Λ_{N0J} represents the above mentioned unknown parameter and does not depend upon the flavor content of the meson. Therefore, (5.5.1) is, like (4.4.1), also flavor independent and the flavor symmetry in relative space for ground state mesons in Sec. 4.4 also holds for radially excited states of these mesons.

Due to the high masses, large error limits and large widths of these states, corrections to (5.5.1) due to electromagnetic mass, considered in Sec. 5.1, can be neglected. Thus, for each pair of N and J values, (5.5.1) can be used to predict the meson masses if one of them is used as input parameter to determine Λ_{N0J}. The results of such predictions are compared to data in Tables 5.6 and 5.7.

Table 5.6. Radially excited singlet, l=0 meson (§3.4.1) assignments and masses in Mev. The data [P1] are given in brackets. Below these are the predicted masses from (5.5.1) using Table 5.2, assuming (4.7.5a) ff. J denotes the spin of the state, N is the total quantum number in (4.3.1d). For η, (2.4.15) replaces (2.3.27). § refers to that the quark content of η' in Table 5.5, according to the discussion below it, has been changed so that the predicted values are raised to that of data. * denotes a state not present in the quark model assignments in [P1].

Isospin	$N^{2S+1}l_J=1^1S_0$	$=2^1S_0$	$=3^1S_0$
1	π	$\pi(1300)$ [1200-1400] input	$\pi(1800)$ [1812±13] input
1/2	K	$K(1460)$ [1400,1460] 1291-1478	$K(1830)$ [≈1830] 1863±12
0	η	$\eta(1295)$ [1294±4] 1323-1507	$\eta(1760)$ [1760±11] 1885±12
0	§η'	$\eta(1475)$ [1476±4] 1529-1691	*$\eta(2225)$ [2220±18] 2035±12
0	η_c	$\eta_c(2S)$ [3638±5] 3196-3276	- -

Table 5.7. Radially excited triplet l=0 meson (§3.4.3) assignments and masses in Mev. The data [P1] are given in brackets. Beneath are the predicted masses from (5.5.1) using Table 5.2 assuming (4.7.5a) ff.

Isospin	$N^{2S+1}l_J=1^3S_0$	$=2^3S_1$
1	ρ	$\rho(1450)$ [1465±25] input
0	ω	$\omega(1420)$ [1400-1450] 1465±25
1/2	$K^*(892)$	$K^*(1410)$ [1414±15] or $K^*(1680)$ [1717±27] 1539±24
0	ϕ	$\phi(1680)$ [1680±20] 1616±23
0	J/ψ	$\psi(2S)$ [3686.1±0.084] 3305±11
0	$\Upsilon(1S)$	$\Upsilon(2S)$ [10023.26±0.00031] 9578±4

Differences between the predicted and measured masses in Tables 5.6 and 5.7 are still greater than those in Table 5.5 for the ground state vector mesons. These are attributed to the simplification (4.7.5a) ff, the approximation a) of Sec. 5.4, and the uncertain quarks content due to mixing and flavor symmetry breaking in Sec. 5.2. Other possible error sources in Table 5.6 are the large width of 200-600 Mev of $\pi(1300)$ used as input and that predictions refer to states, except for $\eta(1295)$ and $\eta(1475)$, that are not considered as established [P1 p51]. The states in Table 5.7 are not beset with the uncertainties like those in in Table 5.6. Agreement with data is much better. The disscrepancies for $\bar{c}c$ and $\bar{b}b$ mesons, also present to a less extent in Table 5.5, are presently attributed to flavor symmetry breaking and to the conjecture mentioned at the end of Sec. 5.2.

5.6. Nonlinearly Confined Radial Wave Functions

Tables 5.6 and 5.7 presume that confined solutions $\psi_{JN}(r)$ in (5.5.2) really exist. It may also be of interest to see the forms of such radial wave functions; some examples will now be obtained numerically.

Knowing the initial and asymtotic forms of $\psi_J(r)$ discussd in the beginning of Sec. 5.5, (3.4.1, 3) and (3.2.17) are solved numerically by the following iterational procedure. At first, one guesses a trial meson wave function $\psi_{JN(0)}(r)$, which can be taken to be the corresponding linearized solution (4.3.1). Inserting this $\psi_{JN(0)}$ (r) into (3.2.17), one obtains the trial confining potential $\Phi_{cJN(0)}(r)$. Here, g_s^2 is considered to have been absorbed into $\psi_{JN(0)}(r)$. Alternatively, it can take on the hypothetical value (8.3.11) below. Next insert this potential into (3.4.1) or (3.4.3), which can then be solved as an initial value problem using a Runge-Kutta routine on a computer. M_m is the averaged quark masses listed in Table 5.2. E_{J0} and the integration constants d_m and d_{m0} are then adjusted so that the first iteration solution $\psi_{JN(1)}(r)$ as well as its derivative with respect to r vanishes at large r. Practically, this is the case for a finite range of large r values, beyond which $\psi_{JN(1)}(r)$ diverges. This is due to the finite step length in the Runge-Kutta integration routine.

$\psi_{JN(1)}(r)$ can differ significantly from $\psi_{JN(0)}(r)$. In the next iteration, $\psi_{JN(0)}$ is replaced by $\psi_{JN(1)}$ to obtain $\Phi_{cJN(1)}(r)$ from (3.2.17). Repeating the integration using $\Phi_{cJN(1)}(r)$ instead of $\Phi_{cJN(0)}(r)$, the second iteration solution $\psi_{JN(2)}(r)$ is found and will differ from $\psi_{JN(1)}(r)$ to a less extent. Also, new E_{J0}, d_m and d_{m0} values are obtained. This cycle is repeated until $\psi_{JN(n-1)}(r)\approx\psi_{JN(n)}(r)\rightarrow\psi_{JN}(r)$. The required number of iterations is rather limited due to the weak dependence of $\Phi_{cJ}(r)$ on ψ_J in (3.2.17). It however increases with increasing total quantum number N.

If the amplitude of the trial radial function $\psi_{JN(0)}(r)$ is altered, the amplitude of $\Phi_{cJN(0)}(r)$ will also alter according to (3.2.17). This will allow for other E_{J0}, d_m and d_{m0} values and radial wave function $\psi_{JN(1)}(r)$. This new degree of freedom can be parameterized by Ω_c according to the left of (4.7.1) where ψ_J is replaced by ψ_{JN} here. Ω_c replaces the unknown parameter Λ_{N0J} in Sec 5.5 and corresponds in principle to that these mesons are under some external hadronic influence like that discussed in Sec. 4.5.

The above numerical program has been applied to the ground state mesons of Tables 5.3-5.5 and the first and second radially excited mesons in Tables 5.6 and 5.7, together with Table 5.2. In these calculations, E_{J0} have been taken from data and d_m and d_{m0} have been allowed to vary. The so-obtained radial wave and potential functions for different sets of d_m and d_{m0}, with the accompanying finite Ω_c, can be seen in Figs. 1-3 and Table 9 of [H6], which show that confined solutions do exist.

Here, d_m and d_{m0} are regarded as fundamental constants and cannot be varied. Using E_{J0} data in Tabels 5.6-7, the corresponding Ω_c are

	$\pi(1300)$, 2^1S_0	$\rho(1450)$, 2^3S_1	$\pi(1800)$, 3^1S_0
number of radial nodes	1	1	2
$Q_c^{1/3}$ (fm)	3.6	3.7	3.02

(5.6.1.)

where $g_s^2/4\pi=1/(137)^{1/4}$ of (8.3.11a) has been used. One sees that the so-estimated volume Ω_c for the both mesons with one radial node is about the same and is about twice that of the meson with two radial nodes. These $(\Omega_c)^{1/3}$ lengths are of the same order as the radius of pseudoscalar mesons in the relative space 4.3 fm in (8.4.19) below.

These and (4.7.5b) leads to the conjecture that the meson size $(\Omega_c)^{1/3}$ cannot vary continuously from ∞ to 0 but are somehow "quantized". The only scale available is the fundamental mesonic length $1/d_m$ of Sec. 4.8. Forms of such a conjecture may for insatance be $(\Omega_c)^{-1/3} \propto n_r d_m$ or $\Omega_c^{-1} \propto n_r d_m^{\,3}$. In the latter case, $\Omega_c=\infty$ for the radial quantum number $n_r=0$ and refers to a ground state meson and $\Omega_c \propto 1/d_m^{\,3}$, $1/2d_m^{\,3}$,... for $n_r=1$, 2,... and refer to the first, second,... radially excited mesons.

5.7. Orbitally Excited Mesons

Next, consider the singlet and triplet, $l>0$ mesons of §3.4.2 and §3.4.4, again based upon the assumption (4.7.5a) ff. In these paragraphs, the meson wave functions depend upon both r and ϑ, and perhaps also ϕ. The confining potentials $\Phi_{c0}(\underline{x})$ in (3.2.12) and $\Phi_c(\underline{x})$ in (3.2.8b), putting ψ_0 and χ_0 there to zero, depend therefore also upon r and ϑ. This renders that $\psi_0(\underline{x})$ in (3.210b) cannot be separated, as in (3.4.2a), and that $\underline{\psi}(\underline{x})$ and $\underline{\chi}(\underline{x})$ in (3.2.2) can also not be separated, as in (3.4.4).

5.7.1. Orbitally Excited Singlet Mesons

The nonseparablity of $\psi_0(\underline{x})$ led to that the radial equation (3.4.2b) holds only for small and large r. For general solution, we have to return to (3.2.10b). Multiplying the first of (3.2.10b) by $\chi^*_0(\underline{x})$ and integrating over $\underline{x}$ leads to a form analogous to that of (5.5.1, 2),

$$E_{Nl0}^2 - (m_p + m_r)^2 = 4\Lambda_{Nl0} \tag{5.7.1}$$

$$\Lambda_{Nl0} = -\frac{\int d^3\underline{x}\psi_0(\underline{x})\left(\Delta + \frac{d_m}{r} + d_{m0} - \Phi_{c0}(\underline{x})\right)\psi_0(\underline{x})}{\int d^3\underline{x}\psi_0^2(\underline{x})} \tag{5.7.2}$$

Similarly, (5.7.1) is also flavor independent and can be used for predictions. These states can be approximately classified using the separated solution (3.4.2a) which holds in the small r region. The results are given in Table 5.8.

Table 5.8. Orbitally excited singlet meson (§3.4.2) assignments and masses in Mev. The data are given brackets. Below them are the predicted masses from (5.7.1), assuming (4.7.5a) ff. * in front denotes a state not present in the quark model assignments in [P1]. $N_r = n_r + 1$ where n_r is the radial quatum number in (4.3.1d).

Quark-content	$N_r^{2S+1}l_J = 1^1P_1$ $J^P = 1^+$	$= 1^1D_2$ $= 2^-$	$= 1^1F_3$ $= 3^+$	$= 1^1G_4$ $= 4^-$
$\bar{d}u,\ \bar{d}d,\ \bar{u}u$	$b_1(1235)$ [1229.5±3.2] input	$\pi_2(1670)$ [1672.4±3.2] input		
$\bar{s}u,\ \bar{s}d$	$K_1(1270)$ [1273±7] 1320±3	$K_2(1770)$ [1773±8] 1737±19	*$K_3(2320)$ [2324±24]	*$K_4(2500)$ [2490±20]
$\bar{s}s,\ \bar{d}d,\ \bar{u}u$	$h_1(1170)$ [1170±20] $h_1(1380)$ [1386±19] 1350±3	$\eta_2(1645)$ [1617±5] $\eta_2(1870)$ [1842±8] 1760±19		
$\bar{u}c,\ \bar{d}c$	$D_1(2420)$ [2422.2±1.8] 2232±2			
$\bar{s}c$	$D_{s1}(2536)$ [2535.35±0.34] 2315±2			
$\bar{c}c$	$h_c(1P)$ [3526.21±0.25] 3207±1			

The discrepancies are presently again attributed to the assumption (4.7.5a) ff and the breaking of flavor symmetry, as was mentioned at the end of Sec. 5.5.

As was pointed out above in this section, $\psi_0(\underline{x})$ cannot be separated in r and ϑ, except in the $r\to 0$ and ∞ limits. Thus, the known separable angle-dependent solution (3.4.2a) at $r\to 0$ and the known angle-independent solution (3.2.21) at $r\to\infty$ are connected by an unknown nonseparable wave function $\psi_0(r, \vartheta, \phi)$ in the intermediate r region. The associated radial and orbital quantum numbers n_r and l, meaningful at $r\to 0$ and ∞ are now coupled and possibly lead to a new pair of quantum numbers.

5.7.2. Orbitally Excited Triplet Mesons

In the $r\to 0$ and ∞ limits, the wave functions of these mesons are given by (3.4.6-9). In the intermediate r region, $\Phi_c(\underline{x})$ in (3.2.8b) with ψ_0 and χ_0 there put to zero, again depend upon r and ϑ and possibly also ϕ. This analogously causes that $\underline{\psi}(\underline{x})$ and $\underline{\chi}(\underline{x})$ cannot be put in the separable form (3.4.4). The wave function components are coupled to each other and we have to revert to (3.2.2). The unknown, nonseparable solutions in this region connects the known separable angle-independent solutions at small r (3.4.6-8) to the known angle-independent asymtotic solution (3.4.9).

Equations (3.2.2) and (3.4.5) show that $E_0^2/4$ and M_m^2 no longer appear together as a difference, as in (3.4.1-3), so that flavor symmetry in relative space valid for the last mentioned equations no longer holds in §3.4.4 and here. This loss of flavor symmetry implies that we no longer have a simple mass formula of the type (5.7.1). Instead, (3.2.8b) with ψ_0, χ_0=0 and (3.2.2) are to be solved. The complexities of these equations have however put their solutions beyond reach presently.

According to Sec. 4.6, the confining potential Φ_c provides a significant contribution to the mass of an excited meson, as it does for the radially excited mesons of Sec. 5.5. This potential tends to confine the meson wave functions to a region close to the origin. In this region, the d_m/r term dominates in (3.2.8a) and (3.4.6-8) applies. Therefore, (3.4.8) can be used to approximately classify the triplet, l>0 mesons. For each l, there is a trio of solutions. Each solution generally involves all three spherical harmonics Y_{l+1}, Y_l and Y_{l-1} and is therefore not associated with any definite angular momentum J. It howevr still possesses a definite parity; $\underline{\psi}(\underline{x}) \to \underline{\chi}(-\underline{x})$ under spatial inversion according to (3.3.2b).

The approximate classification is further simplified if one considers only the leading terms in the expansion (3.4.6). To lowest order, (3.4.8a) shows that the wave function is $g_{+}r^{l-1}Y_{l-1}$ and $J=l-1$ is assigned to the associated meson. The wave function associated with (3.4.8b) is $h_{+}r^{l}Y_l$ and $J=l$ for this meson. The wave function associated with (3.4.8c) is $f_{+}(r^{l+1}Y_{l+1}+(l+1)(2l+3)r^{l-1}Y_{l-1})$ and involves both $J=l+1$ and $J=l-1$ components. The parity of these three states is $P=(-)^l$. The meson masses E_0 associated with each of these solutions are in general different for the same quark content and are eigenvalues of the sixth order equation system (3.2.2). Assignments of the corresponding mesons in [P1] according to the wave functions associated with (3.4.8) are given in Table 5.9.

Table 5.9. Orbitally excited triplet meson (§3.4.4) assignments. * on the left denotes states not present in [P1]. C denotes charge conjugation parity where applicable. +... means small admixtures of other $\bar{q}q$ pairs. § denotes possible radially excited mesons. The quantum numbers refer to the meson wave functions (3.4.4) in relative space. The wave functions associated with each λ is given by (3.4.6) and (3.4.8). For each orbital quantum number l, there are three meson states corresponding to $\lambda=l-1$, l and $l+1$ in (3.4.8). In the first two cases, $J=l-1$ and l respectively. For the last case, both $J=l-1$ and $J=l+1$ components are present, with the former dominating over the latter. $N_r = n_r + 1$(see (4.3.1d)).

Quark content	$N_r^{2S+1}l_J=1^3P_0$ $\lambda(PC)$=0(++)	$=1^3P_1$ =1(++)	$=1^3P_2+1^3P_0$ =2(++)	$=1^3D_1$ =1(- -)	$=1^3D_2$ =2(- -)	$=1^3D_3+1^3D_1$ =3(- -)
$\bar{d}u$, $\bar{d}d$, $\bar{u}u$	$a_0(980)$ $a_0(1450)$	$a_1(1260)$	$a_2(1320)$	$\rho(1700)$		$\rho_3(1690)$
$\bar{u}u$, $\bar{d}d$				$\omega(1600)$		$\omega_3(1670)$
$\bar{s}u$, $\bar{s}d$	$K^*_0(1430)$ *§$K^*_0(1950)$	$K_1(1400)$ *§$K_1(1650)$	$K^*_2(1430)$ §$K^*_2(1980)$	$K^*(1680)$	$K_2(1820)$ *§$K_2(2250)$	$K^*_3(1780)$
$\bar{s}s$, $\bar{d}d$, $\bar{u}u$	*$f_0(980)$ $f_0(1370)$ *§$f_0(1500)$ $f_0(1710)$	$f_1(1285)$ $f_1(1420)$	$f_2(1270)$ $f'_2(1525)$ §$f_2(1810)$ §$f_2(2010)$			
$\bar{s}s$						$\phi_3(1850)$
$\bar{u}c$, $\bar{d}c$			$D^*_2(2460)$			

$\bar{c}c+\ldots$	$\chi_{c0}(1P)$	$\chi_{c1}(1P)$	$\chi_{c2}(1P)$	$\psi(3770)$		
$\bar{b}b+\ldots$	$\chi_{b0}(1P)$	$\chi_{b1}(1P)$	$\chi_{b2}(1P)$			
	*§$\chi_{b0}(2P)$	*§$\chi_{b1}(2P)$	*§$\chi_{b2}(2P)$			
Quark content	$N_r^{2S+1}l_J=1^3F_2$ $\lambda(PC)=2(++)$	$=1^3F_3$ $=3(++)$	$=1^3F_4+1^3F_2$ $=4(++)$	$=1^3G_3$ $=3(--)$	$=1^3G_4$ $=4(--)$	$=1^3G_5+1^3G_3$ $=5(--)$
$\bar{d}u,\ \bar{d}d,\ \bar{u}u$			$a_4(2040)$		*$\rho_3(2250)$	*$\rho_5(2350)$
$\bar{s}u,\ \bar{s}d$			$K^*_4(2045)$			*$K^*_5(2380)$
$\bar{s}s,\ \bar{d}d,\ \bar{u}u$	*$f_2(1565)$		$f_4(2050)$			
	*$f_2(1640)$		$f_J(2220)$			
	*§$f_2(2150)$					
	*$f_2(2300)$					
	*$f_2(2340)$					
Quark content	$N_r^{2S+1}l_J=1^3H_4$ $\lambda(PC)=4(++)$	$=1^3H_5$ $=5(++)$	$=1^3H_6+1^3H_4$ $=6(++)$		$=2^3P_2+2^3P_0$ $=2(++)$	
$\bar{d}u,\ \bar{d}d,\ \bar{u}u$			*$a_6(2450)$			
$\bar{s}u,\ \bar{s}d$					$K^*_2(1980)$	
$\bar{s}s,\ \bar{d}d,\ \bar{u}u$	*$f_4(2300)$		*$f_6(2510)$		$f_2(1810)$	
					$f_2(2010)$	
$\bar{b}b+\ldots$					$\chi_{b2}(2P)$	

Some of the f_2 mesons may also be reassigned to $1^3F_4+1^3F_2$ because the J=4 component is smaller than the J=2 component by a factor of 1/36, according to (3.4.8c), and may have eluded observation. For the same reason, the radially excited $2^3P_2+2^3P_0$ states may be modified $2^3F_4+2^3F_2$ states. It is also noted that states corresponding to the middle member of the trio $f(r)Y_{l+1}$, $h(r)Y_l$ and $g(r)Y_{l-1}$ tends to be underrepresented or absent. This may be related to that (3.4.4b), being symmetric in ψ and χ, contains both $f(r)Y_{l+1}$ and $g(r)Y_{l-1}$. While (3.4.4c), being antisymmetric in ψ and χ, contains only $h(r)Y_l$. Except for the last column, only n_r=0 states appear in Table 5.9. For n_r>0, the radial wave functions will extend themselves farther out from the origin. In this region, d_m/r becomes smaller and $\Phi_c(\underline{x})$ becomes greater in (3.2.8a) so that the separable solutions, which led to the classification of Table 5.9, have to be replaced by unknown, nonseparable solutions mentioned in the beginning of this section. This leads to an n_r to l coupling, analogous to that mentioned at the end of §5.7.1, which yields a more complicated and as yet unknown form of classification scheme.

Chapter 6

LAGRANGIAN FORMALISM AND ELECTROMAGNETIC INTERACTIONS

This chapter follows largely Sec. 2-5 of [H8], Sec. 2-4 of [H9] and [H10]. In the last two chapters, the unperturbed or zeroth order solutions to the equations of motion for meson have been found for free mesons at rest. The energy eigenvalues associated with these solutions were applied to account for meson spectra. In this and the following two chapters, first order perturbational problems connected to meson decay will be treated. For this purpose, the meson equations of motion need be cast in the form of actions.

6.1. MESON ACTION

6.1.1. Meson Action And Boundary Conditions

The equations of motion for two quark mesons (2.3.22) can be replaced by the meson action

$$S_m = \int dx_I^4 dx_{II}^4 L_m \tag{6.1.1}$$

$$2L_m = -\left(\partial_I^{a\dot{b}} \chi^*_{\dot{e}a}\right)\left(\partial_{II}^{f\dot{e}} \chi_{\dot{b}f}\right) - \left(\partial_{I\dot{c}a} \psi^{*d\dot{c}}\right)\left(\partial_{II\dot{e}d} \psi^{a\dot{e}}\right) - \left(M_m^2 - \Phi_m\right)\left(\chi^*_{\dot{e}a} \psi^{a\dot{e}} + \psi^{*d\dot{c}} \chi_{\dot{c}d}\right) + c.c. \tag{6.1.2}$$

where

$$\left(\partial_I^{a\dot{b}} \chi^*_{\dot{e}a}\right)\left(\partial_{II}^{f\dot{e}} \chi_{\dot{b}f}\right) = \partial_I^{a\dot{b}} \chi^*_{\dot{e}a} \partial_{II}^{f\dot{e}} \chi_{\dot{b}f} - \chi^*_{\dot{e}a} \partial_I^{a\dot{b}} \partial_{II}^{f\dot{e}} \chi_{\dot{b}f}, \qquad \chi \rightarrow \psi \tag{6.1.3}$$

has been consulted.

The meson wave functions χ and ψ in (6.1.2) must satisfy the following boundary conditions in the relative space x^{μ} and laboratory space X^{μ} via the transformation (3.1.3a). At large $|\underline{x}|$, χ and ψ must vanish if they are to be confined. They and their first derivatives also vanish for free mesons, for which the confining potential is not active, according to (4.3.2, 3). At large X^μ, the conventional boundary conditions for a one particle wave function are taken over here; χ and ψ observe periodic boundary conditions at large $|\underline{X}|$ and are fixed at large times $|X^0|$.

Shortly after the proposal of the Bethe-Salpeter equation, Wick [W4] pointed out that the boundary conditions for the associated wave functions have not been adequately formulated for a general two particle wave function. This status has remained the same to my knowledge. Here, the boundary conditions of χ and ψ at large relative times $|x^0|$ are assumed to be same as those at large $|X^0|$. This is supported by the symmetric character of the dependence of χ and ψ on x_I^0 and x_{II}^0 in (3.1.10b). The linear transformation (3.1.3a) with (3.1.10a) does not alter this form of dependence. Therefore, if χ and ψ depend upon X^0 in a certain way, they are expect to depend upon x^0 in the same way, like their dependences on x_I^0 and x_{II}^0. This assumption is of fundamental importance because it allows us to formulate the present theory in the integral form (6.1.1). The relative time x^0 underlies the normalization condition (4.2.6) as well as the gauge boson mass (see §6.2.2 and (7.4.6) below).

6.1.2. Variational Problem

We now vary S_m with respect to $\chi*$ and $\psi*$ and require it to be an extremum;

$$\delta S_m = 0 \qquad (6.1.4)$$

The boundary conditions for χ and ψ above lead at large $|x^\mu|$ and $|X^\mu|$ to $\delta\chi=\delta\psi=0$ as well as that δ operated on the first derivatives of χ and ψ also vanish. With (3.1.3), the equation of motion for mesons (2.3.22) is recovered from (6.1.1, 2). Variation of S_m with respect to χ and ψ leads to the hermitian conjugate of (2.3.22) describing the corresponding antimeson. It can be seen that (2.3.22) and its complex conjugate are the same except for $x_I \leftrightarrow x_{II}$. This reflects the fact that, apart from internal properties characterized by ξ, the meson is its own antimeson.

In recovering (2.3.22) from (6.1.1, 2), Φ_m has tacitly been assumed to be not affected by the variations of χ and ψ. Applications of the meson action S_m in this book reside in Chapters 6-8 and are limited to free, ground state mesons for which the nonlinear confining term Φ_c in (3.2.8) vanishes according to §4.3.1. The remaining terms in Φ_m are constant functions not affected by variations of χ and ψ. In Chapter 8, Φ_c is a first order perturbed quantity and $\delta\Phi_c$ is therefore a higher order perturbed quantity and can therefore be neglected.

S_m has not been applied to the excited mesons in Sec. 5.5 and 5.7. A possible application of S_m will require a restricted type of variation δ_r instead of δ in (6.1.4). This is due to that Φ_c is no longer small and depends upon variation of χ and ψ according to (3.2.8b). δ_r is assumed to be a subset of the general and arbitrary variation δ such that

$$\int dx_I^4 dx_{II}^4 \delta_r \Phi_m \left(\chi^{*a\dot{e}} \psi_{\dot{e}a} + c.c.\right) = 0 \qquad (6.1.5)$$

where

$$\Box_I \Box_{II} \delta_r \Phi_m = \frac{1}{4} g_s^4 \left(\psi^{e\dot{a}} \delta_r \chi^*_{\dot{a}e} + \chi^*_{\dot{a}e} \delta_r \psi^{e\dot{a}} + c.c. \right) \tag{6.1.6}$$

For the particular case (3.2.8b), $\Phi_m \to \Phi_c$ and

$$\delta_r \Phi_c(\underline{x}) = \frac{g_s^4}{8\pi} \int d^3 \underline{x}' |\underline{x} - \underline{x}'| \mathrm{Re} \begin{pmatrix} \psi(\underline{x}')\delta_r \chi^*(\underline{x}') + \chi^*(\underline{x}')\delta_r \psi(\underline{x}') \\ -\psi_0(\underline{x}')\delta_r \chi_0^*(\underline{x}') - \chi_0^*(\underline{x}')\delta_r \psi_0(\underline{x}') \end{pmatrix} \tag{6.1.7}$$

With (6.1.5), the restricted form of (6.1.4)

$$\delta_r S_m = 0 \tag{6.1.8}$$

together with (6.1.1, 2) will reproduce (2.3.22) with Φ_m containing a confining term Φ_c that can be large.

6.1.3. Inclusion of Electromagnetic Fields

To include electromagnetic fields, (2.3.19) is generalized in the same way the quark wave equations are generalized to (2.3.11);

$$\begin{aligned} &\left(\partial_I^{a\dot{b}} + iq_{op}(z)A^{a\dot{b}}(x_I)\right)\left(\partial_{II}^{f\dot{e}} + iq_{op}(u)A^{f\dot{e}}(x_{II})\right)\chi_{\dot{b}f}(x_I, x_{II})\xi^p{}_r(z,u) \\ &-\left(m_{2op}(z,u) - \Phi_m(x_I, x_{II})\right)\psi^{a\dot{e}}(x_I, x_{II})\xi^p{}_r(z,u) = 0 \end{aligned} \tag{6.1.9a}$$

$$\begin{aligned} &\left(\partial_{I\dot{c}a} + iq_{op}(z)A_{\dot{c}a}(x_I)\right)\left(\partial_{II\dot{e}d} + iq_{op}(u)A_{\dot{e}d}(x_{II})\right)\psi^{a\dot{e}}(x_I, x_{II})\xi^p{}_r(z,u) \\ &-\left(m_{2op}(z,u) - \Phi_m(x_I, x_{II})\right)\chi^*_{\dot{c}a}(x_I, x_{II})\xi^p{}_r(z,u) = 0 \end{aligned} \qquad \text{6.1.9b)}$$

Multiplying these equations by $\xi^{r}{}_p(z,u)$, integrating over the z and u and employing (2.3.14, 24-27) yields the generalized (2.4.3)

$$\left(\partial_I^{a\dot{b}} + iq_p A^{a\dot{b}}(x_I)\right)\left(\partial_{II}^{f\dot{e}} - iq_r A^{f\dot{e}}(x_{II})\right)\chi_{(pr)\dot{b}f}(x_I, x_{II}) - \left(M_m^2 - \Phi_m(x_I, x_{II})\right)\psi^{a\dot{e}}_{(pr)}(x_I, x_{II}) = 0 \tag{6.1.10a}$$

$$\left(\partial_{I\dot{c}a} + iq_p A_{\dot{c}a}(x_I)\right)\left(\partial_{II\dot{e}d} - iq_r A_{\dot{e}d}(x_{II})\right)\psi^{a\dot{e}}_{(pr)}(x_I, x_{II}) - \left(M_m^2 - \Phi_m(x_I, x_{II})\right)\chi_{(pr)\dot{c}d}(x_I, x_{II}) = 0 \tag{6.1.10b}$$

The gauge fields A are introduced at the quark level because $z=z_I$ and x_I in (6.1.9) are coordinates for the same quark I and $u=z_{II}$ and x_{II} are those for quark II.

6.2. U(1) GAUGE INVARIANCE AND THE U(1) PROBLEM

6.2.1. U(1) Gauge Invariance

The equations governing a quantum mechanical particle must be form invariant under Lorentz and appropriate gauge transformations. Lorentz invariance of the Lagrangian density (6.1.2) is manifest due to their spinor form. The simplest local gauge transformations are associated with the U(1) group. To consider such transformations, replace the decomposition (3.1.5) by the more general form

$$\chi^{a\dot{e}}(x_I, x_{II}) = \Psi(X)\chi^{a\dot{e}}(x), \qquad \chi \to \psi \tag{6.2.1}$$

The relative space coordinates x has been considered as "hidden variable" at the end of Sec. 3.1. Only the laboratory coordinates X are observables. The U(1) or Abelian transformation

$$\Psi(X) \to \Psi(X)\exp(ig_1'\varphi_{u1}(X)) \tag{6.2.2}$$

cannot change the physical results entailed by (6.1.2), because only $\Psi^*(X)\,\Psi(X)$, not $\Psi(X)$ itself, is measurable. Here g_I' is a coupling constant and φ_{uI} a real phase function. This necessitates the existence of a U(1) gauge field B in the minimal substituion on the meson level;

$$\partial_I^{a\dot{b}} = \tfrac{1}{2}\partial_X^{a\dot{b}} - \partial^{a\dot{b}} \to \tfrac{1}{2}\left(\partial_X^{a\dot{b}} + ig_1'B^{a\dot{b}}(X)\right) - \partial^{a\dot{b}} \tag{6.2.3}$$

For $\partial_{II}^{a\dot{b}}$, the $\partial^{a\dot{b}}$ term changes sign. Here (3.1.4) and (3.5.7) have been employed. This gauge field $B(X)$ is introduced on the meson level because the coupling constant g_I' is not associated with any specific quark, as q_{op} in (6.1.9) does.

This U(1) gauge field transform like

$$B^{a\dot{b}}(X) \to B^{a\dot{b}}(X) - \partial_X^{a\dot{b}}\varphi_{u1}(X) \tag{6.2.4}$$

Substituting (6.2.3) into (6.1.2), (6.1.1) after switching the upper and lower indices yields

$$S_m \to S_{mu1} = \int dX^4 dx^4 \frac{1}{2}$$

$$\times\begin{Bmatrix} -\left[\left(\tfrac{1}{2}\left(\partial_X^{a\dot{b}}+ig_1'B^{a\dot{b}}\right)-\partial^{a\dot{b}}\right)\Psi^*(X)\chi^*_{\dot{e}a}(\underline{x})\right]\left[\left(\tfrac{1}{2}\left(\partial_X^{f\dot{e}}+ig_1'B^{f\dot{e}}\right)+\partial^{f\dot{e}}\right)\Psi(X)\chi_{\dot{b}f}(\underline{x})\right] \\ -\left[\left(\tfrac{1}{2}\left(\partial_{X\dot{c}a}+ig_1'B_{\dot{c}a}\right)-\partial_{\dot{c}a}\right)\Psi^*(X)\psi^{*d\dot{c}}(\underline{x})\right]\left[\left(\tfrac{1}{2}\left(\partial_{X\dot{e}d}+ig_1'B_{\dot{e}d}\right)+\partial_{\dot{e}d}\right)\Psi(X)\psi^{a\dot{e}}(\underline{x})\right] \\ -\left(M_m^2-\Phi_m(X,x)\right)\left(\Psi^*(X)\chi^*_{\dot{e}a}\Psi(X)\psi^{a\dot{e}}+c.c.\right)+c.c. \end{Bmatrix}$$

(6.2.5)

where use has been made of (3.1.3c, 9) and (3.5.6). The U(1) gauge transformations (6.2.2, 4) leave the meson action (6.2.5) invariant.

The simplest and most common action employed for gauge fields is that for an electromagnetic field,

$$S_{u1}=-\int dX^4\,\tfrac{1}{4}F_B^{\mu\nu}F_{B\mu\nu} \tag{6.2.6a}$$

$$F_B^{\mu\nu}=\partial_X^{\mu}B^{\nu}(X)-\partial_X^{\nu}B^{\mu}(X) \tag{6.2.6b}$$

which is left invariant by the gauge transformation (6.2.4). The four vector B^{μ} is related to the mixed spinor $B^{a\dot{b}}$ by (Cla). The total action of a meson interacting with a U(1) gauge field is

$$S_{Tu1}=S_{mu1}+S_{u1} \tag{6.2.7}$$

6.2.2. Mass And Source of The U(1) Gauge Field

Variation of (6.2.5) with respect to $\chi^*_{\dot{e}a}$ and $\psi^{*d\dot{c}}$ reproduces (2.3.22) with substitutions of the form (6.2.3)

$$\left(\partial_I^{a\dot{b}}+\tfrac{i}{2}g_1'B^{a\dot{b}}(X)\right)\left(\partial_{II}^{f\dot{e}}+\tfrac{i}{2}g_1'B^{f\dot{e}}(X)\right)\chi_{\dot{b}f}(x_I,x_{II})-\left(M_m^2-\Phi_m(x_I,x_{II})\right)\psi^{a\dot{e}}(x_I,x_{II})=0$$

(6.2.8a)

$$\left(\partial_{I\dot{b}c}+\tfrac{i}{2}g_1'B_{\dot{b}c}(X)\right)\left(\partial_{II\dot{e}d}+\tfrac{i}{2}g_1'B_{\dot{e}d}(X)\right)\psi^{c\dot{e}}(x_I,x_{II})-\left(M_m^2-\Phi_m(x_I,x_{II})\right)\chi_{\dot{b}d}(x_I,x_{II})=0$$

(6.2.8b)

together with the subsidiary conditions

$$\frac{i}{2}g_1'\chi_{\dot{b}f}(x_I,x_{II})\left(\partial_X^{a\dot{b}}B^{f\dot{e}}(X)-\partial_X^{f\dot{e}}B^{a\dot{b}}(X)\right)=0 \tag{6.2.9a}$$

$$\frac{i}{2}g_1'\psi^{c\dot{e}}(x_I,x_{II})\left(\partial_{X\dot{b}c}B_{\dot{e}d}(X)-\partial_{X\dot{e}d}B_{\dot{b}c}(X)\right)=0 \tag{6.2.9b}$$

which comes from the last *c.c.* term in (6.2.5) needed to insure the reality of S_{mu1}. They are satisfied at least for plane wave W. For the singlet (6.2.10) below, they specify that $\underline{W}$ must be parallel to its momentum.

Consider at first a pseudoscalar meson at rest

$$\Psi(X) = \exp\left(-iE_0 X^0\right) \tag{6.2.10a}$$

$$\psi^{a\dot{e}}(\underline{x}) = \delta^{a\dot{e}}\psi_0(r) = -\chi^{a\dot{e}}(\underline{x}) \qquad \text{pseudoscalar} \tag{6.2.10b}$$

according to (3.4.1), (3.2.5b) and (3.4.3). Substitute these into (6.2.5) and integrate it over x. The term quadratic in B determining the squared mass of the B field reads

$$\int d^4X M_{u1}^2 \left(B_0(X)^2 - \underline{B}(X)^2\right) \tag{6.2.11}$$

$$M_{u1}^2 = 2g_1'^2 \int d^4x \psi_0^2(r) = 2g_1'^2 \frac{\int dx^0}{\Omega} = 2g_1'^2 \frac{\infty}{\infty} = \text{finite} \tag{6.2.12}$$

where (4.3.2, 3) has been used. The $B(X)$ field is thus massive. The origin of the gauge boson mass will be further discussed in §7.5.4 and §7.6.2 below.

6.2.3. The U(1) Problem

All the observed ground state pseudoscalar mesons belong to internal multiplets of rank two or higher. Thus, the π's belong to an SO(3) isomultiplet. The K's, D's and B's to SU(2) isodoublets and η to an approximative SU(3) isomuliplet. There is no pseudoscalar counterpart to the vector mesons ω($u\bar{u}$+ $d\bar{d}$), ϕ($s\bar{s}$), J/ψ($c\bar{c}$), and Υ($b\bar{b}$), which are U(1) isosinglets. Note that η_c(2980) is not a pure $c\bar{c}$ state but is a mixed state (see end of Sec. 5.2). This is known as the U(1) problem [W5, G4].

There have been many solutions to this problem [see e.g. A2, G4 and references therein]. All these solutions are to my knowledge relatively isolated and not connected to a sufficient number of other low energy phenomena, such as meson spectra, confinement, absence of Higgs bosons, etc.

The U(1) problem is resolved here as follows. If such a U(1) pseudoscalar isosinglet (6.2.10) exists, they must persist for a relatively long time because they cannot decay via strong or electromagnetic interactions. During their life time, they are accompanied by by a U(1) gauge boson field $\underline{B}(X)$. Now, the only gauge boson field known is the massless electromagnetic field and the massive boson fields $W^{\pm}$ and Z^0 arising from the SU(2)×U(1) gauge transformations treated in the next chapter. Thus, the massive U(1) gauge boson $\underline{B}(X)$ generated by any U(1) pseudoscalar isosinglet does not exist. It then follows the U(1) pseudoscalars themselves *cannot exist*; (6.2.10) drop out. This U(1) gauge field then becomes massless and can be identified as the electromagnetic field.

The nonexistence of pseudoscalar $\bar{u}u + \bar{d}d$, $\bar{s}s$, $\bar{c}c$, and $\bar{b}b$ forces them to form approximate isomultiplets of higher rank than one. Thus, $\bar{u}u + \bar{d}d$, and $\bar{s}s$ are combined to become η which is a member of an approximative SU(3) octet. Further, it has been suggested at the end of Sec. 5.4 that $\bar{c}c$ is mixed with lower mass $\bar{q}q$ pairs to form η' and possibly also $\eta_c(2980)$ which are no longer isosinglets.

The U(1) isosinglet vector mesons ω, ϕ, J/ψ, and Y can exist; they can decay via strong or electromagnetic interactions fastly.

6.3. Meson Magnetic Moment

6.3.1. Gauge Invariance and External Magnetic Field

The gauge transfomations at quark level

$$\chi(X) \to \chi(X)\exp(iq_{op}(z)\varphi_{q1}(X)), \qquad \chi \to \psi \tag{6.3.1a}$$

$$A(X) \to A(X) - \partial\varphi_{q1}(X)/\partial X \tag{6.3.1b}$$

leave the quark wave equations (2.3.11) form invariant. Analogously, it is seen that (6.1.9) is invariant under the gauge transformations at quark level

$$\chi_{\dot{b}f}(x_I, x_{II}) \to \chi_{\dot{b}f}(x_I, x_{II})\exp(iq_{op}(z)\varphi_I(x_I) + iq_{op}(u)\varphi_{II}(x_{II})), \quad \chi \to \psi \tag{6.3.2a}$$

$$\chi^*_{\dot{b}f}(x_I, x_{II}) \to \chi^*_{\dot{b}f}(x_I, x_{II})\exp(-iq_{op}(z)\varphi_I(x_I) - iq_{op}(u)\varphi_{II}(x_{II})), \quad \chi \to \psi \tag{6.3.2b}$$

$$A^{a\dot{b}}(x_I) \to A^{a\dot{b}}(x_I) - \partial_I^{a\dot{b}}\varphi_I(x_I) \tag{6.3.2c}$$

$$A^{a\dot{b}}(x_{II}) \to A^{a\dot{b}}(x_{II}) - \partial_{II}^{a\dot{b}}\varphi_{II}(x_{II}) \tag{6.3.2d}$$

The φ's in (6.3.1, 2) are real phase functions. Since (6.1.10) follow from (2.3.11), they can be considered to incorporate these gauge invariance proporties. While χ, ψ and Φ_m in (6.1.10) can be converted to be functions separable in X and x according to (3.1.5, 7a, 11), the gauge fields A in (6.1.10) cannot because they depend only upon x_I or x_{II}. With (3.1.3) and (3.5.7), one can write

$$\begin{aligned} A^{a\dot{b}}(x_I^\mu) &= A^{a\dot{b}}(X^\mu - \tfrac{1}{2}x^\mu) = A^{a\dot{b}}(X^\mu) - \tfrac{1}{2}x^\mu\partial_{X\mu}A^{a\dot{b}}(X^\mu) + \tfrac{1}{8}(x^\mu)^2\partial^2_{X\mu}A^{a\dot{b}}(X^\mu) + \dots \\ A_{\dot{e}d}(x_{II}^\mu) &= A_{\dot{e}d}(X^\mu + \tfrac{1}{2}x^\mu) = A_{\dot{e}d}(X^\mu) + \tfrac{1}{2}x^\mu\partial_{X\mu}A_{\dot{e}d}(X^\mu) + \tfrac{1}{8}(x^\mu)^2\partial^2_{X\mu}A_{\dot{e}d}(X^\mu) + \dots \end{aligned} \tag{6.3.3}$$

$A(X^{\mu})$ is now identified as an external electromagnetic field in laboratory space. To treat the magnetic moment of a meson, $A(X^{\mu})$ can without loss of generality be taken to give rise to a static and homogenous magnetic field,

$$\underline{H}_{ex}(X^{\nu}) = (0,\ 0,\ H_{ex}) = \partial_{\underline{X}} \times \underline{A}(X^{\nu}) \tag{6.3.4a}$$

$$A^{\mu}(X^{\nu}) = (0,\ \underline{A}(X^{\nu})), \qquad \underline{A}(X^{\nu}) = (0,\ H_{ex}X^{1},\ 0) \tag{6.3.4b}$$

Inserting (6.3.4b) into (6.3.3) leads to

$$A^{a\dot{b}}(x_I) = -\underline{\sigma}^{a\dot{b}}(0,\ H_{ex}X^{1},\ 0) + \tfrac{1}{2}x^{1}\underline{\sigma}^{a\dot{b}}(0,\ H_{ex},\ 0) \tag{6.3.5a}$$

$$A_{\dot{e}d}(x_{II}) = \underline{\sigma}_{\dot{e}d}(0,\ H_{ex}X^{1},\ 0) + \tfrac{1}{2}x^{1}\underline{\sigma}_{\dot{e}d}(0,\ H_{ex},\ 0) \tag{6.3.5b}$$

6.3.2. Magnetic Moments of Mesons

Take the first operator in (6.1.10b) and operate on (6.1.10a) and re-apply (6.1.10b). After dropping the subscripts *pr*, one obtai

$$\begin{aligned}(\partial_{I\dot{c}a} + iq_pA_{\dot{c}a}(x_I))(\partial_I^{a\dot{b}} + iq_pA^{a\dot{b}}(x_I))(\partial_{II\dot{e}d} - iq_rA_{\dot{e}d}(x_{II}))(\partial_{II}^{f\dot{e}} - iq_rA^{f\dot{e}}(x_{II}))\chi_{\dot{b}f}(x_I, x_{II}) \\ = (M_m^2 - \Phi_m(x_I, x_{II}))^2\chi_{\dot{c}d}(x_I, x_{II}) + R_M\end{aligned} \tag{6.3.6}$$

$$R_M = [\Phi_m(x_I, x_{II}),\ (\partial_{I\dot{c}a} + iq_pA_{\dot{c}a}(x_I))(\partial_{II\dot{e}d} - iq_rA_{\dot{e}d}(x_{II}))\psi^{a\dot{e}}(x_I, x_{II})] \tag{6.3.7}$$

where $[A,B]=AB-BA$. Note that operators dependent upon x_I and those dependent upon x_{II} commute. Consider now the magnetic moments of ground state mesons at rest. The unperturbed wave functions in (6.1.10) or (6.3.6), obtained by dropping the qA terms, are according to (3.1.5) of the form

$$\psi^{a\dot{e}}(X, x) = \psi^{a\dot{e}}(x)\exp(-iE_0X^0), \qquad \psi \to \chi \tag{6.3.8}$$

The qA terms are considered as first order perturbations and will give rise to first order perturbations of the wave functions, denoted by the subscript *1*. Make the ansatz,

$$\psi_1^{a\dot{e}}(X, x) = (\psi^{a\dot{e}}(x) + \psi_{1s}^{a\dot{e}}(\underline{X}, \underline{x}))\exp(-i(E_0 + E_{1M})X^0) - \psi^{a\dot{e}}(X, x), \qquad \psi \to \chi \tag{6.3.9}$$

where E_{1M} is a first order energy shift and ψ_{1s} a first order wave function perturbation in space caused by the qA terms.

The pseudoscalar meson has only one wave function component (4.3.2) and has therefore *no magnetic moment*. For vector mesons at rest, (6.3.6-9) with (3.1.4-11) and (3.5.7) leads to the first order equation

$$\begin{bmatrix} \left(i\frac{1}{2}\delta_{\dot{c}a}\left(E_0+E_{1M}\right)-\partial_{\dot{c}a}+\frac{1}{2}\partial_{\underline{X}\dot{c}a}+iq_pA_{\dot{c}a}\left(x_I\right)\right) \\ \left(i\frac{1}{2}\delta^{a\dot{b}}\left(E_0+E_{1M}\right)-\partial^{a\dot{b}}+\frac{1}{2}\partial^{a\dot{b}}_{\underline{X}}+iq_pA^{a\dot{b}}\left(x_I\right)\right) \\ \left(i\frac{1}{2}\delta_{\dot{e}d}\left(E_0+E_{1M}\right)+\partial_{\dot{e}d}+\frac{1}{2}\partial_{\underline{X}\dot{e}d}-iq_rA_{\dot{e}d}\left(x_{II}\right)\right) \\ \left(i\frac{1}{2}\delta^{f\dot{e}}\left(E_0+E_{1M}\right)+\partial^{f\dot{e}}+\frac{1}{2}\partial_{\underline{X}\dot{e}d}-iq_rA^{f\dot{e}}\left(x_{II}\right)\right) \\ -\left(i\frac{1}{2}\delta_{\dot{c}a}E_0-\partial_{\dot{c}a}\right)\left(i\frac{1}{2}\delta^{a\dot{b}}E_0-\partial^{a\dot{b}}\right)\left(i\frac{1}{2}\delta_{\dot{e}d}E_0+\partial_{\dot{e}d}\right)\left(i\frac{1}{2}\delta^{f\dot{e}}E_0+\partial^{f\dot{e}}\right) \end{bmatrix}\underline{\sigma}_{\dot{c}d}\,\underline{\chi}\left(\underline{x}\right)-R_{M1}$$

$$=-\partial_{I\dot{c}a}\partial^{a\dot{b}}_{I}\partial_{II\dot{e}d}\partial^{f\dot{e}}_{II}\chi_{1s\dot{b}f}\left(X,x\right)+\left(M_m^2-\Phi_m\left(r\right)\right)^2\chi_{1s\dot{c}d}\left(X,x\right)+R_{M2}$$

$$\partial_{\underline{X}\dot{b}a}=\underline{\sigma}_{\dot{b}a}\partial_{\underline{X}},\qquad \partial^{b\dot{a}}_{\underline{X}}=-\underline{\sigma}^{b\dot{a}}\partial_{\underline{X}} \tag{6.3.10a}$$

$$R_{M1}=\left[\Phi_m\left(r\right),\ \left(iq_pA_{\dot{c}a}\left(x_I\right)\partial_{II\dot{e}d}-iq_rA_{\dot{e}d}\left(x_{II}\right)\partial_{I\dot{c}a}\right)\psi^{a\dot{e}}\left(\underline{x}\right)\right] \tag{6.3.10b}$$

$$R_{M2}=\left[\Phi_m\left(r\right),\ \partial_{I\dot{c}a}\partial_{II\dot{e}d}\psi^{a\dot{e}}_{1s}\left(\underline{X},\underline{x}\right)\right] \tag{6.3.10c}$$

Here, (3.2.8, 12, 17) have been called upon. $R_{M1}+R_{M2}$ is linear in $\partial\Phi_m(r)/\partial r$ and $\partial^2\Phi_m(r)/\partial r^2$. Insert (6.3.5) and (3.2.5b) into (6.3.10), the first term on the left side of (6.3.10a) becomes after some algebra

$$\tfrac{1}{2}\left(\tfrac{1}{4}E_0^2+\Delta\right)\left[\tfrac{1}{2}E_{1M}E_0\underline{\sigma}_{\dot{c}d}\,\underline{\chi}\left(\underline{x}\right)-2q_p\left(\underline{\sigma}_{\dot{c}}{}^{\dot{b}}\,\underline{H}_{ex}\right)\left(\underline{\sigma}_{\dot{b}d}\,\underline{\chi}\left(\underline{x}\right)\right)-2q_r\left(\underline{\sigma}^{f}{}_{d}\,\underline{H}_{ex}\right)\left(\underline{\sigma}_{\dot{c}f}\,\underline{\chi}\left(\underline{x}\right)\right)\right]+R_{MH} \tag{6.3.11a}$$

$$R_{MH}=i\left(q_p+q_r\right)H_{ex}x^1\frac{\partial}{\partial x^2}\left(\tfrac{1}{4}E_0^2+\Delta\right)\underline{\sigma}_{\dot{c}d}\,\underline{\chi}\left(\underline{x}\right) \tag{6.3.11b}$$

The both terms on the right side of (6.3.5a) contribute equally to the q_p term in (6.3.11a). The same obviously holds for (6.3.5b) in regard to the q_r term in (6.3.11a). The last terms in (6.3.5) are in addition the source of R_{MH}.

It is readily seen that certain linear combinations of the components of $\underline{\chi}(\underline{x})=(\chi^{(1)},\ \chi^{(2)},\ \chi^{(3)})=\hat{r}\,\chi_l(r)$ are eigen functions of the magnetic field operators $\underline{\sigma}_{\dot{c}}{}^{\dot{b}}\,\underline{H}_{ex}$ and $\underline{\sigma}^{f}{}_{d}\underline{H}_{ex}$ (see (6.3.15) below) with different eigenvalues that yield real energy splitting. The same components are however not eigenfunctions of the operator in R_{MH}, which gives no energy splitting. The operators in (6.3.10b) are imaginary and the operators in (6.3.10c) do not contain H_{ex}. From the energy splitting viewpoint, (6.3.10a) becomes

$$\tfrac{1}{2}\left(\tfrac{1}{4}E_0^2+\Delta\right)\left[\tfrac{1}{2}E_{1M}E_0\underline{\sigma}_{\dot{c}d}\,\underline{\chi}\left(\underline{x}\right)-2q_p\left(\underline{\sigma}_{\dot{c}}{}^{\dot{b}}\,\underline{H}_{ex}\right)\left(\underline{\sigma}_{\dot{b}d}\,\underline{\chi}\left(\underline{x}\right)\right)-2q_r\left(\underline{\sigma}^{f}{}_{d}\,\underline{H}_{ex}\right)\left(\underline{\sigma}_{\dot{c}f}\,\underline{\chi}\left(\underline{x}\right)\right)\right]=0 \tag{6.3.12}$$

$$\partial_{I\dot{c}a}\partial_{I}^{ab}\partial_{II\dot{e}d}\partial_{II}^{f\dot{e}}\chi_{1s\dot{b}f}(X,x)-\left(M_m^2-\Phi_m(r)\right)^2\chi_{1s\dot{c}d}(X,x)-R_{M2}=R_{M1}-R_{MH} \tag{6.3.13}$$

The right side of (6.3.13) provides a source for the perturbed functions χ_{1s} and ψ_{1s}. Another similar set of equations are obtained by interchanging (6.1.10a) and (6.1.10b) in arriving at equations that are essentially (6.3.6) and (6.3.7) with $\chi\leftrightarrow\psi$. These equations will analogously lead to basically (6.3.12) and (6.3.13) with $\chi\leftrightarrow\psi$. The former is the same as (6.3.12) because $\underline{\chi}(\underline{x})=\underline{\psi}(\underline{x})$ according to (3.2.5b) and (3.4.3). The latter together with (6.3.13) determine the two perturbed wave functions χ_{1s} and ψ_{1s}.

As was mentioned at the end of Sec. 3.1, $\underline{x}$ is considered as "hidden variable". Therefore, physical results in the laboratory space time X^{μ} can be obtained by multiplying (6.3.12) and (6.3.13) by $\left(\underline{\sigma}^{d\dot{c}}\,\underline{\chi}\right)$ and integrating over $\underline{x}$. After some algebra, it can be shown that the right side of (6.3.13) vanishes and a condition is imposed on its left side,

$$\int d^3\underline{x}\left(\underline{\sigma}^{d\dot{c}}\,\underline{\chi}(\underline{x})\right)\left(\partial_{I\dot{c}a}\partial_{I}^{ab}\partial_{II\dot{e}d}\partial_{II}^{f\dot{e}}\chi_{1s\dot{b}f}(X,x)-\left(M_m^2-\Phi_m(r)\right)^2\chi_{1s\dot{c}d}(X,x)-R_{M2}\right)=0 \tag{6.3.14}$$

Such a multiplication and integration over $\underline{x}$ will however leave all terms in (6.3.12) finite. By writing out explicitly

$$\underline{\sigma}_{\dot{c}d}\,\underline{\chi}=\begin{pmatrix}\chi^{(3)} & \chi^{(1)}-i\chi^{(2)}\\ \chi^{(1)}+i\chi^{(2)} & -\chi^{(3)}\end{pmatrix},\qquad \underline{\sigma}_{\dot{c}}^{\ \dot{b}}\,\underline{H}_{ex}=\begin{pmatrix}H_{ex} & \\ & -H_{ex}\end{pmatrix} \tag{6.3.15}$$

Equation (6.3.12) yields

$$\begin{aligned} E_{1M} &= 0 && \text{for} \quad \chi_3\\ E_{1M} &= \pm\mu_M H_{ex} && \text{for} \quad \chi_1 \pm i\chi_2 \end{aligned}$$

where

$$\mu_M=\frac{1}{E_0}\left(q_p-q_r\right)=\frac{q_M}{E_0} \tag{6.3.16}$$

is the magnetic moment for the vector meson and q_M its charge. The magnetic moment is simply inversely proportional to the mass of the vector meson. Neutral vector meson has no magnetic moment.

With the mass data from [P1], the magnetic moments of ρ^{+}, K^{*+}, D^{*+}, D_s^{*+} and B^{*+} are 2.437, 2.105, 0.9336, 0.8883 and 0.3524 proton magnetons, respectively. No data is available to check these results.

6.4. Radiative Decay of Vector Meson $V \to P\gamma$

In this section, the decay of a vector meson V into the corresponding pseudoscalar meson P and a photon γ is treated. The methods developed here serve as prototypes for weak and strong decays considered in Chapters 7 and 8.

6.4.1. Wave Functions of Decaying Meson

For a free meson, the meson equations (2.3.22) with Φ_c=0 in (3.2.8a) hold. If an electromagnetic field A is introduced, (2.3.22) is replaced by (6.1.10). The magnitude of the difference between (6.1.10) and (2.3.22) is small and is of the order of quark charges or e. Therefore, the associated decay $V \to P\gamma$ can be formulated as a first order perturbational problem. Noting §3.5.2, the wave functions of the decaying meson are taken to a modified form of (3.1.1, 5, 6, 9),

$$\psi^{a\dot{b}}(X,x) = \sum_J \sum_K b_{JK}\left(\delta^{a\dot{b}}\psi_{JK}(\underline{x}) - \underline{\sigma}^{a\dot{b}}\underline{\psi}_{JK}(\underline{x})\right)\exp\left(-iE_{JK}X^0 + i\underline{K}_J\underline{X}\right) \quad (6.4.1a)$$

$$\psi \to \chi \quad (6.4.1b)$$

$$b_{JK} = a_{JK} + a_{JK}^{(1)}\left(X^0\right) \quad (6.4.2)$$

The subscript J=0,1 refers to pseudoscalar and vector mesons, respectively. a_{JK} is unity here but is in a quantized case in §6.4.4 below to be elevated to an annihilation operator annihilating an initial meson of spin J and momentum $\underline{K}$. Similarly, its complex conjugate a^*_{JK} is also unity and is elevated to a creation operator creating a final state with the same J and $\underline{K}$. $a^{(1)}_{JK}(X^0)$ is a small first order amplitude that varies slowly with time and, in the quantized case in §6.4.4, becomes an operator that "slowly" transforms the initial meson to some virtual intermediate vacuum state. It is zero at $X^0 = -\infty$. $a^{(1)*}_{JK}(X^0)$ is the complex conjugate of $a^{(1)}_{JK}(X^0)$ and, in the quantized case, becomes an operator that "slowly" creates the same final state as that created by a^*_{JK}. These quantization assignments are phenomenological and rudimentary.

In $V \to P\gamma$, $a^{(1)}_{JK}(X^0)$ is caused by the qA terms in (6.1.10) and is hence of first order in e. For convenience, write (6.4.1) in the form

$$\psi^{a\dot{b}}(X,x) = \psi_0^{a\dot{b}} + \psi_1^{a\dot{b}}, \qquad \psi_1^{a\dot{b}} = \left(a_{JK}^{(1)}\left(X^0\right)/a_{JK}\right)\psi_0^{a\dot{b}}, \qquad \psi \to \chi \quad (6.4.3)$$

where the subscripts 0 and 1 denote orders in e. $\psi_0^{a\dot{b}}$ is simply (6.4.1a) with $a^{(1)}_{JK}(X^0)$=0. Further, only one J value and one $\underline{K}$ value in the summations of (6.4.1) are picked out in (6.4.3).

6.4.2. First Order Relations

Inserting (6.4.3) into (6.1.10a), multiplying it by $\chi^*_{\dot{e}a}$ and integrating over X and x, the first order part reads

$$\int d^4Xdx^4\,\frac{1}{2}\chi^*_{0\dot{e}a}\left[\begin{array}{l}\partial_I^{a\dot{b}}\partial_{II}^{f\dot{e}}\chi_{1\dot{b}f}-\left(M_m^2-\Phi_m\right)\psi_1^{a\dot{e}}\\+iq_pA^{a\dot{b}}(x_I)\partial_{II}^{f\dot{e}}\chi_{0\dot{b}f}-iq_r\partial_I^{a\dot{b}}A^{f\dot{e}}(x_{II})\chi_{0\dot{b}f}\end{array}\right]=0 \tag{6.4.4}$$

Applying (6.1.3) and the second of (6.4.3) and noting that $a^{(1)}{}_{JK}(X^0)$ can with good approximation be moved to the left of ∂_I or ∂_{II} because it varies slowly over X^0, (6.4.4) becomes

$$\int d^4Xdx^4\,\frac{1}{2}\left\{\begin{array}{l}\partial_I^{a\dot{b}}\left(a_{JK}^{(1)}(X^0)/a_{JK}\right)\chi^*_{0\dot{e}a}\partial_{II}^{f\dot{e}}\chi_{0\dot{b}f}+iq_pA^{a\dot{b}}(x_I)\chi^*_{0\dot{e}a}\left(\partial_{II}^{f\dot{e}}\chi_{0\dot{b}f}\right)\\+iq_r\left(\partial_I^{a\dot{b}}\chi^*_{0\dot{e}a}\right)A^{f\dot{e}}(x_{II})\chi_{0\dot{b}f}-iq_r\partial_I^{a\dot{b}}\chi^*_{0\dot{e}a}A^{f\dot{e}}(x_{II})\chi_{0\dot{b}f}\\-\left(a_{JK}^{(1)}(X^0)/a_{JK}\right)\left[\left(\partial_I^{a\dot{b}}\chi^*_{0\dot{e}a}\right)\left(\partial_{II}^{f\dot{e}}\chi_{0\dot{b}f}\right)+\left(M_m^2-\Phi_m\right)\chi^*_{0\dot{e}a}\psi_0^{a\dot{e}}\right]\end{array}\right\}=0$$

(6.4.5a)

Carrying out the analogous operations that led to this equation for (6.1.10b) leads to

$$\int d^4Xdx^4\,\frac{1}{2}\left\{\begin{array}{l}\partial_{I\dot{c}a}\left(a_{JK}^{(1)}(X^0)/a_{JK}\right)\psi_0^{*d\dot{c}}\partial_{II\dot{e}d}\psi_0^{a\dot{e}}+iq_pA_{\dot{c}a}(x_I)\psi_0^{*d\dot{c}}\left(\partial_{II\dot{e}d}\psi_0^{a\dot{e}}\right)\\+iq_r\left(\partial_{I\dot{c}a}\psi_0^{*d\dot{c}}\right)A_{\dot{e}d}(x_{II})\psi_0^{a\dot{e}}-iq_r\partial_{I\dot{c}a}\psi_0^{*d\dot{c}}A_{\dot{e}d}(x_{II})\psi_0^{a\dot{e}}\\-\left(a_{JK}^{(1)}(X^0)/a_{JK}\right)\left[\left(\partial_{I\dot{c}a}\psi_0^{*d\dot{c}}\right)\left(\partial_{II\dot{e}d}\psi_0^{a\dot{e}}\right)+\left(M_m^2-\Phi_m\right)\psi_0^{*d\dot{c}}\chi_{0\dot{c}d}\right]\end{array}\right\}=0$$

(6.4.5b)

The $a^{(1)}{}_{JK}(X^0)$ factor multiplying the brackets in (6.4.5, 6) can with good approximation be moved outside the integral sign because it varies slowly with X^0. Substituitng (6.1.3) once more into (6.4.5), it is seen that the surface term associated with the first term on the right of (6.1.3) vanishes upon integration. The remaining terms in the brackets also drop out by virtue of the equation of motion (2.3.22) for steady state mesons. The next to last terms in (6.4.5), the $-iq_r\partial_I$ terms, are also surface terms and vanish after integration, noting that $A(x_I)$ and $A(x_{II})$ are real and satisfy periodic boundary conditions at large X^μ and x^μ according to (6.4.9, 8) below. Sum of (6.4.5a) and (6.4.5b) can now be written as

$$S'_{mAd}=S'_{mAs} \tag{6.4.6}$$

$$S'_{mAd}=\int d^4Xdx^4\left[\partial_I^{a\dot{b}}\left(a_{JK}^{(1)}(X^0)/a_{JK}\right)\chi^*_{0\dot{e}a}\partial_{II}^{f\dot{e}}\chi_{0\dot{b}f}+\partial_{I\dot{c}a}\left(a_{JK}^{(1)}(X^0)/a_{JK}\right)\psi_0^{*d\dot{c}}\partial_{II\dot{e}d}\psi_0^{a\dot{e}}\right]$$

(6.4.7a)

$$S'_{mAs} = i\int d^4X dx^4 \begin{Bmatrix} q_p\left[A^{a\dot{b}}(x_I)\chi^*_{0\dot{e}a}\left(\partial^{f\dot{e}}_{II}\chi_{0\dot{b}f}\right) + A_{\dot{c}a}(x_I)\psi_0^{*d\dot{c}}\left(\partial_{II\dot{e}d}\psi_0^{a\dot{e}}\right)\right] + \\ -q_r\left[A_{\dot{f}e}(x_{II})\chi_0^{*b\dot{f}}\left(\partial_{I\dot{a}b}\chi_0^{e\dot{a}}\right) + A^{e\dot{d}}(x_{II})\psi^*_{0\dot{a}e}\left(\partial^{c\dot{a}}_I\psi_{0\dot{d}c}\right)\right] \end{Bmatrix} \quad (6.4.7b)$$

6.4.3. Representation of The Electromagnetic Field

The electromagnetic field in (6.4.7b) will mainly manifest itself as the photon in $V \to P\gamma$. For this purpose, the phase functions in (6.3.2) are chosen such that the time component of the A field vanishes. The photon is then represented by

$$\underline{A}(X) = \sum_K \frac{1}{\sqrt{2E_r\Omega}} \sum_T \underline{e}_T a_T(\underline{K}) \exp\left(-iE_r X^0 + i\underline{K}\,\underline{X}\right) + c.c. \quad (6.4.8)$$

where Ω is a large normalization volume. (E_r, $\underline{K}$) is the four momentum of the photon, $\underline{e}_T$ the unit polarization vector in the directions $T=1, 2$ perpendicular to $\underline{K}$. $a_T(\underline{K})$ is the analog of a_{JK} in (6.4.2) and is set to a unity here but is elevated to an annihilation operator in the quantized case. Inserting (6.4.8) into (6.3.3) yields

$$\begin{aligned} A^{a\dot{b}}(x_I) &= -\underline{\sigma}^{a\dot{b}}\,\underline{A}(X)\exp\left(i\tfrac{1}{2}E_r x^0 - i\tfrac{1}{2}\underline{K}\,\underline{x}\right) + c.c. \\ A^{a\dot{b}}(x_{II}) &= -\underline{\sigma}^{a\dot{b}}\,\underline{A}(X)\exp\left(-i\tfrac{1}{2}E_r x^0 + i\tfrac{1}{2}\underline{K}\,\underline{x}\right) + c.c. \end{aligned} \quad (6.4.9)$$

Apply the null relative energy condition $\omega_K=0$ of (3.5.6) to (3.1.10a), the meson wave functions (3.1.5) with (3.1.9) become independent of the relative time x^0. Alternatively, apply (3.1.3a) and (3.5.7) to (3.1.10b), which then contains X^0 but not x^0. This is exemplified by the zeroth order meson wave functions here describing a vector meson at rest according to (6.4.1-3) and (3.2.4a, 5b),

$$\psi_0^{a\dot{b}} = \chi_0^{a\dot{b}} = -\underline{\sigma}^{a\dot{b}}\hat{r}\psi_1(r)\exp\left(-iE_{10}X^0\right) \quad (6.4.10)$$

This is consistent with the interpretation that the both quarks share the same time $x_I^0=x_{II}^0$ so that $x^0=0$. This will now be *assumed* to apply to (6.4.9) which now becomes

$$\begin{aligned} A^{a\dot{b}}(x_I) &= -\underline{\sigma}^{a\dot{b}}\,\underline{A}(X)\exp\left(-\tfrac{i}{2}\underline{K}\,\underline{x}\right) + c.c. \\ A^{a\dot{b}}(x_{II}) &= -\underline{\sigma}^{a\dot{b}}\,\underline{A}(X)\exp\left(\tfrac{i}{2}\underline{K}\,\underline{x}\right) + c.c. \end{aligned} \quad (6.4.11)$$

If this assumption is not made, the appearance of x^0 in (6.4.9) will lead to that the decay amplitude (6.4.18) below, which contains integrals of $A^{a\dot{b}}$ over x^0, vanish altogether. Actually, this $x^0=0$ assumption violates Lorentz invariance and is invalid. Instead, the time-dependent gauge field $A^{a\dot{b}}(X)$ has to be introduced at the meson level analogous to (7.1.4) below. This has been done in [H9 (A6)] which led to smaller decay rates than those in Table 6.1 below. The difference stems largely from the exponential factor in (6.4.11) and a factor of $1-a=a=1/2$ in front of $A(X)$ in [H9 (A6)]..

6.4.4. Quantization and Decay Amplitude

Up to this point, the scalar strong interaction theory is at quantum mechanical level. To account for decays, the fields in this theory need be quantized. A full fledged field theory is beyond the scope of this book. A standard problem arises when quantizing nonlocal fields which enter the present theory. However, the situation is different here because the meson wave function (3.1.5) or (6.2.1) is separated into a part dependent upon the laboratory coordinates X and a part dependent upon the relative coordinates x. According to Sec. 5.4 of Appendix G, x is a hidden variable and this x dependent part does not participate in quantization. This observation can have basic implications when considering a complete field theory.

The present treatment attempts to describe such decays in a quantum mechanical way, analogous to the semiclassical treatment of radiation in quantum mechanics. The justification is that the energies involved are low so that typical field-theoretical effects such as vacuum polarization are small, just like that analogous effects are small in QED at low energies.

The quantization procedures below are accordingly rudimentary. Let the initial and final states in $V \to P\gamma$ be

$$|i\rangle = |V(\underline{K}_1 = 0)\rangle, \qquad \langle f| = \langle P(\underline{K}_0), \gamma_T(\underline{K})| \tag{6.4.12}$$

where V denotes a vector meson at rest which decays into a pseudoscalar meson P with momentum $\underline{K}_0$ and photon γ.

Let |0> and <0| denote vacuum states, one has conventionally,

$$\langle f|i\rangle = \langle 0|i\rangle = \langle f|0\rangle = 0, \qquad \langle 0|0\rangle = \langle f|f\rangle = 1 \tag{6.4.13}$$

Insert (6.4.3) with the zeroth order solutions (6.4.10) into (6.4.7a), make use of (3.1.4, 10a) and (3.5.6), sandwich the result between <f| and |i>. Let the a's considered below (6.4.2) be elevated to operators according to the interpretations there. After carrying out the integration over X^0, the result reads

$$\langle f|S'_{mAd}|i\rangle = -\frac{i}{2} E_{10} S_{fi} \int d^3\underline{X} \int d^4x \psi_1^2(r) \tag{6.4.14}$$

$$S_{fi} = \langle f|a^*_{JK} a^{(1)}_{JK}(X^0 \to \infty)|i\rangle \tag{6.4.15}$$

S_{fi} corresponds to the conventional S-matrix element and is interpreted as the decay amplitude via the assignments of the a's below (6.4.2). The both forms of <f|...|i> characterize the same physical process and differ only by a phase.

Next, place (6.4.7b) between <f| and |i> of (6.4.12), make use of (6.4.11, 8) and the zeroth order part of (6.4.1), and elevate both a_{JK} and a^*_{JK} to annihilation and creation operators, respectively, so that they are on the same level as $a_T(\underline{K})$ in (6.4.8). Applying these operators to (6.4.12) yields

$$a^*_{0K}a^*_T(\underline{K})a_{10}|i\rangle = a^*_{0K}a^*_T(\underline{K})|0\rangle = |P(\underline{K}_0),\gamma_T(\underline{K})\rangle = |f\rangle \qquad (6.4.16)$$

Here, a_{10} picks out the J=1, $\underline{K}_l$=0 component of the zeroth order part of (6.4.1) or simply (6.4.10) describing a vector meson at rest. The other operators in (6.4.16) pick out a photon of polarization T described by $c.c.$ part of (6.4.8) and a pseudoscalar meson having a momentum $\underline{K}_0$ with the wave function

$$\chi^*_{0\dot{e}a} = \left(\delta_{\dot{e}a}\chi_{0K}(\underline{x}) + \underline{\sigma}_{\dot{e}a}\underline{\chi}_{0K}(\underline{x})\right)\exp\left(iE_{0K} - i\underline{K}_0\underline{X}\right), \qquad \chi \to \psi \qquad (6.4.17)$$

Noting (6.4.13) and carrying out the integration over X, one obtains after some algebra

$$\begin{aligned}\langle f|S'_{mAs}|i\rangle &= \frac{i}{\sqrt{2E_r\Omega}}(2\pi)^4\,\delta\left(E_{0K} + E_r - E_{10}\right)\delta\left(\underline{K}_0 + \underline{K}\right) \\ &\times \int dx^0 \int d^3\underline{x}\left(Q_+(\underline{x})q_{p+r}(r,\vartheta) + iQ_-(\underline{x})q_{p-r}(r,\vartheta)\right)\end{aligned} \qquad (6.4.18)$$

$$\begin{aligned}Q_+(\underline{x}) &= \frac{1}{2}\psi_1\left(E_{10} + E_{0K}\right)\left(\underline{e}_T \times \hat{r}\right)\left(\underline{\psi}_{0K} - \underline{\chi}_{0K}\right) \\ &+ \psi_1\left(\underline{e}_T\hat{r}\right)\left(\underline{\partial}\left(\underline{\psi}_{0K} + \underline{\chi}_{0K}\right)\right) - \psi_1\left(\underline{e}_T \times \hat{r}\right)\left(\underline{\partial} \times \left(\underline{\psi}_{0K} + \underline{\chi}_{0K}\right)\right) \\ &- \left(\frac{\partial\psi_1}{\partial r} + \frac{2\psi_1}{r}\right)\underline{e}_T\left(\underline{\psi}_{0K} + \underline{\chi}_{0K}\right) - \frac{1}{2}\psi_1\underline{K}_0\left(\underline{e}_T \times \hat{r}\right)\left(\psi_{0K} - \chi_{0K}\right)\end{aligned} \qquad (6.4.19a)$$

$$\begin{aligned}Q_-(\underline{x}) &= \frac{1}{2}\psi_1\left(E_{10} + E_{0K}\right)\left(\underline{e}_T\hat{r}\right)\left(\psi_{0K} + \chi_{0K}\right) \\ &+ \frac{1}{2}\psi_1\left(\underline{e}_T \times \hat{r}\right)\left(\underline{K}_0 \times \left(\underline{\psi}_{0K} + \underline{\chi}_{0K}\right)\right) - \frac{1}{2}\psi_1\left(\underline{e}_T\hat{r}\right)\left(\underline{K}_0\left(\underline{\psi}_{0K} + \underline{\chi}_{0K}\right)\right)\end{aligned} \qquad (6.4.19b)$$

$$\begin{aligned}&q_{p\pm r}(r,\vartheta) = q_p\exp\left(-i\tfrac{1}{2}Kr\cos\vartheta\right) \pm q_r\exp\left(i\tfrac{1}{2}Kr\cos\vartheta\right) \\ &\left[q_{p\pm r}(r,\vartheta)\right]_{Kr\to 0} \to q_p \pm q_r\end{aligned} \qquad (6.4.20)$$

Here, (3.5.9b) has been adopted so that

$$\underline{e}_1 = (1,\ 0,\ 0), \qquad \underline{e}_2 = (0,\ 1,\ 0) \tag{6.4.21}$$

in (6.4.8, 11). If $Q_\pm$ depends upon r only, the integral over the polar angle ϑ in (6.4.18) can be carried out to yield

$$\int d\vartheta\, q_{p\pm r}(r,\vartheta)\sin\vartheta = (q_p \pm q_r)\frac{2}{Kr}\sin\frac{1}{2}Kr \tag{6.4.22}$$

The factor multiplying $(q_p \pm q_r)$ is a reduction factor obtained by including effects in relative space on the photon and originates in the exponential factors in (6.4.11). It is unity in the limit of vanishing K or r, as in the second of (6.4.20).

Equating (6.4.14) to (6.4.18) according to (6.4.7a), the decay amplitude is found to be

$$S_{fi} = -\frac{2i}{E_{10}\Omega}\frac{1}{\sqrt{2E_r\Omega}}(2\pi)^4\delta(E_{0K}+E_r-E_{10})\delta(\underline{K}_0+\underline{K})$$
$$\times \int d^3\underline{x}\big(Q_+(\underline{x})q_{p+r}(r,\vartheta)+iQ_-(\underline{x})q_{p-r}(r,\vartheta)\big)\Big/\int d^3\underline{x}\,\psi_1^2(r) \tag{6.4.23a}$$

in which the integrals over the relative time x^0 have been cancelled out and

$$\Omega = \int d^3\underline{X} \tag{6.4.23b}$$

is a large normalization type of volume.

6.4.5. Decay Rate

The decay rate is

$$\Gamma(V\to P\gamma) = \sum\nolimits_{\text{final states}}|S_{fi}|^2/T_d = \sum\nolimits_T\sum\nolimits_K|S_{fi}|^2/T_d \tag{6.4.24a}$$

where T_d is a long time during which decay takes place and Σ_T=2 for the both photon polarizations in (6.4.8, 21). Further,

$$\sum\nolimits_K = \frac{\Omega}{(2\pi)^3}\int d^3\underline{K} \to \frac{\Omega}{(2\pi)^3}4\pi\int dK K^2 \tag{6.4.24b}$$

with (6.4.21) or (3.5.9b), S_{fi} survives only if

$$\underline{K}_0 = (0,\ 0,\ K_0), \qquad K_0 = -K \tag{6.4.25}$$

Making use of (3.5.31) and observing $E_r=K=K_0$, one can write

$$\delta(E_{0K}+E_r-E_{10})=\delta\left(\sqrt{E_{00}^2+K^2}+K-E_{10}\right)=\delta(K-K_0)(1-K_0/E_{10}) \quad (6.4.26a)$$

$$K_0=\left(E_{10}^2-E_{00}^2\right)/2E_{10} \quad (6.4.26b)$$

Inserting this expression into the square of the delta functions in (6.4.23a) leads to

$$[\delta(E_{0K}+E_r-E_{10})\delta(\underline{K}_0+\underline{K})]^2=\delta(K-K_0)\left(1-\frac{K_0}{E_{10}}\right)\frac{T_d}{2\pi}\frac{\Omega^2}{(2\pi)^6} \quad (6.4.27)$$

(see for instance [B4 (7.9)]). Combining (6.4.23-27) leads to the decay rate

$$\Gamma(V\to P\gamma)=\frac{4}{\pi}\frac{K_0}{E_{10}^2}\left(1-\frac{K_0}{E_{10}}\right)I_{pr} \quad (6.4.28a)$$

$$I_{pr}=\left|\int d^3\underline{x}\left(Q_+(\underline{x})q_{p+r}(r,\vartheta)+iQ_-(\underline{x})q_{p-r}(r,\vartheta)\right)\Big/\int d^3\underline{x}\,\psi_1^2(r)\right|^2 \quad (6.4.28b)$$

6.4.6. Dimensional Approximation Estimate of Decay Rate

Via (6.4.19), the upper integral in (6.4.28b) contains wave functions for a pseudoscalar meson in motion, (6.4.17), which have so far not been obtained, as was mentioned in the beginning of Sec. 3.5. Therefore, the decay rate will be estimated using the dimensional approximation wave functions for nonrealistic pseudoscalar mesons there.

The radial scale of $Q_\pm$ of (6.4.19) is that of $\psi_1(r)$ and $\psi_{00}(r)$ according to (3.5.11, 19, 23) and is of the magnitude of a few fm according to (4.7.3, 4). The radial scale of $q_{p\pm r}$ of (6.4.20) is $2/K\cos\vartheta$ and is much greater for the K values appearing in Table 6.1 below, including those for the π's. In the upper integral of (6.4.28b), $q_{p\pm r}$ can therefore be approximated by the constants $q_p \pm q_r$. This can be shown quantitatively by considering the effect of the exponential terms in (6.4.20) in the upper integral of (6.4.28b). Representing the radial dependence of $Q_\pm$ by $\psi_1(r)\psi_{00}(r)$, as mentioned above, the ratio

$$\frac{\int d^3\underline{x} f_u(\vartheta,\phi)\psi_1(r)\psi_{00}(r)\exp(\pm\tfrac{i}{2}K\cos\vartheta)}{\int d^3\underline{x} f_u(\vartheta,\phi)\psi_1(r)\psi_{00}(r)} \quad (6.4.29)$$

turns out to deviate insignificantly from unity for $f_u=1$ and K values in Table 6.1 below. Here, (4.3.2, 3) and (5.2.3) have been employed. For

$$f_u(\vartheta,\phi)=\hat{x}^2=\sin^2\vartheta\sin^2\phi$$

(6.4.29) is still near unity for B and D mesons and becomes 0.9 for kaon and 0.85 for pion. This allows for the simplification of (6.4.28b) to

$$I_{pr}=\left(q_p+q_r\right)^2 Q_{1+}^2+\left(q_p-q_r\right)^2 Q_{1-}^2 \tag{6.4.30a}$$

$$Q_{1\pm}=\int d^3\underline{x}Q_{\pm}(\underline{x})\Big/\int d^3\underline{x}\psi_1^2(r) \tag{6.4.30b}$$

Identify ψ_{0K} and $\underline{\psi}_{0K}$ as ψ_0 and $\underline{\psi}$ in (3.5.2). These are inserted into (6.4.19) and (6.4.30b) and the values of $Q_{1\pm}$ are considered for each order in ε_0 of (3.5.1). According to the ε_0 values given in Table 6.1 below, it is seen that the nonrelativistic approximation holds well for B and D, to some degree also for kaons but not for pions. To zeroth order, the only nonvanishing term is the first term on the right of (6.4.19b), which vanishes exactly due to (3.5.3). By virtue of (3.5.4a), which is also exact, it is seen that Q_- is one order higher than Q_+.

To first order in ε_0, insert (3.5.10a, 14, 22a), (3.5.23, 24) and (3.5.3) into (6.4.19a) and (6.4.30b). The result is Q_{1+}=0 to order ε_0 for both heavy (3.5.9a) and light (3.5.21) mesons. Insert the same set of equations into (6.4.19b) and (6.4.30b) and observe the second order wave function relations (3.5.16a) and (3.5.26). The result is Q_{1-}=0 to order ε_0^2 for both heavy and light mesons. In these calculations, some terms in Q_+ vanishes while others contain odd powers of $\hat{x}$ or $\hat{y}$ and vanish upon integration over the angles. Next, Q_{1+}=0 to order ε_0^2 and Q_{1-}=0 to order ε_0^3 because the integrals over $Q_\pm$ vanish due to the exact results (3.5.4a) and (3.5.27), independent of whether the meson is heavy or light.

To obtain nonvanishing decay rates, one has to go to order ε_0^3 for Q_{1+} and ε_0^4 for Q_{1-}. Since meson wave functions to these orders are more complex and have not been dealt with in §3.5.5, additional assumptions must be made in order to obtain some estimate. These assumptions are largely extrapolations of the relations between the lower order wave functions in Sec. 3.5.

References to Sec. 3.5 below (6.4.30) show that the wave functions that differ by order ε_0^2 generally behave similarly. Thus, (3.5.4a), (3.5.27) and the second of (3.5.3) show that the zeroth order and second order triplet functions both vanish, as do the first order and third order singlet functions. The zeroth order singlet function of the first of (3.5.3) behaves similarly to the second order wave functions for heavy mesons (3.5.16a) and (3.5.19) and is consistent with the second order wave function for light mesons (3.5.26). An extension of these results suggests that the first and third order triplet functions behave similarly and that the second and fourth order singlet wave functions behave analogously. Extending (3.5.10a), (3.5.16a), (3.5.22a), and (3.5.26) to third and fourth orders yields

$$\underline{\chi}_{03}(\underline{x})=-\underline{\psi}_{03}(\underline{x}),\qquad \chi_{04}(\underline{x})=-\psi_{04}(\underline{x})\qquad \text{for } E_{00}>>d_m \tag{6.4.31a}$$

$$\underline{\chi}_{03}(\underline{x})=\underline{\psi}_{03}(\underline{x}),\qquad \chi_{04}(\underline{x})=\psi_{04}(\underline{x})\qquad \text{for } E_{00}<<d_m \tag{6.4.31b}$$

The magnitudes of these wave functions will be estimated from the assumption that the magnitude of wave functions of order $i+1$ is $\varepsilon_0=K_0/E_{00}$ times that of order i. This is in agreement with (3.5.11, 14) for heavy mesons. For light mesons, the corresponding factor has been estimated to $\varepsilon_L=3K/2d_m$ according to (3.5.23, 24) and differs in form from ε_0. However, for the only application to kaons in §6.4.7, it will be shown there that these both ε factors are very close to each other numerically so that there is no actual disagreement.

The magnitude of ψ_{03} is accordingly $\varepsilon_0^{\,3}$ that of ψ_{00}. The radial dependence of $\underline{\psi}_{03}$ is also taken to be that of $\psi_{00}(r)$ of (4.3.2), by analogy to (3.5.11, 14, 23). Further, $\underline{\psi}_{03}(\underline{x})$ is assumed to contain an $\hat{x}$ or $\hat{y}$ factor (in general an odd power of them consistent with (3.2.5b)) so that an $\hat{x}^2$ or $\hat{y}^2$ term is formed in (6.4.19) and the angular integrations in (6.4.30b) do not vanish.

In evaluating Q_{1+} to order $\varepsilon_0^{\,3}$, it is first noted that the last term of (6.4.19a) contributes nothing due to (3.5.16a, 19, 26) and (3.1.7b). For heavy mesons, (6.4.31a) applies so that only the first term on the right side of (6.4.19a) survives which will be approximated by

$$Q_+(\underline{x})=\frac{1}{2}\psi_1(r)(E_{10}+E_{0K})(\underline{e}_T\times\hat{r})(\underline{\psi}_{03}-\underline{\chi}_{03})\rightarrow\psi_1(r)(E_{10}+E_{0K})\varepsilon_0^3\psi_{00}(r)$$
$$E_{00}>>d_m \qquad (6.4.32a)$$

Application of the first of (6.4.31a) to (6.4.19b) shows that only the first term on its right side may survive. Comparing this term to the first term on the right side of (6.4.19a) and making use of the assumption below (6.4.31) yields to order $\varepsilon_0^{\,4}$

$$\overline{Q}_-=\varepsilon_0\overline{Q}_+, \qquad E_{00}>>d_m \qquad (6.4.32b)$$

where the bar denotes some average magnitude. If the second of (6.4.31a) is also applied to (6.4.19b), one further obtains the $\varepsilon_0^{\,4}$ order result

$$Q_-=0, \qquad E_{00}>>d_m \qquad (6.4.32c)$$

For light mesons, the first term on the right side of (6.4.19a) vanishes by virtue of (6.4.31b) and the remaining nonvanishing terms are approximated by

$$Q_+=\psi_1(r)(\underline{e}_T\hat{r})(\underline{\partial}(\underline{\psi}_{03}+\underline{\chi}_{03}))-\psi_1(\underline{e}_T\times\hat{r})(\underline{\partial}\times(\underline{\psi}_{03}+\underline{\chi}_{03}))$$
$$-\left(\frac{\partial\psi_1}{\partial r}+\frac{2\psi_1}{r}\right)\underline{e}_T(\underline{\psi}_{03}+\underline{\chi}_{03})\rightarrow\psi_1(r)d_m\varepsilon_0^3\psi_{00}(r), \qquad d_m>>E_{00} \qquad (6.4.33a)$$

where $|\underline{d}|$ has been replaced by $d_m/2$ according to (4.3.2). By (3.5.27), the first term on the right side of (6.4.19b) is of order $\varepsilon_0^{\,4}$, just like (6.4.32b) is. Apply (6.4.31b) to this term and assume that ψ_{04} contains an even power of $\hat{r}$, which is analogous to the assumption for $\underline{\psi}_{03}$

above (6.4.32a) and is consistent with (3.5.19, 26). This term will then not contribute to (6.4.30b). The remaining terms in (6.4.19b) differ from the second and third terms in (6.4.19a) by the replacement of $K_0/2$ by $\underline{\partial}$. They are analogously approximated by

$$Q_- \to (K/d_m)Q_+, \qquad d_m >> E_{00} \tag{6.4.33b}$$

With (4.3.2, 3), one gets

$$\int d^3\underline{x}\psi_1(r)\psi_{00}(r) \Big/ \int d^3\underline{x}\psi_1^2(r) = \sqrt{\frac{3}{2}\frac{32}{81}} \tag{6.4.34}$$

Inserting (6.4.32a), (6.4.32c-34) and (3.5.1) into (6.4.30) and (6.4.28a) yields the decay rate estimates

$$\Gamma(V \to P\gamma) \approx \frac{1}{137}\frac{8192}{2187}\frac{K_0}{E_{10}^2}\left(1-\frac{K_0}{E_{10}}\right)\left(\frac{K_0}{E_{00}}\right)^6\left(\frac{q_p+q_r}{e}\right)^2\left(E_{10}+\sqrt{E_{00}^2+K_0^2}\right)^2 \text{ Gev}$$

$$E_{00} >> d_m \tag{6.4.35a}$$

$$\Gamma(V \to P\gamma) \approx \frac{1}{137}\frac{8192}{2187}\frac{K_0}{E_{10}^2}\left(1-\frac{K_0}{E_{10}}\right)\left(\frac{K_0}{E_{00}}\right)^6\left[\left(\frac{q_p+q_r}{e}\right)^2 d_m^2 + \left(\frac{q_p-q_r}{e}\right)^2 K_0^2\right] \text{ Gev}$$

$$d_m >> E_{00} \tag{6.4.35b}$$

where E_{10} is the mass of the vector meson, E_{00} the mass of the pseudoscalar meson, K_0 its momentum given by (6.4.26b), q_p the quark charge, q_r the antiquark charge, d_m=0.864 Gev in (5.2.3) and $e^2/4\pi=1/137$. Obviously, the numerous approximations intervening the exact decay rate (6.4.28) and the evaluable rate (6.4.35) render that it can at best provide an estimate of the order of magnitude of the decay rates.

6.4.7. Comparison with Data

The results obtained in the section are applied to $V\to P\gamma$ listed in Table 6.1.

Table 6.1. Estimates of order of magnitude of the decay rate $V\to P\gamma$ and associated parameters. The symbols are explained below (6.4.35). The ρ decay rate cannot be estimated because the nonrelativistic requirement $\varepsilon_0 <<1$ of (3.5.1) is strongly violated for pions. The assignment of B to the heavy meson case (6.4.35a) is acceptable due to the relatively small d_m/E_{00}. The assignment of D to the heavy meson case (6.4.35a) is not fully satisfactory because $d_m<E_{00}$ but not $<<E_{00}$. Analogously, the assignment of K to the light meson case (6.4.35b) is also not clear cut because $d_m>E_{00}$ but not $>>E_{00}$.

	B^{*+} $(b\bar{u})$	B^{*0} $(b\bar{d})$	D^{*+} $(c\bar{d})$	D^{*0} $(u\bar{c})$	K^{*+} $(u\bar{s})$	K^{*0} $(s\bar{d})$	ρ^{+} $(u\bar{d})$	ρ^{0} $(u\bar{u}-d\bar{d})$
E_{10}(Gev)	5.325	5.325	2.01	2.0067	0.89166	0.8961	0.769	0.7693
K_0(Gev)	0.0457	0.0454	0.1357	0.1371	0.3092	0.3099	0.3724	0.3732
$\varepsilon_0=K_0/E_{00}$	0.00866	0.0086	0.0726	0.0735	0.6262	0.6226	2.668	2.765
d_m/E_{00}	0.1637	0.1637	0.4622	0.4634	1.75	1.756	6.191	6.402
Equation	(6.4.35a)		(6.4.35a)		(6.4.35b)			
$\Gamma(V\to P\gamma)$ (kev)	2.3×10^{-10}	8.78×10^{-10}	0.21×10^{-3}	3.655×10^{-3}	74.8	133.4	-	-
Γ(data) [P1]	dominant		<2.1	<800	50.3	116.6	67.6	102.1

It is seen that the nonrelativistic condition $\varepsilon_0<<1$ of (3.5.1) underlying the decay rate estimates (6.4.35) holds well for B and D, to some degree for K, but not for π. Therefore, no estimate of $\Gamma(\rho\to\pi\gamma)$ has been given.

For B^*, the heavy meson approximation (3.5.9a) holds well and (6.4.35a) is applicable. The decay rate estimates are very small. This decay has been seen and is the dominant mode because the sister decay $B^*\to B\pi$ is forbidden by energy conservation. The decay rates have not been measured quantitatively.

For D^*, the heavy meson approximation (3.5.9a) barely holds but will be adopted, bearing in mind that some light meson approximation feature indicated in (6.4.35b) may affect the results weakly. The estimated decay rates from (6.4.35a) are consistent with the data but are orders of magnitudes smaller than the measured upper limits.

The data [P1] provide the ratio of the decays

$$\frac{\Gamma(D^{*0}\to D^0\gamma)}{\Gamma(D^{*0}\to D^0\pi^0)}\times\frac{\Gamma(D^{*+}\to D^+\pi^0)}{\Gamma(D^{*+}\to D^0\gamma)}=\frac{38.1\%}{61.9\%}\times\frac{30.7\%}{1.6\pm 0.4\%}=11.81\mp^{2.36}_{3.94} \quad (6.4.36)$$

In Table 8.1 below,

$$\Gamma(D^{*0}\to D^0\pi^0)/\Gamma(D^{*+}\to D^+\pi^0)$$

has been estimated to be 1.13. Inserting this value into (6.4.36) yields

$$R_D=\Gamma(D^{*0}\to D^0\gamma)/\Gamma(D^{*+}\to D^+\gamma)=13.35\mp^{2.67}_{4,45}=10.67 \text{ to } 17.8 \quad (6.4.37)$$

This ratio can be predicted fairly generally and accurately from (6.4.28a, 30, 32b) together with what may be regarded as a relatively safe assumption that Q_{1+} is the same for these both decays. The result is R_D=15.4. If Q_{1+} for the both decays differ only by ratios of the $(E_{10}+E_{0K})\varepsilon_0^3$ values according to (6.4.32a), R_D=16.6. If in addition (6.4.32c) is adopted instead of (6.4.32b), R_D=17.4. All these results fall within the error limits of (6.4.37). That these R_D values can be considered as relatively accurate is due to that they are obtained without specifying the radial dependencies of the meson wave functions, as in the right member of (6.4.32a).

Turning to K^*, Table 6.1 shows that the nonrelativistic condition $\varepsilon_0 <<1$ of (3.5.1) is not well satisfied and the light meson limit (3.5.21) is also not reached. Therefore, the decay rate estimates obtained from (6.4.35b) will involve greater errors than the estimates for D^* decay.

Resuming the consideration following (6.4.31), Table 6.1 yields ε_L=0.5667 and 0.5379 for K^+ and K^0, respectively. These values are rather close to the ε_0 values in Table 6.1. In fact, if some relativistic effects are included in ε_0 by replacing E_{00} in (3.5.1) by E_{0K}, one finds K_0/E_{0K}=0.5308 and 0.5285 for K^+ and K^0, respectively. The ε_L values lie between these ε_0 values and the ε_0 values in Table 6.1 and thus support the assumption on the magnitude of the wave functions below (6.4.31).

If ε_L replaces ε_0 in (6.4.33a), the decay rates in Table 6.1 estimated from (6.4.35b) are reduced to 29.6 kev for K^{*+} and 55.5 kev for K^{*0}. The data then fall between these values and the corresponding ones in Table 6.1. Note that any estimate in Table 6.1 that is within a factor of 10 from data may be considered as acceptable in view of the gross simplifications made between (6.4.30) and (6.4.35). The estimated K^* decay rates and relatively accurate R_D values below (6.4.37) provide so far the only support from $V \to P\gamma$ to the scalar strong interaction hadron theory.

The driving force of $V \to P\gamma$ is the same as that in the sister process $V \to P\pi_0$ in Sec. 8.2 and 8.3 below. The radius of the vector meson, r_1 of (4.7.4), is greater than that for pseudoscalar meson, r_0 of (4.7.3). Therefore, the potential energy of the vector meson, $-d_m/r_1+2/r_1^2$ in (3.4.3), is higher than that of the corresponding pseudoscalar meson, $-d_m/r_0$ in (3.4.1). In $V \to P\gamma$, the quarks at larger radius r_1 move inwards toward r_0. The potential energy thereby released is used to create a photon and impart a momentum to the resulting pseudoscalar meson.

Chapter 7

SEMILEPTONIC DECAY OF KAON AND PION AND KAON → THREE PIONS

The basic ingredients providing the starting points of this chapter have been given in [H11], [H12], Sec. 6 and 7 of [H8], and §3.4 and §9.2 of [H13]. This chapter is the same as Parts II and III of [H18]. The Abelian gauge fields introduced in the last chapter are now extended to non-Abelian gauge fields to treat some basic semi-leptonic decays of pions and kaons and the decay of kaon into three pions. The gauge fields pertaining to the $SU(2)_I \times U(1)_Y$ group in the standard model [P1] turn out to be insufficient and are extended to those of the SU(3) group below. A comparison to the standard model is made.

7.1. ACTIONS FOR SEMI-LEPTONIC DECAY

7.1.1. Total and Gauge Boson Action

The weak decays considered here consist of two stages. In stage I, the pseudoscalar meson decays into an intermediate gauge boson alone or together with a lighter pseudoscalar meson. In stage II, the intermediate gauge boson can decay into a lepton-antilepton pair and this will be treated first below. It can also decay into a pair of pions and this will be treated last in Sec. 7.7 and 7.8. The total action S_{T3} covering these processes must contain actions for each of these particles,

$$S_{T3} = S_{GB} + S_{m3} + S_L + S_{Lm} + \kappa_\pi S_{m\pi} \tag{7.1.1}$$

where S_{m3} is the action for the decaying meson, S_L is the action for the leptons and S_{Lm} that for the finite lepton mass. $S_{m\pi}$ is the action for a pion-antipion pair regarded as decay products and will be given in §7.7.1 below. All these actions contain the gauge boson fields, whose action is given by S_{GB}. κ_π is a scale constant.

According to the present theory, the s quark is only slightly heavier than the d quark, as is seen in Table 5.2. Based upon this result, the equations of motion for mesons containing only the u, d and s quarks are approximately invariant under global SU(3) transformations, as was considered in §2.4.2. Therefore, these equations are also approximately invariant under local

SU(3) transformations, achieved by replacing ϑ_l in (2.4.19b) by $\vartheta_l(X)$. As is well known, this transformation calls for the introduction of SU(3) gauge fields, analogous to the introduction of U(1) gauge field in §6.2.1.

In the remainder of this section, S_{GB} and S_{m3} are given and their approximate SU(3) gauge invariance shown. S_L is also given and its approximate SU(2)×U(1) gauge invariance considered. The fields in these three actions are coupled via the total action S_{T3} (7.1.1), which now appears to be inconsistent because there are eight SU(3) gauge fields in S_{GB} and S_{m3} but only four SU(2)×U(1) gauge fields in S_L. This inconsistency is removed by the degeneration mechanism given in §7.2.3 below.

The U(1) action (6.2.6) is now replaced by

$$S_{GB} = -\frac{1}{4}\int d^4X \sum_{l=1}^{8} G_l^{\mu\nu} G_{l\mu\nu} \tag{7.1.2a}$$

$$G_l^{\mu\nu}(X) = \partial_X^{\mu} W_l^{\nu} - \partial_X^{\nu} W_l^{\mu} - f_{jkl}\, g W_j^{\mu} W_k^{\nu} \tag{7.1.2b}$$

where f_{ijk} has appeared in (2.4.7). Using (C5), S_{GB} can be shown to be invariant under SU(3) gauge transformations (7.1.4) and (7.1.7b, 7d) below. Variation of S_{GB} with respect to the eight gauge boson fields W_l yields Yang-Mills type of equations of motions for these eight gauge bosons with zero mass(see (7.4.2) below).

7.1.2. Meson Action and Approximate SU(3) Gauge Invariance

The starting point is the single meson action (6.1.1), which is to be generalized to account for meson multiplets in the same way that led (2.3.22) to (2.4.3). Thus, the first term on the right side of (6.1.2) together with its complex conjugate is generalized according to

$$\left(\partial_I^{a\dot{b}} \chi_{\dot{e}a}^{*}\right)\left(\partial_{II}^{f\dot{e}} \chi_{\dot{b}f}\right) + c.c. \rightarrow \left(\partial_I^{a\dot{b}} \chi_{(rp)\dot{e}a}^{+}\right)\left(\partial_{II}^{f\dot{e}} \chi_{(pr)\dot{b}f}\right) + c.c. \tag{7.1.3a}$$

$$\chi_{(rp)\dot{e}a}^{+} = \chi_{(pr)\dot{e}a}^{*} = \left(\chi_{(pr)\dot{a}e}\right)^{+} \qquad r, p = 1, 2, 3 \tag{7.1.3b}$$

where the superscript + denotes hermitian adjoint and interchanges the flavor indices. Consistent with the convention given above (C3) for space time spinor indices, the first internal or flavor index r or p in (7.1.3b) refers to the row number in a matrix of the form (2.4.18) and the second index to the column number.

The minimal substitution with a U(1) gauge field (6.2.3) is generalized to one with gauge fields associated with the SU(3) group,

$$\begin{aligned}
\partial_I^{a\dot{b}} = \tfrac{1}{2}\partial_X^{a\dot{b}} - \partial^{a\dot{b}} &\rightarrow D_{I\,ps}^{a\dot{b}} = \tfrac{1}{2}\left(\partial_X^{a\dot{b}}\delta_{ps} + i\tfrac{1}{2} g(\lambda_l)_{ps} W_l^{a\dot{b}}(X)\right) - \partial^{a\dot{b}}\delta_{ps} = \partial_I^{a\dot{b}}\delta_{ps} + i\tfrac{1}{4} g W_{ps}^{a\dot{b}}(X) \\
\partial_{II}^{f\dot{e}} = \tfrac{1}{2}\partial_X^{f\dot{e}} + \partial^{f\dot{e}} &\rightarrow D_{II\,ps}^{f\dot{e}} = \partial_{II}^{f\dot{e}}\delta_{ps} + i\tfrac{1}{4} g(\lambda_l)_{ps} W_l^{f\dot{e}}(X) = \partial_{II}^{f\dot{e}}\delta_{ps} + i\tfrac{1}{4} g W_{ps}^{f\dot{e}}(X)
\end{aligned} \tag{7.1.4}$$

$$\frac{ig}{2}(\lambda_l)_{ps} W_l^{a\dot{b}}(X) = \frac{ig}{2}\begin{pmatrix} W_3 + \frac{1}{\sqrt{3}}W_8 & \sqrt{2}W_I^- & \sqrt{2}W_V^- \\ \sqrt{2}W_I^+ & -W_3 + \frac{1}{\sqrt{3}}W_8 & \sqrt{2}W_U \\ \sqrt{2}W_V^+ & \sqrt{2}\,\overline{W}_U & -\frac{2}{\sqrt{3}}W_8 \end{pmatrix}^{a\dot{b}} \tag{7.1.5a}$$

$$\begin{aligned} \sqrt{2}W_I^{\pm} &= W_1 \pm iW_2, & \sqrt{2}W_V^{\pm} &= W_4 \pm iW_5 \\ \sqrt{2}W_U &= W_6 + iW_7 & \sqrt{2}\,\overline{W}_U &= W_6 - iW_7 \end{aligned} \tag{7.1.5b}$$

where W_l are the wave functions of the eight gauge bosons and λ_l the eight Gell-Mann matrices. The expression (7.1.3a) becomes

$$\left(D_{I\,ps}^{a\dot{b}+}\chi^{+}_{(rp)\dot{e}a}\right)\left(D_{II\,sq}^{f\ddot{e}}\chi_{(qr)\dot{b}f}\right) + c.c. \tag{7.1.6}$$

The global SU(3) transformations (2.4.19) are now replaced by the local SU(3) transformations

$$\chi_{(qr)\dot{b}f} \to \chi'_{(qr)\dot{b}f} = U_{3qs}(X)\chi_{(sp)\dot{b}f}U_{3pr}^{-1}(X) \tag{7.1.7a}$$

$$U_{3qs}(X) = \exp\left(i\tfrac{1}{2}g(\lambda_l)_{qs}\vartheta_l(X)\right) \tag{7.1.7b}$$

This and (7.1.3b) give

$$\left(\chi'_{(qr)\dot{b}f}\right)^{+} = U_{3rp}(X)\chi^{+}_{(ps)f\dot{b}}U_{3sq}^{-1}(X) = \left(\chi^{+}_{(rq)f\dot{b}}\right)' \tag{7.1.7c}$$

where $\vartheta_l(X)$ are eight gauge functions.

The accompanying transformation of the gauge fields in (7.1.5) is [L4 §14.3.2]

$$\begin{aligned} (\lambda_l)_{qr}W_l^{a\dot{b}}(X) \to (\lambda_l)_{qr}W_l'^{a\dot{b}}(X) = U_{3qs}(X)(\lambda_l)_{sp}W_l^{a\dot{b}}(X)U_{3pr}^{-1}(X) \\ + \frac{2i}{g}\left(\partial_X^{a\dot{b}}U_{3qs}(X)\right)U_{3sr}^{-1}(X) \end{aligned} \tag{7.1.7d}$$

The expression (7.1.6) is invariant under the SU(3) gauge transformations (7.1.7).

The second term on the right side of (6.1.2) is similarly generalized to a form analogous to (7.1.6). The resulting meson action is

$$S_{m3} = \int d^4X d^4x L_{m3} \tag{7.1.8a}$$

$$\begin{aligned}2L_{m3} &= -\left(D_{I\,ps}^{a\dot{b}+}\chi_{(rp)\dot{e}a}^{+}\right)\left(D_{II\,sq}^{f\dot{e}}\chi_{(qr)\dot{b}f}\right)-\left(D_{I\,ps\,\dot{a}b}^{+}\psi_{(rp)}^{+e\dot{a}}\right)\left(D_{II\,sq\,\dot{f}e}\psi_{(qr)}^{b\dot{f}}\right)\\ &- \chi_{(rp)\dot{e}a}^{+}\left(M_{mpr}^{2}-\Phi_{m}\right)\psi_{(pr)}^{a\dot{e}}-\psi_{(rp)}^{+a\dot{e}}\left(M_{mpr}^{2}-\Phi_{m}\right)\chi_{(pr)\dot{e}a}+c.c.\end{aligned} \tag{7.1.8b}$$

In the SU(3) symmetry limit, the small quark mass differences are neglected so that $M_{mpr}=M_m$ according to §2.4.2. The M_m^2 terms in (7.1.8b) are also invariant under the SU(3) transformations (7.1.6a, 6b). Therefore, S_{m3} is invariant under the transformations (7.1.7) in this limit. During stage I, the last three actions in (7.1.1) drop out and the total action S_{T3} is approximately SU(3) gauge invariant.

Generalizing (2.4.3) to include gauge fields according to (7.1.4) yields

$$D_{I\,ps}^{a\dot{b}}D_{II\,sq}^{f\dot{e}}\chi_{(qr)\dot{b}f}-\left(M_{mpr}^{2}-\Phi_{m}\right)\psi_{(pr)}^{a\dot{e}}=0 \tag{7.1.9a}$$

$$D_{I\,ps\dot{c}a}D_{II\,sq\dot{e}d}\psi_{(qr)}^{a\dot{e}}-\left(M_{mpr}^{2}-\Phi_{m}\right)\chi_{(pr)\dot{c}d}=0 \tag{7.1.9b}$$

These equations of motion are also recovered by varying (7.1.8) with respect to χ^+ and ψ^+ together with the subsidiary conditions,

$$\begin{aligned}&(i/4)g\chi_{\dot{b}f}(x_I,x_{II})\left(\partial_X^{a\dot{b}}W^{f\dot{e}}(X)-\partial_X^{f\dot{e}}W^{a\dot{b}}(X)\right)=0\\ &(i/4)g\psi^{c\dot{e}}(x_I,x_{II})\left(\partial_{X\dot{c}a}W_{\dot{e}d}(X)-\partial_{X\dot{e}d}W_{\dot{c}a}(X)\right)=0\end{aligned} \tag{7.1.10}$$

for a given member of the SU(3) multiplet for χ, ψ and for W. These conditions stem from the *c.c.* term in (7.1.8b) and are satisfied at least for a plane wave W. For the singlet meson (7.3.14) below, (7.1.10) specifies that $\underline{W}$ must be parallel to its momentum.

7.1.3. Lepton Action and Approximate SU(2)×U(1) Gauge Invariance

The internal indices appearing in the lepton action are limited to 1 and 2, because the leptons appear in doublets. In the standard model [P1, L5 §22.2], the lepton wave functions are left handed doublets and right handed singlets and the action in spinor form noting (C13) reads

$$S_{L,SM}=S_{Lr,SM}+S_{Ll,SM} \tag{7.1.11}$$

$$S_{Lr,SM}=-i\tfrac{1}{4}\int d^4X\chi_{La}\left(\partial_X^{a\dot{b}}+i\tfrac{1}{2}g'Y_rC^{a\dot{b}}\right)\chi_{L\dot{b}}+c.c.$$

$$S_{Ll,SM}=-i\frac{1}{4}\int d^4X\left(\psi_{\nu L}^{\dot{a}},\,\psi_{L}^{\dot{a}}\right)D_{\dot{a}b}\begin{pmatrix}\psi_{\nu L}^{b}\\ \psi_{L}^{b}\end{pmatrix}+c.c. \tag{7.1.12a}$$

$$D_{\dot{a}b} = \partial_{X\dot{a}b} + \frac{i}{2} g \begin{pmatrix} W_3 + \frac{g'}{g} Y_l C & \sqrt{2} W_l^- \\ \sqrt{2} W_l^+ & -W_3 + \frac{g'}{g} Y_l C \end{pmatrix}_{\dot{a}b} \qquad (7.1.12b)$$

The accompanying action for lepton mass, which is different from that of the standard model (see §7.6.1 below), is

$$S_{Lm} = -\tfrac{1}{2} \int d^4 X m_L \left(\psi^{\dot{a}}_L \chi_{L\dot{a}} + c.c. \right) \qquad (7.1.13)$$

Here L denotes the leptons, electron and muon, r and l right and left handedness, respectivly, ν_L the associated neutrino, ψ^b the left handed spinor and $\chi_{\dot{b}}$ the corresponding right handed spinor according to (C13), g' the charge associated with the U(1) gauge field like g_l' in (6.2.3), $C^{a\dot{b}}$ the U(1) gauge field analogous to $B^{a\dot{b}}(X)$ in (6.2.3), and Y the weak hypercharge. Using the Gell-Mann-Nishijima relation $Q=I_3+Y/2$, where Q is the charge and I_3 the third component of the weak isospin, one finds

$$Y_l = -1, \qquad Y_r = -2 \qquad (7.1.14)$$

$S_{L,SM}$ is invariant under the $SU(2)_I \times U(1)_Y$ group of local gauge transformations. Variation of (7.1.11-13) with respect to the lepton wave functions produces the corresponding equations of motion for the leptons.

The gauge fields in the lepton action S_L must be compatible with those in the other actions in (7.1.1). Therefore, S_L cannot contain quantities like g' and $C_{\dot{a}b}(X)$ absent in the other actions. The quantity corresponding to hypercharge Y in (7.1.5a) is 1 because it can couple to the SU(3) specific doublet $\psi_{(13)}$ and $\psi_{(23)}$ in (2.4.18) associated with K^+ and K^0. Comparison of the parentheses in (7.1.12b) containing the gauge fields to (7.1.5a), with its third row and column removed according to the above compatibility, suggests the identification

$$\frac{g'}{g} C_{\dot{a}b}(X) \rightarrow \frac{1}{\sqrt{3}} W_{8\dot{a}b}(X) \qquad (7.1.15)$$

Thus,

$$S_L = S_{Ll} + S_{Lr} \qquad (7.1.16)$$

$$S_{Lr} = -i \frac{1}{4} \int d^4 X \chi_{La} \left(\partial_X^{a\dot{b}} - ig \frac{1}{\sqrt{3}} W_8^{a\dot{b}} \right) \chi_{L\dot{b}} + c.c. \qquad (7.1.17a)$$

$$S_{Ll} = -\frac{i}{4}\int d^4X\left(\psi_{\nu L}^{\dot a},\,\psi_L^{\dot a}\right)\left[\partial_{X\dot ab} + \frac{i}{2}g\begin{pmatrix} W_3 - \frac{1}{\sqrt3}W_8 & \sqrt2 W_l^- \\ \sqrt2 W_l^+ & -W_3 - \frac{1}{\sqrt3}W_8\end{pmatrix}_{\dot ab}\right]\begin{pmatrix}\psi_{\nu L}^b \\ \psi_L^b\end{pmatrix} + c.c. \tag{7.1.17b}$$

S_L is invariant under the SU(2)$_I$×U(1)$_Y$ group of transformations,

$$\chi_{L\dot a} \to \chi_{L\dot a}\exp\left(-i\tfrac{1}{\sqrt3}g\vartheta_8(X)\right) \tag{7.1.18}$$

$$\begin{pmatrix}\psi_{\nu L}^b \\ \psi_L^b\end{pmatrix} \to U_{21-}(X)\begin{pmatrix}\psi_{\nu L}^b \\ \psi_L^b\end{pmatrix} \tag{7.1.19}$$

$$(\sigma_l)_{qs}W_{l\dot ab} \to U_{21+qs}(X)(\sigma_l)_{sp}W_{l\dot ab}U^{-1}_{21+pr}(X) + \frac{2i}{g}\left(\partial_{X\dot ab}U_{21+qs}(X)\right)U^{-1}_{21+sr}(X) \tag{7.1.20a}$$

$$W_{8\dot ab} \to W_{8\dot ab} - \partial_{X\dot ab}\vartheta_8(X) \tag{7.1.20b}$$

$$U_{21\pm qs}(X) = \exp\left\{i\tfrac12 g\left[(\sigma_l)_{qs}\vartheta_l(X) \pm \delta_{qs}\tfrac{1}{\sqrt3}\vartheta_8(X)\right]\right\} \tag{7.1.21}$$

Here, l runs from 1 to 3 and δ is the usual delta function, as is in (A11). This group of transformations is a subset of the SU(3) transformations (7.1.7) with the meson wave functions χ replaced by the lepton wave functions (7.1.17b).

In S_{Lm} of (7.1.13), the difference between the muon and electron masses on the one hand and the vanishing mass of the neutrinos on the other breaks the SU(2) symmetry, much like that differences between different quark masses break the SU(3) symmetry considered in §2.4.2. In the so-called Glashow model [H11 Sec. 2] adopted here, this noninvariance of S_{Lm} is disregarded.

The SU(2)×U(1) gauge fields in (7.1.17) manifest themselves as the $W^{\pm}$, Z and A bosons in (7.2.2) below. Their masses M_W and M_Z, hence energies, according to (7.4.9) and (7.4.18c) below are very large and these energies are transferred to the decay products, electron or muon and their neutrinos indicated in (7.1.17b). Contributions from the derivative terms there are therefore much greater than S_{Lm} due to the relative smallness of the lepton masses therein. In this high energy limit, S_{Lm} can be neglected and SU(2)×U(1) gauge invariance holds.

At energies much lower than the M_W and M_Z, S_{Lm} is no longer small and the SU(2)×U(1) gauge invariance breaks down. But then, the $W^{\pm}$ and Z bosons cannot be present and enter as virtual states in the decays treated here; there is no contradiction. In this manner, the Glashow model is justified.

An analogous procedure will be consistently applied to the SU(3) gauge symmetry, which is an extension of the SU(2)×U(1) symmetry here, for the light mesons in regard to the quark mass differences. This has been mentioned above (2.4.5) and will be considered at the end of Sec. 7.2.

7.2. THE WEINBERG ANGLE AND DEGENERACY OF SU(3) GAUGE FIELDS

7.2.1. Origin of the Weinberg Angle

Before proceeding with the decay problems, a prediction of the Weinberg angle is made.

In the standard model, there are two independent electroweak coupling constants g and g'. It is customary to introduce the unitary Weinberg transformation

$$Z_{\dot{a}b} = W_{3\dot{a}b} \cos \vartheta_W - C_{\dot{a}b} \sin \vartheta_W, \qquad A_{\dot{a}b} = W_{3\dot{a}b} \sin \vartheta_W + C_{\dot{a}b} \cos \vartheta_W \tag{7.2.1a}$$

$$\tan \vartheta_W = g'/g, \qquad e = g \sin \vartheta_W \tag{7.2.1b}$$

into (7.1.12). The Weinberg angle ϑ_W and g are the two unknown constants to be determined by data. The matrix operator in (7.1.12b) now becomes

$$\frac{i}{2} g \begin{pmatrix} Z/\cos \vartheta_W & \sqrt{2} W_I^- \\ \sqrt{2} W_I^+ & -2A \sin \vartheta_W - Z(\cos 2\vartheta_W / \cos \vartheta_W) \end{pmatrix}_{\dot{a}b} \tag{7.2.2}$$

In the total action S_{T3} (7.1.1), which includes the lepton action (7.1.16), however, there is only one weak coupling constant g; there is no need to fix an extra constant $\sin \vartheta_W$. Conversely, the origin of the Weinberg angle may be inferred by identifying

$$C_{\dot{a}b}(X) \rightarrow W_{8\dot{a}b}(X) \tag{7.2.3a}$$

in (7.1.15). Assuming this, (7.1.15) yields

$$\tan \vartheta_W = 1/\sqrt{3}, \qquad \vartheta_W = 30^\circ, \qquad \sin^2 \vartheta_W = 0.25 \tag{7.2.3b}$$

$$e = g/2 \tag{7.2.3c}$$

which is close to the measured value of $\sin^2 \vartheta_W \approx 0.23$ [L4 §14.5.1].

In the limit of SU(3) symmetry limit, (7.2.3) suggests that the Weinberg angle in the scalar strong interaction hadron theory is just another form of $1/\sqrt{3}$ arsising from normalization of the eighth Gell-Mann matrix λ_8. In other words, g' just stands for $g/\sqrt{3}$. Deviation from (7.2.3) is attributed to that the SU(3) symmetry is broken to first order, mentioned in §2.4.2 and treated in the following §7.2.2.

Nature is sparesome in dealing out fundamental constants. To hand out two natural constants g' and g or e within a factor of √3 appears to be wasteful. This aesthetical and practical defect is removed in the present theory by means of (7.2.3).

7.2.2. Broken Local SU(3) Symmetry and Weinberg's Angle

As was mentioned above (2.4.5), the SU(3) symmetry is broken slightly by the somewhat higher mass m_3 of the s quark relative to m_1 and m_2, the masses of the u and d quarks. This symmetry breaking is reflected in the $M_{mpr}^{\ 2}\psi_{(pr)}$ type of expressions in (2.4.3) and in the meson action (7.1.8). Neglecting the small differences between m_1 and m_2, (2.4.1, 13, 15) show that there are only three pseudoscalar meson masses which are put in the form,

$$M_{m\pi}^{\ 2} = M_{m\pi 0}^{\ 2} = m_{1-2}^2, \quad M_{mK}^{\ 2} = m_{1-2}^2\left(1+\frac{\Delta m_3}{2m_{1-2}}\right)^2 = m_{1-2}^2(1+\Delta_3)^2$$

$$M_{m\eta}^{\ 2} = m_{1-2}^2\left(\frac{1}{3}+\frac{2}{3}\left(1+\frac{\Delta m_3}{m_{1-2}}\right)^2\right) = m_{1-2}^2(1+\Delta_8) \tag{7.2.4a}$$

$$m_1,\ m_2 \to m_{1-2} = (m_1+m_2)/2 = 0.6603 \text{ Mev} \tag{7.2.4b}$$

$$\Delta m_3 = m_3 - m_{1-2} = 0.08283 \text{ Mev} \tag{7.2.4c}$$

where Table 5.2 has been consulted. Similarly,

$$\psi_{(12)} = \psi_{(21)} = \psi_{(\pi 0)} = \psi_{(\pi)}, \quad \psi_{(13)} = \psi_{(31)} = \psi_{(23)} = \psi_{(32)} = \psi_{(K)}, \quad \psi \to \chi \tag{7.2.5}$$

in (2.4.18). These wave functions in rest frame have been obtained via Chapter 3 and §4.3.2 and differ only by three phases containing the meson masses $E_{0\pi}$, E_{0K} and $E_{0\eta}$ according to (2.4.10). Thus, the SU(3) symmetry breaking expression $M_{mpr}^{\ 2}\psi_{(pr)}$ can be put in the form,

$$M_{mpr}^{\ 2}\psi_{(pr)} = \begin{pmatrix} M_{m\pi}^{\ 2}\psi_{(\pi)} + M_{m\eta}^{\ 2}\psi_{(\eta)} & M_{m\pi}^{\ 2}\psi_{(\pi)} & M_{mK}^{\ 2}\psi_{(K)} \\ M_{m\pi}^{\ 2}\psi_{(\pi)} & -M_{m\pi}^{\ 2}\psi_{(\pi)} + M_{m\eta}^{\ 2}\psi_{(\eta)} & M_{mK}^{\ 2}\psi_{(K)} \\ M_{mK}^{\ 2}\psi_{(K)} & M_{mK}^{\ 2}\psi_{(K)} & -2M_{m\eta}^{\ 2}\psi_{(\eta)} \end{pmatrix}$$

$$= \begin{pmatrix} 1+(1+\Delta_8)P_{\eta-\pi} & 1 & (1+\Delta_3)^2 P_{K-\pi} \\ 1 & -1+(1+\Delta_8)P_{\eta-\pi} & (1+\Delta_3)^2 P_{K-\pi} \\ (1+\Delta_3)^2 P_{K-\pi} & (1+\Delta_3)^2 P_{K-\pi} & -2(1+\Delta_8)P_{\eta-\pi} \end{pmatrix} m_{1-2}^2\psi_{(\pi)} \tag{7.2.6a}$$

$$P_{K-\pi} = \exp\left(-i\left(E_{0K} - E_{0\pi}\right)X^0\right), \qquad P_{\eta-\pi} = \exp\left(-i\left(E_{0\eta} - E_{0\pi}\right)X^0\right) \tag{7.2.6b}$$

This mass term is of zeroth order in g. It is expected to affect the originally SU(3) symmetry preserving terms of first order in g, $g(\lambda_l)_{pr}W_l\psi$ in (7.1.8b) via (7.1.4) such that they too loose this symmetry. How these terms are modified to break the symmetry is unknown, except that the degree of symmetry breaking should be of the same magnitude as $\Delta m_3/m_{1-2}$ in (7.2.6). There are infinitely many ways to break the SU(3) symmetry of the g and g^2 terms in (7.1.8). In general, this symmetry breaking will involve i) new unknown functions or ii) new unknown constants. In these cases, the results to be obtained will contain these unknowns and can hence not be evaluated.

Therefore, the criterion "*no new unknown is allowed in the symmetry breaking*" will be adopted in the following. If i) is not allowed, then the only possibility is to modify g to g_l, i. e., the weak charge associated with the different gauge fields W_l are different. In view of (7.2.4-6) and (7.1.5a), the generalization takes the form

$$g(\lambda_l)_{ps}W_l\psi_{(sr)} \rightarrow g_l(\lambda_l)_{ps}W_l\psi_{(sr)}$$
$$= \begin{pmatrix} gW_3 + g_8W_8/\sqrt{3} & g\sqrt{2}W_I^- & g_{4-7}\sqrt{2}W_V^- \\ g\sqrt{2}W_I^+ & -gW_3 + g_8W_8/\sqrt{3} & g_{4-7}\sqrt{2}W_U \\ g_{4-7}\sqrt{2}W_V^+ & g_{4-7}\sqrt{2}W_U & -2g_8W_8/\sqrt{3} \end{pmatrix} \begin{pmatrix} 1+P_{\eta-\pi} & 1 & P_{K-\pi} \\ 1 & -1+P_{\eta-\pi} & P_{K-\pi} \\ P_{K-\pi} & P_{K-\pi} & -2P_{\eta-\pi} \end{pmatrix} \psi_{(\pi)}$$
$$g = g_1 = g_2 = g_3, \qquad g_{4-7} = g_4 = g_5 = g_6 = g_7 \tag{7.2.7}$$

If ii) above is not allowed, Δg_{4-7} and Δg_8 can only depend upon the known SU(3) symmetry breaking parameter Δm_3 in a way not containing any new constant. Such a dependence, apart from a sign, can be determined from a comparison of (7.2.6) and (7.2.7). The phases P_{K}-π and P_{η}-π are of magintude unity. Te dimension of gW in (7.2.7) is (length)$^{-1}$ while the dimension of M_m^2 in (7.2.6) is (length)$^{-2}$. Therefore, such a comparison gives the identification

$$\frac{g_{4-7}}{g} = 1 + \Delta_3, \qquad \frac{g_8}{\sqrt{3}g} = \sqrt{1+\Delta_8} \tag{7.2.8}$$

which together with (7.2.4) yields

$$\frac{g_{4-7}}{g} = 1 \pm \frac{\Delta m_3}{2m_{1-2}} = 1 \pm 0.0627, \quad \frac{g_8}{\sqrt{3}g} = \sqrt{1 \pm \frac{4\Delta m_3}{3m_{1-2}} + \frac{2}{3}\left(\frac{\Delta m_3}{m_{1-2}}\right)^2} = \begin{pmatrix} 1+0.0852 \\ 1-0.0817 \end{pmatrix} \tag{7.2.9}$$

The lower minus sign is introduced because one does not know at this stage whether Δm_3 will give rise to higher or lower weak charges g_{4-7} and g_8. The forms (7.2.8-9) appear to satisfy the above criterion uniquely and are also the simplest possible. It is noted that (7.2.6a) and (7.2.7) still preserves SU(2)$_I$ symmetry and, in the limit of $\Delta m_3 \rightarrow 0$, the original SU(3) symmetry is restored.

With the SU(3) symmetry preserving (7.1.5a) replaced by the SU(3) symmetry breaking parentheses containing gW in (7.2.7), it is seen that the identification (7.1.15) is replaced by

$$\frac{g'}{g} C_{\dot{a}b}(X) \rightarrow \frac{1}{\sqrt{3}} \frac{g_8}{g} W_{8\dot{a}b}(X) \tag{7.2.10}$$

Consequently, the first of (7.2.1b) with the first of (7.2.3b) now becomes

$$\tan \vartheta_W = \frac{g_8}{\sqrt{3}g} = \frac{1}{\sqrt{3}} \begin{pmatrix} 1+0.08124 \\ 1-0.08172 \end{pmatrix} \tag{7.2.11}$$

The lower case is chosen because it gives

$$\sin^2 \vartheta_W = 0.2194, \qquad \vartheta_W = 27.93^0 \tag{7.2.12}$$

in good agreement with data 0.2224 [P1]. This value will also later lead to a lower mass of the Z gauge boson in (7.4.18c) and is therefore preferred. It replaces the SU(3) symmetry preserving value 0.25 given in (7.2.3b). The lower case in the first of (7.2.9) also shows that the weak charge $g_{4\text{-}7}$ of (7.2.7) associated with $W_4,\ldots W_7$ is about 6% lower than g associated with pions.

The empirical link (7.2.9) between the weak charge, which is related to the electric charge via the last of (7.2.1b), and the quark masses is another indication that the strong and electromagnetic interactions are interconnected, as is indicated by (8.3.11) below.

With (7.2.10), the lepton action (7.1.17) together with (7.2.1a, 2, 3a) becomes

$$S_{Lr} = -i\tfrac{1}{4} \int d^4X \chi_{La} \left(\partial_X^{a\dot{b}} + ie\left(-A^{a\dot{b}} + \tan \vartheta_W Z^{a\dot{b}}\right)\right) \chi_{L\dot{b}} + c.c. \tag{7.2.13a}$$

$$S_{Ll} = -\frac{i}{4} \int d^4X \left(\psi_{\nu L}^{\dot{a}}, \psi_L^{\dot{a}}\right) \left[\partial_{X\dot{a}b} + i\frac{g}{2} \begin{pmatrix} Z/\cos\vartheta_W & \sqrt{2}W_l^- \\ \sqrt{2}W_l^+ & -2A\sin\vartheta_W - Z\left(\cos 2\vartheta_W / \cos\vartheta_W\right) \end{pmatrix}_{\dot{a}b} \right] \begin{pmatrix} \psi_{\nu L}^b \\ \psi_L^b \end{pmatrix}$$
$$+ c.c. \tag{7.2.13b}$$

The charge of the charged lepton spinors $\psi_L{}^b$ and $\chi_{L\dot{b}}$ is correctly represented in the electromagnetic field terms.

Similarly, the SU(3) symmetry preserving (7.1.5a) now becomes

$$i\frac{g}{2}(\lambda_l)_{ps} W_{l\dot{a}b}(X) \rightarrow i\frac{g_l}{2}(\lambda_l)_{ps} W_{l\dot{a}b}(X)$$

$$= i\frac{g}{2}\begin{pmatrix} 2A\sin\vartheta_W + Z\cos 2\vartheta_W/\cos\vartheta_W & \sqrt{2}W_I^- & (g_{4-7}/g)\sqrt{2}W_V^- \\ \sqrt{2}W_I^+ & -Z/\cos\vartheta_W & (g_{4-7}/g)\sqrt{2}W_U \\ (g_{4-7}/g)\sqrt{2}W_V^+ & (g_{4-7}/g)\sqrt{2}\overline{W}_U & -2A\sin\vartheta_W + 2Z\sin^2\vartheta_W/\cos\vartheta_W \end{pmatrix}_{\dot{a}b} \tag{7.2.14}$$

In the limit of SU(3) symmetry $\Delta m_3 \to 0$ and (7.2.14) becomes

$$i\frac{g}{2}(\lambda_l)_{ps} W_{l\dot{a}b}(X) = i\frac{g}{2}\begin{pmatrix} A + \frac{1}{\sqrt{3}}Z & \sqrt{2}W_I^- & \sqrt{2}W_V^- \\ \sqrt{2}W_I^+ & -\frac{2}{\sqrt{3}}Z & \sqrt{2}W_U \\ \sqrt{2}W_V^+ & \sqrt{2}\overline{W}_U & -A + \frac{1}{\sqrt{3}}Z \end{pmatrix}_{\dot{a}b} \tag{7.2.15}$$

7.2.3. Degeneracy of SU(3) Gauge Fields and Extended Glashow Model

Lorentz invariance of the total action (7.1.1) is guaranteed by the tensor or spinor forms of the member actions. Approximative gauge invariances for each of these actions have been demonstrated in Sec. 7.1($S_{m\pi}$ will be given by (7.7.3) below). Gauge invariance is a requirement of quantum mechanics saying that the local phases of the lepton and meson wave functions do not affect the physical results. This necessitates the introduction of four gauge fields in (7.1.17) and eight gauge fields in (7.1.8, 9) via (7.1.4-5).

Having served these purposes, gauge symmetries place no further constraint on the development and specialization of these actions but may still be helpful in some cases. This is analogous to that one no longer needs to consider Lorentz invariance when specializing Dirac's equation for application to the hydrogen atom. As Wigner remarked: "Once the equation of motion is known, we do not need symmetry anymore". The equations of motion of the meson, lepton and gauge boson fields have been obtained by varying the invdividual actions in the total action S_{T3} (7.1.1) in Sec. 7.1.

All the fields interact with the gauge boson fields in S_{T3}, which serves the purpose of coupling of these fields in stage II of §7.1.1. Variations of S_{T3} with respect to the gauge fields in §7.4.2 below put these couplings in form of decay of these bosons into lepton pairs. As was mentioned above, there are four known gauge fields $W_I^{\pm}$, Z and A in (7.2.2) or $W_I^{\pm}$, W_3 and W_8 in the lepton action (7.1.17) but eight $W_I^{\pm}$, W_3, W_8, $W_V^{\pm}$, W_U, and $\overline{W}_U$ in the meson action S_{m3} (7.1.8) and gauge boson action S_{GB} (7.1.2) via (7.2.14-15). The last four gauge fields are new, have not been observed and are converted into the four known ones as follows.

Grossly speaking, W_I^- converts a d quark to a u quark and W_V^- converts an s quark to a u quark. Because the s and d quarks have the same charge, so do $W_I^{\pm}$ and $W_V^{\pm}$. In the limit of SU(3) symmetry, the s and d quarks have the same mass and are hence indistinguishable. Therefore, the degeneracy

$$\sqrt{2}W_{V\dot{a}b}^{\pm}(X) \to \sqrt{2}W_{I\dot{a}b}^{\pm}(X) \qquad \text{SU(3) symmetry} \tag{7.2.16}$$

holds in (7.2.15). Analogously, W_U converts an s quark into a d quark and Z can be viewed as converting a d or $\bar{d}$ quark into itself. In the limit of SU(3) symmetry, both conversions become the same. However, W_U is complex and Z is real. Therefore, only the degeneration forms,

$$\sqrt{2}\,\mathrm{Re}(W_U)=\sqrt{2}\,\mathrm{Re}(\overline{W}_U)=W_6\to -2Z/\sqrt{3}, \qquad iW_7\to 0 \qquad (7.2.17)$$

can take place in (7.2.15). There is however an exception. The neutral kaon wave functions entering (7.1.8-9) do not represent physical kaons and are conventionally decomposed according to [L5 (15.22)]

$$\begin{aligned}
&\psi_{(23)}=\left(\psi_{(KS)}+\psi_{(KL)}\right)/\sqrt{2} \quad \text{or symbolically} \quad K^0=\left(K_S^0+K_L^0\right)/\sqrt{2}\\
&\psi_{(32)}=\left(\psi_{(KS)}-\psi_{(KL)}\right)/\sqrt{2} \quad \text{or symbolically} \quad \overline{K}^0=\left(K_S^0-K_L^0\right)/\sqrt{2}\\
&\psi\to\chi
\end{aligned} \qquad (7.2.18)$$

where $\psi_{(KS)}$ and $\psi_{(KL)}$ are eigenstates of the discrete CP operator. These relations show that $\psi_{(KL)}$ is imaginary and the roles of W_6 and W_7 are reversed and (7.2.17) is changed to

$$\sqrt{2}\,\mathrm{Im}(W_U)=-\sqrt{2}\,\mathrm{Im}(\overline{W}_U)=-W_7\to -2Z/\sqrt{3}, \quad W_6\to 0 \quad \text{for } K_L^0 \text{ decay only} \qquad (7.2.19)$$

In the SU(3) symmetry breaking case, the masses of the s and d quarks differ by a small amount. This difference affects the strong interaction part of S_{m3} but not directly the weak interaction part, which is related to charges. The charges of the quarks and gauge bosons are not affected. The direct effect of this symmetry breaking can only take place via the associated weak charges in (7.2.7). Accordingly, the degenerations (7.2.16, 17, 19) for (7.2.15) are replaced by the corresponding ones in the more general (7.2.14);

$$\frac{g_{4-7}}{g}\sqrt{2}W^{\pm}_{V\dot{a}b}(X)\to\sqrt{2}W^{\pm}_{I\dot{a}b}(X) \qquad (7.2.20)$$

$$\frac{g_{4-7}}{g}\sqrt{2}\,\mathrm{Re}(W_U)=\frac{g_{4-7}}{g}\sqrt{2}\,\mathrm{Re}(\overline{W}_U)=\frac{g_{4-7}}{g}W_6\to -Z/\cos\vartheta_W, \qquad iW_7\to 0 \qquad (7.2.21)$$

$$\frac{g_{4-7}}{g}\sqrt{2}\,\mathrm{Im}(W_U)=-\frac{g_{4-7}}{g}\sqrt{2}\,\mathrm{Im}(\overline{W}_U)=-\frac{g_{4-7}}{g}W_7\to -Z/\cos\vartheta_W, \quad W_6\to 0 \qquad (7.2.22)$$

With these degenerations, S_{m3} and S_{GB} only contain the four known gauge boson present in the lepton action S_L (7.2.17) and the total action S_{T3} (7.1.1) is now self consistent in this regard. Thus, in spite of the SU(3) internal symmetry called for to represent kaons, there are no new, unobserved gauge bosons.

The approximation to be adopted in the rest of this chapter is to put $\Delta_{m3}=0$ at first so that the meson action (7.1.8) is invariant under SU(3) transformations (7.1.7). After having obtained results of first and second order in g utilizing this symmetry, Δ_{m3} of (7.2.4c) is restored as if it had that value all along. This procedure may therefore be called the "extended Glashow model" because it is entirely analogous to the Glashow model for lepton action in which the mass difference between the electron or muon and neutrino has been neglected when establishing SU(2)×U(1) gauge invariance. Δ_{m3} plays the role of the lepton masses in S_{Lm} and the mesons in (7.1.8) play the roles of the leptons in (7.1.17b). Justification of this "extended Glashow model" is analogously based upon the large masses M_W and M_Z and energies associated with the $W^{\pm}$ and Z bosons and follow the same reasoning as that given at the end of Sec. 7.1.

7.3. Decay of Pion and Kaon to Gauge Bosons

7.3.1. Ordering of Wave Functions

Turn now to stage I in §7.1.1 and derive expressions for the decay amplitudes. The methods developed for $V \to P\gamma$ in Sec. 6.4 will be followed where applicable. Analogous to (6.4.1-3), where the index J can be dropped here, the meson wave functions in S_{m3} (7.1.8) are written in the form

$$\begin{aligned}
\psi^{a\dot{b}}_{(pr)}(X,x) &= \left(a_{pr} + a^{(1)}_{pr}(X^0)\right)\left(\delta^{a\dot{b}}\psi_{0(pr)K}(\underline{x}) - \underline{\sigma}^{a\dot{b}}\underline{\psi}_{(pr)K}(\underline{x})\right)\exp\left(-iE_{prK}X^0 + i\underline{K}_{pr}\underline{X}\right) \\
&= \psi^{a\dot{b}}_{0(pr)}(X,x) + \psi^{a\dot{b}}_{1(pr)}(X,x) = \left(1 + \frac{a^{(1)}_{pr}(X^0)}{a_{pr}}\right)\psi^{a\dot{b}}_{0(pr)}(X,x) \\
\psi &\to \chi
\end{aligned} \tag{7.3.1}$$

where $a_{pr}=1$ and $a_{pr}^{(1)}(X^0)$ is a first order quantity varying slowly with the time X^0. The subscripts pr have also been attached to E_K and $\underline{K}$ analogous to M_m^2 in (2.4.1).

The terms in the actions that make up S_{T3} of (7.1.1), i. e., (7.1.2, 8, 16, 17), can now be grouped in powers of the small parameter g or e. Only the lowest order and independent quantities are listed in the two alternatives below:

$$\text{First order:} \quad g, \quad a^{(1)}_{pr}(X^0), \quad \partial_X \text{ in } S_{GB}, \quad \psi^{\dot{a}}_L\psi^{b}_L, \quad \chi_{La}\chi_{L\dot{b}} \tag{7.3.2a}$$

$$\begin{aligned}
&\text{First order:} \quad g, \quad W, \quad \partial_X \text{ in } S_{GB}, \quad \psi_L, \quad \chi_L \\
&\text{Second order:} \quad a^{(1)}_{pr}(X^0)
\end{aligned} \tag{7.3.2b}$$

These both alternatives are equivalent except that the nonlinear gWW term in (7.1.2b) is of the same order as the other ones there for the first alternative (7.3.2a) but is of one order higher in the second alternative (7.3.2b). The latter is more consistent with the standard model considered in §7.6.1 below and will be adopted. However, the former alternative will be formally used in the exposition below, bearing the difference on the gWW term in mind.

7.3.2. First Order Relations

Insert (7.3.1) into (7.1.9), multiply (7.1.9a) by $\chi^{*}_{0\dot{e}a(rp)}$ and (7.1.9b) by $\psi^{*d\dot{c}}_{0(rp)}$, add them together, and integrate over X and x. The first order quantities can be put in the form

$$S'_{m3d} = S'_{m3s} \tag{7.3.3}$$

S'_{m3d} is a surface term corresponding to the first right side term in (6.1.3)

$$S'_{m3d} = \int d^4X dx^4 \left[\partial_I^{a\dot{b}}\left(a^{(1)}_{pr}\left(X^0\right)/a_{pr}\right)\chi^{+}_{0(rp)\dot{e}a}\partial_{II}^{f\dot{e}}\chi_{0(pr)\dot{b}f} + \partial_{I\dot{c}a}\left(a^{(1)}_{pr}\left(X^0\right)/a_{pr}\right)\psi^{+d\dot{c}}_{0(rp)}\partial_{II\dot{e}d}\psi^{a\dot{e}}_{0(pr)}\right] \tag{7.3.4}$$

The source part S'_{m3s} of the first order terms contains the gW terms in (7.1.8), ignoring the $c.c.$ term there. It together with (7.1.4) and the left of (7.2.7) reads

$$S'_{m3s} = -\int d^4X dx^4 \frac{ig_l}{4}\left\{\begin{array}{l}\chi^{+}_{0(rp)\dot{e}a}\left(\lambda_l\right)_{ps}W_l^{a\dot{b}}\left(\partial_{II}^{f\dot{e}}\chi_{0(sr)\dot{b}f}\right) - \left(\partial_I^{a\dot{b}}\chi^{+}_{0(rp)\dot{e}a}\right)\left(\lambda_l\right)_{ps}W_l^{f\dot{e}}\chi_{0(sr)\dot{b}f} \\ + \psi^{+d\dot{c}}_{0(rp)}\left(\lambda_l\right)_{ps}W_{l\dot{c}a}\left(\partial_{II\dot{e}d}\psi^{a\dot{e}}_{0(sr)}\right) - \left(\partial_{I\dot{c}a}\psi^{+d\dot{c}}_{0(rp)}\right)\left(\lambda_l\right)_{ps}W_{l\dot{e}d}\psi^{a\dot{e}}_{0(sr)} + R_s\end{array}\right\} \tag{7.3.5a}$$

$$\frac{ig_l}{4}R_s = \partial_I^{a\dot{b}}\chi^{+}_{0(rp)\dot{e}a}\left(\lambda_l\right)_{ps}W_l^{f\dot{e}}\chi_{0(sr)\dot{b}f} + \partial_{I\dot{c}a}\psi^{+d\dot{c}}_{0(rp)}\left(\lambda_l\right)_{ps}W_{l\dot{e}d}\psi^{a\dot{e}}_{0(sr)} \tag{7.3.5b}$$

The surface term R_s corresponds to the $iq_r\partial_I$ terms in (6.4.5).

7.3.3. Quantization

The treatment of §6.4.4 is taken over here. Let the initial and final states of stage I of §7.1.1 be

$$|i\rangle = \left|P_{i(pr)}\left(E_{pr}\right)\right\rangle, \qquad p \neq r \tag{7.3.6}$$

$$\langle f| = \left\langle P_{f(rs)}\left(K^{\mu}_{f1rs}\right), W_{sp}\right| \qquad r, s = 1, 2 \tag{7.3.7a}$$

where P stands for a pseudoscalar meson, i denotes the initial state and f the final state pion in stage I. The subscript K in (7.3.1) is zero for the initial state and is suppressed. The initial state is a kaon or a charged pion characterized by pr with $p \neq r$ and the final states pion by one of

$$(rs) = (12) \text{for } \pi^+ = \pi+, \; (rs) = (21) \text{for } \pi^- = \pi-, \; (rs) = (11-22) \text{for } \pi^0 = \pi 0 \tag{7.3.8}$$

where the corresponding pion symbols given after the parentheses will be used synonymously. When rs=11 or 22, it represents a $\overline{u}u$ or a $\overline{d}d$ state which does not exist. These final states are interpreted as that the quark and antiquark annihilate each other leaving behind a *vacuum pseudoscalar meson state* to be discussed further below (7.3.19). In this case, (7.3.7a) becomes

$$\langle f| = \langle P_{f(rr)}(0), W_{rp}| = \langle 0, W_{rp}| \tag{7.3.7b}$$

W_{sp} is the intermediary gauge bosom field representing the other final state of stage I and is related to (7.2.7) by

$$g_l(\lambda_l)_{sp} W_{l\dot{a}b} = g_l \sqrt{2} W_{sp\dot{a}b}, \qquad s \neq p \tag{7.3.9a}$$

which can with (7.2.7) be written as

$$\begin{aligned} gW_{12} &= gW_I^-, \quad g_{4-7}W_{13} = g_{4-7}W_V^-, \quad g_{4-7}W_{23} = g_{4-7}W_U, \\ g\,W_{21} &= gW_I^+, \quad g_{4-7}W_{31} = g_{4-7}W_V^+, \quad g_{4-7}W_{32} = g_{4-7}\overline{W}_U \end{aligned} \tag{7.3.9b}$$

In stage II, the intermediary W boson decays into a pair of leptons according to (7.4.6) below. It is therefore implicitly understood the W_{sp} in (7.3.7) also picks out these final lepton wave functions in (7.4.7) ff below.

a_{pr} in (7.3.1) is now elevated to bcome an annihilation operator according to the assignment below (6.4.2) so that, analogous to (6.4.16),

$$a_{pr}|P_{i(pr)}(E_{pr})\rangle = |0\rangle \tag{7.3.10a}$$

Similarly, a^*_{rs} is interpreted as an equivalent creation operator acting on |0> or an annihilation operator acting on <f|;

$$a^*_{rs}|0\rangle = |P_{f1(rs)}(K^\mu_{f1rs})\rangle, \qquad \langle P_{f1(rs)}(K^\mu_{f1rs})|a^*_{rs} = \langle 0| \tag{7.3.10b}$$

Obviously,

$$a_{sq}\left|i\right\rangle = a_{sq}\left|P_{i(pr)}\left(E_{pr}\right)\right\rangle = 0, \qquad sq \neq pr \tag{7.3.11}$$

It is possible to choose the negative energy solutions characterized by the lower sign in (5.1.1) for mesons. In this case, the interpretations

$$\begin{aligned} a^*_{pr}\left|i = P_{i(pr)}\left(-\left|E_{pr}\right|\right)\right\rangle &= a_{rp}\left|i = P_{i(rp)}\left(\left|E_{pr}\right|\right)\right\rangle \\ \left\langle f = P_{f1(pr)}\left(-\left|E_{pr}\right|\right)\right| a_{pr} &= \left\langle f = P_{f1(rp)}\left(\left|E_{pr}\right|\right)\right| a^*_{rp} \end{aligned} \tag{7.3.12}$$

are adopted. To illustrate their physical meaning, let pr=13 representing K^- in (2.4.18). These relations mean that creating a negative energy $-E_{13}$ K^- meson is the same as annihilating a positive energy $E_{13}=E_{31}$ K^+ meson.

Consistent with (6.4.15), one writes the decay amplitude

$$S_{fi} = \left\langle f\left|a^+_{rp}a^{(1)}_{pr}\left(X^0 \to \infty\right)\right|i\right\rangle \tag{7.3.13}$$

Following §6.4.4, the first order part S_{T31} of S_{T3} in (7.1.1) is placed between <f| and |i>. Consider at first the contribution stemming from S_{m3d} in (7.3.4). The zeroth order wave functions for a pseudoscalar meson at rest are according to (7.3.1) and (3.4.1)

$$\psi^{a\dot{b}}_{0(pr)}\left(X, x\right) = \delta^{a\dot{b}}\psi_{o(pr)}(r)\exp\left(-iE_{pr}X^0\right), \qquad \psi \to \chi \tag{7.3.14a}$$

$$\psi_{o(pr)}(r) = -\chi_{o(pr)}(r) = \psi_o(r) \tag{7.3.14b}$$

The last quantity is given by (4.3.2). With (7.3.13-14), (7.3.4) yields

$$\left\langle f\left|S'_{m3d}\right|i\right\rangle = -iE_{pr}S_{fi}\int d^3\underline{X}dx^0 d^3\underline{x}\,\psi_0^2(r) = -iE_{pr}S_{fi}\tau_0 \tag{7.3.15}$$

analogous to (6.4.14). Here, $\int d^3\underline{X}=\Omega$ according to the second of (4.2.5) and $\to\infty$ here and

$$\tau_o = \int dx^0 \to \infty \tag{7.3.16}$$

is a relative time. Combining (7.3.15) with (7.3.3) leads to

$$S_{fi} = i\frac{1}{E_{pr}\tau_0}\left\langle f\left|S'_{m3s}\right|i\right\rangle \tag{7.3.17}$$

where E_{pr} is the mass of the initial state pseudoscalar meson present in (7.3.6, 14a).

7.3.4. Amplitude for Decay to Gauge Boson, Vacuum Meson State

In the source matrix element $<f|S'_{m3s}|i>$, the surface term R_s in (7.3.5) turns out not to contribute similar to that the surface term $iq_r\partial_I$ in (6.4.5) does not contribute. After elevating the amplitudes of the wave functions entering S'_{m3s} to annihilation and creation operators according to the assignments below (6.4.2), the quantities operated by ∂_I in (7.3.5b) do not depend upon X^μ. This is due to that the initial and final state four momenta must cancel each other. Further, the meson wave functions vanish at large r due to confinement, as was mentioned in §6.1.1. With (3.1.4), it is readily seen that the R_s term of (7.3.5) vanishes in the source matrix element after integration over $\underline{x}$. With (7.3.5, 6, 7, 9, 10, 11, 14), one obtains for positive energy solutions denoted by the superscript (+),

$$\langle f|S'_{m3s}|i\rangle^{(+)} = -\frac{ig_l}{4}\sqrt{2}\int d^4X dx^4$$
$$\times\begin{Bmatrix} \chi^{+}_{f1(rp)\dot{e}a} W^{a\dot{b}}_{ps}\left(\partial_{II\dot{b}}{}^{\dot{e}}\chi_0(r)\exp(-iE_{sr}X^0)\right) - \left(\partial^{a\dot{b}}_{I}\chi^{+}_{f1(rp)\dot{e}a}\right) W_{ps\dot{b}}{}^{\dot{e}}\chi_0(r)\exp(-iE_{sr}X^0) \\ +\psi^{+d\dot{c}}_{f1(rp)} W_{ps\dot{c}a}\left(\partial^{a}_{II\,d}\psi_0(r)\exp(-iE_{sr}X^0)\right) - \left(\partial_{I\,\dot{c}a}\psi^{+d\dot{c}}_{f1(rp)}\right) W^{\,a}_{ps\,d}\psi_0(r)\exp(-iE_{sr}X^0) \end{Bmatrix}$$

(7.3.18)

where raising and lowering of spinor indices follow those in (C9). For negative energy meson wave functions corresponding to the lower sign in (5.1.1), the braced expression in (7.3.5a) is to be replaced by its complex conjugate. With (7.3.12), it becomes

$$\langle f|S'_{m3s}|i\rangle^{(-)} = -\frac{ig_l}{4}\sqrt{2}\int d^4X dx^4 \times$$
$$\times\begin{Bmatrix} -\chi^{(-)e\dot{a}}_{f1(rp)} W_{sp\dot{a}b}\partial_{IIe}{}^{b}\chi_0(r)\exp(-i|E_{rs}|X^0) + \left(\partial_{I\dot{a}b}\chi^{(-)e\dot{a}}_{f1(rp)}\right) W_{spe}{}^{b}\chi_0(r)\exp(-i|E_{rs}|X^0) \\ -\psi^{(-)}_{f1(rp)\dot{d}c} W^{c\dot{a}}_{sp}\partial_{II\dot{a}}{}^{\dot{d}}\psi_0(r)\exp(-i|E_{rs}|X^0) + \left(\partial^{c\dot{a}}_{I}\psi^{(-)}_{f1(rp)\dot{d}c}\right) W_{sp\dot{a}}{}^{\dot{d}}\psi_0(r)\exp(-i|E_{rs}|X^0) \end{Bmatrix}$$

(7.3.19)

The superscript (−) denotes negative energy meson wave functions.

Consider at first the final states (7.3.7b). The initial state pseudoscalar meson at rest is a singlet having mass E_{sr}, as has been represented in (7.3.18). The meson final state denoted by the subscript $f1$ is a vacuum meson state which will be taken here as the neutral pion π^0 that has lost its four momentum to the intermediate gauge boson $W(X)$. The relative space part of the meson wave functions is unchanged, inasmuch as this part depends only upon the "hidden variable" $\underline{x}$ and is not directly observable [H12 Sec. 1]. One of the open flavors s or r of the initial meson state is thus converted by this $W(X)$ to the other in the final meson state, in which the quark and antiquark have the same flavor and are regarded to annihilate each other virtually;

$$\psi^{*d\dot{c}}_{f1(pr)}(X,x) = \psi^{*}_{0wvac(pr)}(r)\delta^{d\dot{c}} \qquad (7.3.20a)$$

in (7.3.18). Here, the subscript w refers to weak decay and vac to vacuum meson state. Accordingly, (4.3.2) with the normalization (4.2.8) becomes

$$\psi_{0wvac}(r)=\frac{1}{\sqrt{2}}\sqrt{\frac{d_m^3}{8\pi\Omega_c}}\exp(-d_m r/2)=\frac{1}{\sqrt{2}}\sqrt{\frac{\Omega}{\Omega_{wvac}}}\psi_0(r), \qquad \Omega_c=\Omega_{wvac}=\text{finite} \tag{7.3.20b}$$

where $1/\sqrt{2}$ stems from (7.3.23) below. The departure from (4.3.2) where $\Omega_c\to\Omega\to\infty$ is due to the following. In (4.3.2), which describes a real meson, $\Omega\to\infty$ is needed to accommodate the different de Broglie wave lengths so that wave packets representing the real meson can be built and observed. In the vacuum meson state, no meson is observed and there is no need for $\Omega_c\to\Omega\to\infty$; $\Omega_c\to\Omega_{wvac}$ in (7.3.20b) suffices. The interpretation is that the vacuum meson state is not spread out widely and thinly but is concentrated in a finite volume Ω_{wvac}, which can be small.

With (7.1.3b), (7.3.20), (7.3.14b), (3.1.4), and (3.5.7), (7.3.18) is evaluated to be

$$\left\langle 0, W_{rp}\left|S'_{m3s}\right|i\right\rangle=\frac{g_l}{2}E_{pr}\sqrt{\frac{\Omega}{\Omega_{wvac}}}\int d^4Xd^4xW_{rp}^0(X)\psi_0^2(r)\exp\left(-iE_{pr}X^0\right) \tag{7.3.21}$$

The same result is also obtained using (7.3.19) instead. In obtaining this result, it is noted that in (7.3.18) $\underline{W}_{rp}$ only couples to $\underline{\partial}$ which operates on $\psi_0^2(r)$. Such terms vanish upon integration over $\underline{x}$ because $\psi_0(r)$ vanishes at large $\underline{x}$. The decay amplitude (7.3.17) becomes

$$S_{fi}=i\frac{g_l}{2}\frac{1}{\tau_0}\sqrt{\frac{\Omega}{\Omega_{wvac}}}\int d^4Xd^4xW_{rp}^0(X)\psi_0^2(r)\exp\left(-iE_{pr}X^0\right) \tag{7.3.22}$$

Owing to (7.3.6) and (2.4.18), this relation holds for $\pi^\pm$ and kaons only. This result can also be obtained starting from either (7.1.9a) or (7.1.9b), without the addition operation mentioned above (7.3.3).

7.3.5. Amplitude for Decay to Pion and Gauge Boson

Turning to (7.3.7a), rs=11 and 22 each contains a π^0 according to (2.4.18) and (7.3.8). Consider at first $\pi^+\to\pi^0W_l^+$, which is the only possibility. The π^0 and W_l^+ can now move in opposite directions and both the singlet and triplet components of the wave functions of the π^0 enter, according to the introduction of Sec. 3.5. With (7.3.8), (2.4.18), (7.1.3b), (3.1.1), and (3.1.5) adapted to the present case, the final states in (7.3.18) can be written as

$$\chi^{+}_{f1(11)\dot{e}a} = -\chi^{+}_{f1(22)\dot{e}a} = \frac{1}{\sqrt{2}}\chi^{+}_{f1(\pi 0)\dot{e}a} = \frac{1}{\sqrt{2}}\left(\chi^{*}_{0f1\pi 0}\delta_{\dot{e}a} - \underline{\sigma}_{\dot{e}a}\underline{\chi}^{*}_{f1\pi 0}\right)$$
$$= \frac{1}{\sqrt{2}}\left(\chi^{*}_{0\pi 0x}(\underline{x})\delta_{\dot{e}a} - \underline{\sigma}_{\dot{e}a}\underline{\chi}^{*}_{\pi 0x}(\underline{x})\right)\exp\left(iE_{f1\pi 0}X^{0} - i\underline{K}_{f1\pi 0}\underline{X}\right), \quad \chi \to \psi \tag{7.3.23}$$

Note that final meson state is a real meson; $\Omega \to \infty$ in (4.3.2), and (7.3.20) does not apply. Inserting these expressions into (7.3.18) and make use of (3.1.4) and (3.5.7), (7.3.17) for $\pi^{+} \to \pi^{0} W_{I}^{+}$ becomes after some calculation

$$S_{fi} = \frac{i2}{E_{\pi+}\tau_{0}}\left\langle \pi^{0}, W_{I}^{-}\left|S'_{m3s}\right|\pi^{+}\right\rangle = \frac{ig}{8E_{\pi+}\tau_{0}}\int d^{4}Xd^{4}x\exp\left(i\left(E_{f1\pi 0} - E_{\pi+}\right)X^{0} - i\underline{K}_{f1\pi 0}\underline{X}\right)$$

$$\times\begin{cases}\psi_{0}(r)W_{I}^{0+}\left[\left(E_{\pi+} + E_{f1\pi 0}\right)\left(\chi^{*}_{0\pi 0x} - \psi^{*}_{0\pi 0x}\right) - \underline{K}_{f1\pi 0}\left(\underline{\chi}^{*}_{\pi 0x} - \underline{\psi}^{*}_{\pi 0x}\right) - i2\underline{\partial}\left(\underline{\chi}^{*}_{\pi 0x} - \underline{\psi}^{*}_{\pi 0x}\right)\right] \\ -\psi_{0}(r)\underline{W}_{I}^{+}\begin{bmatrix}\left(E_{\pi+} + E_{f1\pi 0}\right)\left(\underline{\chi}^{*}_{\pi 0x} - \underline{\psi}^{*}_{\pi 0x}\right) - \underline{K}_{f1\pi 0}\left(\chi^{*}_{0\pi 0x} - \psi^{*}_{0\pi 0x}\right) - i2\underline{\partial}\left(\chi^{*}_{0\pi 0x} - \psi^{*}_{0\pi 0x}\right) \\ + 2\underline{\partial}\times\left(\underline{\chi}^{*}_{\pi 0x} + \underline{\psi}^{*}_{\pi 0x}\right) + i\left(\underline{\chi}^{*}_{\pi 0x} + \underline{\psi}^{*}_{\pi 0x}\right)\times\underline{K}_{f1\pi 0}\end{bmatrix} \\ + 2\left(\underline{\partial}\psi_{0}(r)\right)\left[\underline{W}_{I}^{+}\times\left(\underline{\chi}^{*}_{\pi 0x} + \underline{\psi}^{*}_{\pi 0x}\right) + i\underline{W}_{I}^{+}\left(\chi^{*}_{0\pi 0x} - \psi^{*}_{0\pi 0x}\right) - iW_{I}^{0+}\left(\underline{\chi}^{*}_{\pi 0x} - \underline{\psi}^{*}_{\pi 0x}\right)\right]\end{cases} \tag{7.3.24}$$

The same result is obtained using (7.3.19) instead. Because the mass difference of π^{+} and π^{0} is small compared to the π^{0} mass, π^{0} will move very slowly so that $\underline{K}_{f1\pi 0}$ is of first order in ε_{0} of (3.5.1). To zeroth order in ε_{0}, (3.5.3) holds for π^{0}. Further, (3.5.21) holds for π^{0} so that the first order results (3.5.22a) and (3.5.23) apply. Equation (7.3.24) can be much simplified if terms to order ε_{0} and higher are neglected. To zeroth order in ε_{0}, it becomes

$$S_{fi} = -\frac{ig}{4\Omega E_{\pi+}}\int d^{4}X\exp\left(i\left(E_{f1\pi 0} - E_{\pi+}\right)X^{0} - i\underline{K}_{f1\pi 0}\underline{X}\right)$$
$$\times\left[\left(E_{\pi+} + E_{f1\pi 0}\right)W_{I}^{0+}(X) + 2\underline{K}_{f1\pi 0}\underline{W}_{I}^{+}(X)\right] \tag{7.3.25}$$

The first order term $\underline{KW}_{I}^{+}$ is retained here to facilitate comparison to the results of the literature in §7.4.5 below.

There are four kaon decays in (7.3.18), which together with (7.3.17) yields the following decay amplitudes analogous to (7.3.24);

$$K^{-} \to \pi^{0}W_{V}^{-}: \quad S_{fi} = -\frac{ig_{4-7}}{8E_{K-}}\int d^{4}Xd^{3}\underline{x}\exp\left(i\left(E_{f1\pi 0} - E_{K-}\right)X^{0} - i\underline{K}_{f1\pi 0}\underline{X}\right)$$

$$\times\begin{Bmatrix} \psi_0(r)W_V^{0-}\left[\left(E_{K-}+E_{f1\pi 0}\right)\left(\chi^*_{0\pi 0x}-\psi^*_{0\pi 0x}\right)-\underline{K}_{f1\pi 0}\left(\underline{\chi}^*_{\pi 0x}-\underline{\psi}^*_{\pi 0x}\right)-i2\underline{\partial}\left(\underline{\chi}^*_{\pi 0x}-\underline{\psi}^*_{\pi 0x}\right)\right] \\ -\psi_0(r)\underline{W}_V^-\begin{bmatrix}\left(E_{K-}+E_{f1\pi 0}\right)\left(\underline{\chi}^*_{\pi 0x}-\underline{\psi}^*_{\pi 0x}\right)-\underline{K}_{f1\pi 0}\left(\chi^*_{0\pi 0x}-\psi^*_{0\pi 0x}\right)-i2\underline{\partial}\left(\chi^*_{0\pi 0x}-\psi^*_{0\pi 0x}\right) \\ +2\underline{\partial}\times\left(\underline{\chi}^*_{\pi 0x}+\underline{\psi}^*_{\pi 0x}\right)+i\left(\underline{\chi}^*_{\pi 0x}+\underline{\psi}^*_{\pi 0x}\right)\times\underline{K}_{f1\pi 0}\end{bmatrix} \\ +2\left(\underline{\partial}\psi_0(r)\right)\left[\underline{W}_V^-\times\left(\underline{\chi}^*_{\pi 0x}+\underline{\psi}^*_{\pi 0x}\right)+i\underline{W}_V^-\left(\chi^*_{0\pi 0x}-\psi^*_{0\pi 0x}\right)-iW_V^{0-}\left(\underline{\chi}^*_{\pi 0x}-\underline{\psi}^*_{\pi 0x}\right)\right]\end{Bmatrix}$$

(7.3.26a)

$K^- \to \pi^- W_U$: $S_{fi} = \sqrt{2}\times$ right of (7.3.26a) with $W_V^- \to W_U$ and subscript ${}_{\pi 0}\!\!\to_{\pi -}$

(7.3.26b)

$\overline{K}^0 \to \pi^- W_V^+$: $S_{fi} = \sqrt{2}\times$ right of (7.3.26a) with subscripts ${}_{K-}\!\!\to_{\overline{K}0}$ and ${}_{\pi 0}\!\!\to_{\pi -}$

(7.3.27a)

$\overline{K}^0 \to \pi^0 W_U$: $S_{fi} = (-)\times$ right of (7.3.26a) with $W_V^- \to W_U$ and subscript ${}_{K-}\!\!\to_{\overline{K}0}$

(7.3.27b)

where $\overline{K}^0$ is a linear combination of the physical $K^0{}_S$ and $K^0{}_L$ in (7.2.18). The amplitudes of the conjugate decays are analogously obtained from (7.3.19) and read

$K^+ \to \pi^0 W_V^+,\ \pi^+ \overline{W}_U$:

$S_{fi} = (7.3.26)\ with\ {}_{K-}\!\!\to_{K+}$ and $W_V^- \to W_V^+$ (26a), $W_U \to \overline{W}_U$, ${}_{\pi -}\!\!\to_{\pi +}$ (26b)

(7.3.28)

$K^0 \to \pi^+ W_V^-,\ \pi^0 \overline{W}_U$:

$S_{fi} = (7.3.27)\ with\ {}_{\overline{K}0}\!\!\to_{K0}$ and $W_V^+ \to W_V^-$, ${}_{\pi -}\!\!\to_{\pi +}$ (27a), $W_U \to \overline{W}_U$ (27b)

(7.3.29)

7.3.6. Amplitude for Decay to Two Gauge Bosons, Second Order Relations

The amplitudes obtained in §7.3.4-5 are of first order in g according to (7.3.2a) and contain one gauge boson. The treatment there can be modified to yield second order amplitudes containing two gauge bosons. First let

$$a_{pr}^{(1)}\left(X^0\right) \to a_{pr}^{(1)}\left(X^0\right) + a_{pr}^{(2)}\left(X^0\right) \tag{7.3.30}$$

in (7.3.1), where the superscript (2) denotes second order perturbation. Next, (7.3.3) is replaced by

$$S'_{m3d2} = S'_{m3s2} \tag{7.3.31}$$

where the subscript *2* denotes second order contributions. The left side is (7.3.4) with the superscripts (*1*)→(*2*). The right side of (7.3.31) is obtained from (7.3.5) in which

$$\partial_{II}, \partial_I \to i\tfrac{1}{4}g(\lambda_{l'})_{ps} W_{l'}(X) \tag{7.3.32}$$

according to (7.1.4), supressing the internal indices. The results is

$$S'_{m3s2} = \int d^4Xdx^4 \frac{g_l g_{l'}}{16} \begin{Bmatrix} \overset{+}{\chi}_{0(rp)\dot{e}a}(\lambda_l)_{ps} W_l^{a\dot{b}} (\lambda_{l'})_{sq} W_{l'}^{f\dot{e}} \chi_{0(qr)\dot{b}f} \\ +\psi_{0(rp)}^{+d\dot{c}}(\lambda_l)_{ps} W_{l\dot{c}a} (\lambda_{l'})_{sq} W_{l'\dot{e}d} \psi_{0(qr)}^{a\dot{e}} \end{Bmatrix} \tag{7.3.33}$$

In general, there is also a second order source term of the form

$$i\tfrac{1}{4}g(\lambda_l)W_l(X)a^{(1)}(X^0) \tag{7.3.34}$$

that contributes to (7.3.33). For reasons given below, this term is dropped because it will not contribute to π^+, $K^+ \to \mu^+ \nu_\mu \gamma$ to be considered here.

The initial state (7.3.6) is the same. The final state (7.3.7b) is modified to

$$\langle f| = \langle P_{f(rr)}(0), W_{rq}, W_{qp}| = \langle 0, W_{rq}, W_{qp}| \tag{7.3.35}$$

The treatments of §7.3.3-4 and §7.2.3 can be taken over and (7.3.17) and (7.3.22) are modified to

$$\begin{aligned} S_{fi} &= i\frac{1}{E_{pr}\tau_0}\langle f|S'_{m3s2}|i\rangle \\ &= -\frac{ge}{2E_{pr}\tau_0}\sqrt{\frac{\Omega}{\Omega_{wvac}}} \int d^4Xd^4x\psi_0^2(r)\underline{A}_{rq}(X)\underline{W}_{qp}(X)\exp(-iE_{pr}X^0) \end{aligned} \tag{7.3.36}$$

where (7.1.4), (7.2.14) and (7.2.1b) have been employed.

By (7.3.13) and (7.3.22), $a^{(1)}(X)$ in (7.3.34) is proportional to $W^0{}_{qp}(X)$. $(\lambda_l)W_l(X)$ in (7.3.34) becomes $\underline{A}_{rq}(X)$ in (7.3.36) which does not couple to the singlet $W^0{}_{qp}(X)$. Therefore, (7.3.34) does not contribute here.

7.4. DECAY AMPLITUDE FOR GAUGE BOSON → LEPTON PAIR

7.4.1. General Considerations

Turn now to stage II of the decay mentioned in §7.1.1, i.e., the decay of the gauge bosons present in the decay amplitudes S_{fi} in the last two paragraphs. The gauge bosons can also decay into two types of end products, namely, a pair of leptons contained in S_L of (7.2.13) via (7.2.20-22) and a pair of pions contained in $S_{m\pi}$ in (7.1.1) which will be considered briefly in §7.7.4 below.

Further, (7.1.2) can by means of (7.2.1a, 3a) and (C4, C5) be transformed to

$$S_{GB} = -\frac{1}{4}\int d^4X \left\{ \begin{array}{l} \partial_{X\dot{a}b} W_I^{-c\dot{d}} \left(\partial_X^{b\dot{a}} W_{I\,\dot{d}c}^{+} - \partial_{X\dot{d}c} W_I^{+b\dot{a}} \right) + \partial_{X\dot{a}b} W_V^{-c\dot{d}} \left(\partial_X^{b\dot{a}} W_{V\,\dot{d}c}^{+} - \partial_{X\dot{d}c} W_V^{+b\dot{a}} \right) \\ + \partial_{X\dot{a}b} W_U^{c\dot{d}} \left(\partial_X^{b\dot{a}} \overline{W}_{U\,\dot{d}c} - \partial_{X\dot{d}c} \overline{W}_U^{b\dot{a}} \right) \\ + \tfrac{1}{2} \partial_{X\dot{a}b} Z^{c\dot{d}} \left(\partial_X^{b\dot{a}} Z_{\dot{d}c} - \partial_{X\dot{d}c} Z^{b\dot{a}} \right) + \tfrac{1}{2} \partial_{X\dot{a}b} A^{c\dot{d}} \left(\partial_X^{b\dot{a}} A_{\dot{d}c} - \partial_{X\dot{d}c} A^{b\dot{a}} \right) \end{array} \right\}$$
$$+ g(.....) \qquad (7.4.1)$$

The last term denotes terms containing g, which by (7.2.7) should actually be generalized to g_l. A term containing g_l will break the SU(3) symmetry of (7.4.1). Such a generalization will complicate the calculations but will have no effect on the results below because these g terms will contribute to order g^2(see (7.4.2) below) and can be neglected.

The gauge boson wave functions are obtained by varying the second order part of the total action S_{T3} with respect to W, Z or A.

7.4.2. Generation of Mass of Charged Gauge Boson

There are four charged bosons, $W_I^{\pm}$ and $W_V^{\pm}$. The W_V^{-} case will be worked out below. The results obtained also hold for the W_V^{+} case by appropriate complex conjugation. The $W_I^{\pm}$ cases are the same as the $W_V^{\pm}$ cases by virtue of (7.2.7, 20).

Variations of (7.4.1) with respect to $W_I^{+a\dot{b}}$ defined in (7.1.5b) yields

$$\delta S_{GB} / \delta W_I^{-a\dot{b}} = \tfrac{1}{2} \Box W_{I\,\dot{b}a}^{+} - \tfrac{1}{4} \partial_{X\dot{b}a} \left(\partial_X^{c\dot{d}} W_{I\,\dot{d}c}^{+} \right) + \tfrac{1}{2} g^2 V_{C\dot{b}a} \left(W_I^{+} \right) \qquad (7.4.2a)$$

$$g^2 V_{C\dot{b}a} \left(W_I^{+} \right) = g^2 \left(\delta_{\dot{b}a} V_C^0 \left(W_I^{+} \right) + \underline{\sigma}_{\dot{b}a} \underline{V}_C \left(W_I^{+} \right) \right) \qquad (7.4.2b)$$

$$\begin{array}{l} V_C^0 \left(W_I^{+} \right) = \left(\underline{W}_I^{+} \right)^2 W_I^{0-} - \left(\underline{W}_I^{-} \underline{W}_I^{+} \right) W_I^{0+} \\ \underline{V}_C \left(W_I^{+} \right) = W_I^{0+} W_I^{0-} \underline{W}_I^{+} - \left(W_I^{0+} \right)^2 \underline{W}_I^{-} + \left(\underline{W}_I^{+} \right)^2 \underline{W}_I^{-} - \left(\underline{W}_I^{-} \underline{W}_I^{+} \right) \underline{W}_I^{+} \end{array} \qquad (7.4.2c)$$

With (7.1.4) and (7.2.7), the second order part of S_{m3} of (7.1.8) containing W_V^+ and W_V^- are coupled to a rest frame $K^+(K^-)$ characterized by the flavor indices 13(31) in the next equation. Applying (7.2.20) and vary this part with respect to $W_I^-(W_I^+)$ yields

$$\delta S_{m3}/\delta W_I^{\mp a\dot{b}} = -\frac{g^2}{16}\int d^4x W_I^{\pm f\dot{e}}\begin{bmatrix}\left(\chi^*_{0(13)\dot{e}a}\chi_{0(13)\dot{b}f} + \chi^*_{0(31)\dot{b}f}\chi_{0(31)\dot{e}a} + \chi^*_0 \leftrightarrow \chi_0\right) \\ +\left(\psi_{0(31)\dot{e}a}\psi^*_{0(31)\dot{b}f} + \psi_{0(13)\dot{b}f}\psi^*_{0(13)\dot{e}a} + \psi^*_0 \leftrightarrow \psi_0\right)\end{bmatrix}$$

(7.4.3a)

If $K^+(K^-)$ is replaced by $\pi^+(\pi^-)$, this relation becomes

$$\delta S_{m3}/\delta W_I^{\mp a\dot{b}} = (7.4.3a) \text{ with index } 3 \to 2 \tag{7.4.3b}$$

Inserting (7.3.14) and (4.3.2) into (7.4.3) yields

$$\delta S_{m3}/\delta W_I^{\mp a\dot{b}} = -\frac{g^2}{2}\frac{\tau_0}{\Omega}W_I^{\pm b\dot{a}} \tag{7.4.4}$$

The same variation of (7.2.13) yields

$$\delta S_L/\delta W_I^{-a\dot{b}} = \frac{g}{2\sqrt{2}}\psi_{\nu L\dot{b}}\psi_{La} \tag{7.4.5}$$

In (7.4.3-5), source terms not containing $W_I^\pm$ have been dropped.

Variation of (7.1.1) with respect to $W_I^{-a\dot{b}}$ gives the equation of motion for $W^+_{I\dot{b}a}$. With (7.4.2, 4, 5), this variation yields,

$$\Box W^+_{I\dot{b}a} - \tfrac{1}{2}\partial_{X\dot{b}a}\left(\partial_X^{c\dot{d}}W^+_{I\dot{d}c}\right) + g^2 V_{C\dot{b}a}\left(W_I^+\right) - M_W^2 W_I^{+b\dot{a}} = -\frac{g}{\sqrt{2}}\psi_{\nu L\dot{b}}\psi_{La} \tag{7.4.6a}$$

where

$$M_W^2 = g^2\frac{\tau_0}{\Omega} = \frac{\infty}{\infty} \tag{7.4.6b}$$

is the square of the mass of the charged gauge boson [H13 §3.4] and is by (7.3.16) a ratio ∞/∞=finite similar to M_{uI}^2 in (6.2.12). It is seen that the origin of the gauge boson mass lies in the presence of the relative time x^0 in a Lorentz covariant formulation. This relative time is absent in conventional, local theories, which call for the unseen Higgs bosons for mass generation. This will be discussed further in §7.5.4 and §7.6.2 below. That Higgs boson is not needed to generate M_W was first shown in [H8 Sec. 6].

Contracting (7.4.6a) by $\delta^{a\dot{b}}$ and $\underline{\sigma}^{a\dot{b}}$ yields

$$-\frac{\partial}{\partial X^0}\left(\frac{\partial}{\partial \underline{X}}\underline{W}_I^+\right)+g^2V_C^0\left(W_I^+\right)-M_W^2W_I^{0+}=-\frac{g}{2\sqrt{2}}\psi_{\nu L\dot{a}}\psi_{La} \quad (7.4.7a)$$

$$\Box\underline{W}_I^+ +\frac{\partial}{\partial \underline{X}}\left(\frac{\partial}{\partial X^0}W_I^{0+}+\frac{\partial}{\partial \underline{X}}\underline{W}_I^+\right)+g^2\underline{V}_C\left(W_I^+\right)+M_W^2\underline{W}_I^+=-\frac{g}{2\sqrt{2}}\underline{\sigma}^{a\dot{b}}\psi_{\nu L\dot{b}}\psi_{La} \quad (7.4.7b)$$

Choose $\vartheta_4(X)-i\vartheta_5(X)$ in (7.1.7b) such that (7.1.7d) via (7.2.20) leads to the Coulomb gauge

$$\left(\partial/\partial\underline{X}\right)\underline{W}_I^+=0$$

Further, the ordering (7.3.2b) adopted relegates the nonlinear g^2V_C terms in (7.4.6-7) to higher order, by two, relative to the rest of the terms there. In the absence of the lepton source terms on the right of (7.4.7), it yields to lowest order

$$W_I^{0+}=0 \quad (7.4.8a)$$

$$\Box\underline{W}_I^+ + M_W^2\underline{W}_I^+=0 \quad (7.4.8b)$$

$\underline{W}_I^-$ is identified with the observed charged gauge boson $\underline{W}$ [P1] with the mass

$$M_W=80.42 \text{ Gev} \quad (7.4.9)$$

which may be considered as a fundamental constant instead of the Fermi constant in (7.4.29) below. The time component W_I^{0+} associated with $\underline{W}_I^+$ in (7.4.8) vanishes in agreement with the nonobservation of such a singlet charged gauge boson W^0 accompanying the observed triplet $\underline{W}$. If Higgs boson were used to generate the gauge boson mass, such a singlet W^0 with same mass (7.4.9) should also be seen, contrary to observation, as will be discussed in §7.6.2 below.

If the Lorentz gauge

$$\partial_X^{c\dot{d}}W_{I\dot{d}c}^+=0$$

were employed, (7.4.8b) remains unchanged and (7.4.8a) becomes

$$\Box W_I^{0+}-M_W^2W_I^{0+}=0$$

This implies that W_I^{0+} has an imaginary mass of (7.4.9) and therefore must vanish and (7.4.8a) remains in effect valid.

The energy and momentum of the virtual gauge bosons in (7.4.7) are determined by those of the lepton pair and are generally much smaller than the mass (7.4.9) and can therefore be dropped. Hence, (7.4.7) reduces to

$$M_W^2 W_I^{0+} = \frac{g}{2\sqrt{2}} \psi_{La}^{(-)} \psi_{\nu L\dot{a}}^{(+)} \tag{7.4.10a}$$

$$M_W^2 \underline{W}_I^{+} = -\frac{g}{2\sqrt{2}} \underline{\sigma}^{b\dot{a}} \psi_{\nu L\dot{a}}^{(+)} \psi_{Lb}^{(-)} \tag{7.4.10b}$$

While the triplet $\underline{W}_I^+$ can exist freely and hence be seen, as is shown in (7.4.8b), it can also be a virtual intermediate state as is seen in (7.4.10b). On the other hand, the singlet W_I^{0+} cannot be observed as is shown in (7.4.8a), but can only be a charged, virtual intermediate singlet as is seen in (7.4.10a).

As was pointed out in the beginning of §7.4.2, the above results for the charged kaons also holds for the charged pions which enter (7.4.3b). One can see that (7.4.10) is not obtainable with the ordering (7.3.2a) for which the g^2V_C terms are of the same order as the other terms in (7.4.7) and cannot be neglected. Retaining this nonlinear term practically excludes any explicit and simple solution of the form (7.4.10).

Note that (7.4.4, 6a) are no longer Lorentz covariant because the triplet part of the meson wave functions vanish for rest frame mesons (7.3.14). This has led to that the signs in front of M_W^2 in (7.4.7a) and (7.4.7b) are different. Thus, M_W^2 there may be considered as a 4×4 diagonal matrix with the element $-M_W^2$ for the time component and $+M_W^2$ for the spatial components. For a meson in motion, both the singlet and triplet wave functions in (3.5.8) would be present in (7.4.3) and the simple forms of (7.4.4, 6a) are lost. The 4×4 matrix for M_W^2 above is no longer diagonal and mixes the four W_I^- components, leading to higher order equations for them. Further, the elements of this nondiagonal matrix cannot be evaluated presently because wave functions for mesons in motion are unknown, except approximately for slowly moving mesons, as was considered in Sec. 3.5.

The present situation therefore differs basically fronm that in which the massive gauge bosons are generated by Higgs boson. The so-generated M_{WH}^2 is a scalar given by (7.6.6b) below and is independent of an eventual motion of the Higgs boson.

7.4.3. Decay of Charged Gauge Boson into Lepton Pair

The plane wave solutions of freely moving leptons satisfy (A3) with I_B=0 and the first of (A15). The solutions (A13, A14) can be put in terms of the spinors of (C11),

$$\psi_{\nu L\dot{b}}^{(+)} = \frac{1}{\sqrt{\Omega_L}} u_{\nu L\dot{b}}^{(+)} \exp\left(iE_{\nu L} X^0 - i\underline{K}_{\nu L}\underline{X}\right)$$

$$\begin{pmatrix} u^{(+)}_{\nu L\dot{1}} \\ u^{(+)}_{\nu L\dot{2}} \end{pmatrix} = \frac{1}{\sqrt{2E_{\nu L}}}\left[\begin{pmatrix} K_{\nu L1} - iK_{\nu L2} \\ E_{\nu L} - K_{\nu L3} \end{pmatrix}, \begin{pmatrix} -E_{\nu L} - K_{\nu L3} \\ -K_{\nu L1} - iK_{\nu L2} \end{pmatrix}\right] \quad (7.4.11a)$$

$$\psi^{(-)}_{Lb} = \frac{1}{\sqrt{\Omega_L}} u^{(-)}_{Lb} \exp\left(iE_L X^0 + i\underline{K}_L \underline{X}\right)$$

$$\begin{pmatrix} u^{(-)}_{L1} \\ u^{(-)}_{L2} \end{pmatrix} = \frac{1}{\sqrt{2E_L(E_L + M_L)}}\left[\begin{pmatrix} E_L + M_L - K_{L3} \\ -K_{L1} + iK_{L2} \end{pmatrix}, \begin{pmatrix} -K_{L1} - iK_{L2} \\ E_L + M_L + K_{L3} \end{pmatrix}\right] \quad (7.4.11b)$$

where the superscripts (+) and (−) denote positive and negative energy, respectively. Negative energy solutions are chosen for the lepton with mass M_L. Ω_L is a large normalization volume for the leptons. The comma in the brackets separates the spin up and spin down solutions.

Inserting (7.4.6b, 11) into (7.4.10a) yields

$$gW_I^{0+} = \frac{1}{2\sqrt{2}}\frac{\Omega}{\tau_0}\frac{1}{\Omega_L} u^{(-)}_{La} u^{(+)}_{\nu L\dot{a}} \exp\left(i(E_{\nu L} + E_L)X^0 - i(\underline{K}_{\nu L} - \underline{K}_L)\underline{X}\right) \quad (7.4.11c)$$

Noting (7.3.9b), (7.4.11c) is inserted into (7.3.22). Carrying out the integration over X^0, it leads to

$$\langle f| = \langle 0, W| = \langle 0, L^{(-)}\nu_L^{(+)}|, \qquad |i\rangle = |\pi^+ \text{ or } K^+\rangle \quad (7.4.12)$$

$$S_{fi} = -i\frac{\pi}{2\sqrt{2}}\frac{\Omega}{\tau_0}\frac{1}{\tau_0 \Omega_L}\delta\left(E_L + E_{\nu L} - E_{pr}\right) u^{(-)}_{La} u^{(+)}_{\nu L\dot{a}} I_{L\nu} \quad (7.4.13a)$$

$$I_{L\nu} = \int d^3\underline{X} d^4 x \sqrt{\frac{\Omega}{\Omega_{wvac}}}\psi_0^2(r)\exp\left(i(\underline{K}_L - \underline{K}_{\nu L})\underline{X}\right) \quad (7.4.13b)$$

These results stem from the c.c. term in (7.1.8b). Here, $L^{(-)}$ and $\nu_L^{(+)}$ refer to the indices in (7.4.11) characterizing the final lepton states. Further, the ψ's of (7.4.11) have been tacitly quantized so that the final state of (7.4.12) picks out the appropriate lepton wave functions in (7.4.10, 13). This is analogous to that the meson wave function (7.3.1) has been quantized in connection with (7.3.10a).

7.4.4. Neutral Gauge Boson Mass and Decay Into Lepton Pair

Consider next the decay of the W_U and $\overline{W}_U$ bosons via the degeneracy (7.2.21) to the Z boson. By means of (7.4.1) and (7.2.13), the equivivalent of (7.4.2) and (7.4.5) become

$$\frac{\delta S_{GB}}{\delta Z^{a\dot{b}}} = \frac{1}{4}\Box Z_{\dot{b}a} - \frac{1}{8}\partial_{X\dot{b}a}\left(\partial_X^{c\dot{d}} Z_{\dot{d}c}\right) - g^2(\ldots) \tag{7.4.14}$$

$$\frac{\delta S_L}{\delta Z^{a\dot{b}}} = \frac{g}{8\cos\vartheta_W}\left(2\psi_{\nu La}\psi_{\nu L\dot{b}} - \psi_{La}\psi_{L\dot{b}} + \chi_{La}\chi_{L\dot{b}}\right) \tag{7.4.15}$$

where the last term in (7.4.14) denotes terms of order g^2.

Contribution from S_{m3} is similar to that of (7.4.3a). With (7.1.4) and (7.2.7), the second order part of S_{m3} of (7.1.8) containing W_U and $\bar{W}_U$ are coupled to K^0 or $\bar{K}^0$ characterized by the subscripts (23) and (32) to χ and ψ there. Apply the degenerations (7.2.21, 22) to this part and vary it with respect to $Z^{a\dot{b}}$ yields the analog of (7.4.3a)

$$\frac{\delta S_{m3}}{\delta Z^{a\dot{b}}} = -\frac{g^2}{32\cos^2\vartheta_W}\int d^4x Z^{f\dot{e}}\left[\begin{array}{l}\left(\chi^*_{0(23)\dot{e}a}\chi_{0(23)\dot{b}f} + \chi^*_{0(32)\dot{b}f}\chi_{0(32)\dot{e}a} + \chi^*_0 \leftrightarrow \chi_0\right) \\ +\left(\psi_{0(32)\dot{e}a}\psi^*_{0(32)\dot{b}f} + \psi_{0(23)\dot{b}f}\psi^*_{0(23)\dot{e}a} + \psi^*_0 \leftrightarrow \psi_0\right)\end{array}\right]$$

$$= -\frac{g^2}{4\cos^2\vartheta_W}\frac{\tau_0}{\Omega}Z^{\dot{b}a} \tag{7.4.16}$$

where the procedure intervening (7.4.3b) and (7.4.4) has been applied. Similarly, source terms in (7.4.14-15) not containing Z have been dropped. Inserting (7.4.14-16) into the variation of (7.1.1) with respect to $Z^{a\dot{b}}$ yields the analog of (7.4.6),

$$\Box Z_{\dot{b}a} - \tfrac{1}{2}\partial_{X\dot{b}a}\left(\partial_X^{c\dot{d}} Z_{\dot{d}c}\right) + g^2 V_{Z\dot{b}a}(Z) - M_Z^2 Z^{\dot{b}a}$$

$$= -\frac{g}{2\cos\vartheta_W}\left(2\psi^{(+)}_{\nu L\dot{b}}\psi^{(-)}_{\nu La} - \psi^{(+)}_{L\dot{b}}\psi^{(-)}_{La} + \chi^{(-)}_{L\dot{b}}\chi^{(+)}_{La}\right) \tag{7.4.17a}$$

$$M_Z^2 = \frac{1}{\cos^2\vartheta_W}g^2\frac{\tau_0}{\Omega} = \frac{M_W^2}{\cos^2\vartheta_W} \tag{7.4.17b}$$

where $g^2 V_Z$ corresponds to (7.4.2b, 2c) and is a modified form of it.

Following the steps intervening (7.4.7) and (7.4.10), (7.4.17a) leads to the analog of (7.4.8),

$$Z^0 = 0 \tag{7.4.18a}$$

$$\Box \underline{Z} + M_Z^2 \underline{Z} = 0 \tag{7.4.18b}$$

Here, $\underline{Z}$ is identified with the neutral gauge boson Z [P1] with the mass

$$M_Z = \frac{M_W}{\cos \vartheta_W} = 91.02 \text{ Gev} \tag{7.4.18c}$$

This mass is close to the measured 91.2 Gev [P1], differing from it by only −0.19%. If the SU(3) symmetry breaking is removed, (7.2.3b) replaces (7.2.12a) and $M_Z=2M_W/\sqrt{3}=92.86$ Gev which is +1.8% off.

The time component Z^0 is also neutral and has the same mass M_Z. Analogous to W_I^{0+} in (7.4.8a) and (7.4.10a), it cannot exist freely in agreement with the nonobservation of such a singlet neutral gauge boson Z^0. Just like W_I^{0+} in (7.4.10a), however, Z^0 can be present as a virtual intermediary neutral gauge boson.

In passing, it is noted that varaiation of S_{m3} (7.1.8) using (7.2.14) with respect to the electromagnetic field A, analogous to (7.4.16), will give rise to a mass M_A to the A field of the same magnitude as M_Z, M_W, contrary to experience. This apparent difficulty is removed if we note that only the fields entering the decay amplitudes S_{fi} of (7.3.17-19) are of interst. Since A does not enter these S_{fi}, it may be dropped in S_T (7.1.1) and variation of it with respect to A gives zero; there is no contradiction to data.

In (7.4.17a), $\psi_{La}^{(-)}$ and $\psi_{\nu L\dot{b}}^{(+)}$ have been given in (7.4.11). The remaining wave functions in (7.4.17a) are obtained from (A13), (A14) and (C11) in the same way as (7.4.11) was. They read

$$\psi_{\nu La}^{(-)} = \frac{1}{\sqrt{\Omega_L}} u_{\nu La}^{(-)} \exp\left(iE_{\nu L} X^0 + i\underline{K}_{\nu L}\underline{X}\right)$$

$$\begin{pmatrix} u_{\nu L1}^{(-)} \\ u_{\nu L2}^{(-)} \end{pmatrix} = \frac{1}{\sqrt{2E_{\nu L}}}\left[\begin{pmatrix} E_{\nu L} - K_{\nu L3} \\ -K_{\nu L1} + iK_{\nu L2} \end{pmatrix}, \begin{pmatrix} -K_{\nu L1} - iK_{\nu L2} \\ E_{\nu L} + K_{\nu L3} \end{pmatrix}\right] \tag{7.4.19a}$$

$$\psi_{L\dot{b}}^{(+)} = \frac{1}{\sqrt{\Omega_L}} u_{L\dot{b}}^{(+)} \exp\left(iE_L^{(+)} X^0 - i\underline{K}_L^{(+)}\underline{X}\right)$$

$$\begin{pmatrix} u_{L\dot{1}}^{(+)} \\ u_{L\dot{2}}^{(+)} \end{pmatrix} = \frac{1}{\sqrt{2E_L^{(+)}\left(E_L^{(+)} + M_L\right)}}\left[\begin{pmatrix} K_{L1}^{(+)} - iK_{L2}^{(+)} \\ E_L^{(+)} + M_L - K_{L3}^{(+)} \end{pmatrix}, \begin{pmatrix} -E_L^{(+)} - M_L - K_{L3}^{(+)} \\ -K_{L1}^{(+)} - iK_{L2}^{(+)} \end{pmatrix}\right] \tag{7.4.19b}$$

$$\chi_{L\dot{b}}^{(-)} = \frac{1}{\sqrt{\Omega_L}} v_{L\dot{b}}^{(-)} \exp\left(iE_L X^0 + i\underline{K}_L\underline{X}\right)$$

$$\begin{pmatrix} v_{L\dot{1}}^{(-)} \\ v_{L\dot{2}}^{(-)} \end{pmatrix} = \frac{1}{\sqrt{2E_L\left(E_L + M_L\right)}}\left[\begin{pmatrix} -K_{L1} + iK_{L2} \\ E_L + M_L + K_{L3} \end{pmatrix}, \begin{pmatrix} -E_L - M_L + K_{L3} \\ K_{L1} + iK_{L2} \end{pmatrix}\right] \tag{7.4.20a}$$

$$\chi_{La}^{(+)} = \frac{1}{\sqrt{\Omega_L}} v_{La}^{(+)} \exp\left(iE_L^{(+)} X^0 - i\underline{K}_L^{(+)} \underline{X}\right)$$

$$\begin{pmatrix} v_{L1}^{(+)} \\ v_{L2}^{(+)} \end{pmatrix} = \frac{1}{\sqrt{2E_L^{(+)}\left(E_L^{(+)} + M_L\right)}} \left[\begin{pmatrix} E_L^{(+)} + M_L + K_{L3}^{(+)} \\ K_{L1}^{(+)} - iK_{L2}^{(+)} \end{pmatrix}, \begin{pmatrix} K_{L1}^{(+)} + iK_{L2}^{(+)} \\ E_L^{(+)} + M_L - K_{L3}^{(+)} \end{pmatrix}\right] \quad (7.4.20b)$$

A superscript (+) has been introduced into the right sides of (7.4.19b) and (7.4.20b) in order to distinguish these E and K from the corresponding quantities in (7.4.11b) and (7.4.20a), respectively. It is also tacitly attached to $E_{\nu L}$ and $\underline{K}_{\nu L}$ in (7.4.11a) for applications here.

Following the steps of the last paragraph and dropping the first three terms in (7.4.17a), contracting it by $\delta^{a\dot{b}}$ and making use of (7.4.19-20) yields the equivalent of (7.4.11c),

$$gZ^0 = \frac{\cos\vartheta_W}{2}\frac{\Omega}{\tau_0}\frac{1}{\Omega_L}\begin{pmatrix} 2u_{\nu L\dot{a}}^{(+)} u_{\nu La}^{(-)} \exp\left(i\left(E_{\nu L}^{(+)} + E_{\nu L}\right)X^0 - i\left(\underline{K}_{\nu L}^{(+)} - \underline{K}_{\nu L}\right)\underline{X}\right) \\ + \left(v_{L\dot{a}}^{(-)} v_{La}^{(+)} - u_{La}^{(-)} u_{L\dot{a}}^{(+)}\right)\exp\left(i\left(E_L^{(+)} + E_L\right)X^0 - i\left(\underline{K}_L^{(+)} - \underline{K}_L\right)\underline{X}\right)\end{pmatrix} \quad (7.4.21)$$

where the choices of positive and negative energy solutions are the same as those given in (7.4.11c). Note that the superscript *0* here refers to the time component of a four vector and not charge.

For *pr*=23 and 32, (7.3.22) refers to the amplitudes for K^0 coupled to $\bar{W}_U^{\,0}$ and $\bar{K}^0$ to $W_U^{\,0}$, respectively. With (7.2.21-22) and (7.4.21), it is found to be

$$S_{fi}\left(K^0, \overline{K}^0 \to Z^0 \to L^+L^-\right)$$
$$= -i\frac{\pi}{2\sqrt{2}}\frac{\Omega}{\tau_0}\frac{1}{\tau_0\Omega_L}\delta\left(E_L^{(+)} + E_L - E_{Ks}\right)\left(v_{La}^{(+)} v_{L\dot{a}}^{(-)} - u_{La}^{(-)} u_{L\dot{a}}^{(+)}\right)I_{LL} \quad (7.4.22)$$

$$I_{LL} = \int d^3\underline{X} d^4x \sqrt{\frac{\Omega}{\Omega_{wvac}}}\,\psi_0^2(r)\exp\left(i\left(\underline{K}_L - \underline{K}_L^{(+)}\right)\underline{X}\right) \quad (7.4.23)$$

which is the analog of (7.4.13).

For Z^0 decaying into a pair of neutrinos, simplification can already be achieved here if one anticipates

$$\underline{K}_{\nu L}^{(+)} = \underline{K}_{\nu L} \quad (7.4.24)$$

which is obvious for this two body final state but is formally obtained first in the next section after integrations over $\underline{K}_{\nu L}^{(+)}$ and $\underline{K}_{\nu L}$. It then follows that $E_{\nu L}^{(+)} = E_{\nu L} = |\underline{K}_{\nu L}|$. With these results, it is easily seen that $u_{\nu L\dot{a}}^{(+)} u_{\nu La}^{(-)}$ in (7.4.21) vanishes for any spin combination. Therefore,

$$S_{fi}\left(K^0, \overline{K}^0 \to Z^0 \to \nu_L \overline{\nu}_L\right) = 0 \tag{7.4.25}$$

7.4.5. Decay Amplitude for $\pi^+ \to \pi^0 e^+ \nu_e$

This decay, also called pion beta decay, has been treated long ago in the literature based upon the so-called conserved vector current (CVC) hypothesis. Källén [K3 (14.94)] gives the phenomenological decay amplitude

$$\begin{aligned} S_{fi(K)}\left(\pi^+ \to \pi^0 e^+ \nu_e\right) &= -i \int dX^0 \left\langle \pi^0, e^+, \nu_e \left| H_1 \right| \pi^+ \right\rangle = i \frac{G}{2\Omega\Omega_L} \frac{1}{\sqrt{E_{\pi+} E_{f1\pi 0}}} \\ &\times \left\{\left(K^\mu_{\pi+} + K^\mu_{f1\pi 0}\right) \overline{u}^{(+)}_{\nu e}\left(\underline{K}_{\nu e}\right)\left(1 - i\gamma_5\right)\gamma_\mu u^{(-)}_e\left(-\underline{K}_e\right)\right\} \\ &\times \int d^4X \exp\left(i\left(K^\mu_{f1\pi 0} + K^\mu_e + K^\mu_{\nu e} - K^\mu_{\pi+}\right)X_\mu\right) \end{aligned} \tag{7.4.26}$$

where <...> is given in the notation of [K3], G is the Fermi constant, γ the Dirac matrices of (A2-2)-(A2-6) of [K3], and the u's are the four component wave functions of (A2-2)-(A2-24) in [K3], of the category (A14) times $\sqrt{\Omega_L}$.

In the present theory, no additional assumption like the CVC hypothesis above is required. For $L=e$, (7.4.10) become

$$W_I^{0+} = \frac{g}{2\sqrt{2}} \frac{1}{M_W^2} \psi^{(-)}_{ea} \psi^{(+)}_{\nu e\dot{a}} \tag{7.4.27a}$$

$$\underline{W}_I^{+} = -\frac{g}{2\sqrt{2}} \frac{1}{M_W^2} \underline{\sigma}^{b\dot{a}} \psi^{(-)}_{eb} \psi^{(+)}_{\nu e\dot{a}} \tag{7.4.27b}$$

Noting (7.4.7), let

$$\begin{aligned} \psi^{(+)}_{\nu e\dot{a}} &= \frac{1}{\sqrt{\Omega_L}} u^{(+)}_{\nu e\dot{a}}\left(\underline{K}_{\nu e}\right) \exp\left(iE_{\nu e} X^0 - i\underline{K}_{\nu e}\underline{X}\right) \\ \psi^{(-)}_{eb} &= \frac{1}{\sqrt{\Omega_L}} u^{(-)}_{eb}\left(-\underline{K}_e\right) \exp\left(iE_e X^0 - i\underline{K}_e \underline{X}\right) \end{aligned} \tag{7.4.28}$$

where the sign of $\underline{K}_e$ has been chosen to be opposite to that of $\underline{K}_L$ in (7.4.13) for the final state <f1|=<0, W|. The Fermi constant is related to M_W^2 by [L5 (22.83), P1]

$$G = \frac{1}{4\sqrt{2}} \frac{g^2}{M_W^2} = 1.1664 \times 10^{-5}\ \text{Gev}^{-2} \tag{7.4.29a}$$

in the present theory; this expression can be combined with (7.4.6b) to give

$$G = \frac{1}{4\sqrt{2}}\frac{\Omega}{\tau_0} \tag{7.4.29b}$$

Thus, the Fermi constant is a ratio ∞/∞ = finite and is a fundamental constant. Alternatively, M_W of (7.4.6b, 9) can be taken as a fundamental constant instead. Inserting (7.4.28, 29a) into (7.4.27) and the resulting expressions into (7.3.25), one gets

$$S_{fi}\left(\pi^+ \to \pi^0 e^+ \nu_e\right) = -\frac{i}{2}\frac{G}{\Omega\Omega_L}\frac{1}{E_{\pi+}}\begin{Bmatrix}\left(E_{\pi+} + E_{f1\pi 0}\right)u_{ve\dot{a}}^{(+)}\left(\underline{K}_{ve}\right)u_{ea}^{(-)}\left(-\underline{K}_e\right) \\ -\underline{K}_{f1\pi 0}\underline{\sigma}^{a\dot{b}}u_{ea}^{(-)}\left(-\underline{K}_e\right)u_{ve\dot{b}}^{(+)}\left(\underline{K}_{ve}\right)\end{Bmatrix}$$
$$\times \int d^4X \exp\left(i\left(E_{ve} + E_e + E_{f1\pi 0} - E_{\pi+}\right)X^0 - i\left(\underline{K}_{ve} + \underline{K}_e + \underline{K}_{f1\pi 0}\right)\underline{X}\right) \tag{7.4.30}$$

By working out the braced expression in (7.4.30) by means of (7.4.11) and compared it to the braced expression in (7.4.26) evaluated by means of (A2-5, A2-6, A2-24, A2-25) of [K3], it turns out that the both results are equal. With this result, (7.4.30) and (7.4.26) are related by

$$S_{fi}\left(\pi^+ \to \pi^0 e^+ \nu_e\right) = -\sqrt{\frac{E_{f1\pi 0}}{E_{\pi+}}}S_{fi(K)}\left(\pi^+ \to \pi^0 e^+ \nu_e\right) \tag{7.4.31}$$

to zeroth order in ε_0. Apart from the small difference between the π^0 energy and the π^+ mass and a phase, S_{fi} of (7.4.30) is the same as $S_{fi(K)}$ of (7.4.26).

7.4.6. Decay Amplitudes for *K* → *π* + Lepton Pair

These amplitudes have been given by (7.3.26-29) in terms of the *W*'s. Consider at first the kaon decay via $W_V^{\pm}$ in (7.3.26a) and the first of (7.3.28). The conversion of W_V^{+} into an equivalent positron and an electron neutrino via (7.2.16, 20) has been given by (7.4.27). Conversion into a muon and a muon neutrino is obtained by

$$(7.4.27) \text{ with } e \to \mu \tag{7.4.32}$$

From (7.3.26-29), one sees that

$$\begin{aligned} &S_{fi}\left(K^+ \to \pi^0 L^+ \nu_L\right) = S_{fi}\left(K^- \to \pi^0 L^- \overline{\nu}_L\right) \\ &\approx \frac{1}{\sqrt{2}}S_{fi}\left(\overline{K}^0 \to \pi^- L^+ \nu_L\right) = \frac{1}{\sqrt{2}}S_{fi}\left(K^0 \to \pi^+ L^- \overline{\nu}_L\right) \end{aligned} \tag{7.4.33}$$

where the ≈ sign becomes = in the limit of $SU(2)_I$ symmetry in which the masses of $K^{\pm}$, K^0 and $\bar{K}^0$ are the same, as do the masses of $\pi^{\pm}$ and π^0.

Unlike π^0 in $\pi^+ \to \pi^0 e^+ \nu_e$, which is nearly stationary, the pions in the kaon decays are generally relativistic. Thus, ε_0 of (3.5.1) is not a small quantity and the approximate meson wave functions of §3.5.4 derived for slow mesons cannot be used in S_{fi} of (7.3.26-29). Therefore, the absolute magnitudes of the decay amplitudes in (7.4.33), as well as those associated with the W_U and $\bar{W}_U$ decays, using (7.2.21-22), cannot be evaluated presently.

7.4.7. Decay Amplitudes for K^+, $\pi^+ \to \gamma$ + Lepton Pair

Anlogous to (7.4.6a), the total action (7.1.1) is now varied with respect to the electromagnetic field $A^{a\dot{b}}$. The last two terms in (7.1.1) do not contain $A^{a\dot{b}}$ and do not contribute. For the present decay, variation of S_{m3} (7.1.8) using (7.1.4) and (7.2.14) also vanishes. Variation of S_L (7.2.13) leads to a constant source representing a static magnetic field generated by the charged lepton moving with a constant speed which field is decoupled from the photon. Finally, variation of S_{GB} (7.4.1) takes the form (7.4.14) with Z replaced by A. Choosing the Lorentz gauge, one find to order g the usual

$$\Box A_{\dot{b}a} = 0 \qquad (7.4.34)$$

Its vector part has the solution (6.4.8).

Inserting this solution and (7.4.10b) into (7.3.36) and carrying out the X^0 integration give the equivalent of (7.4.13),

$$S_{fi2} = \frac{\pi}{4} \frac{g^2 e}{E_{pr}\sqrt{E_r \tau_0 M_W^2}} \frac{1}{\sqrt{\Omega_r \Omega_L}} \delta(E_r + E_L + E_{\nu L} - E_{pr}) \underline{e}_T \underline{\sigma}^{b\dot{a}} u^{(+)}_{\nu L\dot{a}} u^{(-)}_{Lb} I_{Lr} \qquad (7.4.35a)$$

$$I_{Lr} = \int d^3\underline{X} d^4x \sqrt{\frac{\Omega}{\Omega_{wvac}}} \psi_0^2(r) \exp(i(\underline{K}_r + \underline{K}_L - \underline{K}_{\nu L})\underline{X}) \qquad (7.4.35b)$$

where K_r stand for K in (6.4.8).

7.5. Decay Rates and Detachment of Weak and Electromagnetic Couplings

The total decay rate is related to the decay amplitude by

$$\Gamma = \sum\nolimits_{\text{final spins}} \sum\nolimits_{\text{final momenta}} \left|S_{fi}\right|^2 / T_d \qquad (7.5.1)$$

analogous to (6.4.24a). Similar to (6.4.24b), summation over the final momenta $\underline{K}_f$ for each of the decay products is replaced by a integral of the form

$$\sum_{Kf} = \frac{\Omega_f}{(2\pi)^3}\int d^3\underline{K}_f \tag{7.5.2}$$

where Ω_f is a large normalization volume for the wave function of this final state particle denoted by *f*.

7.5.1. $\pi^+ \to \pi^0 e^+ \nu_e$

For the amplitude (7.4.30), (7.5.1) with (7.5.2) becomes

$$\Gamma(\pi^+ \to \pi^0 e^+ \nu_e) = \frac{\Omega\Omega_L^2}{(2\pi)^9}\sum_{\text{final spins}}\frac{1}{T_d}\int d^3\underline{K}_{\pi 0} d^3\underline{K}_{e+} d^3\underline{K}_{\nu}\left|S_{fi}(\pi^+ \to \pi^0 e^+ \nu_e)\right|^2 \tag{7.5.3}$$

This decay rate has been obtained and evaluated long ago and has been given by [K3 (14.95)], using $S_{fi(K)}$ of (7.4.26) instead of S_{fi} of (7.4.30). The calculation has been outlined in [K3 §14.7]. Let

$$\Delta_\pi = E_{\pi+} - E_{\pi 0} \quad \text{and} \quad \varepsilon_\pi = \mathrm{M}_e^2/\Delta_\pi^2$$

and neglect Δ_π^2 next to $4E_\pi^2$, [K3 (14.109)] gives

$$\frac{1}{\tau_{\beta(K)}} = \Gamma(\pi^+ \to \pi^0 e^+ \nu_e) = \frac{G^2}{192\pi}\left(1+\frac{E_{\pi 0}}{E_{\pi+}}\right)^3 \Delta_\pi^5$$
$$\times\left[\frac{2}{5}\sqrt{1-\varepsilon_\pi}\left(2-9\varepsilon_\pi-8\varepsilon_\pi^2\right)+6\varepsilon_\pi^2\ln\frac{1+\sqrt{1-\varepsilon_\pi}}{\sqrt{\varepsilon_\pi}}\right] = \frac{1}{2.457\ \text{sec.}} \tag{7.5.4}$$

where the E's stand for the masses. $\tau_{\beta(K)}$ is just outside the error limit of the measured 2.54 sec.±3.3% [P1]. If (7.5.3) with (7.4.31, 32) is used, the decay time $\tau_{\beta(K)}$ is to be enhanced by the factor $E_{\pi+}/E_{fl\,\pi 0} \approx E_{\pi+}/E_{\pi 0}$ and becomes $\tau_\beta = 2.54$ sec., in agreement with the measured value. The present derivation has been taken from [H17].

Actually, (7.4.30) is based upon (7.3.25) in which terms of order ε_0 (3.5.1) and higher in the more exact (7.3.24) have been neglected. The energy Δ_π less M_e is available to impart the three decay products with momenta. Most of this energy goes to e^+ and ν_e. The upper limit of the π^0 momentum $|\underline{K}_{fl\,\pi 0}|$ is obtained when both e^+ and ν_e move in the same direction. Let the average $|\underline{K}_{f1\pi0}|$ be half of this limit, one finds $\varepsilon_0 \approx 1.49\%$. Thus, τ_β above can be off by about ±3%, which is of the same order as the measured error limit above.

7.5.2. $\pi^+, K^+ \to L^+\nu_L$

The decay rate is obtained by inserting (7.4.13) into (7.5.1) and making use of (7.5.2). This procedure is well known [K3 §15.2] but will be modified for the present application. One obtains

$$\Gamma\left(\pi^+, K^+ \to L^+\nu_L\right) = \frac{\pi^2}{8}\frac{\Omega^2}{\tau_0^2}\frac{1}{\tau_0^2}\frac{1}{(2\pi)^7}\int d^3\underline{K}_L I_{uu} \tag{7.5.5a}$$

$$I_{uu} = \sum\nolimits_{\text{spins}} \int d^3\underline{K}_{\nu L}\frac{\Omega}{\Omega_{wvac}} I_{L\nu1} I^*_{L\nu1}\left|u^{(+)}_{\nu L\dot{a}} u^{(-)}_{La}\right|^2 \delta\left(E_L + E_{\nu L} - E_{pr}\right) \tag{7.5.5b}$$

where

$$\left|\delta\left(E_L + E_{\nu L} - E_{pr}\right)\right|^2 = \frac{T_d}{2\pi}\delta\left(E_L + E_{\nu L} - E_{pr}\right) \tag{7.5.6}$$

is similar to (6.4.27) and has been employed. Performing the integration over $\underline{X}$ in (7.4.13b) yields

$$I_{L\nu1} = (2\pi)^3\delta\left(\underline{K}_L - \underline{K}_{\nu L}\right)\int d^4x\,\psi_0^2(r) = I^*_{L\nu1} \tag{7.5.7}$$

Substituting this expression for one of the two $I_{L\nu l}$'s in (7.5.5b) and carrying out the integration over $\underline{K}_{\nu L}$ leads to

$$I_{uu} = (2\pi)^3\left(\int d^4x\,\psi_0^2(r)\right)\frac{\Omega}{\Omega_{wvac}} I_{Nd}\Sigma_{uu}\delta\left(E_L + E_{\nu L} - E_{pr}\right), \quad \underline{K}_L = \underline{K}_{\nu L} \tag{7.5.8}$$

$$\Sigma_{uu} = \sum\nolimits_{\text{spins}}\left|u^{(+)}_{\nu L\dot{a}} u^{(-)}_{La}\right|^2_{\underline{K}_L = \underline{K}_{\nu L}} \tag{7.5.9a}$$

$$I_{Nd} = \int d^3\underline{X}\,d^4x\,\psi_0^2(r) \tag{7.5.9b}$$

Insert (7.4.11a, 11b) into (7.5.9a) and carry out the sum over all possible spin combinations. The result is

$$\Sigma_{uu} = \left(1+\frac{M_L}{E_L}\right)\left(1-\frac{|\underline{K}_L|}{E_L+M_L}\right)^2 \tag{7.5.10}$$

Digress for a moment to the normalization of meson wave functions and let

$$-iN_d = 2\int dx^0 N_{mJ} \tag{7.5.11}$$

in (4.2.4). The imaginary sign comes obviously from the left side of (4.2.4). Since N_{mJ} is a conserved quantity of dimension energy, N_d must be a scalar quantity and expected to transform as a dimensionless Lorentz scalar, which can however be large. One finds from (7.5.9b) and (4.2.5)

$$N_d = E_{00}\int d^3\underline{X}d^4x\psi_0^2(r) = E_{pr}I_{Nd} = \text{large dimensionless Lorentz scalar} \tag{7.5.12}$$

Here, the meson mass E_{00} has been identified as the mass E_{pr} of the initial state meson in (7.3.1).

Return now to (7.5.8) and apply the formula

$$\delta(f(x)) = \frac{\delta(x-x_0)}{|\partial f(x)/\partial x|_{x=x_0}}, \qquad f(x_0)=0 \tag{7.5.13}$$

to the δ function therein. We obtain

$$\delta\left(E_L+E_{\nu L}-E_{pr}\right)_{\underline{K}_L=\underline{K}_{\nu L}} = \delta\left(\sqrt{K_L^2+M_L^2}+K_L-E_{pr}\right) = \frac{E_L}{E_{pr}}\delta(K_L-K_{L0}) \tag{7.5.14a}$$

$$K_{L0} = \frac{1}{2E_{pr}}\left(E_{pr}^2-M_L^2\right) \tag{7.5.14b}$$

Insert (7.5.10, 12, 14) into (7.5.8), which in its turn is substituted into (7.5.5a). After performing the integration over $\underline{K}_L$, the decay rate takes the form

$$\Gamma\left(\pi^+,K^+\to L^+\nu_L\right) = \frac{1}{64\pi}\frac{1}{E_{pr}}\frac{N_d}{\tau_0^4}\frac{\Omega^3}{\Omega_{wvac}}\int d^4x\psi_0^2(r)M_L^2\left(1-\frac{M_L^2}{E_{pr}^2}\right)^2 \tag{7.5.15}$$

Employing (4.2.8) and (7.4.6b), (7.5.15) becomes

$$\Gamma\left(\pi^+, K^+ \to L^+\nu_L\right) = \frac{F_W^2}{16\pi}\frac{1}{E_{pr}}M_L^2\left(1-\frac{M_L^2}{E_{pr}^2}\right)^2 \tag{7.5.16a}$$

$$F_W^2 = \frac{N_d}{4\tau_0^3}\frac{\Omega^2}{\Omega_{wvac}} = \frac{1}{4}\frac{g^4}{M_W^4}\frac{N_d}{\Omega_{wvac}\tau_0} \tag{7.5.16b}$$

$$M_W^2 = \frac{g^2\tau_0}{\Omega} \tag{7.5.16c}$$

Here, pr=12 and 13 and $E_{12}=E_{\pi+}$ and $E_{13}=E_{K+}$. For $L=\mu$ and e, (7.5.16) yields the ratio

$$\frac{\Gamma\left(\pi^+ \to e^+\nu_\mu\right)}{\Gamma\left(\pi^+ \to \mu^+\nu_e\right)} = \frac{M_e^2}{M_\mu^2}\left(\frac{E_{\pi+}^2 - M_e^2}{E_{\pi+}^2 - M_\mu^2}\right)^2 \tag{7.5.17}$$

in agreement with that of the literature [L5 (21.22)], apart from radiative corrections. Further, (7.5.16) yields

$$\frac{\Gamma\left(K^+ \to \mu^+\nu_\mu\right)}{\Gamma\left(\pi^+ \to \mu^+\nu_\mu\right)} = \left(\frac{E_{\pi+}}{E_{K+}}\right)^5\left(\frac{E_{K+}^2 - E_\mu^2}{E_{\pi+}^2 - E_\mu^2}\right)^2 \approx 1.41 \tag{7.5.18}$$

which is not too far from 1.34 obtained from data [P1]. F_W is related to the conventional charged pion decay constant F_π which enter the decay rates according to [L5 (21.20) and p606],

$$\Gamma\left(\pi^+ \to \mu^+\nu_\mu\right) = \left(F_\pi \cos\vartheta_C\right)^2 \frac{G^2 M_\mu^2}{8\pi E_{\pi+}^3}\left(E_{\pi+}^2 - M_\mu^2\right)^2 \tag{7.5.19a}$$

$$\Gamma\left(K^+ \to \mu^+\nu_\mu\right) = \left(F_\pi \sin\vartheta_C\right)^2 \frac{G^2 M_\mu^2}{8\pi E_{K+}^3}\left(E_{K+}^2 - M_\mu^2\right)^2 \tag{7.5.19b}$$

where ϑ_C is the Cabbibo angle to be determined by data. Such an angle is however absent in (7.5.16), which therefore has a greater predictive power. Comparison of (7.5.18) to the ratio of (7.5.19a) and (7.5.19b) leads to the prediction

$$\tan\vartheta_C = \frac{E_{\pi+}}{E_{K+}} = 0.2827 \tag{7.5.20}$$

which is close to 0.275 determined from data [P1]. From (7.5.16, 19, 20), one finds

$$F_W^2 = 2F_\pi^2 \frac{G^2}{\left(E_{\pi+}^{-2} + E_{K+}^{-2}\right)} \tag{7.5.21}$$

Using the value $F_\pi cos\vartheta_C$=0.128 Gev [L5 (21.21)] and (7.4.29), (7.5.21, 20) give the redefined dimensionless weak coupling constant

$$F_W^2 = 8.684 \times 10^{-14} \tag{7.5.22a}$$

Inserting this value into (7.5.16b) and making use of (7.4.29a), (7.5.12) and (4.2.8) gives

$$\rho_v = \frac{N_d}{\Omega_{wvac}\tau_0} = \frac{E_{00}}{\Omega_{wvac}} = 4798 \times 10^{-5}\ \text{Gev}^4 = \text{inverse of a 4 - volume} \tag{7.5.22b}$$

where ρ_v is a Lorentz invariant "virtual energy density" constant. The magnitude of the rates π, $K\to$lepton pair is not predicted but has to be fixed by data (7.5.22). The ratios of the rates are correctly predicted. This is due to the presence of the vacuum meson state here. When the vacuum meson state is replaced by a real meson state, as in $\pi^+ \to \pi^0 e^+ \nu_e$, $\Omega_{wvac}\to\Omega\to\infty$ and the magnitude of the rate itself (7.5.4) is correctly predicted.

7.5.3. Detachment of Weak and Electromagnetic Couplings

In the standard electroweak model, the strength of weak interactions is given by the square of the Fermi constant G, as examplfied by (7.5.19). In its turn, G is proportional to the conventional weak coupling constant g^2 according to (7.4.29a), where M_W^2 is regarded as a constant given in (7.4.9). The strength of electromagnetic interactions is determined by the fine structure constant $\alpha=e^2/4\pi$, as is examplified by (6.4.35). These both types of charges, g and e, are coupled by a new parameter, the Weinberg angle ϑ_W, according to (7.2.1).

This parameter is predicted in (7.2.12b). The rate of the same decay as that of (7.5.19), given by (7.5.16), contains however no g. Consistently, the Fermi constant G itself is also independent of g according to (7.4.29b) because M_W is proportional to g according to (7.4.6b). In fact, the weak charge g can be eliminated altogether and be replaced by $e/sin\,\vartheta_W$ accoding to the last of (7.2.1b), where ϑ_W has been predicted to be 27.93° in (7.2.12b) and is not an independent parameter or universal constant to be fixed by data.

Here, the strength of weak interactions given by the *weak coupling constant* F_W^2 of (7.5.16b) or G of (7.4.29b) is thus *detached* from the corresponding strength of electromagnetic interactions given by the *fine structure constant* $\alpha=e^2/4\pi$. Using (7.5.22a), F_W^2 is about 1.2×10^{-11} times weaker than $e^2/4\pi=1/137$. Note however that the formation of the weak gauge boson field Z and the electromagnetic field A according to the SU(2)xU(1) formalism remains intact. Such a detachment has however not been investigated generally involving other weak decays.

7.5.4. Gauge Boson Mass, ∞/∞ and Cabbibo Angle

As has been mentioned below (7.4.6), the squared mass the gauge boson, M_W^2 in (7.5.16c) or (7.4.6b), is a ratio of a large relative time τ_0 and a large volume Ω associated with the pseudoscalar meson wave functions that generate it. The physics underlying their finite ratio is not understood presently, because they are ostensibly independent of each other. A finite, universal τ_0/Ω indicates that there is some connection between a possible weak relative time dependence and a weak laboratory space dependence of the rest frame meson wave function. Let

$$\Omega = l_X l_Y l_Z \tag{7.5.23}$$

where l stands for length and X, Y and Z the Cartesian axes in laboratory space. There are infinitely many ways to form a finite ratio between τ_0 and Ω, each approaching infinity. The three prototypes are

$$\tau_0 \to \infty, \qquad l_Z = c_\tau \tau_0, \qquad l_X l_Y = \text{finite}, \qquad l_Z \propto \tau_0 \tag{7.5.24a}$$

$$\tau_0 = \frac{a_\tau}{l_a}, \qquad a_\tau \to \infty, \qquad l_X l_Y = c_a a_\tau, \qquad l_Z = \text{finite}, \qquad l_X,\ l_Y \propto \sqrt{\tau_0} \tag{7.5.24b}$$

$$\tau_0 = \frac{\Omega}{l_\Omega^2}, \qquad \Omega = l_X l_Y l_Z \to \infty, \qquad l_X,\ l_Y,\ l_Z \propto (\tau_0)^{1/3} \tag{7.5.24c}$$

where l_a and l_Ω are finite length constants, a_τ an area, the c's finite dimensionless constants, and (7.5.16c) has been used. In (7.5.24a), the volume Ω is limited to an infinite long tube of crossection $c_\tau l_X l_Y$. In (7.5.24b), Ω is limited to an infinitely extended slab of thickness l_X. In (7.5.25c), Ω is allowed to go to infinity in all three directions. Only this case gives that correct dimension for M_W^2.

The weak coupling constant F_W^2 is also a ∞/∞ ratio according to (7.5.16b), where $N_d \to \infty$ according to (7.5.11, 12). Corresponding to (7.5.24a, 24b), F_W^2 turns out to depend asymmetrically upon one or two of l_X, l_Y, l_Z. This asymmetry is removed in the case of (7.5.24c).

It can also be seen that the most straight forward choice l_X, l_Y, $l_Z \propto \tau_0 \to \infty$ would lead to difficulties. Inserting this into (7.5.16b, 16c, 23) yields

$$\tau_0 \to \infty, \qquad \Omega \propto \tau_0^3, \qquad M_W^2 \propto \tau_0^{-2} \to 0 \tag{7.5.25}$$

which contradicts (7.4.9).

Finally, prediction of the Cabbibo angle ϑ_C in (7.5.20) depends upon the presence of the $1/E_{pr}$ factor in (7.5.16a), which in its turn depends upon the application of the large normalization constant N_d in (7.5.12). If is this normalization condition is not utilized and I_{Nd}

of (7.5.9b) is computed using the meson wave functions (7.3.14b) and (4.3.2), one finds simply

$$I_{Nd} = \tau_0 \tag{7.5.26}$$

Inserting this relation into (7.5.12) and substituting the result into (7.5.15) converts (7.5.16b) to

$$\frac{F_W^2}{E_{pr}} = \frac{g^4}{4M_W^4}\frac{1}{\Omega_{wvac}} \tag{7.5.27}$$

This causes that the $1/E_{pr}$ factor in (7.5.16a) drops out so that the correct Cabbibo angle cannot be predicted and, in addition, the relatively good agreement between the second order predictions of the ratio $\Gamma(K^+)/\Gamma(\pi^+)$ with data in (7.5.49) below will be ruined.

Ω_{wvac} is not a constant but depends upon the initial meson mass E_{pr} in (7.5.22b) where the "virtual energy density" ρ_v is a constant.

7.5.5. K^0, $\overline{K}^0$ → Lepton Pairs

From (7.4.25), it follows that

$$\Gamma\left(K^0, \overline{K}^0 \to \nu_\mu \overline{\nu}_\mu, \nu_e \overline{\nu}_e\right) = 0 \tag{7.5.28}$$

in agreement with data [P1]. The data however are given only for $K_S{}^0$ and $K_L{}^0$ [P1]. By (7.2.18), the decays of $K_S{}^0$ and $K_L{}^0$ are composed of K^0 and $\overline{K}^0$ decays in equal amount. Since the form of the amplitudes (7.4.22) and (7.4.13) are the same, the derivation of the decay rate for the latter in §7.5.2 can be mostly taken over for the former. The main difference is that (7.5.9a) is to be replaced by

$$\Sigma_{uu} = \sum\nolimits_{\text{spins}} \left| v_{L\dot{a}}^{(-)} v_{La}^{(+)} - u_{L\dot{a}}^{(+)} u_{La}^{(-)} \right|^2_{\underline{K}_L^{(+)} = \underline{K}_L} \tag{7.5.29}$$

Insert (7.4.11b, 19b, 20) into (7.5.29) and note that $\underline{K}_L{}^{(+)}$ and $\underline{K}_L$, hence $E_L{}^{(+)} = E_L$, are anticipated analogous to (7.4.24). It turns out that the vv and uu terms in (7.5.29) cancel out so that

$$\Sigma_{uu} = 0 \tag{7.5.30}$$

Analogous to (7.5.5, 8, 9a), (7.5.30) leads to that the equivalent of (7.5.16a) becomes

$$\Gamma\left(K^0, \overline{K}^0 \to L^+ L^-\right) = \Gamma\left(K^0, \overline{K}^0 \to \mu^+\mu^-, e^+e^-\right) = 0 \tag{7.5.31}$$

in agreement with data [P1].

These results show that the virtual time component of the neutral gauge boson Z^0 described by (7.4.21) does not decay into any lepton pair. Finally, it is remarked that π^0 cannot decay weakly via Z to for example e^+e^-. This is due to that, unlike $\pi^\pm$ and kaons, which have open flavors and must decay via weak interaction, π^0 has closed flavor and can decay via strong interaction, as will be considered in Sec. 8.4 below.

7.5.6. $K \to \pi +$ Lepton Pairs

The decay rates for (7.4.33) are

$$\begin{aligned} &2\Gamma\left(K^+ \to \pi^0 L^+ \nu_L\right) = 2\Gamma\left(K^- \to \pi^0 L^- \overline{\nu}_L\right) \\ &\approx \Gamma\left(\overline{K}^0 \to \pi^- L^+ \nu_L\right) = \Gamma\left(K^0 \to \pi^+ L^- \overline{\nu}_L\right) \end{aligned} \tag{7.5.32}$$

In accordance with (7.2.18),

$$\begin{aligned} &\Gamma\left(K_S^0 \to \pi^\pm L^\mp, \overline{\nu}_L, \nu_L\right) = \Gamma\left(K_L^0 \to \pi^\pm L^\mp, \overline{\nu}_L, \nu_L\right) \\ &= \Gamma\left(\overline{K}^0 \to \pi^- L^+ \nu_L\right) = \Gamma\left(K^0 \to \pi^+ L^- \overline{\nu}_L\right) \end{aligned} \tag{7.5.33}$$

The left of this relation agrees with data [P1] to 0.11% for $L=\mu$ and 0.02% for $L=e$. These numbers are far less than the experimental errors ranging from 0.7 to 2.5 % for these decays.

Comparison of the measured decay rates for charged and neutral kaons shows that the predicted factor 2 [L5 (11.119)] in (7.5.32) is to be replaced by 2.05 for $L=\mu$ and 1.93 for $L=e$. These discrepancies are presently attributed to the neglect of the u and d quark mass differences mentioned below (7.4.33).

Finally, (7.4.25) and (7.5.29, 30) also leads to

$$\Gamma\left(K^\pm \to \pi^\pm L^+ L^-, \pi^\pm \overline{\nu}_L \nu_L\right) = \Gamma\left(K^0, \overline{K}^0 \to \pi^0 L^\pm L^\mp, \pi^0 \overline{\nu}_L \nu_L\right) = 0 \tag{7.5.34}$$

This also agrees with data [P1], which show no such decays. For the right part of (7.5.34), agreement with data is inferred from that neither $K_S{}^0$ nor $K_L{}^0$ decays into such final states.

7.5.7. K^+, $\pi^+ \to \mu^+ \nu_\mu \gamma$

Analogous to §7.5.2, the rate of this decay is found by inserting (7.4.35) into (7.5.1) using (7.5.2) and (7.5.6-7) type of expression,

$$\Gamma\left(\pi^{+}, K^{+} \rightarrow L^{+}\nu_{L}\gamma\right)=\frac{1}{64(2\pi)^{5}}\frac{g^{4}e^{2}}{\tau_{0}^{2}M_{W}^{4}}\frac{1}{E_{pr}^{2}E_{r}}\frac{\Omega}{\Omega_{wvac}}\int d^{3}\underline{K}_{r}\int d^{3}\underline{K}_{L}$$
$$\int d^{3}\underline{K}_{\nu L}\delta\left(E_{r}+E_{L}+E_{\nu L}-E_{pr}\right)\delta\left(\underline{K}_{r}+\underline{K}_{L}-\underline{K}_{\nu L}\right)\left(\int d^{4}x\psi_{0}^{2}(r)\right)\Sigma_{uu\gamma} \tag{7.5.35a}$$
$$\int d^{3}\underline{X}\int d^{4}x\psi_{0}^{2}(r)\exp\left(i\left(\underline{K}_{r}+\underline{K}_{L}-\underline{K}_{\nu L}\right)\underline{X}\right)$$

$$\Sigma_{uu\gamma}=\sum\nolimits_{spins}\underline{e}_{T}\underline{\sigma}^{b\dot{a}}\left|u_{\nu L\dot{a}}^{(+)}u_{Lb}^{(-)}\right|^{2} \tag{7.5.35b}$$

Carrying out the $\underline{K}_{\nu L}$ integration leads to the equivalent of (7.5.8-9),

$$\Gamma\left(\pi^{+}, K^{+} \rightarrow L^{+}\nu_{L}\gamma\right)=\frac{F_{W}^{2}}{16(2\pi)^{5}}\frac{e^{2}}{E_{pr}^{3}}\int d^{3}\underline{K}_{r}I_{K} \tag{7.5.36a}$$

$$I_{K}=\int d^{3}\underline{K}_{L}\frac{1}{E_{r}}\delta\left(f\left(\cos\vartheta_{Lr}\right)\right)\left(\Sigma_{uu\gamma}\right)_{\left(\underline{K}_{\nu L}=\underline{K}_{L}+\underline{K}_{r}\right)} \tag{7.5.36b}$$

$$\delta\left(f\left(\cos\vartheta_{Lr}\right)\right)=\delta\left(K_{r}+\sqrt{K_{L}^{2}+M_{L}^{2}}+\sqrt{K_{r}^{2}+K_{L}^{2}-2\cos\vartheta_{Lr}}-E_{pr}\right) \tag{7.5.36c}$$

where (7.5.16b) and (4.2.8) have been employed. ϑ_{Lr} is the angle between $\underline{K}_{L}$ and $\underline{K}_{r}$ which has been put to (0, 0, K_r) in (7.5.36b). This δ function vanishes except for

$$\cos\vartheta_{Lr}=\cos\vartheta_{Lr0}=\pm\frac{1}{2K_{r}K_{L}}\left(E_{pr}^{2}+M_{L}^{2}-2E_{pr}K_{r}-2\left(E_{pr}-K_{r}\right)\sqrt{K_{L}^{2}+M_{L}^{2}}\right) \tag{7.5.37}$$

Thence,

$$\int d^{3}\underline{K}_{L}\delta\left(f\left(\cos\vartheta_{Lr}\right)\right)=2\pi\int dK_{L}\frac{K_{L}}{K_{r}}\left(E_{pr}-K_{r}-\sqrt{K_{L}^{2}+M_{L}^{2}}\right) \tag{7.5.38}$$

Making use of (7.4.11), summing over the spins, as in (7.5.10), and over the photon polarizations T=1, 2, the equivalent of (7.5.10) becomes

$$\left(\Sigma_{uu\gamma}\right)_{\left(\underline{K}_{\nu L}=\underline{K}_{L}+\underline{K}_{r}\right)}=4\left[1+\frac{K_{r}K_{L}\cos\vartheta_{Lr0}+K_{L}^{2}\cos^{2}\vartheta_{Lr0}}{\left(E_{pr}-K_{r}\right)\sqrt{K_{L}^{2}+M_{L}^{2}}-K_{L}^{2}-M_{L}^{2}}\right] \tag{7.5.39}$$

Inserting (7.5.38-39) into (7.5.36b) leads to

$$I_K = \frac{2\pi}{K_r^2}\int dK_L K_L\left(E_{pr} - K_r - \sqrt{K_L^2 + M_L^2}\right)\left(\Sigma_{uu\gamma}\right)_{(\underline{K}_{\nu L} = \underline{K}_L + \underline{K}_r)} \tag{7.5.40}$$

Insert (7.5.37) into (7.5.39) which in its turn is inserted into (7.5.40). Changing the integration variable to

$$E_L = \sqrt{K_L^2 + M_L^2} \tag{7.5.41}$$

(7.5.40) can be integrated to yield

$$I_K = \frac{8\pi}{K_r^2}\Big|_{E_{L-}}^{E_{L+}}\left\{\frac{B_0}{2}\left(\pm 1 + \frac{B_0}{K_r^2}\right)E_L + \left(\begin{pmatrix}0\\ B_1/2\end{pmatrix} - \frac{B_0 B_1}{4K_r^2}\right)E_L^2 + \left(\frac{B_1^2}{12K_r^2} - \frac{1}{3}\right)E_L^3\right\} \tag{7.5.42a}$$

$$B_0 = E_{pr}^2 + M_L^2 - 2E_{pr}K_r, \qquad B_1 = 2\left(E_{pr} - K_r\right) \tag{7.5.42b}$$

Here, the integration limits $E_{L\pm}$ are obtained from (7.5.41) together with the boundary condition

$$\cos^2 \vartheta_{Lr} = 1 \tag{7.5.43}$$

With (7.5.37) and (7.5.41, 42b, 43), one finds

$$E_{L\pm} = \frac{B_0 B_1 \pm \sqrt{B_0^2 B_1^2 - 4\left(B_0^2 + 4K_r^2 M_L^2\right)E_{pr}\left(E_{pr} - 2K_r\right)}}{4E_{pr}\left(E_{pr} - 2K_r\right)} \tag{7.5.44}$$

The upper and lower signs in (7.5.42a) refer to $\cos\vartheta_{Lr} = +1$ and -1, respectively.

Inserting (7.5.42a) into (7.5.36a) gives

$$\Gamma\left(\pi^+, K^+ \to L^+\nu_L\gamma\right) = \frac{F_W^2}{2(2\pi)^3}\frac{e^2}{E_{pr}^3}\int dK_r\left\{\frac{B_0}{2}\left(\pm 1 + \frac{B_0}{K_r^2}\right)\left(E_{L+} - E_{L-}\right)\right.$$
$$\left. + \left(\begin{pmatrix}0\\ B_1/2\end{pmatrix} - \frac{B_0 B_1}{4K_r^2}\right)\left(E_{L+}^2 - E_{L-}^2\right) + \left(\frac{B_1^2}{12K_r^2} - \frac{1}{3}\right)\left(E_{L+}^3 - E_{L-}^3\right)\right\} \tag{7.5.45}$$

The boundaries limiting the allowed K_r and K_L values in (7.5.45) are determined by putting the argument in (7.5.36c) to 0 together with (7.5.43),

$$E_{pr} = K_r + \sqrt{K_L^2 + M_L^2} + \left|K_r \pm K_L\right| \tag{7.5.46}$$

which is illustrated in Figure 7.1.

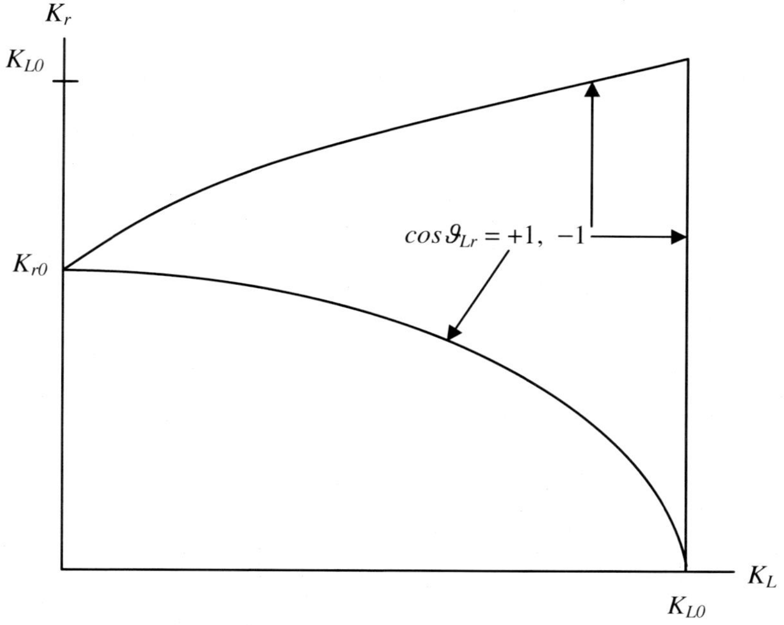

Figure 7.1. The allowed K_r and K_L values in the integral of (7.5.45) are limited to the triangle-like area.

In this figure,

$$K_{L0} = \left(E_{pr}^2 - M_L^2\right)/2E_{pr}, \qquad K_{r0} = \left(E_{pr} - M_L\right)/2 \tag{7.5.47}$$

For the boundaries in Figure 7.1, (7.5.45) becomes

$$\begin{aligned}\Gamma\left(\pi^+, K^+ \to L^+\nu_L\gamma\right) &= \frac{F_W^2}{2(2\pi)^3}\frac{e^2}{E_{pr}^3}\left[\int_0^{K_{r0}} dK_r\left\{\left(\frac{B_0}{2}+\frac{B_0^2}{4K_r^2}\right)\left(E_{L+0}-E_{L-}\right)\right.\right.\\ &\left.-\frac{B_0B_1}{4K_r^2}\left(E_{L+0}^2-E_{L-}^2\right)+\left(\frac{B_1^2}{12K_r^2}-\frac{1}{3}\right)\left(E_{L+0}^3-E_{L-}^3\right)\right\}\\ &+\int_{K_{r0}}^{K_{L0}} dK_r\left\{\left(\frac{B_0^2}{4K_r^2}-\frac{B_0}{2}\right)\left(E_{L+0}-E_{L-}\right)+\frac{B_1}{2}\left(1-\frac{B_0}{2K_r^2}\right)\left(E_{L+0}^2-E_{L-}^2\right)\right.\\ &\left.\left.+\left(\frac{B_1^2}{12K_r^2}-\frac{1}{3}\right)\left(E_{L+0}^3-E_{L-}^3\right)\right\}\right], \qquad E_{L+0}=\sqrt{K_{L0}^2+M_L^2}\end{aligned} \tag{7.5.48}$$

The integrand is a measure of the photon spectrum which shows discontinuous jump at $K_r=K_{r0}$. The ratio R_{disc} of these two amplitudes is the ratio between the two braces evaluated at $K_r=K_{r0}$ and $E_{L-}=M_L$ according to Figure 7.1.

With $E_{pr}=0.4937$ Gev for K^+ and 0.1396 for π^+ and $M_L=0.10566$ for μ^+ [P1], (7.5.22a) and $e^2/4\pi=1/137$, (7.5.48) has been integrated on a computer to yield

$$\begin{aligned} &\Gamma\left(K^+\to\mu^+\nu_\mu\gamma\right)=1.22\times10^{-10}\text{ ev} && \text{data[P1]}=2.93\times10^{-10}\text{ ev}\\ &\Gamma\left(\pi^+,\to\mu^+\nu_\mu\gamma\right)=0.98\times10^{-12}\text{ev} && \text{data[P1]}=3.14\times10^{-12}\text{ ev}\\ &\Gamma\left(K^+\right)/\Gamma\left(\pi^+\right)=125 && \text{data[P1]}=93.3 \end{aligned} \quad (7.5.49)$$

The agreements in magnitude with data may be accepted presently at the quantum mechanical level. The computer results also show that the photon intensity rises from $K_r=0$ to a maximum before reaching $K_r=K_{r0}$, where it jumps up by a factor R_{disc} and then decreases to 0 at $K_r=K_{L0}$. The predictions

$$R_{disc}=5.26\text{ for }K^+\to\mu^+\nu_\mu\gamma\ \text{ and }\ 1.31\text{ for }\pi^+\to\mu^+\nu_\mu\gamma \quad (7.5.50)$$

can be tested.

When μ^+ is replaced by e^+ with $M_L=0.00511$ Gev [P1], (7.5.49) is turned into

$$\begin{aligned} &\Gamma\left(K^+\to e^+\nu_\mu\gamma\right)=1.03\times10^{-10}\text{ ev} && \text{data[P1]}=\text{no data}\\ &\Gamma\left(\pi^+,\to e^+\nu_\mu\gamma\right)=0.296\times10^{-10}\text{ev} && \text{data[P1]}=4.67\times10^{-15}\text{ ev} \end{aligned} \quad (7.5.51)$$

The huge disagreements with data are not understood presently. The small ratio $\Gamma(\pi^+\to e^+\nu_e)/\Gamma(\pi^+\to\mu^+\nu_\mu)$ in (7.5.17) stems from the M_L^2 factor in (7.5.16a) which in its turns comes from W^0 in (7.3.22) via (7.5.10). This W^0 is here replaced by $\underline{W}$ in (7.3.36) which led to (7.5.39) and the suppressing factor M_L^2 factor in (7.5.16a) is lost in (7.5.48). However, a qualitative argument may be mentioned. When a photon is introduced into K^+, $\pi^+\to L^+\nu_L$ of §7.5.2, it may interact dynamically with L^+. This interaction, apart from the static inteaction mentioned in §7.4.7, has not been included here. Viewing the photon as a virtual e^+–e^- pair, it will hardly change the motion of the heavy μ^+ but will do so for the light e^+. This may lead to reduced phase space and hence the decay rate.

If there is also a dynamic interaction between the photon and the meson, it is expected to affect the lighter π^+ more than K^+. This trend agrees with (7.5.49) where the decay rate of π^+ is farther from data the that of K^+.

7.6. Purely Leptonic Interactions and Comparison with Standard Model

Purely leptonic interactions have been treated rather successfully in the standard model. In this case, it will be shown that the total action (7.1.1) reduces to a form close to that of the standard model so that some of the results obtained from this model can be taken over here.

7.6.1. Standard Model and Purely Leptonic Interactions

In the standard model [P1, L5 §22.2 §22.3], S_{m3} of (7.1.1) is replaced by two actions, S_H for Higgs and S_q for quarks. The equivalent of (7.1.1) is

$$S_{SM} = S_{GB,SM} + \sum_{\mu,e} S_{L,SM} + S_{LmH} + S_H + S_q \tag{7.6.1}$$

Here, S_L and S_M have been given by (7.1.11). S_{LmH} is the lepton mass term and is an integral over trilinear products of the left- and right-handed lepton fields and the Higgs field. The action for the Higgs field is

$$S_H = \int d^4X \frac{1}{4}\left[\left(\sigma_\mu^{a\dot{b}} D_{\dot{b}a}\Phi_H\right)^+\left(\sigma^{\mu a\dot{b}} D_{\dot{b}a}\Phi_H\right) - \frac{M_H^2}{\rho_H^2}\left(\Phi_H^+\Phi_H - \rho_H^2\right)^2\right] \tag{7.6.2a}$$

$$\Phi_H = \begin{pmatrix}\Phi_{H+}\\ \Phi_{H0}\end{pmatrix} \tag{7.6.2b}$$

where ρ_H is a constant, M_H the mass of the Higgs boson and $D_{\dot{b}a}$ has been given in (7.1.12b) with $Y_l=1$. Φ_{H+} and Φ_{H0} transform in the same way as the χ members in (7.1.19) do;

$$\Phi_H \to U_{21-}(X)\Phi_H \tag{7.6.3}$$

$S_{GB,SM}$ is essentially a subset of S_{GB} (7.1.2),

$$S_{GB,SM} = -\frac{1}{4}\int d^4X\left(\sum_{l=1}^{3} G_l^{\mu\nu}G_{l\mu\nu} + F_C^{\mu\nu}F_{C\mu\nu}\right)$$
$$G_l^{\mu\nu}(X) = \partial_X^\mu W_l^\nu - \partial_X^\nu W_l^\mu - \varepsilon_{jkl}\, g W_j^\mu W_k^\nu, \quad F_C^{\mu\nu}(X) = \partial_X^\mu C^\nu - \partial_X^\nu C^\mu \tag{7.6.4}$$

S_{SM} is a invariant under $SU(2)_I \times U(1)_Y$ gauge transformations (7.1.19-21) with W_8 reverted back to C according to the reverse of (7.1.15). In passing, it is noted that the invariance of S_{LmH} is achieved by its construction. This is in contrast to S_{Lm} of (7.1.13) which is not invariant under SU(2) gauge transformations. This noninvariance has been disregarded according to the approximation below (7.1.21).

In the SU(2) part of the transformations, there are three phase functions, ϑ_1, ϑ_2 and ϑ_3 in (7.1.21) that can arbitrarily chosen. Using two of them, (7.6.2b) can be transformed to

$$\Phi_H = \begin{pmatrix}0\\ \rho_H + \chi_H(X)\end{pmatrix} \tag{7.6.5}$$

where $\chi_H(X)$ is a Higgs field with mass M_H.

For the purpose of mass generation or for purely leptonic interactions, S_q does not enter and can be dropped. Following §7.4.2 and varying (7.6.1) with respect to $W_I^{-a\dot{b}}$ yields the analog of (7.4.6),

$$\Box W^{+}_{I\dot{b}a} - \tfrac{1}{2}\partial_{X\dot{b}a}\left(\partial^{c\dot{d}}_{X} W^{+}_{I\dot{d}c}\right) + g^2 V_{C\dot{b}a}\left(W_I^{+}\right) + M^2_{WH} W^{+}_{I\dot{b}a} = -\frac{g}{\sqrt{2}}\psi_{\nu L\dot{b}}\psi_{La} \qquad (7.6.6a)$$

$$M_{WH} = \frac{g\rho_H}{\sqrt{2}} \qquad (7.6.6b)$$

where the g^2 term is the same as (7.4.2b). Note that ρ_H not only generates the W_I mass M_{WH} but also the lepton mass $M_L = f_H\rho_H$ associated with $\chi_{L\dot{b}}$, where f_H is a proportionality constant in S_{LmH} [L5 (22.58), (22.70)].

Choosing the Lorentz gauge below (7.4.9) and dropping the g^2V_C term according to the ordering above (7.4.8), (7.6.6a) in component form becomes

$$\Box W_I^{0+} + M^2_{WH} W_I^{0+} = -\frac{g}{2\sqrt{2}}\psi_{\nu L\dot{a}}\psi_{La} \qquad (7.6.7a)$$

$$\Box \underline{W}_I^{+} + M^2_{WH} \underline{W}_I^{+} = -\frac{g}{2\sqrt{2}}\underline{\sigma}^{a\dot{b}}\psi_{\nu L\dot{b}}\psi_{La} \qquad (7.6.7b)$$

Follow the consideration above (7.4.10a), neglect the gauge boson energy and momentum relative to its mass. These steps turn (7.6.7) to the equivalent of (7.4.10),

$$M^2_{WH} W_I^{0+} = -\frac{g}{2\sqrt{2}}\psi_{\nu L\dot{a}}\psi_{La} \qquad (7.6.8a)$$

$$M^2_{WH} \underline{W}_I^{+} = -\frac{g}{2\sqrt{2}}\underline{\sigma}^{b\dot{a}}\psi_{\nu L\dot{a}}\psi_{Lb} \qquad (7.6.8b)$$

7.6.2. Comparison to Standard Model and Replacement of Higgs Bosons by Kaons

Comparison of (7.6.6) to (7.4.6) shows that both are basically same apart from for the origin of the gauge boson mass and the transformation properties of the mass term. In (7.6.6a), this term transforms like the other terms do; so this equation is Lorentz covariant. In (7.4.6a), the mass term transforms as a covariant vector while the others as contravariant vectors. This is due to that the Higgs field (7.6.2b) generating M_{WH} transforms like a Lorentz scalar while M_W of (7.4.6) is generated by the decaying pseudoscalar meson whose wave function $\psi_{0(pr)}$ of (7.3.14) is the time component of a Lorentz four vector so that (7.4.6a) is not Lorentz covariant. This has been considered in §7.4.2.

In the absence of the source terms on the right of (7.6.7), (7.6.7b) becomes the same as (7.4.8b) and can account for the observed vector gauge boson $\underline{W}$. However, the singlet W_l^{0+} in (7.6.7a) also obeys a similar equation and has the same mass M_{WH}, contrary to (7.4.8a). But such a singlet has not been observed; yet it is needed as an intermediary in the decay of the pseudoscalar mesons in §7.4.3. Such a conceptual difficulty is absent in §7.4.2 according to the interpretation above (7.4.10). Further, Higgs boson has not been found.

Comparison of the actions (7.6.1) and (7.1.1) shows that the first two terms on the right sides correspond to each other. The third terms both refer to lepton masses and have been briefly considered below (7.6.4). The basic difference between these two total actions is that the Higg's action S_H and the quark action S_q of the standard model are replaced by the meson action S_{m3} (7.1.8). As has been shown in this chapter, S_{m3} alone gives rise to the decay amplitude for the mesons in terms of massive gauge fields. In the standard model, however, S_H is introduced solely for the purpose of generating masses for the gauge bosons. Still, one has not been able to obtain such a decay amplitude from S_q. This is one of many examples showing that the incapabilities of the standard model are removed in the present scalar strong interaction hadron theory.

The action S_{m3} involves the whole pseudoscalar octet mesons so that it can account for decays such as $K \rightarrow \pi +$ lepton pairs. For purely leptonic interactions, $S_q=0$ and the role of S_{m3} becomes simply that of S_H, i.e., generation of gauge boson masses. For this purpose, we do not need to include all eight mesons represented by $\psi_{(pr)}$ in (7.1.8) or (2.4.18); an isodoublet or the isotriplet inside the octet would suffice. These are provided by the kaon doublets and the pion triplet, which is preferred because it has lower mass. But the simplest choice, which also directly correspond to the Higgs doublet (7.6.2b) is the kaon doublet

$$\psi_{(pr)} \rightarrow \psi_{(p3)} = \begin{pmatrix} \psi_{(13)} \\ \psi_{(23)} \end{pmatrix} = \begin{pmatrix} \psi_{K+} \\ \psi_{K0} \end{pmatrix}, \qquad \psi \rightarrow \chi \tag{7.6.9}$$

Inserting this expression into (7.1.8) leads to that the D operators there are to be modified by removing the third row and column in the gW matrix in (7.2.7). In this matrix, W_8 is further reverted back to C by the reverse of (7.2.11), similar to that done for the standard model below (7.6.4). Thus, (7.1.4) is modified to

$$D_{l\,ps}^{a\dot{b}} \rightarrow D_{l\,ps}^{\prime a\dot{b}} = \partial_l^{a\dot{b}} \delta_{ps} + i\tfrac{1}{4} g(\sigma_l)_{ps} W_l^{a\dot{b}}(X) + i\tfrac{1}{4} g' \delta_{ps} C^{a\dot{b}}(X) \tag{7.6.10}$$

where the gauge field part is the same as that in (7.1.12b). Consistent with this change, S_{GB} and S_L on (7.1.1) also turns to $S_{GB,SM}$ and $S_{L,SM}$, respectively. For purely leptonic interaction, therefore, (7.1.1) becomes

$$S_{T3} \rightarrow S_{T21} = \kappa_{u3} S_{GB,SM} + S_{L,SM} + S_{Lm} + S_{m21} \tag{7.6.11}$$

$$S_{m3} \to S_{m21} = \int d^4 X d^4 x \frac{1}{2} \left\{ \begin{array}{l} \chi^{+}_{(3p)\dot{e}a} D'^{a\dot{b}}_{I\,ps} D'^{f\dot{e}}_{II\,sq} \chi_{(q3)\dot{b}f} + \psi^{+e\dot{a}}_{(3p)} D'_{I\,ps\,\dot{a}b} D'_{II\,sq\,\dot{f}e} \psi^{b\dot{f}}_{(q3)} - \\ - \chi^{+}_{(3p)\dot{e}a} \left(M^2_{mp3} - \Phi_m \right) \psi^{a\dot{e}}_{(p3)} - \psi^{+e\dot{a}}_{(3p)} \left(M^2_{mp3} - \Phi_m \right) \chi_{(p3)\dot{e}a} + c.c. \end{array} \right\} \tag{7.6.12}$$

Noting the remarks about S_{Lm} made below (7.6.4), S_{T21} is invariant under the same $SU(2)_I \times U(1)_Y$ gauge transformations as S_{SM} is. Analogous to (7.6.5), two of the three phase functions can be used to convert (7.6.9) to

$$\psi_{(p3)} \leftarrow \begin{pmatrix} 0 \\ \psi_{(23)} \end{pmatrix} = \begin{pmatrix} 0 \\ \psi_{K0} \end{pmatrix} \tag{7.6.13}$$

Comparison of the (7.6.1) with $S_q=0$ and (7.6.11) shows that the both actions are basically the same apart from the mechanism of the mass generation in S_H and S_{m21}. The role of the wave function $\rho_H+\chi_H(X)$ in (7.6.5) is taken over by an integral over relative space time x involving the wave functions for the neutral kaon (7.6.13). From the viewpoint of W_I^+ mass generation via K^0, (7.6.12) and (7.1.8) with $W_V^+ \to W_I^+$ are the same. Variation of (7.6.11) with respect to $W_I^{+a\dot{b}}$ gives (7.4.6) with $W_V \to W_I$. Consequently, (7.4.10) goes over to

$$M_W^2 \kappa W_I^{0+} = \frac{g}{2\sqrt{2}} \psi^{(+)}_{\nu L\dot{a}} \psi^{(-)}_{La} \tag{7.6.14a}$$

$$M_W^2 \underline{W}_I^+ = -\frac{g}{2\sqrt{2}} \underline{\sigma}^{b\dot{a}} \psi^{(+)}_{\nu L\dot{a}} \psi^{(-)}_{Lb} \tag{7.6.14b}$$

A comparison with the standard model result (7.6.8) can now be made among the following three points.

1) *Intermediate* $W_I^{\pm}$. Comparision of (7.6.8b) with (7.6.14b), noting (7.4.32), leads to the identification

$$M^2_{WH} = M^2_W \tag{7.6.15a}$$

which by (7.6.6b) and (7.4.6b) becomes

$$\rho_H^2 = 2\tau_0 / \Omega \tag{7.6.15b}$$

The constant ρ_H in the Higgs wave function corresponds to the ratio of two large quantities each approaching infinity. Comparison of (7.6.8a) and (7.6.14a) gives rise to minus signs on the right of (7.6.15a, 15b). This however does not affect the observable quantities which depend upon the square of the quantities in (7.6.15), as is shown in (7.5.1).

These comparisons show that *standard model results for purely leptonic interactions mediated by* the charged gauge boson fields *can in essence be derived here.* This is demonstrated by muon decay in the next paragragh. It is understood that the right sides of (7.6.8) or (7.6.14) are interpreted to include higher order effects and the associated consequences of application of QED.

2) *Intermediate Z.* Here, (7.4.17) replaces (7.4.6). The equations corresponding to (7.6.8) is the same equations with

$$W_I^{0-} \to Z^0, \qquad \underline{W}_I^- \to \underline{Z}, \qquad M_Z \to M_{ZH} = M_{WH}/\cos\vartheta_W \tag{7.6.16}$$

where (7.4.18c) has been consulted. Comparison of the so-obtained standard model equation with (7.4.17) follows the same steps given in 1) above. This is possible because (7.4.17a) and (7.4.6a) differ essentially in the source terms on their right sides. The results is therefore entirely analogous to that of 1) with the same provisions. Thus, *standard model results for purely leptonic interaction mediated by* the neutral gauge boson fields *can in essence be derived here* in the limit of SU(3) symmetry.

3) *Virtual K^0 and π^0.* So far, K^0 in S_{m21} has been regarded as a motionless real meson. However, its mass E_{K0} does not appear in the mass generation process. This is due to that the gauge boson mass terms in (7.6.12) are of the form $\psi^* W^+ W^- \psi$ and the energy dependence of ψ in (7.3.14a) is cancelled by that of ψ^*. Therefore, E_{K0} can be less than the meson mass in (7.6.12) without affecting mass generation. This virtual meson state differs from the vacuum meson state of (7.3.7b), for which the energy momentum of the meson vanishes definitely according to discussion below (7.3.19).

The same result can also be obtained if K^0 is replaced by π^0. This can be achieved by using an isotriplet action given in [H8 §7, H11 §9, H14 §2], analogous to the kaon doublet action of (7.6.12). It can be constructed from the isotriplet meson equations in §8.4.1 below.

7.6.3. Muon Decay

The successful standard model results for $\mu \to \bar{\nu}_e \nu_\mu e$ are in essence derived below from the present theory.

1) *Standard Model Results.* The decay amplitude of [K3 14.22, 8] is

$$S_{fi(K)}\left(\mu \to e\bar{\nu}_e \nu_\mu\right) = -i\int dX^0 \left\langle e, \bar{\nu}_e, \nu_\mu \middle| H_1 \middle| \mu\right\rangle = -\frac{i}{\Omega_e \Omega_\mu} \sum_i \sum_{\lambda=1}^4 J_{(e\mu)i\lambda} J_{(\nu\mu\nu e)i\lambda}$$

$$\times \int d^4 X \exp\left(i\left(K_e^\sigma + K_{\nu e}^\sigma + K_{\nu\mu}^\sigma - K_\mu^\sigma\right) X_\sigma\right) \tag{7.6.17a}$$

$$J_{(e\mu)i\lambda} = \bar{u}_e^{(+)}\left(\underline{K}_e\right) O_{i\lambda} u_\mu^{(+)}\left(\underline{K}_\mu\right) \tag{7.6.17b}$$

$$J_{(\nu\mu\nu e)i\lambda} = \overline{u}_{\nu\mu}^{(+)}(\underline{K}_{\nu\mu})O_{i\lambda}\frac{1}{\sqrt{2}}(g_i + ig_i'\gamma_5)u_{\nu e}^{(-)}(-\underline{K}_{\nu e}) \tag{7.6.17c}$$

where the left of (7.6.17a) and $O_{i\lambda}$ are in the notation of [K3] and i denotes the five interaction types S, V, T, A, and P given in Table 13-1 of [K3]. This decay amplitude leads to correct predictions.

2) *Muon Decay Actions.* In the present theory, the starting point is a modified form of (7.1.1),

$$S_{T2} = S_{GB3} + S_{m2} + S_\mu + S_{\mu m} + S_e + S_{em} \tag{7.6.18}$$

where S_{GB3} is S_{GB} of (7.1.2) with l=1,2,3 and S_{m2} is the SU(2) part of S_{m3} in (7.1.8). The remaining lepton actions are shown in (7.1.13, 16, 17). The amplitude for muon decay will now be derived from (7.6.18) and shown to be essentially the same as the standard model amplitude (7.6.17). The procedure is analogous to that employed for $\pi^+ \rightarrow \pi^0 e^+ \nu_e$ in §7.4.5.

Treatment of (7.6.18) that follows is basically the same as that for kaon decay in Sec. 7.3 and 7.4. The role of the meson action S_{m3} in (7.1.1, 8) is taken over by S_μ+$S_{\mu m}$ in § 7.1.3 for the decay of a muon into a ν_μ and a gauge boson W_I^- together with S_{m2} for the generation of the mass of this W_I^-. The roles of the remaining actions S_e+S_{em} for the electron and $\overline{\nu}_e$ and S_{GB3} for the gauge boson are the same as in (7.1.1).

3) *Muon Decay into a Gauge Boson.* The ordering of the quantities contained in (7.6.18) is the same as that of (7.3.2b) with L limited to e. The muon now takes place of the kaon or pion in Sec. 7.3 and the equivalent of (7.3.1) becomes

$$\psi_\mu^b(X) = \left(1 + \frac{a_\mu^{(1)}(X^0)}{a_\mu}\right)\psi_{\mu 0}^b(X), \quad \psi_{\mu 0}^b(X) = \frac{1}{\sqrt{\Omega_\mu}} u_\mu^{b(+)} \exp\left(-iE_\mu X^0 + i\underline{K}_\mu \underline{X}\right) \tag{7.6.19a}$$

$$u_\mu^{b(+)} = \begin{pmatrix} u_\mu^{1(+)} \\ u_\mu^{2(+)} \end{pmatrix} = \frac{1}{\sqrt{2E_\mu(E_\mu + M_\mu)}}\left[\begin{pmatrix} E_\mu + M_\mu - K_{\mu 3} \\ -K_{\mu 1} - iK_{\mu 2} \end{pmatrix}, \begin{pmatrix} -K_{\mu 1} + iK_{\mu 2} \\ E_\mu + M_\mu + K_{\mu 3} \end{pmatrix}\right] \tag{7.6.19b}$$

$$\psi \rightarrow \chi \tag{7.6.19c}$$

analogous to (7.4.19b).

The initial and final states of (7.3.6, 7) become here

$$|i\rangle = |\mu(E_\mu = M_\mu, \underline{K}_\mu = 0)\rangle, \qquad \langle f| = \langle \nu_\mu, W_I^-| \rightarrow \langle \nu_\mu, e, \overline{\nu}_e| \tag{7.6.20}$$

Putting the first order part of S_{m2} to 0, which is equivalent to (7.3.3), and sandwiching it between <f| and |i>, one obtains the analog of (7.3.13, 22),

$$S_{fi} = \langle f | a_\mu^+ a_\mu^{(1)} (X^0 \to \infty) | i \rangle$$
$$= i \frac{g}{16} \frac{\sqrt{2}}{\Omega_\mu} \langle f | \int d^4 X u_{\nu\mu}^{(+)\dot{a}} W_{I\dot{a}b}^- (X) u_\mu^{(+)b} \exp \left(i (K_{\nu\mu\sigma} - K_{\mu\sigma}) X^\sigma \right) | i \rangle \tag{7.6.21}$$

4) *Decay of W_I^+ into and $\bar{\nu}_e$ and Deacy Amplitude.* This decay has been treated in §7.4.1-3, except for the interchange of W_I^+ and W_I^- and for the replacement of kaon by π^0 for mass generation. The last point has been considered at the end of §7.6.2. S_{m2} in (7.6.18) refers to the triplet of pions. A virtual π^0 with zero energy-momentum is chosen because it is the lowest energy meson.

Variation of the fourth order part of (7.6.18) with respect to W_I^- leads to the analog of (7.4.11c),

$$W_{I\dot{a}b}^- = \frac{g}{\sqrt{2} M_W^2} \frac{1}{\Omega_e} u_e^{(+)\dot{b}} (\underline{K}_e) u_{\nu e}^{(-)a} (-\underline{K}_{\nu e}) \exp \left(i (E_e + |E_{\nu e}|) X^0 - i (\underline{K}_e + \underline{K}_{\nu e}) \underline{X} \right) \tag{7.6.22}$$

Inserting this result into (7.6.21) leads to the decay amplitude

$$S_{fi} (\mu \to e \bar{\nu}_e \nu_\mu) = i \frac{g^2}{8 M_W^2} \frac{1}{\Omega_\mu \Omega_e} u_\mu^{(+)b} u_e^{(+)\dot{b}} u_{\nu\mu}^{(+)\dot{a}} u_{\nu e}^{(-)a}$$
$$\times \int d^4 X \exp \left(i (E_{\nu\mu} + |E_{\nu e}| + E_e - M_\mu) X^0 + i (\underline{K}_{\nu\mu} + \underline{K}_e + \underline{K}_{\nu e}) \underline{X} \right) \tag{7.6.23}$$

5) *Comparison with Standard Model.* Return to (7.6.17) and keep only the V and A interactions. Table 13-1 of [K3] defines

$$O_V^\lambda = \gamma^\lambda, \qquad O_A^\lambda = -\gamma^\lambda \gamma_5 \tag{7.6.24}$$

Because there is only one coupling constant, the Fermi constant G, put

$$|g_V| = |g_V'| = |g_A| = |g_A'| = G \tag{7.6.25}$$

in (14.51a) of [K3]. Since the interactions affect only the left handed parts of the lepton wave functions,

$$g_V = g_V' = -g_A = -g_A' = G \tag{7.6.26}$$

is set. With the muon at rest in (7.6.17), one finds

$$S_{fi}\left(\mu \to e\bar{v}_e v_\mu\right) = S_{fi(K)}\left(\mu \to e\bar{v}_e v_\mu \text{ with } i = V,\ A \text{ and } \lambda = 4\right) \tag{7.6.27}$$

analogous to (7.4.31) for $\pi^+ \to \pi^0 e^+ v_e$. The decay amplitude S_{fi} derived from the actions (7.6.18) agrees with that of the standard model $S_{fi(K)}$ limited to the time components λ =4 of the currents in (7.6.17b, 17c) and to V-A interactions. The absence of contributions from the spatial components λ =1,2,3 in (7.6.23) is due that only the time component of the meson wave function (7.3.14) generates the gauge boson mass. This is reflected in that (7.4.4) and (7.6.22) are no longer covariant under the SL(2,C) group of transformations.

The general case in which the decaying kaon is moving, cooresponding to that the virtual π^0 has a momentum in 4) above, should involve more terms that correspond to the λ =1,2,3 contributions in (7.6.17). The simple form (7.4.4) is lost and will involve integrals over the vector parts of the meson wave functions $\underline{\psi}$ and $\underline{\chi}$ in (3.1.7a) leading to very complicated forms of $W_I^{(\pm)}$ in (7.4.10). Further, (3.1.8) determining these $\underline{\psi}$ and $\underline{\chi}$ has not been solved, except for some dimensional analyses in some limits in Sec. 3.5. This general case has not been treated.

On the other hand, it may in principle be possible to transform away the spatial part of the material cuurent in (7.6.17b) so that $S_{fi}=S_{fi}(_K)$ here.

6) *On the Concept of Purely Leptonic Interactions.* In the standard model, muon decay is a purely leptonic process independent of the existence of hadrons. However, this model requires the unseen Higgs boson.

In the present approach, the role of the Higgs is played by the virtual π^0 in 4) above. Muon decay, hence weak interactions in general, therefore necessitates the existence of mesons, here the virtual π^0, hence quarks. *Leptons cannot exist alone but must coexist with quarks.* The usual conception of purely leptonic processes therefore no longer holds.

7.7. NONLEPTONIC DECAY AMPITUDE FOR $K\to 3\pi$

Nonleptonic decays of kaons are of the types $K^\pm \to \pi\pi$, $\pi\pi\pi$, $\pi\pi\mu v_\mu$, $\pi\pi e\, v_e$, $K^0{}_S \to \pi\pi$ and $K^0{}_L \to \pi\pi\pi$, $\pi\pi e\, v_e$ [P1]. Now, these pions, like the decay product pion in §7.4.6, generally move rather fast and their wave functions in relative space differ considerably from that for a pseudoscalar meson at rest and are unknown, as was mentioned below (7.4.33). Analogous to the case of §7.4.6, the associated decay amplitudes can likewise not be calculated generally.

In $K_L{}^0 \to 3\pi$ however, the pions move relatively slowly and their wave functions are approximated by those given in §3.5.4. Only this case is treated below where K^0 and $\bar{K}^0$ are related via (7.2.18) to $K_L{}^0$ and $K_S{}^0(\to 0$ here).

7.7.1. Variation of Pion Action

In the beginning of §7.1.1, it was implied that $K \to \pi L v_L$ consists of two stages. In stage I, the K decays into a π and an intermediate gauge boson. In stage II, the intermediate gauge boson no longer decays into a lepton-antilepton pair treated in Sec. 7.4 and 7.5 but into a pair of π's. In (7.1.1), $\kappa_\pi S_{m\pi}$ takes over the role of S_L+S_{Lm}. This transition is accomplished by modifying (7.4.5, 6) and (7.4.15, 17) according to

$$M_W^2 W_I^{\mp e\dot{f}} = M_W^2 (W_I^{\mp 0}\delta^{e\dot{f}} - \underline{\sigma}^{e\dot{f}}\underline{W}_I^{\mp}) = 2\delta S_L/\delta W_I^{\pm f\dot{e}} \rightarrow 2\kappa_\pi\, \delta S_{m\pi}/\delta W_I^{\pm f\dot{e}} \tag{7.7.1}$$

$$M_Z^2 Z^{e\dot{f}} = M_Z^2 (Z^0\delta^{e\dot{f}} - \underline{\sigma}^{e\dot{f}}\underline{Z}) = 4\delta S_L/\delta Z^{f\dot{e}} \rightarrow 4\kappa_\pi\, \delta S_{m\pi}/\delta Z^{f\dot{e}} \tag{7.7.2}$$

where the terms next to the gauge boson mass terms have been neglected.

$S_{m\pi}$ is taken to be the SU(2) part of S_{m3}. The kaon and η wave functions drop out because they are not the decay products of $W^\pm$ and Z. In stage II of §7.1.1, this action replaces the lepton action in (7.1.1). Making use of (7.1.8), (7.1.4) and (7.2.14), the pion action is written out as

$$S_{m\pi} = \frac{1}{2}\int d^4X \int d^4y$$

$$\times\left\{\begin{array}{l} \begin{pmatrix} \frac{1}{\sqrt{2}}\chi^*_{f3(\pi 0)} & \chi^*_{f3(\pi -)} \\ \chi^*_{f3(\pi +)} & -\frac{1}{\sqrt{2}}\chi^*_{f3(\pi 0)} \end{pmatrix}_{\dot{e}a} \left[\partial_{III}^{a\dot{b}} + i\frac{g}{4}\begin{pmatrix} Z\frac{\cos 2\vartheta_W}{\cos\vartheta_W} & \sqrt{2}W_I^- \\ \sqrt{2}W_I^+ & -\frac{Z}{\cos\vartheta_W} \end{pmatrix}^{a\dot{b}}\right] \\ \times\left[\partial_{IV}^{f\dot{e}} + i\frac{g}{4}\begin{pmatrix} Z\frac{\cos 2\vartheta_W}{\cos\vartheta_W} & \sqrt{2}W_I^- \\ \sqrt{2}W_I^+ & -\frac{Z}{\cos\vartheta_W} \end{pmatrix}^{f\dot{e}}\right]\begin{pmatrix} \frac{1}{\sqrt{2}}\chi^{(-)}_{f2(\pi 0)} & \chi^{(-)}_{f2(\pi +)} \\ \chi^{(-)}_{f2(\pi -)} & -\frac{1}{\sqrt{2}}\chi^{(-)}_{f2(\pi 0)} \end{pmatrix}_{\dot{b}f} \\ +\psi \text{ terms} + \left(M^2_{mpr} - \Phi_m\right)\text{terms} + c.c. \end{array}\right\} \tag{7.7.3}$$

The ψ terms are obtained from the χ terms by $\chi\rightarrow\psi$, upper(lower) indices→lower(upper) indices and the dotted(undotted) indices→undotted(dotted) indices, according to (7.1.8b). The last two terms in the braces refer to the last three terms in (7.1.8b). Summation over the flavor indices r, p, s... in (7.1.8b) has been written out in matrix multiplication form here and the flavor contents of the fields therein given by these indices have been replaced by particle symbols according to (7.1.5b) and (7.2.1a, 3a) together with (2.4.18).

The pion wave functions χ and ψ in (7.7.3), associated with the decay of the gauge bosons, are functions of the quark coordinates x_{III} and x_{IV}. These differ from the quark coordinates x_I and x_{II} for the initial kaon in S_{m3}, much like the strong interaction case of the decay of a vector meson into two pseudoscalar mesons such as $D^* \to D\pi$ of [H13 Sec. 6.1] or (8.1.2). In S_{m3}, the transformation

$$x^\mu = x_{II}^\mu - x_I^\mu, \qquad X^\mu = \tfrac{1}{2}(x_I^\mu + x_{II}^\mu) \tag{7.7.4}$$

has been made according to (3.1.3a, 10a) and the null energy condition [H10 Sec. 4.1] or (3.5.6). In (7.7.3), (7.7.4) is replaced by

$$y^\mu = x_{IV}^\mu - x_{III}^\mu, \qquad X^\mu = \tfrac{1}{2}(x_{III}^\mu + x_{IV}^\mu) \tag{7.7.5}$$

The relative coordinates x^μ and y^μ are hidden variables [H12 Sec. 1] and are integrated over in S_{m3} and (7.7.3).

The subscripts *f2* and *f3* in (7.7.3) actually refer to the same meson therein and are anticipated here and will later, upon quantization of the χ's and ψ's in §7.7.4, refer to final state *2* meson and final state *3* meson, respectively. These mesons will then be different and also differ from the final state *1* meson which is not a decay product of the gauge boson and is denoted by the subscripts *f1* in (7.3.26a). Similarly, the superscript (−) is to be ignored in (7.7.3) but will upon quantization later refer to meson with negative energy.

The action (7.7.3) is now to be varied with respect to $W_I^{\pm}(X)$ and $Z(X)$ as required by (7.7.1, 2). For simplicity of notation, only expressions for $W_I^{+}(X)$ will be written down. The $W_I^{-}(X)$ results are obtained by letting the superscript $+ \to -$ and vice versa in the $W_I^{+}(X)$ results. Thus,

$$\kappa_\pi \frac{\delta S_{m3}}{\delta W_I^{+f\dot{e}}} = i\frac{g}{8}\kappa_\pi$$
$$\times \int d^4y \begin{bmatrix} \chi^*_{f3(\pi-)\dot{b}f}\partial_{IV}^{a\dot{b}}\chi_{f2(\pi0)\dot{e}a} + \chi^*_{f3(\pi-)\dot{e}a}\partial_{III}^{a\dot{b}}\chi^{(-)}_{f2(\pi0)\dot{b}f} - \partial_{III}^{a\dot{b}}\chi^*_{f3(\pi-)\dot{e}a}\chi^{(-)}_{f2(\pi0)\dot{b}f} \\ -\chi^*_{f3(\pi0)\dot{b}f}\partial_{IV}^{a\dot{b}}\chi^{(-)}_{f2(\pi+)\dot{e}a} - \chi^*_{f3(\pi0)\dot{e}a}\partial_{III}^{a\dot{b}}\chi^{(-)}_{f2(\pi+)\dot{b}f} + \partial_{III}^{a\dot{b}}\chi^*_{f3(\pi0)\dot{e}a}\chi^{(-)}_{f2(\pi+)\dot{b}f} \\ +\psi\ \text{terms}(=\chi\to\psi,\ \dot{e}a \leftrightarrow \dot{b}f) + c.c. \end{bmatrix} \tag{7.7.6}$$

Similarly, variation of (7.7.3) with respect to $Z(X)$ yields

$$\kappa_\pi \frac{\delta S_{m3}}{\delta Z^{f\dot{e}}} = i\frac{g}{8}\frac{\kappa_\pi}{\cos\vartheta_W}\times$$

$$\int d^4y \begin{bmatrix} (\cos 2\vartheta_W - 1)\frac{1}{2}(\chi^*_{f3(\pi0)\dot{b}f}\partial^{a\dot{b}}_{IV}\chi^{(-)}_{f2(\pi0)\dot{e}a} + \chi^*_{f3(\pi0)\dot{e}a}\partial^{a\dot{b}}_{III}\chi^{(-)}_{f2(\pi0)\dot{b}f} - \partial^{a\dot{b}}_{III}\chi^*_{f3(\pi0)\dot{e}a}\chi^{(-)}_{f2(\pi0)\dot{b}f}) \\ + \cos 2\vartheta_W (\chi^*_{f3(\pi+)\dot{b}f}\partial^{a\dot{b}}_{IV}\chi^{(-)}_{f2(\pi+)\dot{e}a} + \chi^*_{f3(\pi+)\dot{e}a}\partial^{a\dot{b}}_{III}\chi^{(-)}_{f2(\pi+)\dot{b}f} - \partial^{a\dot{b}}_{III}\chi^*_{f3(\pi+)\dot{e}a}\chi^{(-)}_{f2(\pi+)\dot{b}f}) \\ - \chi^*_{f3(\pi-)\dot{b}f}\partial^{a\dot{b}}_{IV}\chi^{(-)}_{f2(\pi-)\dot{e}a} + \chi^*_{f3(\pi-)\dot{e}a}\partial^{a\dot{b}}_{III}\chi^{(-)}_{f2(\pi-)\dot{b}f} - \partial^{a\dot{b}}_{III}\chi^*_{f3(\pi+)\dot{e}a}\chi^{(-)}_{f2(\pi-)\dot{b}f} \\ + \psi \text{ terms}(= \chi \to \psi,\ \dot{e}a \leftrightarrow \dot{b}f) + c.c. \end{bmatrix} \tag{7.7.7}$$

The ψ terms and *c.c.* in (7.7.6, 7) derive from the ψ terms and *c.c.* in (7.7.3); the $(M_{mpr}{}^2 - \Phi_m)$ term therein drop out because they do not contain $W_I^{\pm}$ and Z. The ψ terms in (7.7.6-7) are obtained from the χ terms by letting $\chi \to \psi$ and switching the index pairs $\dot{e}a \leftrightarrow \dot{b}f$.

7.7.2. Pion Wave Functions and Ordering of the Pion Action

The pion wave functions in (7.7.6, 7) are decomposed analogous to [H10 (3.1)] or (7.3.1) complemented by a relative time factor. For instance,

$$\begin{aligned} &\chi^{(-)}_{f2(\pi+)\dot{e}a}(x_{III}, x_{IV}) = \chi^{(-)}_{f2(\pi+)\dot{e}a}(X, y) = a^{(-)}_{f2(\pi+)}\left(\delta_{\dot{e}a}\chi_{02+}(\underline{y}) + \underline{\sigma}_{\dot{e}a}\chi_{2+}(\underline{y})\right) \\ &\times \exp\left(i\omega^{(-)}_{2+}y^0 - iE^{(-)}_{2+}X^0 + i\underline{K}_{2+}\underline{X}\right), \qquad \chi \to \psi \end{aligned} \tag{7.7.8}$$

where $a_{f2(\pi+)}{}^{(-)}$ is unity here but is later in §7.7.4 elevated to an operator annihilating a π^+ with negative energy $E_{2+}{}^{(-)} < 0$ and momentum $\underline{K}_{2+}$ which is equivalent to the creation of a π^- with positive energy

$$-E^{(-)}_{2+} = \left|E^{(-)}_{2+}\right| = E_{2-} \tag{7.7.9}$$

This quantization procedure is anticipated here. y^0 is the relative time between the quarks constituting the π^- and $\omega_{2+}{}^{(-)}$ the associated relative energy. To be consistent with (7.7.5), the null energy condition mentioned above it holds so that

$$\omega^{(-)}_{2+} = 0 \tag{7.7.10}$$

At rest, $\underline{K}_{2+}=0$ and $\chi_{02+}(\underline{y})=0$ [H10 (B7)]. The normalized meson wave function is given by [H13 (2.2a, 2.3a)] or (4.3.2),

$$\chi_{02+}(\underline{y}) = \chi_0(\underline{y}) = -\psi_0(\underline{y}) = \sqrt{\frac{d_m^3}{8\pi\Omega_\pi}}\exp\left(-d_m r_y/2\right), \quad d_m = 0.864 \text{ Gev} \tag{7.7.11}$$

where reference has been made to

$$\underline{y} = (y^1,\ y^2,\ y^3) = r_y(\sin\vartheta_y \cos\varphi_y,\ \sin\vartheta_y \sin\varphi_y,\ \cos\vartheta_y) \tag{7.7.12}$$

Ω_π is a large normalization volume for the pions decayed from the gauge boson and d_m the fundamental meson energy scale [H13 Sec. 2.4] or (5.2.3).

For a freely moving meson, the meson wave function is unknown generally because the general meson wave equations [H10 (B1)] or (3.1.8) have not been solved. In $K \rightarrow 3\pi$, the pions are light relative to d_m and are nonrelativistic or $K^2 << E^2$. Under these circumstances, these general wave equations can be treated perturbatively in powers of K. To first order in K, these general equations are simplified to give [H10 (B8)] or (3.5.22) which consists of a second order vector differential equation coupled to a first order such. Even these simplified equations have not been solved. However, a dimensional approximation of these equations gives an approximative solution (3.5.23, 24). In terms of the quantities in (7.7.8), the last equations give for the vector part of the meson wave function to first order in K,

$$\underline{\chi}_{2+}(\underline{y}) \rightarrow \underline{\chi}(\underline{y}) = \underline{\psi}(\underline{y}) = -\frac{3}{d_m^2}\underline{K} \times \frac{\partial}{\partial \underline{y}}\chi_0(r_y) = \frac{3}{2d_m}\underline{K} \times \hat{y}\,\chi_0(r_y) \tag{7.7.13}$$

$\chi_{f3}{}^*$, $\psi_{f3}{}^*$, $\psi_{f2}{}^{(-)}$ etc in (7.7.6, 7) are decomposable in entirely the same fashion as was done in (7.7.8-13).

The operators ∂_{III} and ∂_{IV} in (7.7.6, 7) have similar forms as those in (3.1.4). Noting (7.7.4, 5), these read

$$\begin{aligned} \partial_{III}^{a\dot{b}} &= -\delta^{a\dot{b}}\frac{1}{2}\frac{\partial}{\partial X^0} - \underline{\sigma}^{a\dot{b}}\frac{1}{2}\frac{\partial}{\partial \underline{X}} + \delta^{a\dot{b}}\frac{\partial}{\partial y^0} + \underline{\sigma}^{a\dot{b}}\frac{\partial}{\partial \underline{y}} \\ \partial_{IV}^{a\dot{b}} &= -\delta^{a\dot{b}}\frac{1}{2}\frac{\partial}{\partial X^0} - \underline{\sigma}^{a\dot{b}}\frac{1}{2}\frac{\partial}{\partial \underline{X}} - \delta^{a\dot{b}}\frac{\partial}{\partial y^0} - \underline{\sigma}^{a\dot{b}}\frac{\partial}{\partial \underline{y}} \end{aligned} \tag{7.7.14}$$

The expressions (7.7.8-13) can now be inserted into (7.7.6, 7). By (7.7.10) type of conditions, terms containing $\partial/\partial y_0$ in (7.7.14) drop out. The same conditions however leads to that the other terms in (7.7.6, 7) becomes $\int dy_0 \rightarrow \infty$. To arrive at a finite result, therefore, the scaling factor κ_π in (7.7.6, 7) must $\rightarrow 0$ such that $\kappa_\pi \int dy_0$ is finite. The magnitude of this quantity is fixed by requiring that in (7.7.1, 2) variations of S_L and of $\kappa_\pi S_{m\pi}$ are of the same magnitude. From (7.4.5, 15, 19, 20),

$$\left|\frac{\delta S_L}{\delta W_I^{\pm}}\right| \approx \frac{g}{4\sqrt{2}}\frac{1}{\Omega_L}, \qquad \left|\frac{\delta S_L}{\delta Z}\right| \approx \frac{g}{8\cos\vartheta_W}\frac{1}{\Omega_L} \tag{7.7.15}$$

where Ω_L is the normalization volume of the leptons. Using (7.7.14) in (7.7.6, 7), the dominating term is seen to be the $\partial/\partial \underline{y}$ term which by (7.7.8, 11, 12) produces a factor of $d_m/2$. Further, $\chi_{f3}{}^*$ in (7.7.6, 7) has a similar form as (7.7.8), differing from it mainly in the phase factor. Thus, (7.7.11) also holds for $\chi_{f3}(\underline{y})$ so that

$$\int d^3\underline{y}\,\chi_{03}^*(\underline{y})\chi_{02}(\underline{y}) = \frac{1}{\Omega_\pi} \tag{7.7.16}$$

which is simply the normalization condition (4.2.8). Neglecting the small vector function (7.7.13), a typical term in the brackets of (7.7.6, 7) leads to

$$\kappa_\pi\left|\frac{\delta S_{m\pi}}{\delta W_I^\pm}\right| \approx \kappa_\pi \frac{g}{8}\frac{d_m}{2}\frac{1}{\Omega_\pi}\int dy^0, \quad \kappa_\pi\left|\frac{\delta S_{m\pi}}{\delta Z}\right| \approx \kappa_\pi \frac{g}{8\cos\vartheta_W}\frac{d_m}{2}\frac{1}{\Omega_\pi}\int dy^0 \tag{7.7.17}$$

Equating these with (7.7.15) and putting $\Omega_\pi = \Omega_L$ give

$$\kappa_\pi \int dy^0 \approx \frac{2\sqrt{2}}{d_m}, \qquad \frac{2}{d_m} \to c_\pi \frac{2}{d_m} \tag{7.7.18}$$

From a dimensional viewpoint, there is only one scale constant in the relative space time y for an integrand in (7.7.6, 7), namely $d_m/2$ in (7.7.11). Therefore, (7.7.18) is the only possibility. c_π is a constant that may be adjusted but should not deviate very much from the magnitude unity;

$$c_\pi \approx 1 \tag{7.7.19}$$

7.7.3. Evaluation of the Gauge Boson Fields

The expressions (7.7.8-14) and their modification to hold for ψ and for final state 3 meson wave function are inserted into (7.7.6, 7) in order to obtain expressions of W_I^+ and Z in (7.7.1, 2). A typical term in (7.7.6, 7), exemplified by the first term in the brackets of (7.7.6), becomes

$$\begin{aligned}
&\chi^*_{f3(\pi-)\dot{b}f}(X,y)\partial^{a\dot{b}}_{IV}\chi_{f2(\pi0)\dot{e}a}(X,y) \\
&= \frac{d_m^3}{8\pi\Omega_\pi}\exp(i(E_{3-} + |E_{20}^{(-)}|)X^0 + i(\underline{K}_{20} - \underline{K}_{3-})\underline{X} - d_m r_y)\delta_{\dot{e}a} \\
&\times\left\{\begin{matrix} -\frac{i}{2}|E_{20}^{(-)}|\delta^{a\dot{b}} - \frac{i}{2}\underline{\sigma}^{a\dot{b}}\underline{K}_{20} + \frac{d_m}{2}\underline{\sigma}^{a\dot{b}}\hat{y} - \frac{3}{4d_m}\underline{\sigma}^{a\dot{b}} \\ \times\begin{bmatrix} i|E_{20}^{(-)}|\underline{K}_{20}\times\hat{y} - K_{20}^2\hat{y} + \underline{K}_{20}(\underline{K}_{20}\hat{y}) \\ + i\left(d_m - \frac{2}{r}\right)\underline{K}_{20} - i\left(d_m + \frac{2}{r}\right)(\underline{K}_{20}\times\hat{y})\hat{y} \end{bmatrix} \end{matrix}\right\}\left\{\delta_{\dot{b}f} - \frac{3}{2d_m}\underline{\sigma}_{\dot{b}f}(\underline{K}_{3-}\times\hat{y})\right\}
\end{aligned} \tag{7.7.20}$$

The odd powered $\hat{y}$ terms drop out when operated by

$$\int d^3\underline{y} = \int_0^\infty r_y^2 dr_y \int_0^\pi \sin\vartheta_y d\vartheta_y \int_0^{2\pi} d\varphi_y \tag{7.7.21}$$

in (7.7.6), where (7.7.12) has been consulted. Such an integral over the first two terms in the brackets of (7.7.6) yields

$$\int d^3\underline{y}\left(\chi^*_{f3(\pi-)\dot{b}f}\partial^{a\dot{b}}_{IV}\chi_{f2(\pi0)\dot{e}a} + \chi^*_{f3(\pi-)\dot{e}a}\partial^{a\dot{b}}_{III}\chi^{(-)}_{f2(\pi0)\dot{b}f}\right)$$
$$= -\frac{i}{\Omega_\pi}\exp(i(E_{3-} + \left|E^{(-)}_{20}\right|)X^0 + i(\underline{K}_{20} - \underline{K}_{3-})\underline{X} - d_m r_y)$$
$$\times\left\{\delta_{\dot{e}f}\left|E^{(-)}_{20}\right|\left(1 - \frac{3}{2d_m^2}\underline{K}_{20}\underline{K}_{3-}\right) + \underline{\sigma}_{\dot{e}f}\left[\left(1 + \frac{3}{4d_m^2}K_{20}^2\right)\underline{K}_{3-} + \frac{3}{4d_m^2}(\underline{K}_{20}\underline{K}_{3-})\underline{K}_{20}\right]\right\} \tag{7.7.22}$$

A similar integral over the third term in the brackets of (7.7.6) gives

$$\int d^3\underline{y}(-)\partial^{a\dot{b}}_{III}\chi^*_{f3(\pi-)\dot{e}a}\chi^{(-)}_{f2(\pi0)\dot{b}f} = \frac{i}{\Omega_\pi}\exp(i(E_{3-} + \left|E^{(-)}_{20}\right|)X^0$$
$$+ i(\underline{K}_{20} - \underline{K}_{3-})\underline{X} - d_m r_y)$$
$$\times\left\{\begin{aligned}&\delta_{\dot{e}f}\frac{1}{2}\left(E_{3-} + \left|E^{(-)}_{20}\right|\right)\left(1 - \frac{3}{2d_m^2}\underline{K}_{20}\underline{K}_{3-}\right) + \\ &\underline{\sigma}_{\dot{e}f}\left[\frac{1}{2}(\underline{K}_{20} - \underline{K}_{3-}) - \frac{3}{8d_m^2}\left[K_{3-}^2\underline{K}_{20} - K_{20}^2\underline{K}_{3-} + (\underline{K}_{3-} - \underline{K}_{20})(\underline{K}_{3-}\underline{K}_{20}) + i\left(E_{3-} + \left|E^{(-)}_{20}\right|\right)(\underline{K}_{3-}\times\underline{K}_{20})\right]\right]\end{aligned}\right\} \tag{7.7.23}$$

The pion momentum ranges from 0 to about 130 Mev/c, which is about 15% of d_m of (7.7.11). Therefore, the K^2/d_m^2 terms in (7.7.22, 23) can be neglected. The sum of (7.7.22) and (7.7.23) then gives the integral over the first three terms in the brackets of (7.7.6) associated with the $\chi_{f3(\pi-)}$ and $\chi_{f2(\pi0)}^{(-)}$ final state,

$$\int d^3\underline{y}\left(\chi^*_{f3(\pi-)\dot{b}f}\partial^{a\dot{b}}_{IV}\chi_{f2(\pi0)\dot{e}a} + \chi^*_{f3(\pi-)\dot{e}a}\partial^{a\dot{b}}_{III}\chi^{(-)}_{f2(\pi0)\dot{b}f}{}' - \partial^{a\dot{b}}_{III}\chi^*_{f3(\pi-)\dot{e}a}\chi^{(-)}_{f2(\pi0)\dot{b}f}\right) =$$
$$-\frac{i}{\Omega_\pi}\exp(i(E_{3-} + \left|E^{(-)}_{20}\right|)X^0 + i(\underline{K}_{20} - \underline{K}_{3-})\underline{X} - d_m r_y)\left\{\delta_{\dot{e}f}\frac{1}{2}\left(\left|E^{(-)}_{20}\right| - E_{3-}\right) + \underline{\sigma}_{\dot{e}f}\left(-\frac{1}{2}\underline{K}_{20} + \frac{3}{2}\underline{K}_{3-}\right)\right\} \tag{7.7.24}$$

The approximation $K^2 << d_m^2$ employed here does not mean that the K^2 terms can safely be neglected. This is due to that the vector part of the meson wave function (7.7.13) is the result of a coarse dimensional approximation to first order in K and higher order K terms may contribute significantly. This is indicated by the second order energy correction $E_{02}=K^2/2E_{00}$ of (3.5.25d). Putting $K=130$ Mev/c and E_{00} to the average pion energy, $E_{02}/E_{00} \approx 0.31$ which indicates that the effect due to K^2 terms may not be neglected. This will impact upon the quadratic slope parameters in the Dalitz plot to be considered in §7.8.2-5 below.

Furthermore, the omission of the K^2/d_m^2 terms in (7.7.22, 23) to arrive at (7.7.24) does not mean that the vector part of the meson wave function (7.7.13) does not enter. This is testified by the presence of the $\underline{\sigma K}_{3-}$ terms in (7.7.24). This term is the result of (7.7.13) modified to apply to χ_{f3} coupled to the $\underline{\sigma\partial/\partial y}$ term in (7.7.14). This last term produces a factor $-d_m/2$ which cancels the $3/2d_m$ factor in (7.7.13).

It is this $\underline{\sigma K}_{3-}$ term that leads to the asymmetry of (7.7.24) with respect to $\underline{K}_{20}$ and $\underline{K}_{3-}$ which gives rise to the linear Dalitz diagram slope parameter g in §7.8.2 2) below. Physically, this term represents the interaction of the momentum or current of the final state 3 pion with the quark current $\partial/\partial\underline{y}$ in the relative space $\underline{y}$ of the final state 2 pion. The final state 1 pion is not a decay product of the gauge boson and does not contribute.
Inspection of (7.7.6) shows that the 4th-6th terms differ from the first three terms in the brackets only in the sign and pion indices. Therefore, (7.7.24) can simply be modified to arrive at the corresponding expressions for these 4th-6th terms. The ψ terms in (7.7.6) are found via (7.7.8) with $\chi\rightarrow\psi$ and (7.7.13) with $\chi\rightarrow\psi$ and a minus sign because $\chi_{0f2}(r_y)=-\psi_{0f2}(r_y)$ [H5 (6.7-10)] or (3.2.10b). The result turns out to be the same as that for the χ terms in (7.7.24). The $c.c.$ term in (7.7.3) are obatined from the other functions via

$$\left(W_I^{\pm a\dot{b}}\right)^* = W_I^{\mp b\dot{a}}, \quad \left(\chi^*_{f3(\pi\pm)}\right)^* \rightarrow \chi^{(-)}_{f3(\pi\pm)}, \quad \left(\chi^{(-)}_{f2(\pi\pm)}\right)^* \rightarrow \chi^*_{f2(\pi\pm)} \tag{7.7.25}$$

where complex conjugation is accompanied by a change in the sign of the energy in the meson wave functions. The result corresponding to (7.7.24) reads

$$\int d^3\underline{y}\left(\chi^{(-)}_{f3(\pi+)\dot{e}a}\partial^{a\dot{b}}_{IV}\chi^*_{f2(\pi0)\dot{b}f} + \chi^{(-)}_{f3(\pi+)\dot{b}f}\partial^{a\dot{b}}_{III}\chi^*_{f2(\pi0)\dot{e}a} - \partial^{a\dot{b}}_{III}\chi^{(-)}_{f3(\pi+)\dot{b}f}\chi^*_{f2(\pi0)\dot{e}a}\right)=$$
$$-\frac{i}{\Omega_\pi}\exp(i(E_{20}+\left|E^{(-)}_{3+}\right|)X^0 - i(\underline{K}_{20}-\underline{K}_{3+})\underline{X} - d_m r_y)\left\{\delta_{\dot{e}f}\frac{1}{2}(E_{20}-\left|E^{(-)}_{3+}\right|)+\underline{\sigma}_{\dot{e}f}\left(-\frac{3}{2}\underline{K}_{20}+\frac{1}{2}\underline{K}_{3+}\right)\right\} \tag{7.7.26}$$

The remaining $c.c.$ contributions in (7.7.6) follow analogously. With the results in this section and (7.7.17), (7.7.6) is readily evaluated. This result together with (7.7.17) reads

$$\kappa_\pi \frac{\delta S_{m\pi}}{\delta W_I^{+f\dot{e}}} = \kappa_\pi \frac{gc_\pi}{2d_m\Omega_\pi}$$

$$\times\begin{Bmatrix} \exp(i(E_{3-}+|E_{20}^{(-)}|)X^0+i(\underline{K}_{20}-\underline{K}_{3-})\underline{X})\left[\delta_{\dot{e}f}\frac{1}{2}(|E_{20}^{(-)}|-E_{3-})-\underline{\sigma}_{\dot{e}f}\left(\frac{1}{2}\underline{K}_{20}-\frac{3}{2}\underline{K}_{3-}\right)\right] \\ -\exp(i(E_{30}+|E_{2+}^{(-)}|)X^0+i(\underline{K}_{2+}-\underline{K}_{30})\underline{X})\left[\delta_{\dot{e}f}\frac{1}{2}(|E_{2+}^{(-)}|-E_{30})-\underline{\sigma}_{\dot{e}f}\left(\frac{1}{2}\underline{K}_{2+}-\frac{3}{2}\underline{K}_{30}\right)\right] \\ +\exp(i(E_{20}+|E_{3+}^{(-)}|)X^0-i(\underline{K}_{20}-\underline{K}_{3+})\underline{X})\left[\delta_{\dot{e}f}\frac{1}{2}(E_{20}-|E_{3+}^{(-)}|)-\underline{\sigma}_{\dot{e}f}\left(\frac{3}{2}\underline{K}_{20}-\frac{1}{2}\underline{K}_{3+}\right)\right] \\ -\exp(i(E_{2-}+|E_{30}^{(-)}|)X^0-i(\underline{K}_{2-}-\underline{K}_{30})\underline{X})\left[\delta_{\dot{e}f}\frac{1}{2}(E_{2-}-|E_{30}^{(-)}|)-\underline{\sigma}_{\dot{e}f}\left(\frac{3}{2}\underline{K}_{2-}-\frac{1}{2}\underline{K}_{30}\right)\right] \end{Bmatrix} \tag{7.7.27}$$

In an entirely analogous manner, (7.7.7) is evaluated to be

$$\kappa_\pi\frac{\delta S_{m\pi}}{\delta Z^{f\dot{e}}}=\kappa_\pi\frac{gc_\pi}{2d_m\cos\vartheta_W\Omega_\pi}\times$$

$$\begin{Bmatrix} \exp(i(E_{30}+|E_{20}^{(-)}|)X^0+i(\underline{K}_{20}-\underline{K}_{30})\underline{X})\frac{1}{2}(\cos 2\vartheta_W-1))\left[\delta_{\dot{e}f}\frac{1}{2}(|E_{20}^{(-)}|-E_{30})-\underline{\sigma}_{\dot{e}f}\left(\frac{1}{2}\underline{K}_{20}-\frac{3}{2}\underline{K}_{30}\right)\right] \\ -\exp(i(E_{3-}+|E_{2-}^{(-)}|)X^0+i(\underline{K}_{2-}-\underline{K}_{3-})\underline{X})\left[\delta_{\dot{e}f}\frac{1}{2}(|E_{2-}^{(-)}|-E_{3-})-\underline{\sigma}_{\dot{e}f}\left(\frac{1}{2}\underline{K}_{2-}-\frac{3}{2}\underline{K}_{3-}\right)\right] \\ +\exp(i(E_{3+}+|E_{2+}^{(-)}|)X^0+i(\underline{K}_{2+}-\underline{K}_{3+})\underline{X})\cos 2\vartheta_W\left[\delta_{\dot{e}f}\frac{1}{2}(E_{2+}-|E_{3+}^{(-)}|)-\underline{\sigma}_{\dot{e}f}\left(\frac{1}{2}\underline{K}_{2+}-\frac{3}{2}\underline{K}_{3+}\right)\right] \\ +\exp(i(E_{20}+|E_{30}^{(-)}|)X^0-i(\underline{K}_{20}-\underline{K}_{30})\underline{X})\frac{1}{2}(\cos 2\vartheta_W-1))\left[\delta_{\dot{e}f}\frac{1}{2}(E_{20}-|E_{30}^{(-)}|)-\underline{\sigma}_{\dot{e}f}\left(\frac{3}{2}\underline{K}_{20}-\frac{1}{2}\underline{K}_{30}\right)\right] \\ -\exp(i(E_{2+}+|E_{3+}^{(-)}|)X^0-i(\underline{K}_{2+}-\underline{K}_{3+})\underline{X})\left[\delta_{\dot{e}f}\frac{1}{2}(E_{2+}-|E_{3+}^{(-)}|)-\underline{\sigma}_{\dot{e}f}\left(\frac{3}{2}\underline{K}_{2+}-\frac{1}{2}\underline{K}_{3+}\right)\right] \\ +\exp(i(E_{2-}+|E_{3-}^{(-)}|)X^0-i(\underline{K}_{2-}-\underline{K}_{3-})\underline{X})\cos 2\vartheta_W\left[\delta_{\dot{e}f}\frac{1}{2}(E_{2-}-|E_{3-}^{(-)}|)-\underline{\sigma}_{\dot{e}f}\left(\frac{3}{2}\underline{K}_{2-}-\frac{1}{2}\underline{K}_{3-}\right)\right] \end{Bmatrix} \tag{7.7.28}$$

In the braces of (7.7.27) and (7.7.28), the second half of the terms are derived from the *c.c.* terms in (7.7.6) and (7.7.7), respectively.

7.7.4. Decay Amplitudes

In the decay amplitude S_{fi} of (7.3.26, 27), the amplitudes of the wave functions therein, exemplified by $a_{f2}^{(-)}$ in (7.7.8), have been elevated to annililation and creation operators. These also hold implicitly for the gauge boson fields W_I and hence also the pion fields in (7.7.6, 7) and the equations that follow via (7.7.1, 2). This has been anticipated in §7.7.1-3. The rudimentary quantization procedure used here has been considered in §7.3.3.

From (7.7.1, 2) and (7.7.6, 7), it is seen that the gauge bosons $W^+ \to \pi^+\pi^0$, $W^- \to \pi^-\pi^0$ and $Z \to \pi^0\pi^0$ and $\pi^+\pi^-$. With (7.2.20, 22), the amplitudes (7.3.26, 27) can symbolically refer to the following decay modes

$$\alpha = 1: \quad K^- \to \pi^0 W_I^- \to \pi^0\pi^-\pi^0 \tag{7.7.29a}$$

$$K^- \to \pi^- Z \to \pi^-\pi^0\pi^0 \tag{7.7.29b}$$

$$\alpha = 2: \quad \overline{K}^0 \to \pi^- W_I^+ \to \pi^-\pi^+\pi^0 \tag{7.7.30a}$$

$$\overline{K}^0 \to \pi^+ W_I^- \to \pi^+\pi^-\pi^0 \tag{7.7.30b}$$

$$\overline{K}^0 \to \pi^0 Z \to \pi^0\pi^+\pi^- \tag{7.7.30c}$$

$$\alpha = 0: \quad \overline{K}^0 \to \pi^0 Z \to \pi^0\pi^0\pi^0 \tag{7.7.31}$$

$$\alpha = 3: \quad K^- \to \pi^- Z \to \pi^-\pi^+\pi^- \tag{7.7.32}$$

Here, α denotes the number of charged pions with mass m_+.
Insert (7.2.20) into (7.3.26a) and make use of (7.7.8-14). This yields for (7.7.29a)

$$S_{fi}\left(K^- \to \pi^0 W_I^-\right) = -\frac{ig}{8m_{K-}\Omega}\int d^4X \exp\left(i\left(E_{10} - m_{K-}\right)X^0 - i\underline{K}_{10}\underline{X}\right)$$
$$\times \left\{2W_I^{0-}\left(m_{K-} + E_{10}\right) + 8\underline{W}_I^-\underline{K}_{10}\right\} \tag{7.7.33}$$

where m_{K-} is the mass of K^-, $E_{10}=E_{f1\pi0}$, $\underline{K}_{10}=\underline{K}_{f1\pi0}$, and Ω a large normalization volume associated with K^- and π^0 but not with the pions decayed from W_I^-. Insert now (7.7.27) into (7.7.1) to obtain expressions for W_I^{-0} and $\underline{W}_I^-$, which are then substituted into (7.7.33). Carrying out the integrations over X^0 and $\underline{X}$ turns (7.7.33) into the decay amplitude

$$S_{fi}\left(K^- \to \pi^0 W_I^-\right) = A_1\left\{\begin{aligned} &\delta\left(-m_{K-} + E_{10} + \left|E_{20}^{(-)}\right| + E_{3-}\right)\delta\left(-\underline{K}_{10} + \underline{K}_{20} - \underline{K}_{3-}\right)M_{10-} \\ &+ \delta\left(-m_{K-} + E_{10} + \left|E_{2+}^{(-)}\right| + E_{30}\right)\delta\left(-\underline{K}_{10} + \underline{K}_{2+} - \underline{K}_{30}\right)M_{1-0} \\ &+ \delta\left(-m_{K-} + E_{10} + E_{20} + \left|E_{3+}^{(-)}\right|\right)\delta\left(-\underline{K}_{10} - \underline{K}_{20} + \underline{K}_{3+}\right)M_{10-*} \\ &+ \delta\left(-m_{K-} + E_{10} + E_{2-} + \left|E_{30}^{(-)}\right|\right)\delta\left(-\underline{K}_{10} - \underline{K}_{2-} + \underline{K}_{30}\right)M_{1-0*} \end{aligned}\right\} \tag{7.7.34}$$

where

$$A_\alpha M_{\alpha\beta\gamma}\,(\text{no summation over }\alpha) = \text{decay matrix element} \tag{7.7.35a}$$

$$A_1 = A_3 = i\frac{g^2}{M_W^2}\frac{c_\pi}{\Omega\Omega_\pi}\frac{4\pi^4}{m_{K-}d_m} \tag{7.7.35b}$$

$$2M_{10-} = -\left[(m_{K-} + E_{10})\left(\left|E_{20}^{(-)}\right| - E_{3-}\right) + 4\underline{K}_{10}\left(\underline{K}_{20} - 3\underline{K}_{3-}\right)\right] \quad (7.7.36a)$$

$$2M_{1-0} = \left[(m_{K-} + E_{10})\left(\left|E_{2+}^{(-)}\right| - E_{30}\right) + 4\underline{K}_{10}\left(\underline{K}_{2+} - 3\underline{K}_{30}\right)\right] \quad (7.7.36b)$$

$$2M_{10-*} = -\left[(m_{K-} + E_{10})\left(E_{20} - \left|E_{3+}^{(-)}\right|\right) + 4\underline{K}_{10}\left(3\underline{K}_{20} - \underline{K}_{3+}\right)\right] \quad (7.7.36c)$$

$$2M_{1-0*} = \left[(m_{K-} + E_{10})\left(E_{2-} - \left|E_{30}^{(-)}\right|\right) + 4\underline{K}_{10}\left(3\underline{K}_{2-} - \underline{K}_{30}\right)\right] \quad (7.7.36d)$$

The subscripts β and γ stand for the charges of the final state *2* and *3* pions, respectively, and the subscript $*$ refers to the corresponding *c.c.* terms in (7.7.3). The signs of the charges in $|E_{\pm}^{(-)}|$ actually designate the opposite charges. This is due to that such a final state pion has negative energy and the operator associated with the pion wave function is interpreted to annihilate a negative energy pion of a given charge, which is equivalent to the creation of a positive energy pion of the opposite charge.

The case (7.7.29b) together with (7.7.32) is treated analogously. Applying (7.2.21-22) to (7.3.26b) and making use of (7.7.8-14) leads to

$$S_{fi}\left(K^- \to \pi^- Z\right) = \frac{ig}{8m_{K-}\Omega}\int d^4X \exp\left(i(E_{1-} - m_{K-})X^0 - i\underline{K}_{1-}\underline{X}\right)$$
$$\times \frac{1}{\cos\vartheta_W}\left\{Z^0\left(m_{K-} + E_{1-}\right) + 2\underline{K}_{1-}\underline{Z}\right\} \quad (7.7.37)$$

Insert (7.7.28) into (7.7.2) to obtain Z^0 and $\underline{Z}$ and substitute the results into (7.7.37). Carrying out the integrations gives

$$S_{fi}\left(K^- \to \pi^- Z\right) = S_{fi}\left(K^- \to \pi^- Z \to \pi^-\pi^0\pi^0\right) + S_{fi}\left(K^- \to \pi^- Z \to \pi^-\pi^+\pi^-\right) \quad (7.7.38)$$

$$S_{fi}\left(K^- \to \pi^- Z \to \pi^-\pi^0\pi^0\right)$$
$$= A_1\begin{Bmatrix} \delta\left(-m_{K-} + E_{1-} + \left|E_{20}^{(-)}\right| + E_{30}\right)\delta\left(-\underline{K}_{1-} + \underline{K}_{20} - \underline{K}_{30}\right)M_{100} \\ + \delta\left(-m_{K-} + E_{1-} + E_{20} + \left|E_{30}^{(-)}\right|\right)\delta\left(-\underline{K}_{1-} - \underline{K}_{20} + \underline{K}_{30}\right)M_{100*} \end{Bmatrix} \quad (7.7.39)$$

$$2M_{100} = (\cos 2\vartheta_W - 1)\left[(m_{K-} + E_{1-})\left(\left|E_{20}^{(-)}\right| - E_{30}\right) + 4\underline{K}_{1-}\left(\underline{K}_{20} - 3\underline{K}_{30}\right)\right] \quad (7.7.40a)$$

$$2M_{100*} = (\cos 2\vartheta_W - 1)\left[(m_{K-} + E_{1-})\left(E_{20} - \left|E_{30}^{(-)}\right|\right) + 4\underline{K}_{1-}\left(3\underline{K}_{20} - \underline{K}_{30}\right)\right] \quad (7.7.40b)$$

$$S_{fi}\left(K^- \to \pi^- Z \to \pi^-\pi^+\pi^-\right)$$
$$= A_3 \begin{Bmatrix} \delta\left(-m_{K-} + E_{1-} + \left|E_{2-}^{(-)}\right| + E_{3-}\right)\delta\left(-\underline{K}_{1-} + \underline{K}_{2-} - \underline{K}_{3-}\right)M_{3+-} \\ + \delta\left(-m_{K-} + E_{1-} + \left|E_{2+}^{(-)}\right| + E_{3+}\right)\delta\left(-\underline{K}_{1-} + \underline{K}_{2+} - \underline{K}_{3+}\right)M_{3-+} \\ + \delta\left(-m_{K-} + E_{1-} + E_{2+} + \left|E_{3+}^{(-)}\right|\right)\delta\left(-\underline{K}_{1-} - \underline{K}_{2+} + \underline{K}_{3+}\right)M_{3+-*} \\ + \delta\left(-m_{K-} + E_{1-} + E_{2-} + \left|E_{3-}^{(-)}\right|\right)\delta\left(-\underline{K}_{1-} - \underline{K}_{2-} + \underline{K}_{3-}\right)M_{3-+*} \end{Bmatrix} \tag{7.7.41}$$

$$M_{3+-} = -\left[\left(m_{K-} + E_{1-}\right)\left(\left|E_{2-}^{(-)}\right| - E_{3-}\right) + 4\underline{K}_{1-}\left(\underline{K}_{2-} - 3\underline{K}_{3-}\right)\right] \tag{7.7.42a}$$

$$M_{3+-} = \cos 2\vartheta_W \left[\left(m_{K-} + E_{1-}\right)\left(\left|E_{2+}^{(-)}\right| - E_{3+}\right) + 4\underline{K}_{1-}\left(\underline{K}_{2+} - 3\underline{K}_{3+}\right)\right] \tag{7.7.42b}$$

$$M_{3+-*} = -\left[\left(m_{K-} + E_{1-}\right)\left(E_{2+} - \left|E_{3+}^{(-)}\right|\right) + 4\underline{K}_{1-}\left(3\underline{K}_{2+} - \underline{K}_{3+}\right)\right] \tag{7.7.42c}$$

$$M_{3-+*} = \cos 2\vartheta_W \left[\left(m_{K-} + E_{1-}\right)\left(E_{2-} - \left|E_{3-}^{(-)}\right|\right) + 4\underline{K}_{1-}\left(3\underline{K}_{2-} - \underline{K}_{3-}\right)\right] \tag{7.7.42d}$$

Here, $M_W = M_Z \cos\vartheta_W$ of (7.4.18c) has been noted.

The amplitudes of the four types of decay in (7.7.29-32) are

$$\alpha = 1: \quad S_{fi}\left(K^- \to \pi^-\pi^0\pi^0\right) = S_{fi}\left(K^- \to \pi^0 W_I^-\right) + S_{fi}\left(K^- \to \pi^- Z \to \pi^-\pi^0\pi^0\right) \tag{7.7.43a}$$

$$\begin{aligned} \alpha = 2: \quad & S_{fi}\left(\overline{K}^0 \to \pi^+\pi^-\pi^0\right) = S_{fi}\left(\overline{K}^0 \to \pi^+ W_I^-\right) \\ & + S_{fi}\left(\overline{K}^0 \to \pi^- W_I^+\right) + S_{fi}\left(\overline{K}^0 \to \pi^0 Z \to \pi^0\pi^+\pi^-\right) \end{aligned} \tag{7.7.43b}$$

$$\alpha = 0: \quad S_{fi}\left(\overline{K}^0 \to \pi^0\pi^0\pi^0\right) = S_{fi}\left(\overline{K}^0 \to \pi^0 Z \to \pi^0\pi^{+0}\pi^0\right) \tag{7.7.43c}$$

$$\alpha = 3: \quad S_{fi}\left(K^- \to \pi^-\pi^+\pi^-\right) = S_{fi}\left(K^- \to \pi^- Z \to \pi^-\pi^+\pi^-\right) \tag{7.7.43d}$$

Here, the $\overline{K}^0$ decay amplitudes are related to the K^- amplitudes and (7.7.34, 39, 41) via (7.3.26, 27), noting that

$$A_2 = A_0 = A_1\, m_{K-}/m_{K+} \tag{7.7.44}$$

The K^+ and K^0 decays are also mentioned in the beginning of §7.7.4.

7.8. Dalitz Slope Parameters and Estimate of Decay Rates

7.8.1. Phase Space: Present and Conventional

Following (7.5.1, 2), the decay rate is

$$\Gamma(\alpha) = \sum_{\mathrm{fsc}} \sum_{\underline{K}} \left| S_{fi}(\alpha) \right|^2 / T_d \tag{7.8.1}$$

where α refers to those in (7.7.43) and T_d to a long decay time. The subscripts *fsc* stand for final state combinations. Further,

$$\sum_{\underline{K}} = \frac{\Omega}{(2\pi)^3} \int d^3\underline{K}_1 \frac{\Omega_\pi^2}{(2\pi)^6} \int d^3\underline{K}_2 \int d^3\underline{K}_3 \tag{7.8.2}$$

According to (7.5.6) and [H11 (6.4)],

$$\left|\delta(-m_K + E_1 + E_2 + E_3)\right|^2 = \frac{T_d}{2\pi} \delta(-m_K + E_1 + E_2 + E_3) \tag{7.8.3a}$$

$$\left|\delta(-\underline{K}_1 + \underline{K}_2 - \underline{K}_3)\right|^2 = \frac{\Omega}{(2\pi)^3} \delta(-\underline{K}_1 + \underline{K}_2 - \underline{K}_3) \tag{7.8.3b}$$

Inserting (7.7.34, 39, 41) type of expressions into (7.8.1) and making use of (7.8.2, 3) and (7.7.35b) leads to

$$\Gamma(\alpha) = \sum_{\mathrm{fsc}} B_\alpha I_\alpha \sum_{\beta\gamma} \left| M_{\alpha\beta\gamma} \right|^2 \tag{7.8.4a}$$

$$B_1 = B_3 = \frac{1}{2(2\pi)^5} \frac{c_\pi^2 G^2}{m_{K-}^2 d_m^2} = 3.818 \times 10^{-14} c_\pi^2 \ \mathrm{Gev}^{-8}$$

$$B_0 = B_2 = B_1 m_{K-}^2 / m_{K0}^2 = 3.757 \times 10^{-14} c_\pi^2 \ \mathrm{Gev}^{-8}$$

$$G = \frac{1}{4\sqrt{2}} \frac{g^2}{M_W^2} = 1.1664 \times 10^{-5} \ \mathrm{Gev}^{-2} \tag{7.8.4b}$$

$$I_\alpha = \int d^3\underline{K}_1 \int d^3\underline{K}_2 \int d^3\underline{K}_3 \delta(-m_K + E_1 + E_2 + E_3) \delta(-\underline{K}_1 + \underline{K}_2 - \underline{K}_3) \tag{7.8.5}$$

where I_α is considered as an operator. Here, the cross or interference terms in the squares of (7.7.34, 39, 41) generally drop out because the product of two different $\delta(-\underline{K}_1 + \ldots)$ functions is not of the form (7.8.3b) and vanish upon integration over $\underline{K}$. In addition, the two different

$\delta(-\underline{K}_1+...)$ functions involve more $\underline{K}$ variables than the three $\underline{K}$ variables in (7.8.2), leaving the extra $\underline{K}$ variables not integration over. However, there is an exception in §7.8.3 below for which the cross terms do contribute because the both $\delta(-\underline{K}_1+...)$ functions can be regarded to coincide with each other and that the both $M_{\alpha\beta\gamma}$ functions are proportional to each other.

The associated phase space integral $I_{p\alpha}$ is I_α of (7.8.5) operating on a constant. For $\alpha=1$,

$$I_1 \to I_{p1} = \int d^3\underline{K}_{10} \int d^3\underline{K}_{20} \int d^3\underline{K}_{3-}\delta\left(-m_{K-} + E_{10} + \left|E_{20}^{(-)}\right| + E_{3-}\right)\delta\left(-\underline{K}_{10} + \underline{K}_{20} - \underline{K}_{3-}\right)N_1 \tag{7.8.6a}$$

$$N_1 = 1 \tag{7.8.6b}$$

Let the mass of the neutral pion be m_0 and the charged ones be

$$m_+ = m_0(1+\Delta) \tag{7.8.7}$$

and write

$$E_{10} = \sqrt{m_0^2 + K_{10}^2} = m_0 + T_{10} \tag{7.8.8a}$$

$$\left|E_{20}^{(-)}\right| = \sqrt{m_0^2 + K_{20}^2} = m_0 + T_{20} \tag{7.8.8b}$$

$$E_{3-} = \sqrt{m_+^2 + K_{3-}^2} = m_+ + T_{3-} \tag{7.8.8c}$$

where T denotes kinetic energy. Carrying out the $\underline{K}_{3-}$ integration, one obtains

$$\underline{K}_{3-} = \underline{K}_{20} - \underline{K}_{10} \tag{7.8.9}$$

$$I_{p1} = \int d^3\underline{K}_{10} \int d^3\underline{K}_{20}\delta\left(-m_{K-} + E_{10} + \left|E_{20}^{(-)}\right| + \sqrt{m_+^2 + K_{20}^2 + K_{10}^2 - 2\underline{K}_{10}\underline{K}_{20}}\right) \tag{7.8.10}$$

Express now $\underline{K}_{10}$ and $\underline{K}_{20}$ in spherical coordinates and choose $\underline{K}_{10}$ to be parallel to its third component in the $\underline{K}_{20}$ integral and noting its isotropy, (7.8.10) becomes

$$\begin{aligned} I_{p1} &= 8\pi^2 \int dK_{10}K_{10}^2 \int dK_{20}K_{20}^2 \\ &\times \int_0^\pi d\vartheta_{20} \sin\vartheta_{20}\delta\left(-m_{K-} + m_+ + T_{3-} + \sqrt{m_0^2 + K_{10}^2} + \sqrt{m_0^2 + K_{20}^2}\right) \end{aligned} \tag{7.8.11}$$

$$m_+ + T_{3-} = \sqrt{m_+^2 + K_{20}^2 + K_{10}^2 - 2K_{10}K_{20}\cos\vartheta_{20}} \tag{7.8.12}$$

where ϑ_{20} is the angle between $\underline{K}_{10}$ and $\underline{K}_{20}$. Putting the argument of the δ function to 0 gives

$$\cos\vartheta_{20} = \mp\frac{1}{2K_{10}K_{20}}\left[\left(m_{K-} - \sqrt{m_0^2 + K_{10}^2} - \sqrt{m_0^2 + K_{20}^2}\right)^2 - m_+^2 - K_{20}^2 - K_{10}^2\right]$$

$$= \mp\frac{1}{2K_{10}K_{20}}\left[(m_{K-} - 2m_0)^2 - m_+^2 - 2(m_{K-} - m_0)(T_{10} + T_{20}) + 2T_{10}T_{20}\right] \qquad (7.8.13)$$

where (7.8.8) has been consulted, which also gives

$$K_{10}dK_{10} = (m_0 + T_{10})dT_{10}, \qquad K_{20}dK_{20} = (m_0 + T_{20})dT_{20} \qquad (7.8.14)$$

Inserting these into (7.8.11) and integrating over ϑ_{20} using (7.5.13) yields

$$I_{p1} = 8\pi^2 \int dT_{10} \int dT_{20}(m_0 + T_{10})(m_0 + T_{20})(m_{K-} - 2m_0 - T_{10} - T_{20}) \qquad (7.8.15)$$

The boundary of the integration is determined by

$$\cos^2\vartheta_{20} = 1 \qquad (7.8.16)$$

Since the mesons are largely nonrelativistic, we have

$$T_{10} = K_{10}^2/2m_0, \qquad T_{20} = K_{20}^2/2m_0, \qquad T_{3-} = K_{3-}^2/2m_+ \qquad (7.8.17)$$

This and (7.8.12, 13, 7) turn (7.8.16) into

$$4T_{10}T_{20} = [(1+\Delta)Q_1 - (2+\Delta)(T_{10} + T_{20})]^2 \qquad (7.8.18)$$

$$Q_\alpha \to Q_1 = T_{10} + T_{20} + T_{3-} = m_{K-} - (3+\Delta)m_0 \qquad (7.8.19)$$

In the following, only terms up to order Δ will be retained. Let

$$\frac{1}{\sqrt{3}}(T_{20} - T_{10}) = \rho_D \cos\varphi_D = x_D \qquad (7.8.20a)$$

$$\left(\frac{2}{3} + \frac{5}{9}\Delta\right)Q_1 - \left(1 + \frac{2}{3}\Delta\right)(T_{20} + T_{10}) = \rho_D \sin\varphi_D = z_D \qquad (7.8.20b)$$

$$dT_{10}T_{20} = \frac{\sqrt{3}}{2}\left(1 - \frac{2}{3}\Delta\right)\rho_D d\rho_D d\varphi_D \qquad (7.8.21)$$

The boundary (7.8.16) becomes a circle with radius

$$\rho_D \to \rho_{D\max} = \frac{1}{3}\left(1+\frac{\Delta}{3}\right)Q_1 \tag{7.8.22}$$

Substituting (7.8.20, 21) into (7.8.15) yields

$$I_{p1} = 8\pi^2 \frac{\sqrt{3}}{2}\left(1-\frac{2}{3}\Delta\right)\int_0^{\rho_{D\max}} d\rho_D \rho_D \int_0^{2\pi} d\varphi_D \left\{ m_{K-} - 2m_0 - \left(\frac{2}{3}-\Delta\right)Q_1 + \left(1-\frac{2}{3}\Delta\right)\rho_D \sin\varphi_D \right\}$$
$$\times \left\{ \left[m_0 + \left(\frac{1}{3}-\frac{\Delta}{2}\right)Q_1 - \left(\frac{1}{2}-\frac{\Delta}{3}\right)\rho_D \sin\varphi_D \right]^2 - \frac{3}{4}\rho_D^2 \cos^2\varphi_D \right\} F(M_T) \tag{7.8.23a}$$

$$F(M_T)=1 \tag{7.8.23b}$$

The integrand is a polynomial in ρ_D and $sin\varphi_D$ and the integration is readily carried out to give

$$I_{p1} = \frac{4\sqrt{3}}{9}\pi^3 Q_1^2 \left\{ \begin{array}{l} \left[m_0 + \left(\frac{1}{3}-\frac{\Delta}{2}\right)Q_1 \right]^2 \left[m_{K-} - 2m_0 - \left(\frac{2}{3}-\Delta\right)Q_1 \right] \\ -\frac{1}{9}Q_1^2 \left[\frac{1}{8}\left(1+\frac{2}{3}\Delta\right)m_{K-} + \frac{\Delta}{12}\left(m_{K-} - 6.m_0 - 2Q_1\right) \right] \end{array} \right\} = 7.44\times10^{-4}\ \text{Gev}^{-5}$$

(7.8.24)

where the meson masses from [P1] have been used.

The value in (7.8.24) is associated with (7.7.29), in which α=1 indicating that there is one pion with mass m_+. For (7.7.30-32), this number is different. For these cases, (7.8.24) is simply modified to yield

$$\alpha = 2: \ \ \overline{K}^0 \to \pi^-\pi^+\pi^0, \quad m_0 \to m_+, \quad \Delta \to -\Delta, \quad I_{p2} = 7.52\times10^{-4}\ \text{Gev}^{-5} \tag{7.8.25a}$$

$$\alpha = 0: \ \ \overline{K}^0 \to \pi^0\pi^0\pi^0, \quad \Delta \to 0, \quad I_{p0} = 9.25\times10^{-4}\ \text{Gev}^{-5} \tag{7.8.25b}$$

$$\alpha = 3: \ \ K^- \to \pi^-\pi^+\pi^-, \quad m_0 \to m_+, \quad \Delta \to 0, \quad I_{p3} = 5.93\times10^{-4}\ \text{Gev}^{-5} \tag{7.8.25c}$$

In passing, the phase integral for a final state with two pions is also readily obtained. Corresponding to (7.8.5) is

$$I_{2\pi} = \int d^3\underline{K}_1 \int d^3\underline{K}_2 \delta(-m_K + E_1 + E_2)\delta(\underline{K}_1 + \underline{K}_2) = M_{12}\frac{\sqrt{(M_{12}^2 + m_1^2)(M_{12}^2 + m_2^2)}}{\sqrt{(M_{12}^2 + m_1^2)} + \sqrt{(M_{12}^2 + m_2^2)}}$$

$$M_{12}^2 = m_K^2 - 2(m_1^2 + m_2^2), \qquad E_1^2 = m_1^2 + K_1^2, \qquad E_2^2 = m_2^2 + K_2^2 \qquad (7.8.26)$$

Conventionally, the phase space integral corresponding to (7.8.6) is (7.8.6a) together with

$$N_1 = \frac{1}{E_1 E_2 E_3} \qquad (7.8.27)$$

[L5 (15.4]]. The integral is then reduced to an elliptical integral [K3 (7.73]] instead for the explicit result (7.8.24). Only by expanding the integral in powers of two small parameters, could this elliptical integral be approximated by an explicit expression [K3 (16.23]].

The factor (7.8.27) comes from the square of (7.8.28) below, which in its turn is based upon the assumptions that i) the pion wave function $\psi_\pi(X)$ satisfies the Klein-Gordon equation for a point particle and that ii) the amplitude $\psi_\pi(X)$ is a probability amplitude. According to [H13 §3.1] or Sec. 4.1,

$$\psi_\pi(X) = \frac{1}{\sqrt{2E_\pi \Omega_\pi}} \exp(-iE_\pi X) \qquad (7.8.28)$$

for a point pion at rest. However, no scalar or pseudoscalar point particle has so far been found in nature; a pion has extensions. Therefore, (7.8.28) has no experimental support and has to be rejected here. Consequently, the assumption ii) becomes meaningless. Furthermore, the phase space integral is a kinematical one and should not involve the wave function of the particle, as it does in (7.8.6a) and (7.8.27).

In the scalar strong interaction hadron theory, a meson is described by the product of a wave function in the laboratory space and a wave function in the relative space of the quarks [H5 (6.3)] or (7.7.8). The appearance of the relative time together with Lorentz invariance leads to a normalization of the meson wave functions different from that of the conventional probability amplitude interpretation in [H13 Sec. 3.2] or Sec. 4.2. The amplitude of the meson wave functions (3.1.5, 6) with $\underline{K} \rightarrow 0$ and (4.3.2) is a constant independent of energy E, as is seen in (7.7.11, 13) and (7.7.8) where $a_{f2(\pi+)}^{(-)}$ is independent of the energy $E_{2+}^{(-)}$ for $\underline{K}_{2+} \rightarrow 0$ and is assumed to largely remain so for small $\underline{K}_{2+}$.

This independence of energy is now harmonious with that for the amplitude of a free particle obeying the Schrödinger equation. It does not make much difference relative to the conventional phase space integral in the following section because (7.8.27) is nearly a constant here due to that the final state poins are nonrelativistic. In high energy events, however, the final state poins are relativistic and (7.8.6) will produce significant deviations from the conventional results obtained from (7.8.6a) combined with (7.8.27). This may for instance basically bear upon hadronization models.

7.8.2. $K^- \to \pi^- \pi^0 \pi^0$

1) *Decay Matrix Element and Rate.* Pursue the particular case that started with (7.8.6) for α=1. This is equivalent to putting

$$N_1 = \sum_{\beta\gamma} \left|M_{1\beta\gamma}\right|^2 \tag{7.8.29}$$

into (7.8.6a) to obatin $I_1 \Sigma_{\beta\gamma} |M_{1\beta\gamma}|^2$ in (7.8.4a). Carry out the the K_{3-} integration to remove the $\delta(-\underline{K}_{10}+...)$ function in (7.8.6a). Inserting the result (7.8.9) into (7.7.36a) leads to

$$M_{10-} = -\left[\frac{1}{2}\left(m_{K-} + E_{10}\right)\left(\left|E_{20}^{(-)}\right| - E_{3-}\right) + 6K_{10}^2 - 4K_{10}K_{20}\cos\vartheta_{20}\right] \tag{7.8.30}$$

The ϑ_{20} integration is subsequently carried out to remove the remaining δ function in (7.8.11). The result (7.8.13), where the minus sign applies, is inserted into (7.8.30). Noting (7.8.15, 5), one finds

$$I_1\left|M_{10-}\right|^2 = 8\pi^2 \int dT_{10} \int dT_{20} \left(m_0 + T_{10}\right)\left(m_0 + T_{20}\right)\left(m_{K-} - 2m_0 - T_{10} - T_{20}\right)\left|M_{10-}\right|^2 \tag{7.8.31}$$

$$M_{10-} = 2\begin{bmatrix} \frac{1}{4}\left(m_{K-} + m_0 + T_{10}\right)\left(-m_{K-} + 3m_0 + T_{10} + 2T_{20}\right) + 6m_0T_{10} + 3T_{10}^2 \\ +\left(m_{K-} - 2m_0\right)^2 - m_+^2 - 2\left(m_{K-} - m_0\right)\left(T_{10} + T_{20}\right) + 2T_{10}T_{20} \end{bmatrix} \tag{7.8.32a}$$

$$\begin{aligned} &= \frac{1}{8}\begin{bmatrix} 7m_{K-}^2 - 18m_{K-}m_0 - 9m_0^2 - \left(18m_{K-} - 7(2+\Delta)m_0\right)\Delta m_0 - \left(18m_{K-} - 46(1+\Delta)m_0\right)T_{odd} \\ + \left(22m_{K-} - 26(1+\Delta)m_0\right)T_\Delta + 23T_{odd}^2 + 19T_\Delta^2 - 26T_{odd}T_\Delta \end{bmatrix} \\ &= M_{T6}\left(m_{K-},\ m_0,\ \Delta\right) \end{aligned} \tag{7.8.32b}$$

where (7.8.8, 19) have been employed. T_{odd} refers to the odd pion and T_Δ to the difference between the T's of pions of the same kind. Here,

$$T_{odd} = T_{3-}, \qquad T_\Delta = T_{10} - T_{20} \tag{7.8.33}$$

and the dependence of M_{T6} upon the T's are supressed. The transformations (7.8.20, 21) turns (7.8.31, 32a) into the form of (7.8.23). If (7.8.32b) is used, the transformations are

$$T_\Delta = -\sqrt{3}\rho_D \cos\varphi_D, \qquad T_{odd} = \left(1 - \frac{2}{3}\Delta\right)\rho_D \sin\varphi_D - \frac{1}{3}\left(1 - \frac{\Delta}{3}\right)Q_1 \tag{7.8.34}$$

The resulting integrand is again a power series in ρ_D and $\sin\varphi_D$ up to seventh power. The integration can similarly be carried out to arrive at an algebraic expression for $I_1|M_{10-}|^2$ containing only the three masses m_{K-}, m_0 and m_+ of the form (7.8.24). This expression is however somewhat cumbersome and will not be presented here. Instead, its numerical value will enter (7.8.44) below directly.

Next consider (7.7.36b) with $\beta\gamma=-0$. Making use of the associated $\delta(-\underline{K}_{10}+...)$ functions in (7.7.35), the equivalent of (7.8.30) reads

$$M_{1-0} = \left[\frac{1}{2}\left(m_{K-} + E_{10}\right)\left(\left|E_{2+}^{(-)}\right| - E_{30}\right) + 2K_{10}^2 - 4K_{10}K_{30}\cos\vartheta_{30}\right] \tag{7.8.35}$$

$\cos\vartheta_{30}$ is given by (7.8.13) with the indices 20 replaced by 30 and using the lower sign. With the indices $3- \to 20$, the equivalent of (7.8.32b) reads

$$M_{1-0} = \frac{1}{8}\begin{bmatrix} 9m_{K-}^2 - 14m_{K-}m_0 - 39m_0^2 - \left(14m_{K-} - 9(2+\Delta)m_0\right)\Delta m_0 \\ -14\left(m_{K-} + (1+\Delta)m_0\right)T_{odd} + \left(10m_{K-} + 6(1+\Delta)m_0\right)T_\Delta - 7T_{odd}^2 - 3T_\Delta^2 + 6T_{odd}T_\Delta \end{bmatrix}$$
$$= M_{T2}\left(m_{K-},\, m_0,\, \Delta\right) \tag{7.8.36}$$

$I_1|M_{1-0}|^2$ is evaluated in a way entirely analogous to that for (7.8.31). For $\alpha\beta\gamma=10-*$ of (7.7.36c), it can simply be shown that

$$M_{10-*} = M_{1-0} \text{ with indices } {}_{30}\!\rightarrow{}_{20} \quad \text{and} \quad {}_{2+}\!\rightarrow{}_{3+} \tag{7.8.37a}$$

$$I_1\left|M_{10-*}\right|^2 = I_1\left|M_{1-0}\right|^2 = I_1\left|M_{T2}\left(m_{K-},\, m_0,\, \Delta\right)\right|^2 \tag{7.8.37b}$$

Similarly, comparison of (7.7.36d) with $\beta\gamma=-0*$ and (7.7.36a) shows that

$$M_{1-0*} = M_{10-} \text{ with indices } {}_{20}\!\rightarrow{}_{30} \quad \text{and} \quad {}_{3-}\!\rightarrow{}_{2-} \tag{7.8.38a}$$

$$I_1\left|M_{1-0*}\right|^2 = I_1\left|M_{10-}\right|^2 = I_1\left|M_{T6}\left(m_{K-},\, m_0,\, \Delta\right)\right|^2 \tag{7.8.38b}$$

Consider next (7.7.40a) which refers to (7.7.29b). Observing the $\delta(-\underline{K}_{10}+...)$ functions in (7.7.39) leads to

$$M_{100} = \left(\cos 2\vartheta_W - 1\right)$$
$$\times\left[\frac{1}{2}\left(m_{K-} + E_{10}\right)\left(\left|E_{20}^{(-)}\right| - E_{30}\right) + 2K_{20}^2 + 6K_{30}^2 - 8K_{20}K_{30}\cos\vartheta_{2030}\right] \tag{7.8.39}$$

where ϑ_{2030} is the angle between $\underline{K}_{20}$ and $\underline{K}_{30}$ and is obtained from (7.8.13) with $\vartheta_{20}\rightarrow\vartheta_{2030}$ and $_{10\rightarrow 20}$ using the upper sign. In terms of T_{odd} and T_Δ, (7.8.39) becomes

$$M_{100} = (\cos 2\vartheta_W - 1)\frac{1}{2}\begin{bmatrix} 16(1+\Delta)m_0 T_{odd} + (9(m_{K-} - \Delta m_0) - 17m_0)T_\Delta \\ + 8T_{odd}^2 + 8T_\Delta^2 - 9T_{odd}T_\Delta \end{bmatrix}$$
$$= M_{T+}(m_{K-}, m_0, \Delta) \tag{7.8.40}$$

Similarly, (7.7.40b) leads to

$$M_{100*} = (\cos 2\vartheta_W - 1)$$
$$\times\left[\frac{1}{2}(m_{K-} + E_{10})(E_{20} - |E_{30}^{(-)}|) - 6K_{20}^2 - 2K_{30}^2 + 8K_{20}K_{30}\cos\vartheta_{3020}\right] \tag{7.8.41}$$

where ϑ_{3020} is again the angle between $\underline{K}_{20}$ and $\underline{K}_{30}$ and is obtained from (7.8.13) with $\vartheta_{20}\rightarrow\vartheta_{3020}$ and $_{10\rightarrow 30}$ using the lower sign. The equivalent of (7.8.40) becomes

$$M_{100*} = (\cos 2\vartheta_W - 1)\frac{1}{2}\begin{bmatrix} 16(1+\Delta)m_0 T_{odd} + (7(m_{K-} - \Delta m_0) - 31m_0)T_\Delta \\ + 8T_{odd}^2 + 8T_\Delta^2 - 7T_{odd}T_\Delta \end{bmatrix} = M_{T-}(m_{K-}, m_0, \Delta) \tag{7.8.42}$$

In the above $M_{\alpha\beta\gamma}$ expressions, the angles ϑ_{20}, ϑ_{30} etc are those between two pions of the same kind or at least the same mass and the momentum of the odd pion is integrated over in (7.8.5) to remove the $\delta(-\underline{K}_1+...)$ functions. The choice of $+$ or $-$ signs in (7.8.13) is dictated by the signs of the momenta in the $\delta(-\underline{K}_1+...)$ functions in (7.7.34, 39, 41).

$M_{\alpha\beta\gamma}\sqrt{B_\alpha}$ corresponds to the conventional decay matrix element, denoted by A in [L5 (15.3]). There are four basic types M_{T6}, M_{T2}, M_{T+}, and M_{T-}, given by (7.8.32b, 36, 40, 42).

Summing over the two decay modes (7.7.29a, 29b), as in (7.7.43a), and noting (7.7.34, 39) and (7.8.32, 36, 37, 38, 40, 42), (7.8.4) yields

$$\Gamma(K^- \rightarrow \pi^-\pi^0\pi^0) = 2B_1 I_1 \sum_{\beta\gamma} |M_{1\beta\gamma}|^2 \tag{7.8.43a}$$

$$\sum_{\beta\gamma} |M_{1\beta\gamma}|^2 = 2|M_{T6}(\alpha_1)|^2 + 2|M_{T2}(\alpha_1)|^2 + |M_{T+}(\alpha_1)|^2 + |M_{T-}(\alpha_1)|^2$$
$$(\alpha_1) = (m_{K-}, m_0, \Delta) \tag{7.8.43b}$$

The factor 2 in (7.8.43a) comes from Σ_{fsc} in (7.8.4a) because there are two ways to combine the two neutral pions into a final state. Knowing the meson masses [P1] and using (7.2.12) and (7.8.6), the integrals in (7.8.43a) are readily evaluated, as was mentioned below (7.8.34). The result and data Γ_{exp} [P1] are

$$\Gamma\left(K^- \to \pi^-\pi^0\pi^0\right) = c_\pi^2 1.6\times 10^{-9}\ \text{ev} \tag{7.8.44}$$

$$\Gamma_{\exp}\left(K^- \to \pi^-\pi^0\pi^0\right) = 0.919\times 10^{-9}\ \text{ev} \tag{7.8.45}$$

Equating (7.8.44) and (7.8.45) gives

$$c_\pi = 0.757 \tag{7.8.46}$$

which is not far from unity in (7.7.19).

2) *Dalitz Plot Parameters.* Apart from $K_L^0 \to 3\pi^0$, the Dalitz plot distributions have been parametrized in the decay matrix element M_{PDG} according to the charged K meson chapter of [P1],

$$|M_{PDG}|^2 = C_P\left[\begin{array}{l} 1 + gu + jv + hu^2 + kv^2 + fuv + h_3u^3 + k_3v^3 + h_{21}u^2v \\ + h_{12}uv^2 + h_4u^4 + k_4v^4 + h_{13}uv^3 + h_{22}u^2v^2 + h_{31}u^3v \end{array}\right] \tag{7.8.47}$$

$$u = \frac{1}{m_+^2}\left(s_{odd} - s_0\right), \quad v = \frac{1}{m_+^2}\left(s_2 - s_1\right), \quad s_i = \left(m_K - m_i\right)^2 - 2m_K T_i. \quad i = 1,\ 2,\ \text{odd}$$

$$s_0 = \frac{1}{3}\left(m_+^2 + m_1^2 + m_2^2 + m_{odd}^2\right) \tag{7.8.48}$$

$$u = \frac{1}{m_+^2}\left[\left(m_K - m_{odd}\right)^2 - s_0 - 2m_K T_{odd}\right] \qquad v = -\frac{1}{m_+^2}2m_K T_\Delta \tag{7.8.49}$$

where C_P, g, j, f, the h's, and the k's are constants, the subscript *odd* denotes the odd pion, mentioned below (7.8.32b, 42). In [P1], only the first six terms in the brackets of (7.8.47) have been written out and the remaining terms have been introduced here. Inserting (7.8.48, 49) into (7.8.47) gives

$$|M_{PDG}|^2 = C_P \sum\nolimits_{jk} c_{jk} T_{odd}^j T_\Delta^k \tag{7.8.50}$$

where c_{jk} are constants containing the meson masses and the coefficients g, h...

These C_P and c_{jk} constants can now be identified as functions of the C_P, g, j, f, h... constants in (7.8.47) by putting (7.8.43b) equal to (7.8.50);

$$C_P \sum\nolimits_{jk} c_{jk} T_{odd}^j T_\Delta^k = \sum\nolimits_{\beta\gamma} \left|M_{1\beta\gamma}\right|^2 \tag{7.8.51}$$

Making use of (7.8.32b, 36, 40, 42), these coefficients are readily obtained starting from the highest power T terms, which have the simplest forms. g and C_P are the last ones to be determined. Because T_1 in (7.8.51) can be identified with either of the both neutral pions, the sign of T_Δ in (7.8.50) can also be reversed. Taking the sum over these both possible assignments leads to that the odd powered v terms in (7.8.47) vanish. The relevants results are

$$g = 0.597, \qquad h = 0.276 \tag{7.8.52}$$

A direct comparison of (7.8.52) with data is not exact because the phase space used in obtaining (7.8.47) is that of the conventional type (7.8.6a) together with (7.8.27), which can written as

$$\frac{1}{N_1} = E_1E_2E_3 = (m_0 + T_1)(m_0 + T_2)(m_+ + T_3) \tag{7.8.53}$$

Let T_3 refer to the odd pion. Making use of (7.8.49, 33) and (7.8.7, 18) type of relations, (7.8.53) can be put in the normalized form

$$\frac{N_1(T \to 0)}{N_1} = \left(1 + \frac{T_{odd}}{m_0(1+\Delta)}\right)\left(1 - \frac{2T_{odd}}{m_{K-} - m_0(1+\Delta)} + \frac{T_{odd}^2 - T_\Delta^2}{(m_{K-} + m_0(1+\Delta))^2}\right) \tag{7.8.54}$$

The effect of the difference in phase space is compensated for by replacing (7.8.51) by

$$C_P \sum_{j+k=0}^{4} c_{jk} T_{odd}^j T_\Delta^k = \frac{N_1(T \to 0)}{N_1} \sum_{\beta\gamma} |M_{1\beta\gamma}|^2 \tag{7.8.55}$$

The right side now contains T's to seventh power. Because the higher power T terms are small, only T terms up to fourth power will be kept and be compared to those on the left of (7.8.55). Instead of (7.8.52), it is found that

$$g \to g' = 0.599, \qquad h \to h' = 0.264 \tag{7.8.56}$$

These values differ little from those in (7.8.52) because (7.8.53, 54) are essentially constants due to the nonrelativstic motion of the pions. The experimental results [P1] are

$$g_{\exp} = 0.638, \qquad h_{\exp} = 0.051 \tag{7.8.57}$$

3) *Origin of the g Parameters.* Comparison of (7.8.56) and (7.8.57) shows that g' is rather close to data while h' far exceeds the experimental value. This discrepancy is presently attributed to that the higher order pion momenta $\underline{K}$ terms in (7.7.13) and (7.7.24) type of expressions have been neglected, as was considered below (7.7.24). The approximation to keep only terms linear in $\underline{K}$ in (7.7.24) type of relations corresponds to results in (7.8.43b) to order K^2 or T. These correspond to g, which multiplies T_{odd} in (7.8.47). On the other hand, h multiplies $T^2 \propto K^4$ there and this kind of terms has not been included in (7.7.24) type of

expressions. The mentioned approximation also affects the decay rate (7.8.43, 44) to that order. The remaining coefficients in (7.8.47) are associated with still higher powered T terms and are hence unreliable. They are not presented here because there are no data on them.

The g parameter gives a fairly large asymmetry in the Dalitz diagram for $K^- \rightarrow \pi^- \pi^0 \pi^0$. This asymmetry is related to the asymmetry in momenta $\underline{K}_{20}$ and $\underline{K}_{3-}$ of the both pions decayed from the gauge boson W_I^+ shown in (7.7.24) type of expressions, inasmuch as these quantities refer to the same order of expansion in $\underline{K}$.

As was mentioned in the second paragraph following (7.7.24), the last asymmetry is the result of the interaction between the current of the odd pion with momentum $\underline{K}_{3-}$ in laboratory space and the current of the quarks consituting the other pion in the relative space $\underline{y}$ of the quarks. Both of these pions are decay products of the gauge boson. Other expressions equivalent to (7.7.24) are (7.7.26) and the second term in the braces of (7.7.27). The same kind of interaction also prevails there but with the roles of the odd and non-odd pions interchanged. These considerations agree with that g is associated with T_{odd} in (7.8.47, 48).

Putting

$$F(M_T) = \sum_{\beta\gamma} \left| M_{1\beta\gamma} \right|^2 \tag{7.8.58}$$

in (7.8.23a), the integrand in the braces times $F(M_T)$ describes the intensities of the decay as a function of ρ_D and φ_D, which corresponds to a surface above the Dalitz diagram. Normalizing the height of this surface at the origin ρ_D=0 to unity, this surface is sketched in Table 7.8.1.

Table 7.8.1. Decay intensities as function of ρ_D and φ_D in a Dalitz diagram for $K^- \rightarrow \pi^- \pi^0 \pi^0$. ρ_{Dmax} is given by (7.8.22).

ρ_D/ρ_{Dmax}	$\varphi_D=\pi/2$	$\pi/3, 2\pi/3$	$\pi/6, 5\pi/6$	$0, \pm\pi$	$-\pi/6, -5\pi/6$	$-\pi/3, -2\pi/3$	$-\pi/2$
0	1	1	1	1	1	1	1
0.2	0.86	0.887	0.955	1.04	1.12	1.17	1.18
0.4	0.768	0.828	0.983	1.17	1.31	1.4	1.41
0.6	0.723	0.824	1.08	1.38	1.58	1.67	1.68
0.8	0.729	0.875	1.25	1.67	1.93	2.0	1.98
1.0	0.784	0.979	1.43	2.04	2.36	2.38	2.32

The z axis in the Dalitz diagram corresponds to $\varphi_D=\pm\pi/2$ according to (7.8.20b) and depends linearly upon T_{odd}, as does the gu term in (7.8.47). The intensities are symmetric about this axis. The change of the values along this axis clearly shows that the g>0 trend via (7.8.47-49). The intensities rise more steeply near the boundary of the Dalitz diagram and are less reliable. This is due to that ρ_D there corresponds to large $\underline{K}$ or T values for which the neglected higher order pion momenta $\underline{K}$ terms mentioned in the beginning of this subsection become large.

Experimentally, h_{exp} is small compared to g_{exp} in (7.8.57). Dropping the higher order terms in (7.8.47), the surface corresponding to Table 7.8.1 is largely a circular disc with radius ρ_{Dmav} tilted around the x_D axis of (7.8.20b) with a slope given by g_{exp}. This disc is warped slightly by $h_{exp}\neq0$ and this warping is more pronounced at large ρ_D values. For small ρ_D values, this warping is small and the angular variations of the intensities should be sinusoidal. This is verified by the ρ_D/ρ_{Dmav}=0.2 line on Table 7.8.1. For ρ_D/ρ_{Dmav}=0.8 and 1, the effect of h_{exp} becomes greater but more importantly that of the neglected $\underline{K}$ terms mentioned above become large. These greatly distort the sinusoidal angular variations stemming from g_{exp}, as is seen on the last two lines of Table 7.8.1.

7.8.3. $K^0 \to \pi^+\pi^-\pi^0$

This decay consists of the three modes shown in (7.7.30) and the amplitude is given by (7.7.43b). These amplitudes are closely related to the amplitudes (7.7.43a) for $K^- \to \pi^-\pi^0\pi^0$ considered in §7.8.2. By (7.3.27a), the first two terms on the right of (7.7.43b) are each √2 times that of (7.7.34), apart from some changes of indices in (7.7.34, 35).

Similarly, the last term in (7.7.43b) is by (7.3.26b, 27b) the same as −1/√2 times (7.7.41) with appropriate changes of the indices in (7.7.41, 42);

$$S_{fi}\left(\overline{K}^0 \to \pi^0 Z \to \pi^0\pi^+\pi^-\right) = -\frac{1}{\sqrt{2}} S_{fi}\left(K^- \to \pi^- Z \to \pi^-\pi^+\pi^-\right) \tag{7.8.59}$$
$$\text{with } K^- \to \overline{K}^0 \text{ and the indices } 1 - \to 10$$

Here, the cross terms mentioned below (7.8.5) enter due to the following. The parametrization (7.8.47) only contains kinematical quantities; the charges or flavors do not enter. There, the measurable quantities g and C_P, which are proportional to the decay rate (7.8.1), depend only upon the masses and energies of the pions. Therefore, the charge symbols in S_{fi} in (7.8.1) serves only to designate the masses and momenta of the different pion species, namely, $\pi_\pm$ and π_0. The effects of flavors have been taken ínto account in the process from (7.1.1) to S_{fi} in (7.7.34, 39, 41).

From this viewpoint, the + and − subscripts in (7.8.59) can be dropped, bearing in mind that the quantities referred to by these indices are associated with the heavier pion with mass m_+. Inspection of (7.7.41, 42) together with (7.8.59) shows that (7.7.42a) and (7.7.42b) now become proportional to each other. This leads to that the first two terms in the braces of (7.7.41) also become proportional to each other. These both terms can now be summed to form one term containing the factor $(cos2\vartheta_W-1)$. The last two terms in the braces of the so-modified (7.7.41) together with the similarly modified (7.7.42c, 42d) behave entirely analogously. The result is

$$S_{fi}\left(\overline{K}^0 \to \pi^0 Z \to \pi^0\pi^+\pi^-\right) = -\frac{1}{\sqrt{2}} A_2$$
$$\times\begin{Bmatrix} \delta\left(-m_{K0} + E_{10} + \left|E_2^{(-)}\right| + E_3\right)\delta\left(-\underline{K}_{10} - \underline{K}_2 + \underline{K}_3\right)M_2 + \\ + \delta\left(-m_{K0} + E_{10} + E_2 + \left|E_3^{(-)}\right|\right)\delta\left(-\underline{K}_{10} + \underline{K}_2 - \underline{K}_3\right)M_{2*} \end{Bmatrix} \tag{7.8.60}$$

$$M_2 = \left(\cos 2\vartheta_W - 1\right)\left[\left(m_{K0} + E_{10}\right)\left(\left|E_2^{(-)}\right| - E_3\right) + 4\underline{K}_{10}\left(\underline{K}_2 - 3\underline{K}_3\right)\right] \tag{7.8.61a}$$

$$M_{2*} = (\cos 2\vartheta_W - 1)\left[(m_{K0} + E_{10})\left(E_2 - \left|E_3^{(-)}\right|\right) + 4\underline{K}_{10}(3\underline{K}_2 - \underline{K}_3)\right] \tag{7.8.61b}$$

These equations are the same as (7.7.39, 40) apart from a factor of 2 and a switch of the roles of the charged and neutral pions, with the accompanying changes in the subscripts. This is expected because $Z \to \pi^0\pi^0$ and $Z \to \pi^+\pi^-$ behave the same way kinematically, differing only in the masses and momenta.

Knowing the the three amplitudes in (7.7.43b), (7.8.1) and (7.8.4a) for α=2 gives the equivalent of (7.8.43);

$$\Gamma\left(\overline{K}^0 \to \pi^0\pi^+\pi^-\right) = B_2 I_2 \sum_{\beta\gamma} \left|M_{2\beta\gamma}\right|^2 \tag{7.8.62}$$

$$\sum_{\beta\gamma} \left|M_{2\beta\gamma}\right|^2 = 8\left|M_{T6}(\alpha_2)\right|^2 + 8\left|M_{T2}(\alpha_2)\right|^2 + 2\left|M_{T+}(\alpha_2)\right|^2 + 2\left|M_{T-}(\alpha_2)\right|^2$$
$$(\alpha_2) = (m_{K0},\ m_+,\ -\Delta) \tag{7.8.63}$$

The factor 2 in (7.8.43a) is absent in (7.8.62) because Σ_{fsc} in (7.8.4a) is 1 here. Carrying out the computations following the same procedure as in §7.8.2 1) leads to the equivalent of (7.8.44),

$$\Gamma\left(\overline{K}^0 \to \pi^0\pi^+\pi^-\right) = c_\pi^2\, 3.08 \times 10^{-9} \text{ ev} \tag{7.8.64}$$

The amplitude $S_{fi}(K^0 \to \pi^+\pi^-\pi^0)$ comes from the *c.c.* term in S_{m3} of (7.1.8b) and are related to (7.3.27b) via (7.3.29). The result is the same;

$$\Gamma\left(K^0 \to \pi^0\pi^+\pi^-\right) = \Gamma\left(\overline{K}^0 \to \pi^0\pi^+\pi^-\right) \tag{7.8.65}$$

Therefore, (7.8.64) can be compared to [P1]

$$\Gamma_{\exp}\left(K_L^0 \to \pi^+\pi^-\pi^0\right) = 1.6 \times 10^{-9} \text{ ev} \tag{7.8.66}$$

Equating (7.8.64) to (7.8.66) gives

$$c_\pi = 0.721 \tag{7.8.67}$$

which is not too far from (7.8.46).

Replacing the right of (7.8.51) by (7.8.63) gives the equivalent of (7.8.52, 56),

$$g = 0.634, \qquad h = 0.289, \qquad k = 0.189 \tag{7.8.68a}$$

$$g \to g' = 0.636, \qquad h \to h' = 0.275, \qquad k \to k' = 0.184 \tag{7.8.68b}$$

[P1] gives

$$g_{\exp} = 0.678, \qquad h_{\exp} = 0.076, \qquad k_{\exp} = 0.0099 \tag{7.8.69}$$

It's relation to (7.8.68b) is entirely analogous to that between (7.8.57) to (7.8.56) for $K^- \to \pi^- \pi^0 \pi^0$, discussed in §7.8.2 3). These two decays behave remarkbly similar with respect to g in that

$$\frac{g\left(K^- \to \pi^- \pi^0 \pi^0\right)}{g\left(\overline{K}^0 \to \pi^0 \pi^+ \pi^-\right)} \approx 0.941 \approx \frac{g_{\exp}\left(K^- \to \pi^- \pi^0 \pi^0\right)}{g_{\exp}\left(\overline{K}^0 \to \pi^0 \pi^+ \pi^-\right)} \tag{7.8.70}$$

The considerations of the effects of neglected higher order terms $\underline{K}$ terms in §7.8.2 3) also hold here; k and h behave similarly.

7.8.4. $K^0 \to 3\pi^0$

This decay differs from those in §7.8.2-3 in that all the three final state pions have the same mass and charge. This degeneracy leads to that i) the final state *1* pion is interchangeable with the final state *2* pion and the final state *3* pion and ii) there no odd pion.

Consider i) at first. The final state *1* π^0 comes from $\chi_{(rp=11)}{}^*$ in S_{m3} of (7.1.8). The flavor indices *11* refer to a $\overline{u}u$ quark pair. The final state *3* π^0 in $S_{m\pi}$ of (7.7.3) that is multiplied to $Z\cos 2\vartheta_W/\cos\vartheta_W$ is also associated with a $\overline{u}u$ pair originally. On the other hand, the final state *3* π^0 in $S_{m\pi}$ that is multiplied to $-Z/\cos\vartheta_W$ is associated with a $\overline{d}d$ pair. An exchange of the final state *1* π^0 with the originally $\overline{u}u$ flavored final state *3* π^0 leads to no change in the internal or flavor content. But such an exchange with the $\overline{d}d$ flavored final state *3* π^0 leads a different flavor distribution. The same holds for an exchange of the final state *1* π^0 with the corresponding final state *2* π^0.

From this flavor interchangeability viewpoint, the final state *3* or *2* π^0 wave functions associated with $Z\cos 2\vartheta_W/\cos\vartheta_W$ and those with $-Z/\cos\vartheta_W$ are different and can therefore not be simplified to one π^0 wave function multiplied by $(\cos 2\vartheta_W - 1)/\cos\vartheta_W$, as is the case in (7.7.39, 40). This last case refers to $K^- \to \pi^- \pi^0 \pi^0$ in which the final state *1* pion is π^- which is not interchangeable with the π^0's from the Z decay. For the present case, (7.7.39, 40) together with (7.3.26b, 27b) go over to

$$S_{fi}\left(\overline{K}^0 \to \pi^0 Z \to \pi^0 \pi^0 \pi^0\right)$$

$$= A_0 \begin{Bmatrix} \delta\left(-m_{K0} + E_{10} + \left|E_{20}^{(-)}\right| + E_{30}\right)\delta\left(-\underline{K}_{10} + \underline{K}_{20} - \underline{K}_{30}\right)M_{000(\bar{u}u)} \\ + \delta\left(-m_{K0} + E_{10} + \left|E_{20}^{(-)}\right| + E_{30}\right)\delta\left(-\underline{K}_{10} + \underline{K}_{20} - \underline{K}_{30}\right)M_{000(\bar{d}d)} \\ + \delta\left(-m_{K0} + E_{10} + E_{20} + \left|E_{30}^{(-)}\right|\right)\delta\left(-\underline{K}_{10} - \underline{K}_{20} + \underline{K}_{30}\right)M_{000(\bar{u}u)*} \\ + \delta\left(-m_{K0} + E_{10} + E_{20} + \left|E_{30}^{(-)}\right|\right)\delta\left(-\underline{K}_{10} - \underline{K}_{20} + \underline{K}_{30}\right)M_{000(\bar{d}d)*} \end{Bmatrix} \tag{7.8.71}$$

$$M_{000(\bar{u}u)} = -\left(\cos 2\vartheta_W/\sqrt{2}\right)\left[\left(m_{K0} + E_{10}\right)\left(\left|E_{20}^{(-)}\right| - E_{30}\right) + 4\underline{K}_{10}\left(\underline{K}_{20} - 3\underline{K}_{30}\right)\right]$$
$$M_{000(\bar{d}d)} = \left(1/\sqrt{2}\right)\left[\left(m_{K0} + E_{10}\right)\left(\left|E_{20}^{(-)}\right| - E_{30}\right) + 4\underline{K}_{10}\left(\underline{K}_{20} - 3\underline{K}_{30}\right)\right]$$
$$M_{000(\bar{u}u)*} = -\left(\cos 2\vartheta_W/\sqrt{2}\right)\left[\left(m_{K0} + E_{10}\right)\left(E_{20} - \left|E_{30}^{(-)}\right|\right) + 4\underline{K}_{10}\left(3\underline{K}_{20} - \underline{K}_{30}\right)\right]$$
$$M_{000(\bar{d}d)*} = \left(1/\sqrt{2}\right)\left[\left(m_{K0} + E_{10}\right)\left(E_{20} - \left|E_{30}^{(-)}\right|\right) + 4\underline{K}_{10}\left(3\underline{K}_{20} - \underline{K}_{30}\right)\right] \tag{7.8.72}$$

Turning to ii) in the beginning of this section, there are now three possibilities to assign the odd pion to, namely to final state *1*, *2* and *3* pions. Assigning the odd pion to final state *1* leads to M_{T+} and M_{T-} type of expressions in (7.8.40, 42). When the final state *3* or *2* pion is odd, the decay amplitude is then akin to (7.7.34) and to M_{T6} and M_{T2} type of expressions in (7.8.32b, 36).

Following the same procedure that led to (7.8.43) and (7.8.62, 63), one finds

$$\Gamma\left(\overline{K}^0 \to 3\pi^0\right) = 6B_0 I_0 \sum_{\beta\gamma}\left|M_{0\beta\gamma}\right|^2 \tag{7.8.73}$$

$$\sum_{\beta\gamma}\left|M_{0\beta\gamma}\right|^2 = \frac{1}{2}\left(\cos^2 2\vartheta_W + 1\right)\left[2\left|M_{T6}(\alpha_0)\right|^2 + 2\left|M_{T2}(\alpha_0)\right|^2 + \frac{\left|M_{T+}(\alpha_0)\right|^2 + \left|M_{T-}(\alpha_0)\right|^2}{\left(\cos 2\vartheta_W - 1\right)^2}\right]$$
$$(\alpha_0) = (m_{K0},\ m_0,\ 0) \tag{7.8.74}$$

The factor 6 in (7.8.73) comes from Σ_{fsc} in (7.8.4a) because there are 3!=6 ways to combine the three π^0's to a final state. Working out the integral in (7.8.73) gives

$$\Gamma\left(\overline{K}^0 \to 3\pi^0\right) = c_\pi^2 6.69 \times 10^{-9} \text{ ev} \tag{7.8.75}$$

$$\Gamma_{\exp}\left(K_L^0 \to 3\pi^0\right) = 2.69 \times 10^{-9} \text{ ev} \tag{7.8.76}$$

where data [P1] is also shown. Comparison between them yields

$$c_\pi = 0.634 \tag{7.8.77}$$

which is also not too far from those in (7.8.67) and (7.8.46).

The parametrization (7.8.47) does not hold for $K_L^0 \to 3\pi^0$. It has been replaced by

$$|M_S(3\pi^0)|^2 = C_S[1 + hR_S^2] \tag{7.8.78}$$

$$\begin{aligned} R_S^2 &= \frac{4}{m_+^4}\left[s_0^2 - \frac{1}{3}(s_1s_2 + s_2s_3 + s_3s_1)\right] \\ &= \frac{4}{m_+^4}\left[\begin{array}{l}\left(\frac{1}{3}m_{K0}^2 + m_0^2\right)^2 - (m_{K0} - m_0)^4 + \frac{4}{3}m_{K0}(m_{K0} - m_0)^2(m_{K0} - 3m_0) \\ -\frac{2}{3}m_{K0}^2(m_{K0} - 3m_0)^2 + \frac{2}{3}m_{K0}^2(T_1^2 + T_2^2 + T_3^2)\end{array}\right] \end{aligned} \tag{7.8.79}$$

of [S13] where C_S is another constant. The g parameter is absent in this decay. Because no precise result is expected, (7.8.74) is expanded in powers of T and approximated by truncating the series at order T^2. Equating the so-approximated (7.8.74) to (7.8.78) leads to

$$h = 0.114, \qquad h' = 0.092, \qquad h_{\exp} = -0.005 \tag{7.8.80}$$

where data [P1] has been appended. The consideration given in §7.8.2 3) also holds here for h in (7.8.80) and Γ in (7.8.75).

7.8.5. $K^- \to \pi^-\pi^+\pi^-$

This decay is similar to $K^0 \to 3\pi^0$ in that the masses of the three final state pions degenerate but differs from it in that the charges are not. As was pointed out in the beginning of §7.8.3, only masses and momenta of the pions are effectively present in the decay amplitude S_{fi}, the physical effects of the flavor or charges have already been taken care of in the derivations leading up to S_{fi}. This amplitude is employed in (7.8.1) to obtain the decay rate which is then compared to the parametrizartion (7.8.47), which is also of pure kinematical nature and independent of the charges. If the masses of the pions degenerate, there is actually no way to identify the odd pion.

At first this mass degeneracy is ignored and the odd pion is identified by its charge, here the π^+. This decay amplitude has been given by (7.7.41, 42). Inserting this into (7.8.1) and follow the steps leading to (7.8.43) yields

$$\Gamma(K^- \to \pi^-\pi^+\pi^-) = 2B_3I_3\sum\nolimits_{\beta\gamma}|M_{3\beta\gamma}|^2 \tag{7.8.81}$$

$$\begin{aligned} &\sum\nolimits_{\beta\gamma}|M_{3\beta\gamma}|^2 = 4(\cos^2 2\vartheta_W + 1)\left[|M_{T6}(\alpha_3)|^2 + |M_{T2}(\alpha_3)|^2\right] \\ &(\alpha_3) = (m_{K0},\ m_+,\ 0) \end{aligned} \tag{7.8.82}$$

The factor 2 in (7.8.81) again derives from from Σ_{fsc} in (7.8.4a) because there are two ways to combine the two π^-'s into a final state. The equivalent of (7.8.44-46, 52, 56, 57) is

$$\Gamma\left(K^- \to \pi^-\pi^-\pi^+\right) = c_\pi^2\, 2.45\times 10^{-9}\ \text{ev} \tag{7.8.83}$$

$$\Gamma_{\exp}\left(K^- \to \pi^-\pi^-\pi^+\right) = 2.96\times 10^{-9}\ \text{ev} \tag{7.8.84}$$

$$c_\pi = 1.1 \tag{7.8.85}$$

$$\begin{matrix} g = 0.834, & g' = 0.856, & g_{\exp} \approx -0.216 \\ h = 0.31, & h' = 0.297, & h_{\exp} \approx -0.092 \end{matrix} \tag{7.8.86}$$

g' completely disagrees with data. Also, $c_\pi>1$ and differs considerably from c_π ranging from 0.634 to 0.757 for the other three decays.

These disagreements may be regarded as the result of ignoring the mass degeneracy of the three pions. If, however, the charges are completely ignored, as is the case mentioned above (7.8.60), then this case reduces to one of the same type as $\bar{K}^0\to 3\pi^0$. In this case, there is no way to identify an odd pion and the linear slope parameter g drops out or in practice

$$g = g' = 0 \tag{7.8.87}$$

as is seen in (7.8.78). This result also conflicts with data in (7.8.86).

Therefore, a compromise interpretation is adopted here. Let the odd pion be among the decay products of Z, then it must be π^+. This identification led to the form (7.8.82), which in its turn led to (7.8.83-6) above. Kinematically, however, the odd pion can also be assigned to the pion not belonging to the decay products of Z. In this case, it is the final state $1\ \pi^-$. The decay amplitude for such a case is kinematically the same as that given by (7.8.71, 72). The jusitification leading to these expressions, based upon the interchangeability of the three π^0's, is absent and is not needed here, since the charges of the decay products of Z are kept here. In addition, it is noted that the final state $1\ \pi^-$ is associated with a creation operator. The final state $2\ \pi^-$, multiplied by $\cos 2\vartheta_W$ in (7.8.71, 72), adapted to the present case is associated with an annihilation operator and the final state $3\ \pi^-$, not multiplied by $\cos 2\vartheta_W$, adapted to the present case is associated with a creation operator. For the viewpoint of interchangeability of the final state $1\ \pi^-$ with the final state $2\ \pi^-$ and the final state $3\ \pi^-$ are different.

This amplitude will now also contribute to (7.8.82) and enlarge it with the terms of the form given by the last term in the brackets of (7.8.74). The result is

$$\sum_{\beta\gamma}\left|M_{3\beta\gamma}\right|^2 = 4\left(\cos^2 2\vartheta_W + 1\right)\left[\left|M_{T6}(\alpha_3)\right|^2 + \left|M_{T2}(\alpha_3)\right|^2 + \frac{\left|M_{T+}(\alpha_3)\right|^2 + \left|M_{T-}(\alpha_3)\right|^2}{\left(\cos 2\vartheta_W - 1\right)^2}\right] \tag{7.8.88}$$

which replaces (7.8.82). The results ensuing from this expression replacing (7.8.83-6) are

$$\Gamma\left(K^- \to \pi^-\pi^-\pi^+\right) = c_\pi^2 5.11\times 10^{-9}\ \text{ev} \tag{7.8.89}$$

$$c_\pi = 0.761 \tag{7.8.90}$$

$$g = -0.427, \qquad g' = -0.43, \qquad h = 0.572, \qquad h' = 0.565 \tag{7.8.91}$$

c_π is now in line with the three other values 0.634, 0.721 and 0.757. g' has now the correct sign and g_{exp} now lies between this g' value and g'=0 of (7.8.87).

The deviations of g' and h' from data in (7.8.86) are presently again attributed to the neglect of higher powered pion momenta terms discussed in §7.8.2-3. The intensities of this decay as a function of ρ_D and φ_D are obtained from I_3 and (7.8.88) just like that done for $K^- \to \pi^-\pi^0\pi^0$ at the end of §7.8.2-3. The results corresponding to Table 7.8.1 is Table 7.8.2 below.

Table 7.8.2. Decay intensities as function of ρ_D and φ_D in a Dalitz diagram for $K^- \to \pi^-\pi^+\pi^-$. ρ_{Dmax} is given by (7.8.22).

ρ_D/ρ_{Dmax}	φ_D=π/2	π/3,2π/3	π/6,5π/6	0,±π	−π/6,−5π/6	−π/3,−2π/3	−π/2
0	1	1	1	1	1	1	1
0.2	1.14	1.13	1.09	1.03	0.98	0.941	0.927
0.4	1.37	1.33	1.25	1.14	1.03	0.95	0.922
0.6	1.67	1.62	1.49	1.31	1.14	1.02	0.98
0.8	2.05	1.98	1.8	1.55	1.31	1.15	1.1
1.0	2.51	2.43	2.2	1.87	1.55	1.33	1.26

The values along z axis in the Dalitz diagram shows that on the average $g' < 0$ as in (7.8.91). The large h' values is indicated by that actually $g' > 0$ in the lower half plane which trend is overcome by the stronger $g' < 0$ trend in the upper half plane. The very large h' value is also presently attributed to the neglected higher order pion momenta $\underline{K}$ terms mentioned in §7.8.2-3.

The considerations below Table 7.8.1 also hold here. Again, the expected sinusoidal angular variations are seen for the ρ_D/ρ_{Dmav}=0.2 and 0.4 lines in Table 7.8.2. However, such variations are also seen for ρ_D/ρ_{Dmav}=0.8 and 1, contrary to expectation. This may be due to that some of the neglected and large T^2 or K^4 terms happen to cancel out.

Chapter 8

STRONG DECAY OF MESON

The first three sections of this chapter largely reproduce Sec. 4-9 of [H13]. Sec. 8.4 is mostly taken over from [H14], with some assignments and interpretations changed.

In the last chapter, weak interactions are mediated by SU(3) gauge fields in laboratory space. Their strength is determined by the Fermi constant G (7.4.29a) and F_W^2 (7.5.22a). In Chapter 6, electromagnetic interactions are based upon U(1) gauge field in quark coordinate space. The associated strength is given by the fine structure constant $e^2/4\pi$. In this chapter, strong decays will take place via perturbations of the intrameson interaction potential $\Phi_m(x_I,x_{II})$ in for instance (2.3.22). Their strength turns out to be given by g_s^8 below, where g_s^2 is the quark-antiquark scalar strong coupling constant in (2.1.2, 4).

The 0ZI rule obeying decay of a vector meson into two slow pseudoscalar mesons $V\rightarrow PP$ will be considered. The sister processes $VV\rightarrow VV$, $P\rightarrow PP$ and $P\rightarrow VV$ are included at the same time when allowed energetically. However, these PP or VV can be a virtual pair. When coupled to electromagnetic fields, the vector parts VV associated with PP arises. $\pi^0\rightarrow\gamma\gamma$ via intermediary virtual $\pi^+\pi^-$ and $\rho^+\rho^-$ belongs to this category according to the present theory and will be treated.

8.1. MESON EQUATIONS INVOLVING FOUR QUARKS

8.1.1. Coupled Four Quark Equations

In the simplest OZI rule obeying $V\rightarrow PP$ or $P\rightarrow VV$, four quarks are involved. Quark A located at x_I having flavor p in z_I space and antiquark B located at x_{II} with antiflavor $\bar{r}$ in z_{II} space form the initial V or P. The space time equations assigned to them have been given by (2.1.1-4). When internal functions are included, (2.1.1, 3) are replaced by the form (2.3.11) with q_{op} and V_{PB} put to zero. The wave functions of these quarks are then

$$\chi_{A\dot{b}}(x_I)\xi_A^p(z_I),\quad \psi_A^a(x_I)\xi_A^p(z_I),\quad \chi_B^{\dot{f}}(x_{II})\xi_{Br}(z_{II}),\quad \psi_{B\dot{e}}(x_{II})\xi_{Br}(z_{II}) \qquad (8.1.1)$$

In addition, a quark-antiquark pair comprising quark C of flavor q and antiquark D of antiflavor $\bar{q}$ is created from the vacuum. The initial meson V or P, denoted by AB, decays into two mesons PP or VV, denoted by AD and CB or AB and CD. Quasi-classically, one may try to give this created pair CD the same status as the quarks AB in the initial meson. In this case, their wave functions corresponding to (8.1.1) would be

$$\chi_{C\dot{b}}(x_{III})\xi_C^q(z_{III}),\ \psi_C^a(x_{III})\xi_C^q(z_{III}),\ \chi_D^f(x_{IV})\xi_{Dq}(z_{IV}),\ \psi_{D\dot{e}}(x_{IV})\xi_{Dq}(z_{IV}) \quad (8.1.2)$$

which satisfy (2.3.11) type of equations and (2.1.1, 3) with $A\to C$, $B\to D$, $I\to III$, and $II\to IV$. In addition, (2.1.2, 4) needs be further modified to include contributions from the C and D quarks. For instance, (2.1.2) would be replaced by

$$\Box_I V_{SBCD}(x_I)=\tfrac{1}{2}g_s^2\left(\psi_B^b(x_I)\chi_{Bb}(x_I)+\psi_C^b(x_I)\chi_{Cb}(x_I)+\psi_D^b(x_I)\chi_{Db}(x_I)+c.c.\right) \quad (8.1.3)$$

Internal functions are not included on the right side of (8.1.3) for the reasons given below (2.3.19). Here, quarks interact only pair-wise; genuine three quark or four quark interactions have been neglected.

Even with these simplified interactions, the four quark problem posed by four sets of equations of the form (2.1.1 ,2) which are coupled by nonlinear terms like those on the right side of (8.1.3) would be very complicated even before generalizations of the form (2.2.1-3). The complications are mainly due to the increase of the number of independent variables, from x_I and x_{II} to x_I, x_{II}, x_{III}, and x_{IV}.

It will however turn out that such a doubling of the number of independent variables, hence also the ensuing complications, may be side-stepped if one recalls the basic steps in $K^+\to\mu^+\nu_\mu$ in the last chapter and try to taken them over for applications in this chapter.

8.1.2. Intermediary Scalar Potential Perturbation

Ostensibly, $K^+\to\mu^+\nu_\mu$ involves four fermions, the u and $\bar{s}$ quarks, μ^+ and ν_μ and would require four set of coordinates to describe them, similar to the $V\to PP$ case in the previous paragraph. Nevertheless, only two sets, X and x obtained from x_I and x_{II} according to (3.1.3a), appear in the calculations. This is due to the division of a semileptonic decay into two stages in §7.1.1. In stage I, K^+, described in terms of X and x, decays into a gauge boson $W^-(X)$ and a vacuum pseudoscalar meson state explained below (7.3.18). In stage II, $W^-(X)$ in its turn decays into μ^+ and ν_μ. In this last stage, K^+ has already "disappeared" in X space so that this X can now be "reused" by the μ^+ and ν_μ wave functions. This is possible because μ^+ and ν_μ have been treated as different states of the same particle related to each other by SU(2) gauge transformations, as can be seen from (7.1.12, 19). Here, the Glashow model mentioned below (7.1.17c) and (7.1.21) has been assumed. Therefore, only X, the "reused" X, is needed in the wave functions describing them. In this way, extra and independent coordinates for μ^+ and ν_μ do not enter the decay formalism.

This procedure to describe semi-leptonic decay will be analogously taken over here. The difference is largely the replacement of the intermediate gauge boson $W(X)$ by $\Phi_{Im4}(x_I,x_{II})$, a perturbation of the strong scalar intrameson potential $\Phi_m(x_I,x_{II})$ mentioned in the beginning of this chapter. In stage I, V or P decays into a virtual intermediate state Φ_{Im4J}, to be defined in (8.2.5) ff below, and a *vacuum meson state*, analogous to the vacuum pseudoscalar meson state defined below (7.3.18). In stage II, the pertubational Φ_{Im4J} decays into PP or VV whose wave functions depend upon the "reused" x_I and x_{II}. Again PP or VV nsidered as different states of the same meson so they share the same x_I and x_{II}.

In this manner, the extra independent coordinates x_{III} and x_{IV} in (8.1.2) no longer need to enter the formalism below. Since the internal coordinates z_{III} and z_{IV} are on equal footing with x_{III} and x_{IV} as is implied in §2.3.1 (see also (2.3.15, 18)), they also do not appear in the following formalism.

8.1.3. Mixed Quark Wave Functions and Equations

In accordance with the approach put forth in the last paragraph, I shall introduce wave functions of the type (8.1.2, 1) for the $\bar{q}q$ pair DC created from vacuum by the following generalization of (8.1.1),

$$\begin{aligned}
&\chi_{A\dot{b}}(x_I)\xi_A^{p}(z_I) \to \chi_{A\dot{b}}(x_I)\xi_A^{p}(z_I) + \chi_{C\dot{b}}(x_I)\xi_C^{q}(z_I) \\
&\psi_A^{a}(x_I)\xi_A^{p}(z_I) \to \psi_A^{a}(x_I)\xi_A^{p}(z_I) + \psi_C^{a}(x_I)\xi_C^{q}(z_I)
\end{aligned} \tag{8.1.4a}$$

$$\begin{aligned}
&\chi_B^{f}(x_{II})\xi_{Br}(z_{II}) \to \chi_B^{f}(x_{II})\xi_{Br}(z_{II}) + \chi_D^{f}(x_{II})\xi_{Dq}(z_{II}) \\
&\psi_{B\dot{e}}(x_{II})\xi_{Br}(z_{II}) \to \psi_{B\dot{e}}(x_{II})\xi_{Br}(z_{II}) + \psi_{D\dot{e}}(x_{II})\xi_{Dq}(z_{II})
\end{aligned} \tag{8.1.4b}$$

In this ansatz, χ_C, ψ_C, χ_D, and ψ_D are considered to be small during stage I of the decay so that they can be regarded as small perturbations of the quark wave functions for A and B generating the initial meson V or P. Later, in stage II, their amplitudes have grown to be of the same magnitude as those for A and B so that the right sides of (8.1.4a) become mixed quark states and the right sides of (8.1.4b) represent mixed antiquark states.

With (8.1.4), the quark wave equations (2.1.1-4) including internal functions as in (2.3.11) are generalized to

$$\begin{aligned}
&\partial_I^{a\dot{b}}\left(\chi_{A\dot{b}}(x_I)\xi_A^{p}(z_I) + \chi_{C\dot{b}}(x_I)\xi_C^{q}(z_I)\right) - iV_{BD}(x_I)\left(\psi_A^{b}(x_I)\xi_A^{p}(z_I) + \psi_C^{b}(x_I)\xi_C^{q}(z_I)\right) \\
&= im_{Aop}(z_I)\left(\psi_A^{b}(x_I)\xi_A^{p}(z_I) + \psi_C^{b}(x_I)\xi_C^{q}(z_I)\right)
\end{aligned} \tag{8.1.5a}$$

$$\begin{aligned}
&\partial_{I\dot{c}b}\left(\psi_A^{b}(x_I)\xi_A^{p}(z_I) + \psi_C^{b}(x_I)\xi_C^{q}(z_I)\right) - iV_{BD}(x_I)\left(\chi_{A\dot{c}}(x_I)\xi_A^{p}(z_I) + \chi_{C\dot{c}}(x_I)\xi_C^{q}(z_I)\right) \\
&= im_{Aop}(z_I)\left(\chi_{A\dot{c}}(x_I)\xi_A^{p}(z_I) + \chi_{C\dot{c}}(x_I)\xi_C^{q}(z_I)\right)
\end{aligned} \tag{8.1.5b}$$

$$\begin{aligned}
&\partial_{II\dot{e}f}\left(\chi_B^{f}(x_{II})\xi_{Br}(z_{II}) + \chi_D^{f}(x_{II})\xi_{Dq}(z_{II})\right) - iV_{AC}(x_{II})\left(\psi_{B\dot{e}}(x_{II})\xi_{Br}(z_{II}) + \psi_{D\dot{e}}(x_{II})\xi_{Dq}(z_{II})\right) \\
&= im_{Bop}(z_{II})\left(\psi_{B\dot{e}}(x_{II})\xi_{Br}(z_{II}) + \psi_{D\dot{e}}(x_{II})\xi_{Dq}(z_{II})\right)
\end{aligned} \tag{8.1.6a}$$

$$\partial_{II}^{d\dot{e}}\left(\psi_{B\dot{e}}(x_{II})\xi_{Br}(z_{II})+\psi_{D\dot{e}}(x_{II})\xi_{Dq}(z_{II})\right)-iV_{AC}(x_{II})\left(\chi_B^d(x_{II})\xi_{Br}(z_{II})+\chi_D^d(x_{II})\xi_{Dq}(z_{II})\right)$$
$$=im_{Bop}(z_{II})\left(\chi_B^d(x_{II})\xi_{Br}(z_{II})+\chi_D^d(x_{II})\xi_{Dq}(z_{II})\right) \tag{8.1.6b}$$

$$\Box_I V_{BD}(x_I)=\frac{1}{2}g_s^2\left[\left(\psi_B^b(x_I)+\psi_D^b(x_I)\right)\left(\chi_{Bb}(x_I)+\chi_{Db}(x_I)\right)+c.c.\right] \tag{8.1.7}$$

$$\Box_{II} V_{AC}(x_{II})=\frac{1}{2}g_s^2\left[\left(\psi_A^b(x_{II})+\psi_C^b(x_{II})\right)\left(\chi_{Aa}(x_{II})+\chi_{Ca}(x_{II})\right)+c.c.\right] \tag{8.1.8}$$

Consistent with (2.1.2, 4), the internal functions in (8.1.4) have been dropped in (8.1.7, 8). This is due to that the V's are scalar potential in space time and are independent of internal functions which characterize the mass and charge of the quarks. This point has been discussed below (2.3.19).

8.1.4. Construction of Coupled Meson Equations

In Chapter 2, wave equations for a single isolated meson (2.3.19, 23) were constructed by first multiplying the wave equations (2.1.1, 2) for quark A with (2.1.3, 4) for antiquark B. The same procedure will now be repeated using (8.1.5, 7) and (8.1.6, 8) instead. The initial meson AB can decay into either i) AD and CB or ii) AB and CD. In each case, only three of the four mesons AB, AD, CB and CD enter; the fourth can be dropped. In the following, only case i) will be treated explicity. The interchange $B\leftrightarrow D$ among the decay products turn case i) results into those for case ii). Repeating the three steps in Chapter 2 leads to the following equations replacing (2.3.19, 23):

$$\partial_I^{a\dot{b}}\partial_{II\dot{e}f}\begin{pmatrix}\chi_{AB\dot{b}}^f(x_I,x_{II})\xi_r^p(z_I,z_{II})+\chi_{AD\dot{b}}^f(x_I,x_{II})\\ \times\xi_q^p(z_I,z_{II})+\chi_{CB\dot{b}}^f(x_I,x_{II})\xi_r^q(z_I,z_{II})\end{pmatrix}=\left(\Phi_{m4}(x_I,x_{II})-m_{2op}(z_I,z_{II})\right)$$
$$\times\left(\psi_{AB\dot{e}}^a(x_I,x_{II})\xi_r^p(z_I,z_{II})+\psi_{AD\dot{e}}^a(x_I,x_{II})\xi_q^p(z_I,z_{II})+\psi_{CB\dot{e}}^a(x_I,x_{II})\xi_r^q(z_I,z_{II})\right) \tag{8.1.9a}$$

$$\partial_{Ib\dot{c}}\partial_{II}^{d\dot{e}}\begin{pmatrix}\psi_{AB\dot{e}}^b(x_I,x_{II})\xi_r^p(z_I,z_{II})+\psi_{AD\dot{e}}^b(x_I,x_{II})\\ \times\xi_q^p(z_I,z_{II})+\psi_{CB\dot{e}}^b(x_I,x_{II})\xi_r^q(z_I,z_{II})\end{pmatrix}=\left(\Phi_{m4}(x_I,x_{II})-m_{2op}(z_I,z_{II})\right)$$
$$\times\left(\chi_{AB\dot{c}}^d(x_I,x_{II})\xi_r^p(z_I,z_{II})+\chi_{AD\dot{c}}^d(x_I,x_{II})\xi_q^p(z_I,z_{II})+\chi_{CB\dot{c}}^d(x_I,x_{II})\xi_r^q(z_I,z_{II})\right) \tag{8.1.9b}$$

$$V_{BD}(x_I)V_{AC}(x_{II})\to\Phi_{m4}(x_I,x_{II})=\Phi_{mAB}(x_I,x_{II})+\Phi_{1m4}(x_I,x_{II}) \tag{8.1.10a}$$

$$\Box_I \Box_{II} \Phi_{mAB}(x_I, x_{II}) = -\tfrac{1}{2} g_s^4 \operatorname{Re} \psi_{AB}^{a\dot{b}}(x_I, x_{II}) \chi_{AB\dot{b}a}^{*}(x_I, x_{II}) \tag{8.1.10b}$$

$$\Box_I \Box_{II} \Phi_{1m4}(x_I, x_{II})$$
$$= -\frac{1}{4} g_s^4 \begin{bmatrix} \psi_{AD}^{a\dot{b}}(x_I, x_{II}) \chi_{CB\dot{b}a}^{*}(x_I, x_{II}) + \psi_{CB}^{a\dot{b}}(x_I, x_{II}) \chi_{AD\dot{b}a}^{*}(x_I, x_{II}) \\ + \psi_{CB}^{a\dot{b}}(x_I, x_{II}) \chi_{CB\dot{b}a}^{*}(x_I, x_{II}) + \psi_{AD}^{a\dot{b}}(x_I, x_{II}) \chi_{AD\dot{b}a}^{*}(x_I, x_{II}) \\ + \psi_{AB}^{a\dot{b}}(x_I, x_{II}) \chi_{CB\dot{b}a}^{*}(x_I, x_{II}) + \psi_{CB}^{a\dot{b}}(x_I, x_{II}) \chi_{AB\dot{b}a}^{*}(x_I, x_{II}) \\ + \psi_{AB}^{a\dot{b}}(x_I, x_{II}) \chi_{AD\dot{b}a}^{*}(x_I, x_{II}) + \psi_{AD}^{a\dot{b}}(x_I, x_{II}) \chi_{AB\dot{b}a}^{*}(x_I, x_{II}) \end{bmatrix} + c.c. \tag{8.1.10c}$$

Because $\xi^{p}{}_{r}$, $\xi^{p}{}_{q}$ and $\xi^{q}{}_{r}$ are independent of each other, (8.1.9) splits up into three equations, one for each ξ. Upon application of (2.3.21), one obtains

$$\partial_I^{a\dot{b}} \partial_{II}^{f\dot{e}} \chi_{AB\dot{b}f}(x_I, x_{II}) = \left(M_{AB}^2 - \Phi_{m4}(x_I, x_{II})\right) \psi_{AB}^{a\dot{e}}(x_I, x_{II}) \tag{8.1.11a}$$

$$\partial_{I\dot{c}b} \partial_{II\dot{e}d} \psi_{AB}^{b\dot{e}}(x_I, x_{II}) = \left(M_{AB}^2 - \Phi_{m4}(x_I, x_{II})\right) \chi_{AB\dot{c}d}(x_I, x_{II}) \tag{8.1.11b}$$

$$(8.1.11) \text{ with } AB \to AD \tag{8.1.12}$$

$$(8.1.11) \text{ with } AB \to CB \tag{8.1.13}$$

For the simplest mass operator form (2.3.25, 26), $M_{AB}{}^2$ takes on the form (2.4.1). All other product equations drop out in the same manner as the corresponding ones considered in Sec. 2.2.

Equations (8.1.10-13) form a set of seven equations for the seven variables ψ_{AB}, χ_{AB}, $AB \to AD$, $AB \to CB$ and Φ_{m4}. The various meson wave functions for AB, AD and CB are coupled via the Φ's of (8.1.10). Equations (8.1.10b, 11) include the zeroth order terms which account for the initial meson AB. Evidently, only the first two terms in the bracket of (8.1.10c) will contribute to Φ_{1m4}; the remaining terms drop out because the associated mesons do not appear among the decay products.

8.2. Action for Strong Decay and Decay Amplitude

8.2.1. First Order Relations

The development of $V \to P\gamma$ of Sec. 6.4 and $K^+ \to L^+ \nu_L$ of Sec. 7.3 can be largely taken over here. The confinement potential in (8.1.10a) is given by (3.2.8),

$$\Phi_{mAB} = d_m / r + d_{m0} \tag{8.2.1}$$

where (3.2.20) has been employed. Further, (3.2.8c) drops out via (3.2.5b) and (4.2.7, 8). ψ_{AB} and χ_{AB} are expanded in the form (6.4.1-3) modified by the extra subscripts *AB* attached to the wave functions. Insert the so-modified (6.4.1-3) into (8.1.11), multiply (8.1.11a) by $\chi^{*}_{0AB\dot{e}a}$ and (8.1.11b) by $\psi^{*d\dot{c}}_{0AB}$, add them together, and integrate over *X* and *x*. Noting (8.1.10a), the first order terms can by analogy to (6.4.6) and (7.3.3-5) be written as

$$S'_{mABd} = S'_{mABs} \tag{8.2.2}$$

$$S'_{mABd} = \int d^4 X dx^4 \left[\partial_I^{a\dot{b}} \left(a_{J0}^{(1)} \left(X^0 \right) / a_{J0} \right) \chi^{*}_{0AB\dot{e}a} \partial_{II}^{f\dot{e}} \chi_{0AB\dot{b}f} + \partial_{I\dot{c}a} \left(a_{J0}^{(1)} \left(X^0 \right) / a_{J0} \right) \psi^{*d\dot{c}}_{0AB} \partial_{II\dot{e}d} \psi^{a\dot{e}}_{0AB} \right] \tag{8.2.3}$$

$$S'_{mABs} = - \int dX^4 dx^4 \Phi_{1m4} \left(\chi^{*}_{0AB\dot{e}a} \psi^{a\dot{e}}_{0AB} + c.c. \right) \tag{8.2.4}$$

8.2.2. Quantization

Analogous to (7.3.6, 7b), the initial and final states in stage I are taken to be

$$\left| i \right\rangle = \left| AB_J \left(E_{J0}, K_J = 0 \right) \right\rangle, \qquad \left\langle f \right| = \left\langle 0,\, \Phi_{1m4J} \right| \tag{8.2.5}$$

where *AB* stands for the initial meson and *J* its spin. Φ_{1m4J} stands for the virtual intermediate state consisting of $\Phi_{1m4}(x_I, x_{II})$ together with the observable properties, i. e., spin, four momentum and flavors of the initial meson. It plays the same role as the intermediate gauge boson W_{rp} does in (7.3.7b). While W_{rp} decays into a pair of leptons, Φ_{1m4J} decays into a pair of mesons. Analogous to the implication below (7.3.9), Φ_{1m4J} is also implicitly understood to pick out the wave functions of the final pair.

The initial meson AB_J decays into a *vacuum meson state*, represented by <0| in (8.2.5), similar to that described below (7.3.19). The wave function of this state described by (7.3.20) holds for J=0 or an initial pseudoscalar meson. For J=1 or an initial vector meson, this form is to be replaced by that of (4.3.3). Further, Ω_{wvac} in (7.3.20) becomes

$$\Omega_{wvac} \rightarrow \Omega_{svac}$$

representing the volume occupied by the vacuum meson state associated with strong interaction, denoted by the subscript *s*.

Sandwiching (8.2.4a) between <f| and |i> of (8.2.5) and follow the same steps that led to (6.4.14, 15) and (7.3.15) yields

$$\langle f|S'_{mABd}|i\rangle = -\frac{i}{2}E_{J0}S_{fiJ}\Omega\int d^4x\psi_J^2(r) \tag{8.2.6a}$$

$$S_{fi} = \langle f|a^*_{J0}a^{(1)}_{J0}(X^0\to\infty)|i\rangle \tag{8.2.6b}$$

Insert now (8.2.4b) between <f| and |i> of (8.2.5). One obtains the analog of (7.3.21),

$$\langle f|S'_{mABs}|i\rangle = 4\int d^4X\,d^4x\sqrt{\frac{\Omega}{\Omega_{svac}}}\psi_J^2(r)\Phi_{1m4}(X,x)\exp(-iE_{J0}X^0) \tag{8.2.7}$$

where (6.4.1) modified according to §8.2.1, (3.2.5b) and the second of (3.4.2b) and of (3.4.3) have been noted. Combining (8.2.3, 6, 7) leads to the decay amplitude for the initial meson with spin J,

$$S_{fiJ} = i\frac{8}{E_{J0}}\frac{1}{\Omega}\int d^4X\,d^3\underline{x}\sqrt{\frac{\Omega}{\Omega_{svac}}}\psi_J^2(r)\Phi_{1m4}(X,x)\exp(-iE_{J0}X^0)\Big/\int d^3\underline{x}\psi_J^2(r) \tag{8.2.8}$$

This result can also be obtained starting from either (8.1.11a) or (8.1.11b), without the addition operation mentioned above (8.2.2).

8.2.3. Perturbed Scalar Potential Equation

The perturbed potential Φ_{1m4} in the virtual intermediate state Φ_{1m4J} must satisfy the condition

$$\Phi_{1m4} \ll \Phi_{mAB} \tag{8.2.9}$$

in (8.1.10a) to allow for a perturbational calculation. Putting r in (8.2.1) to r_0 and r_1 of (4.7.3, 4) and using Table 5.2, one gets the estimates

$$\Phi_{mAB} \approx d_m/r_J + d_{m0} \approx \begin{pmatrix}0.75+0.24\\0.19+0.24\end{pmatrix}\text{Gev}^2 \quad \text{for } J = \begin{pmatrix}0\\1\end{pmatrix} \tag{8.2.10}$$

During early stage I, (8.2.9) is satisfied by virtue of the perturbational character of χ_C, ψ_C, χ_D, and ψ_D in (8.1.4) because these wave functions take part in the generalizations of the type (2.2.1) to arrive at terms on the right of (8.1.10c) but not (8.1.10b). During late stage II, the pseudoscalar mesons in V decay are sufficiently far apart thet they can be regarded as free mesons having amplitude $1/\sqrt{\Omega}\to 0$ according to (4.3.2) together with (3.5.11, 19, 23, 26). Here, we are limited to slowly moving mesons, as are the cases treated below. With the last mentioned expressions, (8.1.10c) becomes small and (8.2.9) holds. Similarly, the final decay products, the twin photons in $\pi^0\to\rho^+\rho^-\to\gamma\gamma$ mentioned in the beginning of this chapter, are described by wave functions $\propto 1/\sqrt{\Omega_r}\to 0$ according to (8.4.23) below. Therefore, (8.2.9) is analogously satisfied during the late stage II in this case.

In the intermediary stage, i.e., during late stage I and early stage II, the amplitudes of perturbational χ_C, ψ_C, χ_D, and ψ_D may no longer be small. Further, the newly formed *PP* in *V* decay are still close to each other so that they interact strongly and the $1/\sqrt{\Omega}\to 0$ dependence of (4.3.2) is replaced by $1/\sqrt{\Omega_c}$=finite of (4.2.8) according to Sec. 4.5. In this intermediary stage, the bracket of (8.1.10c) may therefore not be small. To conform to (8.2.9), use is made of the hypothesis of unified strong and electromagnetic coupling $g_s^4 \propto e$ of (8.3.11) below and is thus a small parameter.

According to the end of Sec. 8.1, (8.1.10c) becomes

$$\Box_I \Box_{II} \Phi_{1m4} = -\frac{1}{4} g_s^4 \left(\psi_{AD}^{a\dot{b}} \chi_{CB\dot{b}a}^* + \psi_{CB}^{a\dot{b}} \chi_{AD\dot{b}a}^* + \psi_{AD}^{*b\dot{a}} \chi_{CB\dot{a}b} + \psi_{CB}^{*b\dot{a}} \chi_{AD\dot{a}b} \right) \tag{8.2.11}$$

At large times indicated by (8.2.6b), the pseudoscalar mesons *AD* and *CB* in *V* decay are sufficiently apart so that they may be regarded as free meson having wave function of the form (3.1.5, 9). Any other more general form than these would complicate the calculations greatly and practically prohibit any analytical solution.

Analogous to the choice of negative energy for the massive lepton in $K^+\to L^+\nu_L$, shown in (7.4.11b), negative energy solutions will be chosen for *AD*. In this case, (8.2.11) becomes

$$\Box_I \Box_{II} \Phi_{1m4} = -\frac{1}{2} g_s^4 \left(\psi_{0jKAD}(\underline{x}) \chi_{0jKCB}^*(\underline{x}) - \underline{\psi}_{jKAD}(\underline{x}) \underline{\chi}_{jKCB}^*(\underline{x}) + \psi \leftrightarrow \chi \right)$$
$$\times \exp\left(i\left(E_{jKAD} + E_{jKCB}\right) X^0 + i\left(\underline{K}_{jAD} - \underline{K}_{jCB}\right)\underline{X} \right) \tag{8.2.12}$$

where *j* denotes the spin of the decay product mesons. Further, only the first and last terms in the parentheses of (8.2.11) contribute because we anticipate

$$E_{J0} = E_{jKAD} + E_{jKCB} \tag{8.2.13a}$$

from (8.2.8) after integration over X^0. The middle two terms there will produce zero after such an integration. The roles of these two groups of terms in (8.2.8) are reversed if negative energy solutions are chosen for *CB* instead.

Since the right side of (8.2.12) is separated in *X* and $\underline{x}$, the left side must be so also;

$$\Phi_{1m4}(X, \underline{x}) = \Phi_{1jK}(\underline{x}) \exp\left(i\left(E_{jKAD} + E_{jKCB}\right) X^0 + i\left(\underline{K}_{jAD} - \underline{K}_{jCB}\right)\underline{X} \right) \tag{8.2.14}$$

We further anticipate

$$\underline{K}_{jAD} = \underline{K}_{jCB} = \underline{K}_j \tag{8.2.13b}$$

from the integral over $\underline{X}$ in (8.2.8). Inserting (8.2.13, 14) into (8.2.12) and making use of (3.1.3) and (3.5.7), one finds

$$\left(\Delta+\frac{1}{4}E_{J0}^2\right)^2\Phi_{1jK}(\underline{x})=\frac{1}{2}g_s^4\Xi_{jK}(\underline{x}) \tag{8.2.15a}$$

$$\begin{aligned}\Xi_{jK}(\underline{x})&=-\psi_{0jKAD}(\underline{x})\chi_{0jKCB}^*(\underline{x})+\underline{\psi}_{jKAD}(\underline{x})\underline{\chi}_{jKCB}^*(\underline{x})\\&-\chi_{0jKAD}(\underline{x})\psi_{0jKCB}^*(\underline{x})+\underline{\chi}_{jKAD}(\underline{x})\underline{\psi}_{jKCB}^*(\underline{x})\end{aligned} \tag{8.2.15b}$$

8.2.4. Green's Function and Decay Amplitude

Green's function for (8.1.15) satisfies

$$\left(\Delta+\frac{1}{4}E_{J0}^2\right)^2 G(\underline{x},\underline{x}')=\delta(\underline{x}-\underline{x}') \tag{8.2.16a}$$

$$G(\underline{x},\underline{x}')=-\frac{1}{4\pi E_{J0}}\sin\left(\frac{1}{2}E_{J0}|\underline{x}-\underline{x}'|\right) \tag{8.2.16b}$$

In the $E_{J0}\to 0$ limit, (8.2.16b) reduces to

$$G(\underline{x},\underline{x}')=-\frac{|\underline{x}-\underline{x}'|}{8\pi} \tag{8.2.17}$$

which is the confining Green's function (3.2.6b) for the initial meson AB. Note that homogeneous solutions of the type

$$d_m/|\underline{x}-\underline{x}'|,\quad d_{m0}\quad \text{or}\quad d_{m2}|\underline{x}-\underline{x}'|^2$$

in (3.2.8a) cannot be added to the right side of (8.2.16b) due to the presence of E_{J0} in (8.2.16a). Application of (8.2.16) to (8.2.15) yields

$$\Phi_{1jK}(\underline{x})=-\frac{g_s^4}{8\pi E_{J0}}\int d^3\underline{x}'\,\Xi_{jK}(\underline{x}')\sin\left(\frac{1}{2}E_{J0}|\underline{x}-\underline{x}'|\right) \tag{8.2.18}$$

For j=0, we limit ourselves to nonrelativistic pseudoscalar mesons. The products of the triplet wave functions in (8.2.15b) are of order ε_0^2<<1 according (3.5.1, 11, 23) and can be dropped.

The remaining terms in (8.2.15b) depend only upon r and are flavor independent according to (3.5.3) and Sec. 4.4,

$$\Xi_{0K}(\underline{x}) \approx \Xi_{0K}(r) \approx \Xi_{00}(r) = 2\psi_0^2(r) \tag{8.2.19}$$

For j=1, it will be shown in §8.4.5 below that only the vector part of the virtual intermediate $\rho^+\rho^-$ in $\pi^0 \to \gamma\gamma$ will couple to $\gamma\gamma$. Therefore, products of the singlet wave functions in (8.2.15b) can be dropped so that

$$\Xi_{1K}(\underline{x}) = \underline{\psi}_{1KAD}(\underline{x})\underline{\chi}^*_{1KCB}(\underline{x}) + \underline{\chi}_{1KAD}(\underline{x})\underline{\psi}^*_{1KCB}(\underline{x}) \tag{8.2.20a}$$

These wave functions have been estimated in §8.4.7 and are such that (8.2.20a) turns out to be independent of the angles in relative space so that it can be written as

$$\Xi_{1K}(r) = \underline{\psi}_{1KAD}(r)\underline{\chi}^*_{1KCB}(r) + \underline{\chi}_{1KAD}(r)\underline{\psi}^*_{1KCB}(r) \tag{8.2.20b}$$

For these both cases, the angular integrations in (8.2.18) can be carried out to yield

$$\Phi_{1jK}(r) = -\frac{4g_s^4}{E_{J0}^4}\left\{\begin{aligned} &\frac{\cos R}{R}\int_0^R dR'R'(\sin R' - R'\cos R')\Xi_{jK}(r') + \sin R\int_0^R dR'R'\sin R'\Xi_{jK}(r') \\ &-\cos R\int_R^\infty dR'R'\cos R'\Xi_{jK}(r') + \frac{\sin R}{R}\int_R^\infty dR'R'(\cos R' + R'\sin R')\Xi_{jK}(r') \end{aligned}\right\} \tag{8.2.21a}$$

$$R = E_{J0}r/2 \tag{8.2.21b}$$

In the limit of $E_{J0} \to 0$ in (8.2.15, 16),

$$\sin\left(\frac{1}{2}E_{J0}|\underline{x} - \underline{x}'|\right) \to \frac{1}{2}E_{J0}|\underline{x} - \underline{x}'| \tag{8.2.22}$$

and (8.2.18) and (8.2.21) reduce to the forms (3.2.12) and (3.2.17), respectively.

Inserting (8.2.14) into (8.2.8), carrying out the X^0 integration and noting (8.2.21) leads to the decay amplitude

$$S_{fiJ} = i\frac{16\pi}{E_{J0}}\frac{1}{\Omega}\delta(E_{jKAD} + E_{jKCB} - E_{J0})I_{Kj}\overline{\Phi}_{1jK}(r) \tag{8.2.23a}$$

$$I_{Kj} = \int d^3\underline{X}\exp[i(\underline{K}_{jCB} - \underline{K}_{jAD})\underline{X}] \tag{8.2.23b}$$

$$\overline{\Phi}_{1jK}(r) = \int d^3\underline{x}\sqrt{\frac{\Omega}{\Omega_{svac}}}\psi_J^2(r)\Phi_{1jK}(r) \bigg/ \int d^3\underline{x}\psi_J^2(r) \tag{8.2.23c}$$

8.3. Rate For $V \to PP$ and the Unified Strong and Electromagnetic Coupling Hypothesis

8.3.1. Decay Rate for Slow Pseudoscalar Mesons

For $V \to PP$, the decay rate is given by the form (6.4.24) or (7.5.1, 2);

$$\Gamma(V \to PP) = \sum\nolimits_{\text{final states}} \left|S_{fi}\right|^2 / T_d = \frac{\Omega_f^2}{(2\pi)^6} \int d^3\underline{K}_{0AD} d^3\underline{K}_{0CB} \left|S_{fi}\right|^2 / T_d \tag{8.3.1}$$

Inserting (8.2.23) into (8.3.1) leads to the analog of (7.5.5),

$$\Gamma(V \to PP) = \frac{2}{\pi^5}\frac{1}{E_{10}^2}\frac{1}{\Omega^2} C_{CG}^2 \int d^3\underline{K}_{0CB} I_{PP} \tag{8.3.2a}$$

$$I_{PP} = \int d^3\underline{K}_{0AD} I_{K0} I_{K0}^* \left|\Omega_f \overline{\Phi}_{10K}\right|^2 \delta\left(E_{0KAD} + E_{0KCB} - E_{10}\right) \tag{8.3.2b}$$

where the equivalent of (7.5.6) has been used. Because such decays are of strong interaction nature, the branching ratio determined by the isospin Clebsch-Gordan coefficient C_{CG} has been inserted. Carrying out the integration in (8.2.23b) yields

$$I_{K0} = (2\pi)^3 \delta\left(\underline{K}_{0CB} - \underline{K}_{0AD}\right) = I_{K0}^* \tag{8.3.3}$$

Substituting this expression into one of the two I_K's in (8.3.2b) and carrying out the integration yields

$$I_{PP} = (2\pi)^3 \Omega \left|\Omega_f \overline{\Phi}_{10K}\right|^2 \delta\left(E_{0KAD} + E_{0KCB} - E_{10}\right) \tag{8.3.4a}$$

together with (8.2.13b). Ω is the same as I_K of (8.2.23b) observing (8.2.13b) or

$$\Omega = \int d^3\underline{X} \exp\left(i(\underline{0})\underline{X}\right) = \left(2\pi\delta(0)\right)^3 \tag{8.3.4b}$$

which is a large volume.

Next, (4.3.2) with $\Omega=\Omega_f$, is inserted into (8.2.19), which in its turn is inserted into (8.2.21) with $J=1$, $j=0$ and the slow pseudoscalar meson approximation $\underline{K}=0$. The integrations can be evaluated to yield

$$\Phi_{100}(r)=\frac{g_s^4}{\pi}\frac{d_m^3}{E_{10}^4}\frac{1}{\Omega}$$

$$\times\left[2\frac{\exp(-2\beta_1 R)}{\left(1+4\beta_1^2\right)^2}\left(1+\frac{8\beta_1}{1+4\beta_1^2}\frac{1}{R}\right)-\frac{\cos R}{R}\frac{16\beta_1}{\left(1+4\beta_1^2\right)^3}-\sin R\frac{4\beta_1}{\left(1+4\beta_1^2\right)^2}\right] \quad (8.3.5a)$$

$$\beta_1 = d_m/E_{10}, \qquad R = E_{10}r/2 \quad (8.3.5b)$$

In the limit $r\to 0$, Φ_{100} is a constant analogous to (3.2.18). For large r, $\Phi_c(r\to\infty)\to 0$, contrary to the confining potential of (3.2.19).

Inserting (4.3.3) and (8.3.5) into (8.2.23c) and carrying out the integration yields

$$\Omega_f\overline{\Phi}_{100}=\frac{g_s^4}{8\pi E_{10}}\sqrt{\frac{\Omega}{\Omega_{svac}}}I_{100} \quad (8.3.6a)$$

$$I_{100}=\left(\frac{\beta_1}{1+4\beta_1^2}\right)^3\left[\frac{16}{243}\left(1+10\beta_1^2\right)-\frac{32\beta_1^6}{\left(1+\beta_1^2\right)^5}\left(2-11\beta_1^2-40\beta_1^4+21\beta_1^6\right)\right] \quad (8.3.6b)$$

Substituting (3.5.31) together with (8.2.13b) into the δ function in (8.3.4a) and making use of (7.5.13) leads to

$$\delta(E_{0KAD}+E_{0KCB}-E_{10})=\delta(K_0-K_{00})\frac{1}{E_{10}K_0}\sqrt{E_{00AD}^2+K_0^2}\sqrt{E_{00CB}^2+K_0^2} \quad (8.3.7a)$$

$$K_{00}=\frac{1}{2E_{10}}\sqrt{\lambda_3}, \qquad \lambda_3=E_{10}^4+E_{00AD}^4+E_{00CB}^4-2E_{10}^2\left(E_{00AD}^2+E_{00CB}^2\right)-2E_{00AD}^2-2E_{00CB}^2 \quad (8.3.7b)$$

where E_{00AD} and E_{00CB} are the masses of the mesons AD and CB, respectively.

Insert now (8.3.6a, 7a) into (8.3.4a), which in its turn is substituted into (8.3.2a). After performing the integration using (8.2.13b), one obtains

$$\Gamma(V\to PP)=f_s^2C_{CG}^2I_{100}^2\frac{K_{00}}{E_{10}^5}\sqrt{E_{00AD}^2+K_{00}^2}\sqrt{E_{00CB}^2+K_{00}^2} \quad (8.3.8a)$$

$$f_s^2 = \frac{g_s^8}{\pi^3} \frac{1}{\Omega_{svac}} \tag{8.3.8b}$$

for slowly moving pseudoscalar mesons *AD* and *CB*.

8.3.2. Decay Rate and Application to Data

Let

$$\frac{1}{\Omega_{svac}} = M_s^3 \tag{8.3.9}$$

so that (8.3.8b) is written as

$$f_s^2 = \frac{g_s^8}{\pi^3} M_s^3 \tag{8.3.10a}$$

The decay rate (8.3.8) together with (8.3.6b, 7b) and the meson masses E_{10}, E_{00AD} and E_{00CB} can be applied to data [P1]. As was mentioned in the beginning of §8.1.4, the type ii) decay $AB \to AB+CD$, where CD represents π^0 here, becomes the same as the type i) decay $AB \to AD+CB$ treated above if $B \leftrightarrow D$ in the decay products. Further, (8.3.8) holds only for very slowly moving decay products, as was pointed out below (8.2.18).

The strong decay constant f_s^2 of (8.3.10a) is unknown and is to be fixed by $\Gamma(\phi \to K^+K^-)$ in Table 8.1 below, which collects the results obtained. The result is

$$f_s^2 = 5.6 \text{ Gev}^3 \tag{8.3.10b}$$

Table 8.1. Decay rates and parameters for $V \to PP$. ε_0^2 here denotes the product of the ε_0's in column 2 for the both decay products and is a measure of departure from the assumption (8.2.19) in which the motion of the decay products is neglected. C_{CG} is the isospin Clebsch-Gordon coefficient. I_{100} is given by (8.3.6b) and holds strictly for $\varepsilon_0=0$ and will be modified by a factor $1 \pm \varepsilon_0^2$ for small ε_0^2. The predicted decay rates are obtained from (8.3.8, 6b, 7b, 10b) together with the masses from [P1]. They hold for small ε_0^2 only, i.e., for D^* and ϕ. They do not hold for K^* and ρ decays, the small ε_0^2 predictions are given in parentheses.

	$\varepsilon_0=K_0/E_{00}$	ε_0^2	C_{CG}^2	I_{100}	Γ(8.3.8) (kev)	Γ(data) [P1] (kev)
$D^{*+} \to D^+\pi^0$	0.021(D) 0.284(π)	0.00596	1/3	0.00448	0.0115	<40.2
$D^{*+} \to D^0\pi^+$	0.021(D) 0.284(π)	0.00596	2/3	0.00448	0.0245	<88.7

$D^{*0}\to D^0\pi^0$	0.023(*D*) 0.32(π)	0.00736	1/3	0.00452	0.013	<1300
$D^{*0}\to D^+\pi^-$	0.023(*D*) 0.32(π)	0.00736	2/3	0.00452	0	0
$\phi\to K^+K^-$	0.257	0.066	1/2	0.1613	Input	2.19×10^3
$\phi\to K^0K^0$	0.221	0.0448	1/2	0.1613	1.89×10^3	1.51×10^3
$K^{*+}\to K^+\pi^0$	0.586(*K*) 2.14(π)	1.26	1/3			
				0.218	(25.2×10^3)	50.8×10^3
$K^{*+}\to K^0\pi^+$	0.574(*K*) 2.05(π)	1.17	2/3			
$K^{*0}\to K^0\pi^0$	0.583(*K*) 2.15(π)	1.25	2/3			
				0.22	(24.8×10^3)	50.7×10^3
$K^{*0}\to K^+\pi^-$	0.59(*K*) 0.209(π)	1.23	1/3			
$\rho^+\to\pi^+\pi^0$	$\approx$2.6	6.91	1			
				0.238	(62.5×10^3)	150×10^3
$\rho^0\to\pi^+\pi^-$	2.61	6.81	1			

The decay $B^*\to B\pi$ is forbidden by energy conservation and have also not been observed. For decays involving π^0 and ρ^0, only $\overline{u}u$ or $\overline{d}d$ enters in the formalism but each will be treated as a π^0. The effect of small ε_0^2 on the decay rates enters via modification of I_{100}^2 in (8.3.8a) and is of the order of $(1\pm\varepsilon_0^2)^2-1 \approx \pm2\varepsilon_0^2$, as is implied in the approximation leading to (8.2.19). The predictions in Table 8.1 can thus be off by this amount.

Although $\varepsilon_0^2<<1$ for D^* decay, the data [P1] only provides upper limit. However, data [P1] also gives the ratio

$$\Gamma\left(D^{*+}\to D^0\pi^+\right)/\Gamma\left(D^{*+}\to D^+\pi^0\right)=2.205\pm0.052$$

which is fairly close to the predicted ratio 2.13 obtained from the first two lines of Table 8.1. The predicted K^* decay rates for small ε_0^2 is about half the values of data. The same rates for ρ decay is about 40% the value of data. These discrepancies are presently attributed to the breakdown of (8.2.19) and hence (8.3.6b) and (8.3.8a) for sizable ε_0^2 values. The predictions are thus consistent with data but are not conclusive. A determination of the D^* decay rates would provide an important test of the present theory.

8.3.3. Ratio of $V \to P\gamma$ and $V \to P\pi$ Decay Rates

Although the D^* decay rates have not been measured, except for upper limits, data [P1] exists for ratios between the various D^* decays. Ratios of the predicted decay rates for $V \to P\gamma$ in Table 6.1 to those for $V \to P\pi$ in Table 8.1 are given on Table 8.2 below together with data.

Table 8.2. Ratios of $V \to P\gamma$ and $V \to P\pi^0$ decay rates. The estimated values are ratios of the order of magnitude estimates of $\Gamma(V \to P\gamma)$ of Table 6.1 to the predicted $\Gamma(V \to P\pi^0)$ in Table 8.1. The last line gives the experimental data [P1] using the Clebsch-Gordon coefficients of Table 8.1.

	$\frac{\Gamma(D^{*0} \to D^0\gamma)}{\Gamma(D^{*0} \to D^0\pi^0)}$	$\frac{\Gamma(D^{*+} \to D^+\gamma)}{\Gamma(D^{*+} \to D^+\pi^0)}$	$\frac{\Gamma(K^{*0} \to K^0\gamma)}{\Gamma(K^{*0} \to K^0\pi^0)}$	$\frac{\Gamma(K^{*+} \to K^+\gamma)}{\Gamma(K^{*+} \to K^+\pi^0)}$	$\frac{\Gamma(\rho^+ \to \pi^+\gamma)}{\Gamma(\rho^+ \to \pi^+\pi^0)}$
Estimate	0.281	0.01825	8.07×10^{-3}	8.09×10^{-3}	-
Data [P1]	0.6155	0.0391-0.651	3.45×10^{-3}	2.97×10^{-3}	0.45×10^{-3}

The estimated ratios are consistent with data, given that $\Gamma(V \to P\gamma)$ is an order of magnitude estimate and that $\Gamma(K^* \to K\pi^0)$ is also an estimate because of the too large ε_0^2 values for these decays in Table 8.1.

8.3.4. Unified Strong and Electromagnetic Coupling Hypothesis

The driving force of $V \to P\gamma$ and $V \to PP$ lies in the greater radius (4.7.4) of the vector meson wave functions relative to that for the pseudoscalar meson in (4.7.3). The corresponding potential energy term for V is about 0.75–0.19=0.56 Gev^2 greater than that for P according to (8.2.10) or (5.2.2). This difference in the squared energy is released when $V \to P$. For $V \to P\gamma$, it is used to create a photon and impart it and P with momenta. For $V \to PP$, it is used to create a virtual quark-antiquark pair CD, separate them from each so that each of them becomes part of a P, and impart the both P's with momenta.

The driving force is thus the same for $V \to P\gamma$ and $V \to PP$. For $D^* \to D\gamma$ and $D^* \to D\pi$, the π mass is small relative to the D mass and may to some extent be regarded as a massless particle like the γ. Further, the phase space available for these both decays are rather small due to the small differences between the D^* and D masses. From these qualitative reasonings, one may expect that these both decay rates should be of the same order of magnitude. This turns out to be the case, as is seen in Table 8.2 when taking into account the differences in the $V \to P\gamma$ rates due to different quark charge combinations.

For K^* decay, the π mass is not negligible next to the K mass. Further, the available phase space is much greater due to the large difference between K^* and K masses. Therefore, the reasonings applied to D^* decay above do not apply here. Table 8.2 also shows that the decay rates for $K^* \to K\gamma$ and $K^* \to K\pi$ and differ considerably.

Each type of decay or interaction is characterized by its characteristic coupling strength. The coupling strength in the electromagnetic decay rate (6.4.35) is the fine structure constant

α=1/137. The weak coupling strength is F_W^2 in the two particle final state decay rate (7.5.16) dependent upon Ω_{wvac} and is G^2 in the three particle final state decay rate (7.4.29b, 7.5.4). These both coupling strengths are independent of α but are ratios of ∞/∞ dependent upon the relative time scale τ_0, as was discussed in §7.5.4.

The coupling strength for strong decay $V\to PP$ is f_s^2 of (8.3.8) and consists of a combination of two types of coupling strength; a dimensionless coupling constant g_s^8 analogous to α in the electromagnetic decays and $1/\Omega_{svac}$, analogous to $1/\Omega_{wvac}$ in the weak decay case in (7.5.16b). Thus, there appears to be a redundancy in the specification of the strong coupling strength; only one of the two coupling strength is needed and the other one may be removed. Now $1/\Omega_{svac}$ does not contain the relative time scale τ_0 and can therefore not be identified with any combination of ∞/∞ associated with F_W^2 of (7.5.16b) and G^2 of (7.4.29b) which depend upon τ_0. Hence the removal can only be achieved by identifying g_s^8 with α.

That the $D^{*0}\to D^0\gamma$ and $D^0\pi$ decay rates are of the same magnitude supports the identification of g_s^8 with α. This is further aided by noting that the $\eta\to\gamma\gamma$ and 3π decay rates are also of the same magnitude.

On these grounds, the following *hypothesis unifying the strong and electromagnetic couplings* is advanced,

$$\left(g_s^2/4\pi\right)^4 \equiv \alpha_s^4 = \alpha = e^2/4\pi \approx 1/137 \tag{8.3.11a}$$

$$\alpha_s \approx (\pm 1, \pm i)\times 0.2923 \tag{8.3.11b}$$

The factor 1/4π has been attached to g_s^2 for analogy with α. The value of the strong coupling constant $\alpha_s \approx 0.29$ among the quarks in (2.1.2, 4) is of the same magnitude as ≈0.12 cited in the literature [P1].

Another indication to possible interconnection between strong and electromagnetic interactions has been provided by the rather successful link between the electric charge *e* and quark masses considered below (7.2.12) above.

This hypothesis is also in harmony with "*Nature is sparesome in dealing out fundamental constants*" mentioned at the end of §7.2.1. To hand out two natural constants α and α_s within the relatively small factor of 0.12×137≈16.5 likewise appears to be wasteful. This aesthetical defect is removed by (8.3.11).

The hypothesis (8.3.11) eliminates another "fundamental parameter" α_s in the literature, in addition to the removal of the weak charge *g* and the Weinberg angle ϑ_W in §7.2.1 and the Cabbibo angle ϑ_C in (7.5.20). The three charges *g*, *e* and $\sqrt{\alpha_s}$ associated with the three basic types of interactions, the weak, electromagnetic and strong interactions, are thus reduced to one, the electron charge $-e$. This unification further implies that the *existence of electrons requires that quarks or hadrons must also exist.* The current concept of *electroweak* interactions which unifies electromagnetic and weak interactions is, in view of §7.5.3 and (8.3.11), here replaced by what may be called *electrostrong* interactions.

The purely strong interaction coupling strength for meson decay now resides in M_s of (8.3.9). Combining (8.3.10) and (8.3.11a) yields

$$M_s \approx 0.9847 \text{ Gev} \tag{8.3.12}$$

which corresponds to a volume Ω_{swac} of $(1.26 \text{ fm})^3$ for the vacuum meson state. This M_s value is not too far from the fundamental mesonic energy scale d_m=0864 Gev of (5.2.3).

8.4. $\pi^0 \rightarrow \gamma\gamma$

This Section is largely taken over from [H14].

8.4.1. Isotriplet Meson Wave Equations

In the absence of s and heavier quarks, it is sufficient to consider the SU(2)$_I$ part of the approximately SU(3) gauge invariant action S_{m3} of (7.1.8). Under some circumstances, it can be more convenient to work with meson equations transforming according to the regular representation of SU(2), i.e., SO(3) (Special Orthogonal group of rank 3) which is also the group for three dimensional rotations. Here, the basis vector generating the representations of this group is the π or ρ triplet, which is usually written either in Cartesian basis denoted by the subscripts $C=1$, 2 and 3 or in spherical basis by $s= +$, 0 and $-$ (for π^+, π^0, π^-, $\pi \rightarrow \rho$).

The relationship between meson wave functions in (7.1.8b) and those in the isotriplet formalism χ_t has been indicated in (7.3.8) above. In component form,

$$\chi_{ts} = \begin{pmatrix} \chi_{t+} \\ \chi_{t0} \\ \chi_{t-} \end{pmatrix} = \begin{pmatrix} \chi_{(12)} \\ \left(\chi_{(11)} - \chi_{(22)}\right)/\sqrt{2} \\ \chi_{(21)} \end{pmatrix}, \qquad \chi \rightarrow \psi \tag{8.4.1a}$$

where (2.4.18) has been consulted. Alternatively,

$$\chi_{tC} = \begin{pmatrix} \chi_{t1} \\ \chi_{t2} \\ \chi_{t3} \end{pmatrix} = \begin{pmatrix} \left(\chi_{t+} + \chi_{t-}\right)/\sqrt{2} \\ i\left(\chi_{t+} - \chi_{t-}\right)/\sqrt{2} \\ \chi_{t0} \end{pmatrix}, \qquad \chi \rightarrow \psi \tag{8.4.1b}$$

The Gell-Mann matrices λ_l in (7.1.5a), which are generators of the SU(3) group, are now replaced by the generators of the SO(3) group S_{Cl} or S_{sl};

$$S_{s1} = \frac{1}{\sqrt{2}}\begin{pmatrix} 0 & 1 & 0 \\ 1 & 0 & 1 \\ 0 & 1 & 0 \end{pmatrix}, \qquad S_{s2} = \frac{1}{\sqrt{2}}\begin{pmatrix} 0 & -i & 0 \\ i & 0 & -i \\ 0 & i & 0 \end{pmatrix}, \qquad S_{s3} = \begin{pmatrix} 1 & 0 & 0 \\ 0 & 0 & 0 \\ 0 & 0 & -1 \end{pmatrix} \tag{8.4.2a}$$

$$S_{C1}=\begin{pmatrix}0&0&0\\0&0&-i\\0&i&0\end{pmatrix},\qquad S_{C2}=\begin{pmatrix}0&0&i\\0&0&0\\-i&0&0\end{pmatrix},\qquad S_{C3}=\begin{pmatrix}0&-i&0\\i&0&0\\0&0&0\end{pmatrix}\qquad(8.4.2b)$$

S_{Cl} can be turned into S_{sl} and vice versa by a well-known similarity transformation. Both satisfy the same commutation rules as do λ_l, i.e., (2.4.7), but with the index l limited to *1*, *2* and *3*.

The derivative (7.1.4) now takes the form

$$\partial_I^{a\dot{b}}=\tfrac{1}{2}\partial_X^{a\dot{b}}-\partial^{a\dot{b}}\rightarrow D_{Is}^{a\dot{b}}=\tfrac{1}{2}\left(\partial_X^{a\dot{b}}+i\tfrac{1}{2}gS_{sl}W_l^{a\dot{b}}(X)\right)-\partial^{a\dot{b}}$$

$$=\partial_I^{a\dot{b}}+i\frac{g}{4}\begin{pmatrix}W_3&W_I^+&0\\W_I^-&0&W_I^+\\0&W_I^-&-W_3\end{pmatrix}^{a\dot{b}}\qquad(8.4.3a)$$

$$\partial_I^{a\dot{b}}=\tfrac{1}{2}\partial_X^{a\dot{b}}-\partial^{a\dot{b}}\rightarrow D_{IC}^{a\dot{b}}=\tfrac{1}{2}\left(\partial_X^{a\dot{b}}+i\tfrac{1}{2}gS_{Cl}W_l^{a\dot{b}}(X)\right)-\partial^{a\dot{b}}$$

$$=\partial_I^{a\dot{b}}+i\frac{g}{4}\begin{pmatrix}0&-iW_3&iW_2\\iW_3&0&-iW_1\\-iW_2&iW_1&0\end{pmatrix}^{a\dot{b}}\qquad(8.4.3b)$$

where (3.1.4), (3.5.7) and (7.1.5b) have been consulted. The meson wave equations (2.4.3) are now generalized to

$$\begin{aligned}&D_{Is}^{a\dot{b}}D_{IIs}^{f\dot{e}}\chi_{ts\dot{b}f}-\left(M_{ms}^2-\Phi_m\right)\psi_{ts}^{a\dot{e}}=0\\&D_{Is\dot{b}c}D_{IIs\dot{e}d}\psi_{ts}^{c\dot{e}}-\left(M_{ms}^2-\Phi_m\right)\chi_{ts\dot{b}d}=0\end{aligned}\qquad(8.4.4a)$$

$$M_{ms}^2=\begin{pmatrix}M_{m+}^2&0&0\\0&M_{m0}^2&0\\0&0&M_{m-}^2\end{pmatrix}\qquad(8.4.4b)$$

In the limit of SO(3) symmetry, the M_m's in (8.4.4) are the same. In this limit, (8.4.4a) is covariant under the SO(3) gauge transformations

$$\chi_{ts}^{a\dot{e}}\rightarrow\chi_{ts}'^{a\dot{e}}=O_3(X)\chi_{ts}^{a\dot{e}},\qquad\chi\rightarrow\psi\qquad(8.4.5a)$$

$$S_{sl}W_l^{a\dot{b}} \to S_{sl}W_l'^{a\dot{b}} = O_3(X)S_{sl}W_l^{a\dot{b}}O_3^{-1}(X) + \frac{2i}{g}\left(\partial_X^{a\dot{b}}O_3(X)\right)O_3^{-1}(X) \tag{8.4.5b}$$

$$O_3(X) = \exp\left[i\tfrac{1}{2}gS_{sl}\vartheta_l(X)\right] \tag{8.4.5c}$$

Obviously, the spherical basis in (8.4.4, 5) can be replaced by the Cartesian basis using (8.4.1b, 2b, 3b) instead together with an obvious modification of (8.4.4b).

Since $\pi^0 \to \gamma\gamma$ is not a weak decay, the charged gauge bosons $W^{\pm}$ in (8.4.3a) can be dropped. In component form, (8.4.4a) bcomes

$$\begin{aligned} &\partial_I^{a\dot{b}}\partial_{II}^{f\dot{e}}\chi_{t0\dot{b}f} - \left(M_{m0}^2 - \Phi_m\right)\psi_{t0}^{a\dot{e}} = 0 \\ &\partial_{I\dot{b}c}\partial_{II\dot{e}d}\psi_{t0}^{c\dot{e}} - \left(M_{m0}^2 - \Phi_m\right)\chi_{t0\dot{b}d} = 0 \end{aligned} \tag{8.4.6a}$$

$$\begin{aligned} &\left(\partial_I^{a\dot{b}} \pm ig\tfrac{1}{4}W_3^{a\dot{b}}\right)\left(\partial_{II}^{f\dot{e}} \pm ig\tfrac{1}{4}W_3^{f\dot{e}}\right)\chi_{t\pm\dot{b}f} - \left(M_{m\pm}^2 - \Phi_m\right)\psi_{t\pm}^{a\dot{e}} = 0 \\ &\left(\partial_{I\dot{b}c} \pm ig\tfrac{1}{4}W_{3\dot{b}c}\right)\left(\partial_{II\dot{e}d} \pm ig\tfrac{1}{4}W_{3\dot{e}d}\right)\psi_{t\pm}^{c\dot{e}} - \left(M_{m\pm}^2 - \Phi_m\right)\chi_{t\pm\dot{b}d} = 0 \end{aligned} \tag{8.4.6b}$$

8.4.2. Prototype Mechanism for $\pi^0 \to \gamma\gamma$

W_3 has been identified in the literature [L5 p670] to be the electromagnetic field,

$$A^{a\dot{b}}(X) = W_3^{a\dot{b}}(X) \tag{8.4.7}$$

This can be seen here as follows. Choose two of the three ϑ_l's in (8.4.5c) such that $\chi_{t\pm}$, hence also $\psi_{t\pm}$, are gauged away. Equation (8.4.6b) then drops out. Further, (8.4.6a) does not contain W_3. Variation of an action constructed from (8.4.6) in conjunction with the gauge boson action (7.6.4) with respect to W_3 shows that W_3 is massless [H8 §7]. Therefore, the identification of (8.4.7) is possible. If the SU(3) symmetry result $g=2e$ of (7.2.3c) is employed, The correct charges of the $t\pm(\pi^{\pm}, \rho^{\pm})$ wave functions in (8.4.6b) are obtained.

Since g is small, the gW_3 terms in (8.4.6b) are considered as perturbations. In the absecne of W_3, (8.4.6) is considered to be of zeroth order in g. Since π^0 is a pseudoscalar meson, let the zeroth order wave functions in (8.4.6) be singlets. The presence of W_3 in (8.4.6) must be balanced by perturbations superimposed upon the zeroth order wave functions. With (3.1.1), write

$$\chi_{ts}^{a\dot{e}} = \chi_{0ts}\delta^{a\dot{e}} - \underline{\chi}_{1ts}\underline{\sigma}^{a\dot{e}}, \qquad \chi \to \psi \tag{8.4.8}$$

where χ_{0ts} denotes the zeroth order wave functions for the π's and the subscript 1 denotes perturbation of first order in g. The first order singlet wave functions have been absorbed into the zeroth order ones. Higher order terms are not included here.

Inserting (8.4.8) into (8.4.6) leads to the zeroth order meson wave equations

$$\begin{aligned}\partial_I^{a\dot b}\partial_{II}^{f\dot e}\chi_{0ts}\delta_{\dot b f} - \left(M_{ms}^2 - \Phi_m\right)\psi_{0ts}\delta^{a\dot e} = 0 \\ \partial_{I\dot b c}\partial_{II\dot e d}\psi_{0ts}\delta^{c\dot e} - \left(M_{ms}^2 - \Phi_m\right)\chi_{0ts}\delta_{\dot b d} = 0\end{aligned} \tag{8.4.9}$$

and the first order equations

$$\partial_I^{a\dot b}\partial_{II}^{f\dot e}\underline{\chi}_{1t\pm}\underline{\sigma}_{\dot b f} + \left(M_{m\pm}^2 - \Phi_m\right)\underline{\psi}_{1t\pm}\underline{\sigma}^{a\dot e} = \pm ie\tfrac{1}{2}\left(\underline{A\sigma}^{a\dot b}\partial_{II}^{f\dot e} + \partial_I^{a\dot b}\underline{A\sigma}^{f\dot e}\right)\delta_{\dot b f}\chi_{0t\pm} \tag{8.4.10a}$$

$$\partial_{I\dot b c}\partial_{II\dot e d}\underline{\psi}_{1t\pm}\underline{\sigma}^{c\dot e}{}' + \left(M_{m\pm}^2 - \Phi_m\right)\underline{\chi}_{1t\pm}\underline{\sigma}_{\dot e d} = \mp ie\tfrac{1}{2}\left(\underline{A\sigma}_{\dot b c}\partial_{II\dot e d} + \partial_{I\dot b c}\underline{A\sigma}_{\dot e d}\right)\delta^{c\dot e}\psi_{0t\pm} \tag{8.4.10b}$$

Here, (8.4.7) and (7.2.8c) have been introduced. Further, the time component A^0 of the electromagnetic field has been removed by a suitable gauge transformation (8.4.5).

Putting s=0 in (8.4.9), it is seen that π^0 cannot decay to any gauge boson but must decay strongly through some perturbation Φ_{1m} of the confining potential Φ_m, whose source is a product of two meson wave functions according to (2.3.23). On the other hand, the electromagnetic field is only coupled to the charged triplet mesons in (8.4.10). These considerations together with (8.4.9, 10) suggest the prototype of the decay mechanism depicted by the quark line diagram of Figure 8.1.

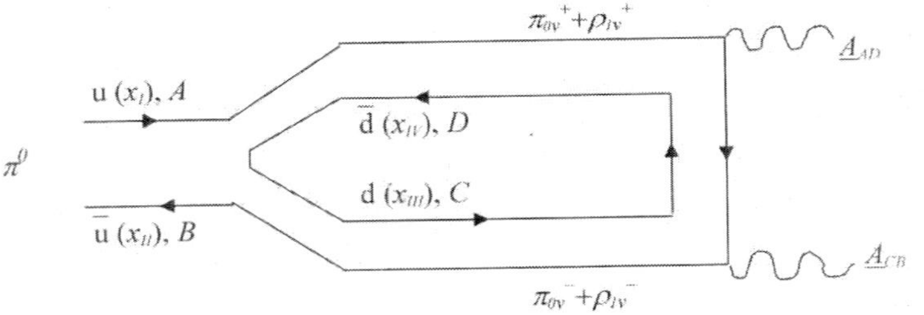

Figure 8.1. Mechanism of $\pi^0 \to \gamma\gamma$. This diagram is to be complemented by a similar diagram with $u \to d$ and subscript $+ \leftrightarrow -$ so that $\pi^0 = (\bar{u}u - \bar{d}d)/\sqrt{2}$ is adequately represented. x_I, x_{II}, x_{III}, and x_{IV} are the coordinates of the four quarks A, C, B and D, respectively, mentioned in §8.1.1. The subscript v denotes virtuality, 0 zeroth order in e and 1 first order in e. A virtual $\bar{d}d$ pair is created from the vacuum. This pair together with $\bar{u}u$ from π^0 form two charged intermediary mesons $\pi_{0v}^{\pm} + \rho_{1v}^{\pm}$. The first order virtual charged vector mesons $\rho_{1v}^{\pm}$ decay into two real photons $\underline{A}_{AD}$ and $\underline{A}_{CB}$ together with a pair of zeroth order virtual charged pseudoscalar mesons $\pi_{0v}^{\pm}$, which in their turn annihilate each other.

This mechanism makes use of the four quarks considered in §8.1.1 and cannot be fully represented by the two quark equations (8.4.9, 10). However, (8.4.9, 10) have their correspondences in Figure 8.1. Thus, π^0 is described by (8.4.9) with s=0 and $\pi_{0v}{}^{\pm}$ are associated with $s=\pm$ in (8.4.9). $\rho_{1v}{}^{\pm}$ are associated with the left sides of (8.4.10). Their decay products $\pi_{0v}{}^{\pm}$ and $\underline{A}_{AD}$ and $\underline{A}_{CB}$ are associated to the right sides of (8.4.10).

8.4.3. Virtual Intermediate Meson Wave Equations

A full four quark treatment is presently not feasible, as was mentioned in §8.1.1. On the other hand, the two quark equations (8.4.9, 10) are inadequate for the four quark problem posed by Figure 8.1. Therefore, the coupled meson equations of §8.1.4, based upon the mixed quark wave equations of §8.1.3, used for strong decay in Sec. 8.2 will be used as a compromise to treat $\pi^0\to\gamma\gamma$ approximately.

To zeroth order in e. the left part of Figure 8.1 represents a form of $P\to PP$ mentioned in §8.1.2 and has been worked out in Sec. 8.1 and 8.2. In addition, there is a first order part VV which arises due to the introduction of the electromagnetic gauge fields, as is indicated by (8.4.10). The mesons AD and CB described by (8.1.12, 13) are now associated with the middle part of Figure 8.1. The treatment of Sec. 8.1 and 8.2 does not include gauge fields and hence does not cover the right half of Figure 8.1. To represent the final state $\gamma\gamma$, electromagnetic field A_{AD} and A_{CB} associated with the U(1) group are introduced into (8.1.12, 13) via the usual minimal substitution

$$\partial_X^{a\dot{b}} \to \partial_X^{a\dot{b}} + ieA_{AD}^{a\dot{b}}(X) = \partial_X^{a\dot{b}} + ie\left(\delta^{a\dot{b}}A_{0AD}(X) - \underline{\sigma}^{a\dot{b}}\underline{A}_{AD}(X)\right)$$
$$AD \to CB, \qquad\qquad e \to -e$$

analogous to (6.2.3). Substituting these expressions into (8.1.12, 13) and consulting (3.1.4) and (3.5.7) leads to

$$\left(\partial_I^{a\dot{b}} + ie\tfrac{1}{2}A_{AD}^{a\dot{b}}\right)\left(\partial_{II}^{f\dot{e}} + ie\tfrac{1}{2}A_{AD}^{f\dot{e}}\right)\chi_{AD\dot{b}f}^{+} - \left(M_{AD}^2 - \Phi_m\right)\psi_{AD}^{+a\dot{e}} = 0 \qquad (8.4.11a)$$

$$\left(\partial_{I\dot{b}c} + ie\tfrac{1}{2}A_{AD\dot{b}c}\right)\left(\partial_{II\dot{e}d} + ie\tfrac{1}{2}A_{AD\dot{e}d}\right)\psi_{AD}^{c\dot{e}} - \left(M_{AD}^2 - \Phi_m\right)\chi_{AD\dot{b}D}^{+} = 0 \qquad (8.4.11b)$$

$$(8.4.11)\text{ with } e\to -e,\ AD\to CB,\text{ and } {}^{+}\to{}^{-} \qquad (8.4.12)$$

The superscripts $+$ and $-$ here indicate the signs of the charges associated with these wave functions; $^{+}$ here does not denote hermitian conjugation, as in (7.1.3b). These equations take over the roles of (8.4.6b). Analogous to (8.4.8), let

$$\chi_{AD\dot{b}f}^{+} = \chi_{0AD}^{+}(X,x)\delta_{\dot{b}f} + \underline{\chi}_{1AD}^{+}(X,x)\underline{\sigma}_{\dot{b}f}, \qquad \chi\to\psi \qquad (8.4.13a)$$

$$\psi_{CB}^{-a\dot{e}} = \psi_{0CB}^{-}(X,x)\delta^{a\dot{e}} - \underline{\psi}_{1CB}^{-}(X,x)\underline{\sigma}^{a\dot{e}}, \qquad \psi \to \chi \tag{8.4.13b}$$

Substituting these expression into (8.4.11, 12) leads to the zeroth order analog of (8.4.9)

$$\begin{aligned} &\partial_I^{a\dot{b}}\partial_{II}^{b\dot{a}}\chi_{0AD}^{+}(X,x) - 2\left(M_{AD}^2 - \Phi_{mAB} - \Phi_{1m4}\right)\psi_{0AD}^{+}(X,x) = 0 \\ &\partial_{I\dot{b}c}\partial_{II\dot{c}b}\psi_{0AD}^{+}(X,x) - 2\left(M_{AD}^2 - \Phi_{mAB} - \Phi_{1m4}\right)\chi_{0AD}^{+}(X,x) = 0 \end{aligned} \tag{8.4.14}$$

$$(8.4.14) \text{ with } AD \to CB \text{ and } {}^{+}\!\to{}^{-} \tag{8.4.15}$$

where (8.1.10a) has been consulted. The first order equations corresponding to (8.4.10) are

$$\begin{aligned} &\partial_I^{a\dot{b}}\partial_{II}^{f\dot{e}}\underline{\chi}_{1AD}^{+}(X,x)\underline{\sigma}_{\dot{b}f} + \left(M_{AD}^2 - \Phi_{mAB} - \Phi_{1m4}\right)\underline{\psi}_{1AD}^{+}(X,x)\underline{\sigma}^{a\dot{e}} \\ &= i\tfrac{1}{2}e\left(\underline{A}_{AD}(X)\underline{\sigma}^{a\dot{b}}\partial_{II}^{f\dot{e}} + \partial_I^{a\dot{b}}\underline{A}_{AD}(X)\underline{\sigma}^{f\dot{e}}\right)\delta_{\dot{b}f}\chi_{0AD}^{+}(X,x) \end{aligned} \tag{8.4.16a}$$

$$\begin{aligned} &\partial_{I\dot{b}c}\partial_{II\dot{e}d}\underline{\psi}_{1AD}^{+}(X,x)\underline{\sigma}^{c\dot{e}} + \left(M_{AD}^2 - \Phi_{mAB} - \Phi_{1m4}\right)\underline{\chi}_{1AD}^{+}(X,x)\underline{\sigma}_{\dot{b}d} \\ &= i\tfrac{1}{2}e\left(\underline{A}_{AD}(X)\underline{\sigma}_{\dot{b}c}\partial_{II\dot{e}d} + \partial_{I\dot{b}c}\underline{A}_{AD}(X)\underline{\sigma}_{\dot{e}d}\right)\delta^{c\dot{e}}\psi_{0AD}^{+}(X,x) \end{aligned} \tag{8.4.16b}$$

$$(8.4.16a,16b) \text{ with } e \to -e,\ AD \to CB \text{ and } {}^{+}\!\to{}^{-} \tag{8.4.17a,17b}$$

8.4.4. Virtual Intermediate Pseudoscalar Meson Wave Functions

The final state of $V \to PP$ in Sec. 8.3 is a pair of real pseudoscalar mesons that move away from each other and no longer interact with each other after a short time. In Figure 8.1, however, $\pi_{0v}^{\pm}$ are not final states but virtual intermediate pseudoscalar mesons. They are at rest because the initial meson AB is at rest and the created pair CD is likewise so. Therefore, π_{0v}^{+} and π_{0v}^{-} are close to each other, as is seen in Figure 8 1, and interact with each strongly prior to their mutual annihilation in the final state. This strong interaction activates the confinement mechanism and limit the volume occupied by π_{0v}^{+} or π_{0v}^{-} in laboratory space to some finite volume Ω_c according to Sec. 4.5.

This strong interaction between π_{0v}^{+} and π_{0v}^{-} requires that they both be located at different points, $(x_I+x_{IV})/2$ and $(x_{II}+x_{III})/2$, respectively, according to §8.1.1 and Figure 8.1. It is however absent in (8.4.14, 15) because these both points have merged into one point $X=(x_I+x_{II})/2$, due to the underlying mixed quark approximation of §8.1.3. Since one cannot treat the full four quark problem presently, the effect of this strong interaction will be incorporated into (8.4.14, 15) in the following phenomenological manner.

In the absence of strong meson-meson interaction, (8.4.14, 15) hold. These equations turn out to be the same as those for pseudoscalar mesons at rest and yield the solutions

$$-\chi_{0AD}^{\pm}(X,x)=\psi_{0AD}^{\pm}(X,x)=\sqrt{\frac{d_m^3}{8\pi\Omega}}\exp\left(-iE_{0AD}X^0-\frac{d_m r}{2}\right),\qquad AD\to CB \tag{8.4.18}$$

Here, (8.2.1, 9), (3.1.5) and (4.3.2) have been employed. The effect of the strong meson-meson interaction is included phenomenologically by the reverse of (4.2.7), i.e., $\Omega\to\Omega_c$ in (8.4.18), according to Sec. 4.5.

The only length scale associated with $\pi^0\to\gamma\gamma$ is the size of π^0 in relative space contained in (4.3.2) or (8.4.18). The average radius of the π^0, hence also of $\pi_{0v}^{\pm}$, is

$$\bar{r}_0=\int d^3\underline{x}\,\psi_0^2(r)r\Big/\int d^3\underline{x}\,\psi_0^2(r)=3/d_m\approx 4.3\ \text{fm} \tag{8.4.19}$$

where (4.3.2) has been used. With Figure 8.1, it will be estimated by (4.7.5b) and indications of (5.6.1) that this is also the radius of the $\pi_{0v}^{\pm}$ wave function in laboratory space or

$$\Omega_c=\frac{4}{3}\pi\,\bar{r}_0^3=\frac{36\pi}{d_m^3} \tag{8.4.20}$$

Whence (8.4.18) becomes

$$-\chi_{0AD}^{\pm}(X,x)=\psi_{0AD}^{\pm}(X,x)=a_{00AD}\psi_{0v}(r)\exp\left(-iE_{0AD}X^0\right),\ AD\to CB \tag{8.4.21a}$$

$$\psi_{0v}(r)=\frac{d_m^3}{12\sqrt{2\pi}}\exp\left(-\frac{d_m r}{2}\right) \tag{8.4.21b}$$

where a_{00AD} is of the type given in (6.4.2). It is unity here but is elevated to an annihilation operator in quantized cases.

8.4.5. Quantization

Let the virtual vector meson wave functions in (8.4.16, 17) be of a form similar to that of (6.4.1, 2),

$$\underline{\psi}_{1AD}^{\pm}(X,x)=b_{1K}\underline{\psi}_{1KAD}^{\pm}(\underline{x})\exp\left(-iE_{1KAD}X^0+i\underline{K}_{1AD}\underline{X}\right) \tag{8.4.22}$$

Note that the subscript *1* to ψ denotes first order in e while the other *1*'s signify $J=1$. Analogous to (6.4.8), let

$$\underline{A}_{AD}(X)=\frac{1}{\sqrt{2E_{rAD}\Omega_r}}\underline{e}_{TAD}a_{TAD}\exp\left(-iE_{rAD}X^0+i\underline{K}_{rAD}\underline{X}\right)+c.c.,\quad AD\to CB \tag{8.4.23}$$

where Ω_r is a large normalization volume for the photon wave functions.

The virtual vector meson wave functions (8.4.22) is in principle obtainable from (8.4.16, 17) in terms of the photon field (8.4.23) and the virtual pseudoscalar meson wave functions (8.4.21). These functions together with (8.4.21) can then be inserted into (8.2.15b,18) to obtain $\Phi_{1jK}(\underline{x})$. This quantity is in its turn substituted into (8.2.14, 8) to obtain the decay amplitude S_{fi0}.

This decay amplitude however depends only upon one meson coordinate X and cannot represent the annihilation of π_{0v}^+ with π_{0v}^-, which have different coordinates according to the right part of Figure 8.1. These circumstances are analogous to the absence of π_{0v}^+ and π_{0v}^- strong interaction in the mixed quark approximation discussed in the beginning of §8.4.4. Analogous to the phenomenological approach there, the effect of π_{0v}^+ and π_{0v}^- annihilation will be represented by that $\pi_{0v}^+(AD)$ and $\pi_{0v}^-(CB)$ are separately vacuum meson states like that mentioned below (8.2.5) and considered below (7.3.18).

The initial state is represented by

$$|i\rangle=\left|\pi^0(\underline{K}_0=0)\right\rangle \tag{8.4.24a}$$

and the final state is represented by

$$\langle f|=\langle 0,\Phi_{1m40}|=\langle 0,00\gamma_{TAD}\gamma_{TCB}| \tag{8.4.24b}$$

where the three zeros represent vacuum meson states of π^0, π_{0v}^+ and π_{0v}^- and γ_T represents a photon of polarization T.

Consider at first the singlet product terms in (8.2.15b), which are products of members of (8.4.21). When these appear between <f| and |i> of (8.4.24) via (8.2.18, 14, 4b, 8), it is seen that they vanish because they do not contain final state photons. This leads to that (8.2.15b) goes over to (8.2.20a). The terms in (8.2.20a) will according to the observation given below (8.4.23) contain the product

$$a_{00AD}a_{00CB}^{*}a_{TAD}a_{TCB}^{*} \tag{8.4.25}$$

Each a is a creation or annihilation operator in (8.4.21, 23). When this prooduct appears to the right of <f| via (8.2.18, 14, 4b, 7), the vacuum meson states represented by the second and third zeros in (8.4.24b) will be operated upon by $a_{00AD}a_{00CB}^*$ and produce

$$E_{0AD}=E_{0CB}=0 \tag{8.4.26}$$

in (8.4.21) that end up in (8.2.7). a_{TCB}^* acting upon γ_{TCB} produces a positive energy photon with energy E_{rCB} in (8.4.23). a_{TAD} operating on γ_{TAD} creates a positive energy photon with energy E_{rAD} by annihilating the negative energy component in (8.4.23) having energy $-E_{rAD}$. This assignment is entirely analogous to the negative energy choice for the real pseudoscalar meson *AD* above (8.2.12).

Since the wave functions of $\pi_{0v}^+(AD)$ and $\pi_{0v}^-(CB)$ will appear in product form in (8.2.20a), the flavors or charges carried by π_{0v}^+ and π_{0v}^- are interpreted to have neutralized each other in the final state. If (8.2.20a) can take the form (8.2.20b), then the decay amplitude (8.2.23) with J=0, j=1 and (8.4.26) holds for $\pi^0 \to \gamma\gamma$.

8.4.6. Virtual Intermediate Vector Meson Wave Functions

Substistute (8.4.21-23) and (8.2.2, 9) into (8.4.16, 17) anticipating the quantization results of §8.4.5 and make use of (3.1.4) and (3.5.7). The result is

$$\begin{Bmatrix} \left(\tfrac{1}{4}E_{1KAD}^2 - \tfrac{1}{4}\underline{K}_{1AD}^2 - \Delta\right)\underline{\chi}_{1KAD}^+(\underline{x}) + \tfrac{1}{2}\underline{K}_{1AD}\left(\underline{K}_{1AD}\underline{\chi}_{1KAD}^+(\underline{x})\right) \\ + 2\underline{\partial}\left(\underline{\partial}\underline{\chi}_{1KAD}^+(\underline{x})\right) + E_{1KAD}\underline{\partial}\times\underline{\chi}_{1KAD}^+(\underline{x}) + \left(d_m/r + d_{m0} - M_{AD}^2\right)\underline{\psi}_{1KAD}^+(\underline{x}) \end{Bmatrix}$$
$$\times\exp\left(-iE_{1KAD}X^0 + i\underline{K}_{1AD}\underline{X}\right) = \frac{1}{\sqrt{2E_{rAD}\Omega_r}}\exp\left(iE_{rAD}X^0 + i\underline{K}_{rAD}\underline{X}\right)$$
$$\times\frac{e}{4}\left\{E_{rAD}\underline{e}_{TAD}\psi_{ov}(r) + i\underline{K}_{rAD}\times\underline{e}_{TAD}\psi_{ov}(r) - 4\underline{\partial}\psi_{ov}(r)\times\underline{e}_{TAD}\right\} \quad (8.4.27a)$$

$$\begin{Bmatrix} \left(\tfrac{1}{4}E_{1KAD}^2 - \tfrac{1}{4}\underline{K}_{1AD}^2 - \Delta\right)\underline{\psi}_{1KAD}^+(\underline{x}) + \tfrac{1}{2}\underline{K}_{1AD}\left(\underline{K}_{1AD}\underline{\psi}_{1KAD}^+(\underline{x})\right) + \\ + 2\underline{\partial}\left(\underline{\partial}\underline{\psi}_{1KAD}^+(\underline{x})\right) - E_{1KAD}\underline{\partial}\times\underline{\psi}_{1KAD}^+(\underline{x}) + \left(d_m/r + d_{m0} - M_{AD}^2\right)\underline{\chi}_{1KAD}^+(\underline{x}) \end{Bmatrix}$$
$$\times\exp\left(-iE_{1KAD}X^0 + i\underline{K}_{1AD}\underline{X}\right) = \frac{1}{\sqrt{2E_{rAD}\Omega_r}}\exp\left(iE_{rAD}X^0 + i\underline{K}_{rAD}\underline{X}\right)$$
$$\times\frac{e}{4}\left\{-E_{rAD}\underline{e}_{TAD}\psi_{ov}(r) - i\underline{K}_{rAD}\times\underline{e}_{TAD}\psi_{ov}(r) - 4\underline{\partial}\psi_{ov}(r)\times\underline{e}_{TAD}\right\} \quad (8.4.27b)$$

$$(8.4.27a, 27b) \text{ with } AD \to CB,\ {}^+ \to {}^-,\ e \to -e, \text{ and } -E_{rAD} \to E_{rCB} \quad (8.4.28a,28b)$$

Comparison of the left and right sides of these equations gives

$$E_{1KAD} = -E_{rAD},\quad E_{1KCB} = E_{rCB},\quad \underline{K}_{1AD} = \underline{K}_{rAD},\quad \underline{K}_{1CB} = \underline{K}_{rCB} \quad (8.4.29)$$

Equations (8.4.27, 28) are akin to (3.5.4b, 4c) for a slowly moving pseudoscalar meson. The presence of the $\underline{K}$ or the right side of (3.5.4b, 4c) breaks the spherical symmetry in relative space of the wave functions of a pseudoscalar meson with $\underline{K}$=0. This symmetry breaking causes that the first order meson wave functions in (3.5.4b, 4c) to be not separable in angular and radial parts. Since these nonseparable equations are not readily solved, the first order meson wave functions were estimated by means of a dimensional approximation in §3.5.3-4.

In (8.4.27, 28), the role of symmetry breaking is taken over by $\underline{K}_r$ and $\underline{e}_T$. Analogously, the nonseparable (8.4.27, 28) are too complicated to solve presently. Therefore, the wave functions $\chi_I^{\pm}$ and $\underline{\psi}_I^{\pm}$ will be estimated by means of the following "directional" approximation, which bears some resemblance to the dimensional approximations in §3.5.3-4.

8.4.7. Directional Approximation

The directional approximation consists of two steps, step 1) and step 2). In step 1), the nonseparable wave functions in (8.4.27, 28) are *assumed* to take on the separated form

$$\underline{\psi}^{+}_{1KAD}(\underline{x}) = \hat{r}_{AD}\psi^{+}_{1AD}(r), \qquad \psi \to \chi \tag{8.4.30a}$$

$$\underline{\chi}^{-}_{1KCB}(\underline{x}) = \hat{r}_{CB}\chi^{-}_{1CB}(r), \qquad \chi \to \psi \tag{8.4.30b}$$

The angular part is contained in the directional cosine vector $\hat{r}$.

Insert (8.4.30a) into (8.4.27). It is seen that the $\underline{\partial}\times$ terms on the left sides vanish, as in (3.2.5). Multiply both sides by $\hat{r}_{AD}$ and note that

$$\underline{\partial}\psi_{0v}(r) = \hat{r}\left(\partial\psi_{0v}(r)/\partial r\right)$$

One sees that the $\underline{\partial}\psi_{0v}(r)\times$ terms on the right sides also drop out so that the sum of the right sides of the so-modified (8.4.27a) and (8.4.27b) vanishes. The sum of the left sides must therefore also vanish and this can only be achieved by

$$\chi^{+}_{1AD}(r) = -\psi^{+}_{1AD}(r) \tag{8.4.31}$$

Inserting this expression into the so-modified (8.4.27) and noting (8.4.29) yields

$$\begin{aligned}&\left(\tfrac{1}{4}E^{2}_{rAD} - \tfrac{1}{4}\underline{K}^{2}_{rAD} - d_m/r - d_{m0} + M^{2}_{AD}\right)\psi^{+}_{1AD}(r) + \tfrac{1}{2}\hat{r}_{AD}\underline{K}_{rAD}\left(\underline{K}_{rAD}\hat{r}_{AD}\psi^{+}_{1AD}(r)\right)\\ &+ \hat{r}_{AD}\left[2\underline{\partial}\left(\underline{\partial}\hat{r}_{AD}\psi^{+}_{1AD}(r)\right) - \Delta\hat{r}_{AD}\psi^{+}_{1AD}(r)\right]\\ &= -\frac{e}{4}\frac{1}{\sqrt{2E_{rAD}\Omega_r}}\hat{r}_{AD}\left(E_{rAD}\underline{e}_{TAD} + i\underline{K}_{rAD}\times\underline{e}_{TAD}\right)\psi_{ov}(r)\end{aligned} \tag{8.4.32}$$

Because both sides must have the same r dependence, let

$$\psi_{1AD}^{+}(r)=\alpha_{AD}\psi_{0v}(r) \tag{8.4.33a}$$

$$\chi_{1CB}^{-}(r)=\alpha_{CB}\psi_{0v}(r) \tag{8.4.33b}$$

where the α's are constants. With (8.4.33a), the bracketed expression in (8.4.32) becomes

$$2\underline{\partial}\left(\underline{\partial}\hat{r}_{AD}\alpha_{AD}\psi_{0v}(r)\right)-\Delta\hat{r}_{AD}\alpha_{AD}\psi_{0v}(r)=\left(\frac{d_m}{r}+\frac{d_m^2}{4}\right)\hat{r}_{AD}\alpha_{AD}\psi_{0v}(r) \tag{8.4.34}$$

where (8.4.21b) has been consulted. For the photon,

$$E_{rAD}=K_{rAD},\quad \underline{K}_{rAD}=\hat{K}_{TAD}K_{rAD},\quad AD\to CB \tag{8.4.35}$$

Substituting (8.4.33-35) into (8.4.32) yields

$$\left(\frac{d_m^2}{4}-d_{m0}+M_{AD}^2\right)\alpha_{AD}+\frac{1}{2}\left(\hat{r}_{AD}\underline{K}_{rAD}\right)^2\alpha_{AD}=-\frac{e}{4}\sqrt{\frac{E_{rAD}}{2\Omega_r}}\hat{r}_{AD}\left(\underline{e}_{TAD}+i\hat{K}_{TAD}\times\underline{e}_{TAD}\right) \tag{8.4.36}$$

Next, insert (8.4.30b) into (8.4.28) and repeat the same procedure that led (8.4.27) to (8.4.36). The basic difference lies in the – signs in (8.4.28). The result corresponding to (8.4.31, 36) is

$$\chi_{1CB}^{-}(r)=-\psi_{1CB}^{-}(r) \tag{8.4.37}$$

$$\left(\frac{d_m^2}{4}-d_{m0}+M_{CB}^2\right)\alpha_{CB}+\frac{1}{2}\left(\hat{r}_{CB}\underline{K}_{rCB}\right)^2\alpha_{CB}=-\frac{e}{4}\sqrt{\frac{E_{rCB}}{2\Omega_r}}\hat{r}_{CB}\left(\underline{e}_{TCB}-i\hat{K}_{TCB}\times\underline{e}_{TCB}\right) \tag{8.4.38}$$

That the α's in (8.4.36, 38) are constants justifies the ansatz (8.4.33).

Insert now (8.4.33, 31, 37) into (8.4.30), which in its turn is substituted into (8.2.20a). This yields

$$\Xi_{1K}(\underline{x})=2\hat{r}_{AD}\hat{r}_{CB}\alpha_{AD}\alpha_{CB}^{*}\psi_{0v}^{2}(r) \tag{8.4.39}$$

The perturbed scalar potential Φ_{Im4} of (8.1.10a) is real by (8.1.10c). Therefore, Φ_{IjK} of (8.2.14, 18) and $\Xi_{IK}(\underline{x})$ of (8.2.20a) and (8.4.39) are also real.

Now, proceed to step 2) of the directional approximation which consists of the *assumptions*

$$\hat{r}_{AD} = \frac{1}{\sqrt{2}}\left(\underline{e}_{TAD} - i\hat{K}_{TAD} \times \underline{e}_{TAD}\right), \qquad \hat{r}_{CB} = \frac{1}{\sqrt{2}}\left(\underline{e}_{TCB} + i\hat{K}_{TCB} \times \underline{e}_{TCB}\right) \tag{8.4.40}$$

These are associated with the emerging photon $\underline{A}_{AD}$ with negative helicity and $\underline{A}_{CB}$ with positive helicity. The polarizations of the emerging photons must be antiparallel to each other,

$$\underline{e}_{TAD} = -\underline{e}_{TCB} \tag{8.4.41}$$

so that there is no net polarization, as is the case for th initial π^0. Further,

$$\underline{e}_{TAD} \perp \underline{K}_{TAD} = \underline{K}_{TCB} \tag{8.4.42}$$

The last relation is anticipated form (8.2.23) and the analog of (8.3.3). Making use of (8.4.40-42), (8.4.39) becomes

$$\Xi_{1K}(\underline{x}) = 2\alpha_{AD}\alpha^*_{CB}\psi^2_{0\nu}(r) = \Xi_{1K}(r) \tag{8.4.43}$$

Subsistuting (8.4.40-42) into (8.4.36, 38) leads to that the $(\hat{r}K_r)^2$ terms drop out and

$$\alpha_{AD} = -\frac{e}{4}\sqrt{\frac{E_{rAD}}{\Omega_r}}\left(\frac{d_m^2}{4} - d_{m0} + M^2_{AD}\right)^{-1}, \qquad AD \to CB \tag{8.4.44}$$

In this way, the α's in (8.4.43) are also real so that the reality requirement of $\Xi_{lK}(\underline{x})$ is fullfilled by the step 2) ansatz. This ansatz provides the only choice for setting the unknown directions $\hat{r}_{AD}$ and $\hat{r}_{CD}$ so that (8.4.43) is real and nontrivial. If $i \to -i$ in (8.4.40), the α's would vanish and there is no decay, contrary to data.

Due to the right relation of (8.4.43), the decay amplitude (8.2.23) applies, as was mentioned at the end of §8.4.5.

8.4.8. Decay Rate Estimate

From (8.4.43, 21b), (8.2.19) and (4.3.2), it is seen that the radial dependence of Ξ_{0K} and Ξ_{lK} are the same. Therefore, $\Phi_{l1K}(r)$ evaluated from (8.2.18) has the same radial form as $\Phi_{l00}(r)$ of (8.3.5). The calculation leading to (8.3.5) can be taken over to yield

$$\Phi_{11K}(r) = -8\frac{g_s^4}{E_{10}^4}\alpha_{AD}\alpha_{CB}\psi^2_{0\nu}(0)$$

$$\times\left[2\frac{\exp(-2\beta_0 R_0)}{\left(1+4\beta_0^2\right)^2}\left(1+\frac{8\beta_0}{1+4\beta_0^2}\frac{1}{R_0}\right)-\frac{\cos R_0}{R_0}\frac{16\beta_0}{\left(1+4\beta_0^2\right)^3}-\sin R_0\frac{4\beta_0}{\left(1+4\beta_0^2\right)^2}\right] \tag{8.4.45a}$$

$$\beta_0 = d_m / E_{00}, \qquad R_0 = E_{00} r/2 \tag{8.4.45b}$$

Inserting this expression into (8.2.23c) with $j=1$, $J=0$ gives the equivalent of (8.3.6),

$$\overline{\Phi}_{11K} = \frac{1}{18432\pi^2} g_s^4 e^2 \sqrt{\frac{\Omega}{\Omega_{svac}}} \frac{d_m^3}{\Omega_r L_\pi} I_{11} \tag{8.4.46a}$$

$$I_{11} = \frac{64\beta_0^6}{\left(1+4\beta_0^2\right)^2}\left[\frac{8\beta_0\left(20\beta_0^2-3\right)}{\left(1+4\beta_0^2\right)^3}-\frac{1}{\beta_0\left(1+4\beta_0^2\right)}-\frac{1}{16\beta_0^3}\right] \tag{8.4.46b}$$

$$L_\pi = M_{\pi\pm}^2 + d_m^4/4 - d_{m0} \tag{8.4.46c}$$

where (4.3.2) has been used. Further,

$$E_{rAD} = E_{rCB} = E_{00}/2 \tag{8.4.47}$$

according to (8.4.35), the right member of (8.4.42) and (8.4.53) below. Also,

$$M_{AD}^2 = M_{CB}^2 = M_{\pi\pm}^2 = (m_1 + m_2)^2/4 \tag{8.4.48}$$

according to (2.3.27).

The decay rate formulae (8.3.1) or (7.5.1) becomes here

$$\Gamma\left(\pi^0 \to \gamma\gamma\right) = \sum\nolimits_{\text{final states}} \left|S_{fi0}\right|^2 / T_d = 2\frac{\Omega_r^2}{(2\pi)^6}\int d^3\underline{K}_{rAD} d^3\underline{K}_{rCB}\left|S_{fi0}\right|^2/T_d \tag{8.4.49}$$

where the factor 2 stems from the two polarization index values $T=1, 2$ in (8.4.23). Inserting (8.2.23) with $j=1$, $J=0$ into (8.4.49) leads to the analog of (8.3.2),

$$\Gamma\left(\pi^0 \to \gamma\gamma\right) = \frac{4}{\pi^5}\frac{1}{E_{00}^2}\frac{1}{\Omega^2}\int d^3\underline{K}_{rCB} I_{VV} \tag{8.4.50a}$$

$$I_{VV} = \int d^3\underline{K}_{rAD} I_{K1} I_{K1}^* \left|\Omega_r \overline{\Phi}_{11K}\right|^2 \delta\left(E_{rAD} + E_{rCB} - E_{00}\right) \tag{8.4.50b}$$

where (8.4.29) has been used. Analogous to (8.3.3, 4a),

$$I_{K1} = (2\pi)^3 \delta(\underline{K}_{rCB} - \underline{K}_{rAD}) = I^*_{K1} \tag{8.4.51}$$

$$I_{VV} = (2\pi)^3 \Omega_s |\Omega_r \overline{\Phi}_{11K}|^2 \delta(E_{rAD} + E_{rCB} - E_{00}) \tag{8.4.52}$$

Because of (8.4.35), the analog of (8.3.7a) is

$$\delta(E_{rAD} + E_{rCB} - E_{00}) = \frac{1}{2}\delta\left(K_{r0} - \frac{E_{00}}{2}\right), \qquad K_{r0} = K_{rAD} = K_{rCB} \tag{8.4.53}$$

Insert (8.4.46, 53) into (8.4.52), which in its turn is substituted into (8.4.50a), and carry out the K_{r0} integration. One finds

$$\Gamma(\pi^0 \to \gamma\gamma) = \frac{1}{(4608\pi)^2} f_s^2 e^4 \frac{d_m^6 I_{11}^2}{L_\pi^4} \text{ Gev} \tag{8.4.54}$$

where f_s^2 has been defined in (8.3.8b).

With the π^0 mass E_{00}=0.135 Gev and d_m from (5.2.3), (8.4.45b, 46b) give I_{11}=1.318. Using m_1, m_2 and d_{m0} values from Table 5.2 in (8.4.48, 46c) yields L_π=0.378 Gev^2. f_s^2 has been given by (8.3.10b) and e^2=4π/137. With these values (8.4.54) yields

$$\Gamma(\pi^0 \to \gamma\gamma) = 7.96 \text{ ev} \tag{8.4.55}$$

which agrees well with the measured value 7.75 ev [P1].

This agreement can however not be taken literally due to the rather coarse assumptions made in the directional approximation of (8.4.30,40) and the estimate of Ω_c in (8.4.20). In view of these approximations, (8.4.55) is considered as an order of magnitude estimate; any estimate that is within a factor of ten from 7.75 ev would be considered as acceptable. The good aggreement can be accidental, since there is only one data point to account for. However, the present theory has also been successful in many other areas of low energy hadronic phenomena, shown in Chapters 4-8 and Chapters 11-12 below. From this viewpoint, the estimate (8.4.55) can be considered to provide an additional support of the scalar strong interaction hadron theory.

This is not the case for the earlier predictions of the rate of this classical and basic decay. More than half a century ago, Steinberger [S6] first predicted the rate to be 13.8 ev. The virtual intermediate state has been assumed to be a proton-antiproton pair. Then came the vector meson dominance (VMD) models [G5, R3, S7, B5] in which the virtual intermediate state was assumed to be dominated by vector mesons. A third type of treatment took place in the frame work of current algebra [A3] and the virtual intermediate state consists of quarks. There are vaiations of these three types of approach.

Most of these earlier predictions have been good. However, the models employed have largely been tailored for this decay and have very narrow application angles. Thus, they bear little or no relation to other low energy phenomena, such as those considered in Chapters 4-8 and Chapters 11-12 below. Without a sufficient number of such relations, the correctness of any decay mechanism assumed by these models is uncertain; to rely on one data point is obviously insufficient.

Compared to these earlier works, Figure 8.1 agrees largely with the vector dominance model [G5] and rejects that the virtual intermediate state consists of a proton-antiproton pair [S6] or quarks fewer then four [A3].

Chapter 9

CONSTRUCTION OF EQUATIONS OF MOTION FOR BARYON

The construction of a set of equations of motion for baryons is likewise guided by the principles of Sec. 1.5 and follows the related steps of [H15]. It is entirely analogous to the construction of the meson wave equations of Chapter 2.

9.1. QUARK WAVE AND POTENTIAL EQUATIONS

A baryon consists of three quarks. In a unified theory, the quark-quark interaction must be the same in both baryons and mesons. Following Sec. 2.1, the first step is to assume that the quarks in a baryon act on each other pairwise via the same massless scalar interaction between the quarks in a meson. Any genuine three body interaction will drop out later in Sec. 9.2 below and are therefore not included here.

Let x_I, x_{II} and x_{III} denote the coordinates of the quarks *A*, *B*, and *C*, respectively. For simplicity, x_I, x_{II} and x_{III} as arguments are replaced by *I*, *II* and *III*, respectively. The starting quark equations corresponding to (2.1.1-4) are

$$\partial_I^{a\dot{b}} \chi_{A\dot{b}}(I) - i(V_{AB}(I) + V_{AC}(I))\psi_A^a(I) = im_A \psi_A^a(I) \tag{9.1.1a}$$

$$\partial_{I\dot{b}c} \psi_A^c(I) - i(V_{AB}(I) + V_{AC}(I))\chi_{A\dot{b}}(I) = im_A \chi_{A\dot{b}}(I) \tag{9.1.1b}$$

$$\partial_{II}^{d\dot{e}} \chi_{B\dot{e}}(II) - i(V_{BC}(II) + V_{BA}(II))\psi_B^d(II) = im_B \psi_B^d(II) \tag{9.1.2a}$$

$$\partial_{II\,\dot{e}f} \psi_B^f(II) - i(V_{BC}(II) + V_{BA}(II))\chi_{B\dot{e}}(II) = im_B \chi_{B\dot{e}}(II) \tag{9.1.2b}$$

$$\partial_{III}^{g\dot{h}} \chi_{C\dot{h}}(III) - i(V_{CA}(III) + V_{CB}(III))\psi_C^g(III) = im_C \psi_C^g(III) \tag{9.1.3a}$$

$$\partial_{III\,\dot{h}k} \psi_C^k(III) - i(V_{CA}(III) + V_{CB}(III))\chi_{C\dot{h}}(III) = im_C \chi_{C\dot{h}}(III) \tag{9.1.3b}$$

$$\square_I \left(V_{AB}(I)+V_{AC}(I)\right)=\frac{1}{2}g_s^2\begin{pmatrix}\psi_B^e(I)\chi_{Be}(I)+\psi_B^{\dot{e}}(I)\chi_{B\dot{e}}(I)+\\+\psi_C^g(I)\chi_{Cg}(I)+\psi_C^{\dot{g}}(I)\chi_{C\dot{g}}(I)\end{pmatrix} \tag{9.1.4a}$$

$$\square_{II} \left(V_{BC}(II)+V_{BA}(II)\right)=\frac{1}{2}g_s^2\begin{pmatrix}\psi_C^g(II)\chi_{Cg}(II)+\psi_C^{\dot{g}}(II)\chi_{C\dot{g}}(II)+\\+\psi_A^a(II)\chi_{Aa}(II)+\psi_A^{\dot{a}}(II)\chi_{A\dot{a}}(II)\end{pmatrix} \tag{9.1.4b}$$

$$\square_{III} \left(V_{CA}(III)+V_{CB}(III)\right)=\frac{1}{2}g_s^2\begin{pmatrix}\psi_A^a(III)\chi_{Aa}(III)+\psi_A^{\dot{a}}(III)\chi_{A\dot{a}}(III)+\\+\psi_B^e(III)\chi_{Be}(III)+\psi_B^{\dot{e}}(III)\chi_{B\dot{e}}(III)\end{pmatrix} \tag{9.1.4c}$$

9.2. Baryon Wave Equations in Space Time

Following the procedure of Sec. 2.2, the left sides of (9.1.1-3) are multiplied together to form triple products containing all three quark wave functions. The corresponding right sides are analogously multiplied together with the operators placed to the left of the wave functions. Consider at first the right side terms, We can form two types of such products. Type a) consists of products with three dotted or three undotted indices, such as $\psi_A{}^a(I)\,\psi_B{}^d(II)\,\psi_C{}^{ag}(III)$. Type b) products contain mixed dotted and undotted indices, such as $\chi_{A\dot{b}}(I)\chi_{C\dot{h}}(III)\psi_B^f(II)$.

Fierz and Pauli [F2] have shown that rank three spinors transforming as type b) can represent spin 3/2 particles. Such a spinor can also include representation of spin 1/2 particles [B6]. These particles can obviously be associated respectively with quartet baryons, such as the Δ baryons, and doublet baryons, such as the nucleons. The *first part* of the *second step* is the generalization to nonseparable baryons wave functions akin to (2.2.1),

$$\chi_{A\dot{b}}(I)\chi_{C\dot{h}}(III)\psi_B^f(II)\rightarrow\chi_{\dot{b}\dot{h}}^f(I,III,II) \tag{9.2.1a}$$

$$\psi_A^c(I)\psi_C^k(III)\chi_{B\dot{e}}(II)\rightarrow\psi^{ck}_{\dot{e}}(I,III,II) \tag{9.2.1b}$$

The subscripts *A, B* and *C* are dropped via the interpretation that it is not possible to distinguish the quarks from each other inside a baryon.

Turning to the left side terms of the product equations mentioned at the beginning of this section. The *second part* of the *second step* by analogy to (2.2.3) consists of the generalization of *VVV* term to the nonseparable intrabaryon potential

$$\left(V_{AB}(I)+V_{AC}(I)\right)\left(V_{CA}(III)+V_{CB}(III)\right)\left(V_{BC}(II)+V_{BA}(II)\right)\rightarrow\Phi_b(I,III,II) \tag{9.2.2}$$

Product terms containing one V drop out on the same ground given above (2.2.4), i.e., there are no free quarks in the baryon system so that the right sides, hence also the left sides, of (9.1.4) vanish. Product terms containing $V(I)V(II)$ type of expressions also drop out for the same reason. If $V(I)V(II)$ is generalized to $\Phi_m(I,II)$ type of form similar to (2.2.3), the products on the right sides of (9.1.4a, 4b) will be analogously generalized as in (2.2.1, 2). The result is the creation of objects transforming like diquarks or mesons. Since such objects are not observed in ground state baryons at least, they are dropped. Therfore, the generalized form of $V(I)V(II)$ on the left side likewise vanish.

The result of the multiplications and generalizations started at the beginning of this section with these two parts of the second step is

$$\begin{aligned} \partial_I^{a\dot b}\partial_{III}^{g\dot h}\partial_{II\,\dot e f}\chi_{\dot b\dot h}^{\;\;f}(I,III,II) &= -i\left(m_A m_B m_C + \Phi_b(I,III,II)\right)\psi^{ag}_{\;\;\dot e}(I,III,II) \\ \partial_{I\,\dot b c}\partial_{III\,\dot h k}\partial_{II}^{d\dot e}\psi^{ck}_{\;\;\dot e}(I,III,II) &= -i\left(m_A m_B m_C + \Phi_b(I,III,II)\right)\chi_{\dot b\dot h}^{\;\;d}(I,III,II) \end{aligned} \quad (9.2.3)$$

This set of baryon wave equations in space time contains 2×8=16 wave function components of (9.2.1), far less than 4×4×4=64, the number of wave function components in the Bethe-Salpeter formalism for baryons. This reduction is number great simplies mathematical handling.

The rank three spinors (9.2.1) have eight components each and can be decomposed according the Clebsch-Gordan series [L2 p225]

$$\underline{2}\otimes\underline{2}\otimes\underline{2} = \underline{4}\oplus\underline{2}_s\oplus\underline{2}_a \quad (9.2.4)$$

By means of the invariant antisymmetric tensors ε^{ab} (B4) and (B15), each member of (9.2.1) can be decomposed into a doublet antisymmetric in the first two indices and a sextet symmetric in these indices;

$$\begin{aligned} &\chi_{\dot b\dot h}^{\;\;f} = \chi_{[\dot b\dot h]}^{\;\;f} + \chi_{\{\dot b\dot h\}}^{\;\;f},\quad \chi_{[\dot b\dot h]}^{\;\;f} = \tfrac{1}{2}\left(\chi_{\dot b\dot h}^{\;\;f} - \chi_{\dot h\dot b}^{\;\;f}\right),\quad \chi_{\{\dot b\dot h\}}^{\;\;f} = \tfrac{1}{2}\left(\chi_{\dot b\dot h}^{\;\;f} + \chi_{\dot h\dot b}^{\;\;f}\right),\quad \varepsilon^{\dot b\dot h}\chi_{\{\dot b\dot h\}}^{\;\;f} = 0 \\ &\psi^{ck}_{\;\;\dot e} = \psi^{[ck]}_{\;\;\dot e} + \psi^{\{ck\}}_{\;\;\dot e},\quad \psi^{[ck]}_{\;\;\dot e} = \tfrac{1}{2}\left(\psi^{ck}_{\;\;\dot e} - \psi^{kc}_{\;\;\dot e}\right),\quad \psi^{\{ck\}}_{\;\;\dot e} = \tfrac{1}{2}\left(\psi^{ck}_{\;\;\dot e} + \psi^{kc}_{\;\;\dot e}\right),\quad \varepsilon_{ck}\psi^{\{ck\}}_{\;\;\dot e} = 0 \end{aligned} \quad (9.2.5)$$

These functions are the same as those in (9.2.3) omitting the arguments $(I,III,II)=(x_I,x_{III},x_{II})$ for clarity. The antisymmetric doublet corresponding to $\underline{2}_a$ in (9.2.4) reads

$$\begin{aligned} \chi^f &= \varepsilon^{\dot b\dot h}\chi_{\dot b\dot h}^{\;\;f} = \chi_{\dot 1\dot 2}^{\;\;f} - \chi_{\dot 2\dot 1}^{\;\;f} = \varepsilon^{\dot b\dot h}\chi_{[\dot b\dot h]}^{\;\;f} \\ \psi_{\dot e} &= \varepsilon_{ck}\psi^{ck}_{\;\;\dot e} = \psi^{21}_{\;\;\dot e} - \psi^{12}_{\;\;\dot e} = \varepsilon_{ck}\psi^{[ck]}_{\;\;\dot e} \end{aligned} \quad (9.2.6)$$

The sextet can be decomposed into a doublet $\underline{2}_s$ in (9.2.4) of mixed symmetry, being symmetric in the indices inside the braces, but antisymmetric when an upper index and a lower index are interchanged,

$$\chi_{0\dot a}=\tfrac{1}{2}\chi_{\{\dot a\dot c\}c},\qquad \chi_{0\dot 1}=\tfrac{1}{2}\left(\chi^{\;1}_{\{\dot 1\dot 2\}}-\chi^{\;2}_{\{\dot 1\dot 1\}}\right),\qquad \chi_{0\dot 2}=\tfrac{1}{2}\left(\chi^{\;1}_{\{\dot 2\dot 2\}}-\chi^{\;2}_{\{\dot 1\dot 2\}}\right)$$
$$\psi^{a}_{0}=-\tfrac{1}{2}\psi^{\{ac\}\dot c},\qquad \psi^{1}_{0}=\tfrac{1}{2}\left(\psi^{\{12\}}_{\;\;\dot 1}-\psi^{\{11\}}_{\;\;\dot 2}\right),\qquad \psi^{2}_{0}=\tfrac{1}{2}\left(\psi^{\{22\}}_{\;\;\dot 1}-\psi^{\{12\}}_{\;\;\dot 2}\right) \tag{9.2.7}$$

and a quartet $\underline{4}$ in (9.2.4) totally symmetric under the interchange of any pair of indices,

$$\chi_{3/2}=\chi^{\;1}_{\{\dot 1\dot 1\}},\qquad\qquad \chi_{-3/2}=-\chi^{\;2}_{\{\dot 2\dot 2\}},$$
$$\chi_{1/2}=-\tfrac{1}{\sqrt3}\left(\chi^{\;1}_{\{\dot 2\dot 1\}}+\chi^{\;1}_{\{\dot 1\dot 2\}}+\chi^{\;2}_{\{\dot 1\dot 1\}}\right),\qquad \chi_{-1/2}=\tfrac{1}{\sqrt3}\left(\chi^{\;2}_{\{\dot 1\dot 2\}}+\chi^{\;2}_{\{\dot 2\dot 1\}}+\chi^{\;1}_{\{\dot 2\dot 2\}}\right)$$
$$\chi\to\psi,\quad \text{lower dotted indices}\leftrightarrow\text{upper undotted indices} \tag{9.2.8}$$

Turning to the potential equations (9.1.4). The left and right sides of (9.1.4a), (9.1.4b) and (9.1.4c) are multiplied together observing (9.2.2). We obtain

$$\Box_I\Box_{III}\Box_{II}\,\Phi_b(I,III,II)=\frac{1}{8}g_s^6$$
$$\times\left(\left\{\left[\left\{\begin{array}{l}\chi_{B\dot b}(I)\chi_{B\dot h}(III)\psi^{a}_{A}(II)\psi^{\dot b}_{B}(I)\psi^{\dot h}_{B}(III)\chi_{Aa}(II)+\chi_{B\dot b}(I)\chi_{B\dot h}(III)\chi_{A\dot a}(II)\psi^{\dot b}_{B}(I)\psi^{\dot h}_{B}(III)\psi^{\dot a}_{A}(II)\\ +\chi_{B\dot b}(I)\psi^{a}_{B}(III)\psi^{h}_{A}(II)\psi^{\dot b}_{B}(I)\chi_{Ba}(III)\chi_{Ah}(II)\\ +\chi_{B\dot b}(I)\psi^{h}_{B}(III)\chi_{A\dot a}(II)\psi^{\dot b}_{B}(I)\chi_{Bh}(III)\psi^{\dot a}_{A}(II)+c.c.\end{array}\right\}\right.\right.\right.$$
$$\left.+\text{same terms with the second B index}\to A\right]$$
$$\left.+\text{same terms with A}\to\text{C}\right\}$$
$$\left.+\text{same terms with the first B index}\to\text{C}\right) \tag{9.2.9}$$

The first term in the inner braces can be generalized via (9.2.1), dropping the *A*, *C*, *B* indices as was mentioned there. The next three terms are similarly generalized into

$$\chi_{B\dot b}(I)\chi_{B\dot h}(III)\chi_{A\dot a}(II)\psi^{\dot b}_{B}(I)\psi^{\dot h}_{B}(III)\psi^{\dot a}_{A}(II)\to\chi_{\dot b\dot h\dot a}(I,III,II)\psi^{\dot b\dot h\dot a}(I,III,II)$$
$$\chi_{B\dot b}(I)\psi^{a}_{B}(III)\psi^{h}_{A}(II)\psi^{\dot b}_{B}(I)\chi_{Ba}(III)\chi_{Ah}(II)\to\psi_{\dot b}^{\;ah}(I,III,II)\chi^{\dot b}_{\;\;ah}(I,III,II)$$
$$\chi_{B\dot b}(I)\psi^{h}_{B}(III)\chi_{A\dot a}(II)\psi^{\dot b}_{B}(I)\chi_{Bh}(III)\psi^{\dot a}_{A}(II)\to\chi^{\;\;h}_{\dot b\;\;\dot a}(I,III,II)\psi^{\dot b\;\;\dot a}_{\;\;h}(I,III,II) \tag{9.2.10}$$

None of the spinors on the right side transforms like those in (9.2.1); they differ in form with respect to the first two indices. They do not enter the equations of motion (9.2.3) and are therefore put to zero. The *c.c.* term and the remainimg terms in (9.2.9) are treated entirely analogously and lead to terms composed of those in (9.2.1) and their *c.c.* In this way, (9.2.9) is simplified to

$$\Box_I\Box_{III}\Box_{II}\,\Phi_b(I,III,II)=\frac{1}{4}g_s^6\left\{\chi_{\dot b\dot h}^{\;\;\;f}(I,III,II)\psi^{\dot b\dot h}_{\;\;\;f}(I,III,II)+c.c.\right\} \tag{9.2.11}$$

which together with (9.2.3) are the baryon wave equations in space time. Note that the ψ term in (9.2.11) is the *c.c.* of ψ on the right of (9.2.1b).

Consider at first the above exposition from the view point of baryon spectra. Classification of known baryon multiplets has shown that a free baryon consists of a quark-diquark pair rather than three separate quarks [L2 (12.79) ff]. To form such a diquark, the form of (9.2.1) calls for that x_{III} merges into x_I so that (9.2.1) becomes

$$\chi_{A\dot{b}}(I)\chi_{C\dot{h}}(III)\psi_B^{\dot{f}}(II) \rightarrow \chi_{A\dot{b}}(I)\chi_{C\dot{h}}(I)\psi_B^{\dot{f}}(II) \rightarrow \chi_{\dot{b}\dot{h}}^{\ \ \dot{f}}(I, II) \tag{9.2.12a}$$

$$\psi_A^{\dot{c}}(I)\psi_C^{\dot{k}}(III)\chi_{B\dot{e}}(II) \rightarrow \psi_A^{\dot{c}}(I)\psi_C^{\dot{k}}(I)\chi_{B\dot{e}}(II) \rightarrow \psi^{\dot{c}\dot{k}}_{\ \ \dot{e}}(I, II) \tag{9.2.12b}$$

The decoposition (9.2.5-8) also holds for (9.2.12).

The *third part* of the *second step* consists of the merge of x_{III} with x_I (9.2.12). In this step, the original three body problem is reduced to a more tractable two body problem. With the merging of quarks A and C into a diquark, the interaction between them vanishes; V_{AC}, V_{CA} in (9.2.2) and the corresponding terms on the right sides of (9.1.4) and (9.2.9) drop out. Noting these and applying (9.2.12), (9.2.11) becomes

$$\Box_I \Box_I \Box_{II} \Phi_b(I, II) = \frac{1}{4} g_s^6 \left\{\chi_{\dot{b}\dot{h}}^{\ \ \dot{f}}(I, II)\psi^{\dot{b}\dot{h}}_{\ \ \dot{f}}(I, II) + c.c.\right\} \tag{9.2.13a}$$

A specialization of this expression is

$$\Box_I \Box_I \Box_{II} \Phi_b(I, II) = \frac{1}{4} g_s^6 \left\{\chi_{\{\dot{b}\dot{h}\}}^{\ \ \ \ \dot{f}}(I, II)\psi^{\{\dot{b}\dot{h}\}}_{\ \ \ \ \dot{f}}(I, II) + c.c.\right\} \tag{9.2.13b}$$

where the braced indices are interchangeable. Correspondlingly, (9.2.3) becomes

$$\begin{aligned} \partial_I^{a\dot{b}}\partial_I^{g\dot{h}}\partial_{II\,\dot{e}f}\chi_{\dot{b}\dot{h}}^{\ \ \dot{f}}(I, II) &= -i(m_A m_B m_C + \Phi_b(I, II))\psi^{ag}_{\ \ \dot{e}}(I, II) \\ \partial_{I\,\dot{b}c}\partial_{I\,\dot{h}k}\partial_{II}^{d\dot{e}}\psi^{ck}_{\ \ \dot{e}}(I, II) &= -i(m_A m_B m_C + \Phi_b(I, II))\chi_{\dot{b}\dot{h}}^{\ \ d}(I, II) \end{aligned} \tag{9.2.14}$$

These are the baryon counterpart of (2.2.4, 5). Since there are only two independent coordinates, x_I for the diquark and x_{II} for the quark, in these equations, the neglect of genuine three quark interaction above (9.1.1) is justified.

As to the triple products containing at least one of (9.1.1, 2, 3) and one of (9.1.4a, 4b, 4c), these likewise drop out. The reasons are the same as those given beneath (9.2.2) which led to that the V and VV terms there drop out. This is anlogous to the corresponding discussion for the meson case beneath (2.2.5).

9.3. BARYON INTERNAL FUNCTIONS AND TOTAL WAVE EQUATIONS

To include flavor dependence, the quark equations (9.1.1-3) are generalized to the form (2.3.11). Analogously, (9.1.1-3) are multiplied by $\xi_A{}^p(z_I)$, $\xi_B{}^q(z_{II})$ and $\xi_C{}^s(z_{III})$, respectively. m_A, m_B and m_C are replaced by mass operators of the form of (2.3.13). Following a procedure similar to that in Sec. 2.3, the so-modified (9.1.1-3) are multiplied together and the generalizations of second step in Sec. 9.2 are analogously repeated.

9.3.1. Baryon Internal Or Flavor Functions

The *first part* of the *third step* is the analog of (2.3.18a) for mesons. Using (2.3.12), the baryon internal function is written as

$$\xi_A^p(z_I)\xi_B^q(z_{II})\xi_C^s(z_{III}) = z_I^p z_{II}^q z_{III}^s \leftrightarrow \xi^{psq}(z_I, z_{III}, z_{II}) \qquad (9.3.1)$$

Consider the SU(3) case, (9.3.1) is decomposed into irreducible multiplets analogous to those in (2.4.8). The corresponding Clebsch-Gordan series reads [C3 (3.32)]

$$\underline{3}\otimes\underline{3}\otimes\underline{3} = \left(\underline{6}\oplus\underline{3}^*\right)\otimes\underline{3} = \left(\underline{6}\otimes\underline{3}\right)+\left(\underline{3}^*\otimes\underline{3}\right) = \left(\underline{10}\oplus\underline{8}_s\right)+\left(\underline{8}_a\oplus\underline{1}\right) \qquad (9.3.2)$$

The corresponding decomposition of (9.3.1) can be achieved by means of the totally antisymmetric operator ε_{sqr}, ε_{123}=1, analogous to (9.2.4, 6-8),

$$\underline{10}: \quad \xi^{\{psq\}}(z_I, z_{III}, z_{II,}) \quad \text{totally symmetric in } p, s, q \qquad (9.3.3a)$$

$$\underline{8}_s: \quad \eta_r^p(z_I, z_{III}, z_{II,}) = \xi^{\{ps\}q}(z_I, z_{III}, z_{II,})\varepsilon_{sqr} \qquad (9.3.3b)$$

$$\underline{8}_a: \quad \eta_{(a)r}^{\;p}(z_I, z_{III}, z_{II,}) = \xi^{[ps]q}(z_I, z_{III}, z_{II,})\varepsilon_{sqr} \qquad (9.3.3c)$$

$$\underline{1}: \quad \eta_{(1)}(z_I, z_{III}, z_{II,}) = \xi^{psq}(z_I, z_{III}, z_{II,})\varepsilon_{psq} \qquad (9.3.3d)$$

As in (9.2.5), $\{ps\}$ denotes symmetry and $[ps]$ antisymmetry under interchange of p and s.

To be consistent with the diquark-quark description of a free baryon, z_{III} is similarly merged into z_I by analogy to (9.2.12) and (9.3.1) becomes

$$\xi_A^p(z_I)\xi_B^q(z_{II})\xi_C^s(z_{III}) \rightarrow \xi_A^p(z_I)\xi_B^q(z_{II})\xi_C^s(z_I) = z_I^p z_{II}^q z_I^s \leftrightarrow \xi^{psq}(z_I, z_{II}) \qquad (9.3.4)$$

Application to (9.3.3) leads to

$$\underline{10}: \quad \xi^{\{psq\}}(z_I, z_{II}) \quad \text{totally symmetric in } p, s, q \qquad (9.3.5a)$$

$$\underline{8}_s: \qquad \eta_r^p(z_I, z_{II}) = \xi^{\{ps\}q}(z_I, z_{II})\varepsilon_{sqr} \tag{9.3.5b}$$

$$\underline{8}_a: \qquad \eta_{(a)r}^{\ p}(z_I, z_{II}) = \xi^{[ps]q}(z_I, z_{II})\varepsilon_{sqr} = 0 \tag{9.3.5c}$$

$$\underline{1}: \qquad \eta_{(1)}(z_I, z_{II}) = \xi^{psq}(z_I, z_{II})\varepsilon_{psq} = 0 \tag{9.3.5d}$$

9.3.2. Total Baryon Wave Functions

The total wave functions of the ground state baryons in the SU(3) scheme are obtained by combining (9.2.1) and (9.3.1),

$$\chi_{\dot{b}\dot{h}}^{\ f}(I, III, II)\xi^{psq}(z_I, z_{III}, z_{II}), \qquad \psi_{\ \dot{e}}^{ck}(I, III, II)\xi^{psq}(z_I, z_{III}, z_{II}) \tag{9.3.6}$$

These can be decomposed into baryon multiplets according to (9.2.6-8) combined with (9.3.3),

$$\underline{4} \otimes \underline{10}, \text{ totally symmetric spin } 3/2 \text{ decuplet}: (\chi_{3/2}, \chi_{1/2}, \chi_{-1/2}, \chi_{-3/2}, \psi_{3/2}, \psi_{1/2}, \psi_{-1/2}, \psi_{-3/2},)\xi^{\{psq\}} \tag{9.3.7a}$$

$$\underline{2}_s \otimes \underline{8}_s, \text{ mixed symmetric spin } 1/2 \text{ octet}: \qquad \chi_{0\dot{a}}\,\eta_r^p, \qquad \psi_0^a\,\eta_r^p \tag{9.3.7b}$$

$$\underline{2}_a \otimes \underline{8}_a, \text{ mixed antisymmetric spin } 1/2 \text{ octet}: \quad \chi^f\eta_{(a)r}^{\ p}, \qquad \psi_{\dot{e}}\,\eta_{(a)r}^{\ p} \tag{9.3.7c}$$

$$0 \otimes \underline{1}, \quad 0 \text{ singlet}: \qquad 0 \tag{9.3.7d}$$

where the arguments (I,III,II) and (z_I,z_{III},z_{II}) have been omitted.

The assignments (9.3.7) are dictated by the following *symmetric quark postulate: "The total baryon wave functions must be totally symmetric under the simultaneous interchange of the space-time index and the internal index associated with one quark with those associated with another quark."* This constitutes the *second part* of the *third step*. The essence of this postulate is supported by the approximately correct calculations of the baryon magnetic moments [L2 p238]. This postulate plays a role akin to that of Pauli's theorem in conventional quantum mechanics in eliminating extraneous solutions. The role of identical particles in Pauli's theorem is taken over here by quarks which, being not observable, are indistinguishable, hence appear as identical inside hadrons.

For free baryons, (9.2.12) and (9.3.4) apply and (9.3.5c, d) leads to that the singlet $\underline{1}$ and mixed antisymmetric octet $\underline{8}_a$ drops out, in agreement with data. In addition, it will be shown in connection with (10.2.1´) and the conclusion beneath (10.2.5) below that (9.2.14) does not have any solution for the $\underline{2}_a$ of (9.2.6) associated with $\underline{8}_a$ in (9.3.7c).

The simplified notations $z_I=z$ and $z_{II}=u$ of (2.3.16, 17) are now introduced and generalized to include $z_{III}=v$. The mass term $m_A m_B m_C$ in (9.2.3) is generalized by analogy to (2.3.18b);

$$m_{1op}(z_I)m_{1op}(z_{III})m_{1op}(z_{II}) = m_{1op}(z)m_{1op}(v)m_{1op}(u) \rightarrow m_{3op}(z,v,u) \qquad (9.3.8)$$

This is the *third part* of the *third step* and is the internal counterpart of the space time generalization (9.2.2). This *third step* generalizes (9.2.3) to the total baryons wave equations

$$\partial_I^{a\dot{b}}\partial_I^{g\dot{h}}\partial_{II\dot{e}f}\chi_{\dot{b}\dot{h}}^{f}(I,III,II)\xi^{psq}(z,v,u) = -i\left(m_{3op}(z,v,u)+\Phi_b(I,III,II)\right)\psi^{ag}{}_{\dot{e}}(I,III,II)\xi^{psq}(z,v,u)$$
$$\partial_{I\dot{b}c}\partial_{I\dot{h}k}\partial_{II}^{d\dot{e}}\psi^{ck}{}_{\dot{e}}(I,III,II)\xi^{psq}(z,v,u) = -i\left(m_{3op}(z,v,u)+\Phi_b(I,III,II)\right)\chi_{\dot{b}\dot{h}}^{d}(I,III,II)\xi^{psq}(z,v,u)$$
$$(9.3.9)$$

The equation for Φ_b (9.2.11) is not affected by the inclusion of internal functions for the same reason given beneath (2.3.19). Equations (9.3.9) and (9.2.11) are the set of total baryon wave equations and may themselves be regarded as a basic postulate, instead of the three steps of assumptions given in this chapter, of the scalar strong interaction hadron theory for baryons. Note that ξ^{psq} is not limited to the SU(3) case (9.3.3); *c* and *b* quarks may be present.

Invariance of (9.3.9) and (9.2.11) under proper Lorentz transformations and space inversion are manifest due to their spinor form, on similar grounds as those mentioned above (2.3.20). Application of (B31) to (9.2.1) leads to

$$\chi_{\dot{b}\dot{h}}^{f}\left(x_I^0,\underline{x}_I,x_{III}^0,\underline{x}_{III},x_{II}^0,\underline{x}_{II}\right) \leftrightarrow \psi^{bh}{}_{\dot{f}}\left(x_I^0,-\underline{x}_I,x_{III}^0,-\underline{x}_{III},x_{II}^0,-\underline{x}_{II}\right)$$

under space inversion.

For a free baryon, (9.2.12) and (9.3.4) apply and (9.3.8, 9) become

$$m_{1op}(z_I)m_{1op}(z_{III})m_{1op}(z_{II}) \rightarrow m_{1op}(z)m_{1op}(z)m_{1op}(u) \rightarrow m_{3op}(z,u) \qquad (9.3.10)$$

$$\partial_I^{a\dot{b}}\partial_I^{g\dot{h}}\partial_{II\dot{e}f}\chi_{\{\dot{b}\dot{h}\}}^{f}(I,II)\xi^{\{ps\}q}(z,u) = -i\left(m_{3op}(z,u)+\Phi_b(I,II)\right)\psi^{\{ag\}}{}_{\dot{e}}(I,II)\xi^{\{ps\}q}(z,u)$$
$$\partial_{I\dot{b}c}\partial_{I\dot{h}k}\partial_{II}^{d\dot{e}}\psi^{\{ck\}}{}_{\dot{e}}(I,II)\xi^{\{ps\}q}(z,u) = -i\left(m_{3op}(z,u)+\Phi_b(I,II)\right)\chi_{\{\dot{b}\dot{h}\}}^{d}(I,II)\xi^{\{ps\}q}(z,u)$$
$$(9.3.11)$$

Observing (9.3.5c, d), only the symmetric (9.3.7a, b) survive here. This and (9.2.13b), giving $\Phi_b(I,II)$, are the total *equations of motion for free, ground state baryons* in diquark-quark configuration.

9.3.3. Internal or Flavor Functions and Mass Operator

Analogous to (2.3.25), (9.3.8) is put to

$$m_{3op}(z,v,u)=m_{1op}^{3}(z,v,u) \tag{9.3.12}$$

Similarly, (2.3.26) is generalized to

$$m_{1op}(z,v,u)=\frac{1}{2}m_q\left(z^q\partial_{zq}+v^q\partial_{vq}+u^q\partial_{uq}\right) \tag{9.3.13}$$

where the flavor index q is summed over. For a free baryon, (9.2.12) and (9.3.4) apply and (9.3.12) becomes

$$m_{3op}(z,u)=m_{1op}^{3}(z,u) \tag{9.3.14}$$

and the v term drops out in (9.3.13).

The total equations of motion for baryons are now specified by (9.2.11), (9.3.9), (9.3.12), and (9.3.13). For baryon spectra considerations, the simpler equations of motion for free baryon or diquark-quark, (9.3.11, 13, 14) can be used. Equation (9.3.11) is separable in x and z if the baryon internal function $\xi^{\{ps\}q}$ is an eigenfunction of $m_{3op}(z,u)$. With (9.3.13, 14), we can write

$$m_{3op}(z,u)\xi^{\{ps\}q}(z,u)=M_b^3\xi^{\{ps\}q}(z,u) \tag{9.3.15}$$

$$\partial_I^{a\dot{b}}\partial_I^{g\dot{h}}\partial_{II\dot{e}f}\chi_{\{\dot{b}\dot{h}\}}^{f}(I,II)=-i\left(M_b^3+\Phi_b(I,II)\right)\psi^{\{ag\}}_{\dot{e}}(I,II) \tag{9.3.16a}$$

$$\partial_{I\dot{b}c}\partial_{I\dot{h}k}\partial_{II}^{d\dot{e}}\psi^{\{ck\}}_{\dot{e}}(I,II)=-i\left(M_b^3+\Phi_b(I,II)\right)\chi_{\{\dot{b}\dot{h}\}}^{d}(I,II) \tag{9.3.16b}$$

where the eigenvalue M_b^3 is the separation constant between functions in space time and in internal space and corresponds to M_m^2 for mesons in (2.3.21, 22). If there exists no eigenvalue, (9.3.16) can still be obtained. Multiplying (9.3.16) by the complex conjugate of the baryon internal function and integrating over the internal space leads to

$$M_b^3=\frac{\int dv_{zq}dv_{uq}\xi_{\{ps\}q}(z,u)m_{3op}(z,u)\xi^{\{ps\}q}(z)}{\int dv_{zq}dv_{uq}\xi_{\{ps\}q}(z,u)\xi^{\{ps\}q}(z)} \tag{9.3.17}$$

which corresponds to (2.3.24) for mesons. This M_b^3 is not the strict separation constant like M_b^3 in (9.3.15) but both represent the quark mass contribution to the baryon mass.

The baryon internal functions are analogous to the corresponding meson internal functions (2.4.8). With (9.3.4) and (9.3.5a, b), they are

$$\eta_r^p(z,u)=z^pz^su^q\varepsilon_{rsq} \tag{9.3.18a}$$

$$\xi^{\{psq\}}(z,u) = z^p z^s u^q + z^p u^s z^q + u^p z^s z^q \tag{9.3.18b}$$

Inserting (9.3.18) into (9.3.15) yields

$$M_b^3 = \frac{1}{8}(m_p + m_s + m_q)^3 \tag{9.3.19}$$

which is the counterpart of (2.3.27). This form will however not distinguish between Σ^0 and Λ, which according to (9.3.18a) have the internal functions

$$\eta_1^1 - \eta_2^2 = 2z^1 z^2 u^3 - (z^2 u^1 + z^1 u^2) z^3 \qquad \text{for } \Sigma^0 \tag{9.3.20a}$$

$$\eta_1^1 + \eta_2^2 - 2\eta_3^3 = 3(z^2 u^1 - z^1 u^2) z^3 \qquad \text{for } \Lambda \tag{9.3.20b}$$

The situation is analogous to that for the η meson at the end of Sec. 5.4. For baryons, (5.4.1) becomes (9.46) of [L2],

$$M_{GMOb} = a_{8b} + b_{8b} Y + c_{8b}\left(I(I+1) - Y^2/4\right) \tag{9.3.21}$$

For the same reason as that mentioned beneath (5.4.1), M_{GMOb} is equated to $2M_b$ of (9.3.19) with $m_1 = m_2$ for the neutron, Σ^0 and Ξ^0. This fixed the three constants

$$a_{8b} = 2m_1 + m_3, \qquad b_{8b} = m_1 - m_3, \qquad c_{8b} = 0 \tag{9.3.22}$$

The last relation shows that the term quadratic in the generators in (9.3.21) is absent in the quark mass summation operator m_{1op} in (9.3.13, 14) using (2.3.26). It is the quadratic I^2 that differentiates Σ^0 from Λ in (9.3.20) and this accounts for that (9.3.19) is the same for Σ^0 and Λ; the simple form (9.3.14) needs be modified. The pros and cons of (9.3.14, 19) versus (9.3.21) are the same as those mentioned at the end of Sec. 5.4.

Chapter 10

REDUCTION OF BARYON WAVE EQUATIONS

The basic set of equations of motion for free baryon in diquark-quark configuration is given by (9.3.16) and (9.2.13b). Reduction of these equations is analogous to that of the meson equations of Sec. 3.1 and follows [H15] closely.

Equations determining the doublet wave functions (9.2.7) can be formally obtained as follows. The e, f and d indices in (9.3.16) are raised and lowered. Multiply the so-modified (9.3.16a) and (9.3.16b) by $\delta_{\dot{e}g}/2$ and $\delta^{d\dot{h}}/2$, respectively, and apply (9.2.7). Equation (9.3.16) is now reduced to

$$\partial_I^{a\dot{b}}\partial_{II}^{f\dot{e}}\partial_I^{e\dot{h}}\tfrac{1}{2}\chi_{\{\dot{b}\dot{h}\}f}(I,II)=-i\left(M_b^3+\Phi_b(I,II)\right)\psi_0^a(I,II) \tag{10.0.1a}$$

$$\partial_{I\dot{b}c}\partial_{II\dot{e}h}\partial_{I\dot{h}k}\tfrac{1}{2}\psi^{\{ck\}\dot{e}}(I,II)=i\left(M_b^3+\Phi_b(I,II)\right)\chi_{0\dot{b}}(I,II) \tag{10.0.1b}$$

The doublet wave functions on the right sides of (10.0.1) are expressed in terms of the left sides which in general contains both the doublet and quartet wave functions (9.2.7, 8).

Analogously, (9.2.14) can be contracted by the ε^{ab} tensor in (9.2.6) and becomes

$$\begin{aligned}\partial_I^{a\dot{b}}\varepsilon_{ag}\partial_I^{g\dot{h}}\partial_{II\dot{e}f}\chi_{\dot{b}\dot{h}}^{\;\;f}(I,II)&=-i\left(m_Am_Bm_C+\Phi_b(I,II)\right)\psi_{\dot{e}}(I,II)\\ \partial_{I\dot{b}c}\varepsilon^{\dot{b}\dot{h}}\partial_{I\dot{h}k}\partial_{II}^{d\dot{e}}\psi^{ck}_{\;\;\dot{e}}(I,II)&=-i\left(m_Am_Bm_C+\Phi_b(I,II)\right)\chi^d(I,II)\end{aligned} \tag{10.0.2}$$

If the symmetric quark postulate [H15] beneath (9.3.7) is employed, (10.0.1) takes a simpler form. After raising and lowering the d, e and f indices and dropping the arguments, the baryon wave functions in (9.3.11) can be decomposed into the form

$$\chi_{\{\dot{b}\dot{h}\}f}\xi^{\{ps\}q}\to\chi_{0\dot{b}}\delta_{\dot{h}f}\left(\xi^{\{ps\}q}-\xi^{\{psq\}}\right)+\chi_{\{\dot{b}\dot{h}f\}}\xi^{\{psq\}} \tag{10.0.3a}$$

$$\psi^{\{ck\}\dot{e}}\xi^{\{ps\}q}\to\psi_0^c\delta^{k\dot{e}}\left(\xi^{\{ps\}q}-\xi^{\{psq\}}\right)+\psi^{\{ck\dot{e}\}}\xi^{\{psq\}} \tag{10.0.3b}$$

Observing (9.2.8), the last terms in (10.0.3) correspond to $\underline{4}\times\underline{10}$ of (9.3.7a). The internal functions in the parentheses have 6×3−10=8 components so that the first terms on the right of (10.0.3) correspond to $\underline{2}_s\times\underline{8}_s$ of (9.3.7b). This is seen more directly by contracting (10.0.3) by ε_{rsq} via (9.3.5b, 5c),

$$\varepsilon_{rsq}\left(\chi_{\{\dot{b}\dot{h}\}f}, \quad \psi^{\{ck\}\dot{e}}\right)\xi^{\{ps\}q} = \left(\chi_{0\dot{b}}\delta_{\dot{h}f}, \quad \psi_0^c\delta^{k\dot{e}}\right)\eta_r^p \tag{10.0.4}$$

The operations (10.0.3a, 4) reduce the first of (9.3.11) to the doublet wave equation

$$\partial_I^{a\dot{b}}\partial_I^{g\dot{f}}\partial_{II}^{f\dot{e}}\chi_{0\dot{b}}(I,II)\eta_r^p(z,u) = -i\left(m_{3op}(z,u)+\Phi_b(I,II)\right)\psi_0^a(I,II)\delta^{g\dot{e}}\eta_r^p(z,u) \tag{10.0.5}$$

Contracting by $\delta_{\dot{e}g}$ and noting (9.3.15), (10.0.5) can be turned into

$$\partial_I^{a\dot{b}}\partial_{II}^{f\dot{e}}\partial_I^{e\dot{f}}\tfrac{1}{2}\chi_{0\dot{b}}(I,II) = -i\left(M_b^3+\Phi_b(I,II)\right)\psi_0^a(I,II) \tag{10.0.6a}$$

Analogously, the second of (9.3.11) reduces to

$$\partial_{I\dot{b}c}\partial_{II\dot{e}h}\partial_{I\dot{h}e}\tfrac{1}{2}\psi_0^c(I,II) = -i\left(M_b^3+\Phi_b(I,II)\right)\chi_{0\dot{b}}(I,II) \tag{10.0.6b}$$

These two equations are the same as (10.0.1) with

$$\chi_{\{\dot{b}\dot{h}\}f} = \chi_{0\dot{b}}\delta_{\dot{h}f}, \qquad \psi^{\{ck\}\dot{e}} = -\psi_0^c\delta^{k\dot{e}} \tag{10.0.7}$$

10.1. Baryon Equations in Relative Space

The coordinate transformations (3.1.3, 4), (3.1.9) with $\varphi(\underline{x})$ representing the baryon relative wave functions, and (3.1.10a) are taken over here. These relations are formally modified for application to baryons via the substitution of the transformation constant a_m by $a_b \to a_{b0}$. Equation (3.1.5) is replaced by

$$\begin{aligned}\chi_{\{\dot{a}\dot{c}\}}^{\ e}(x_I, x_{II}) &= \chi_{\{\dot{a}\dot{c}\}}^{\ e}(\underline{x})\exp\left(-iK_\mu X^\mu + i\omega_K x^0\right)\\ \psi^{\{ac\}}_{\ \ \dot{e}}(x_I, x_{II}) &= \psi^{\{ac\}}_{\ \ \dot{e}}(\underline{x})\exp\left(-iK_\mu X^\mu + i\omega_K x^0\right)\end{aligned} \tag{10.1.1}$$

where $K_\mu=(E_0, -\underline{K})$, as in (3.1.6). Normalization will be considered in Sec. !0.3. As in Sec. 3.2, we consider only the rest frame baryons and put $\underline{K}=0$. The modified relations mentioned above (10.1.1) will remove the unknown relative energy ω_K and allow for a decoupling of the doublet wave functions in (10.2.1) from the quartet wave functions in (10.5.1) from the general six component wave functions in (10.1.2) below. This is entirely analogous to the decoupling of the singlet and triplet mentioned below (3.1.10a).

The basic set (9.3.16) and (9.2.13b) now reduce to

$$\left(i\delta^{a\dot{b}}E_0/2+\underline{\sigma}^{a\dot{b}}\underline{\partial}\right)\left(i\delta^{g\dot{h}}E_0/2+\underline{\sigma}^{g\dot{h}}\underline{\partial}\right)\left(i\delta_{\dot{e}f}E_0/2+\underline{\sigma}_{\dot{e}f}\underline{\partial}\right)\chi_{\{\dot{b}\dot{h}\}}^{\;\;f}(\underline{x})$$
$$=-i\left(M_b^3+\Phi_b(\underline{x})\right)\psi^{\{ag\}}_{\;\;\dot{e}}(\underline{x})$$
$$\left(i\delta_{\dot{b}c}E_0/2-\underline{\sigma}_{\dot{b}c}\underline{\partial}\right)\left(i\delta_{\dot{h}k}E_0/2-\underline{\sigma}_{\dot{h}k}\underline{\partial}\right)\left(i\delta^{d\dot{e}}E_0/2-\underline{\sigma}^{d\dot{e}}\underline{\partial}\right)\psi^{\{ck\}}_{\;\;\dot{e}}(\underline{x})$$
$$=-i\left(M_b^3+\Phi_b(\underline{x})\right)\chi_{\{\dot{b}\dot{h}\}}^{\;\;d}(\underline{x}) \qquad (10.1.2)$$

$$\Delta\Delta\Delta\Phi_b(\underline{x})=\frac{1}{4}g_s^6\left\{\chi_{\{\dot{b}\dot{h}\}}^{\;\;f}(\underline{x})\psi^{*\{\dot{b}\dot{h}\}}_{\;\;f}(\underline{x})+c.c.\right\} \qquad (10.1.3a)$$

$$\psi^{*\{\dot{b}\dot{h}\}}_{\;\;f}(\underline{x})=\left(\psi^{\{bh\}}_{\;\;\dot{f}}\right)^*, \qquad \chi^{*\;\dot{f}}_{\{bh\}}=\left(\chi_{\{\dot{b}\dot{h}\}}^{\;\;f}\right)^* \qquad (10.1.3b)$$

Here, it is seen that the requirement mentioned below (3.1.10a) is analogously fulfilled here. In the absence of the interaction potential Φ_b and quark motion represented by $\underline{\partial}$, (10.1.2) leads to that the baryon mass E_0 reduces to the sum of the quark masses via (9.3.19).

Green's function of (10.1.3) satisfies

$$\Delta\Delta\Delta G_b(\underline{x},\underline{x}')=\delta(\underline{x}-\underline{x}') \qquad (10.1.4a)$$

$$G_b(\underline{x},\underline{x}')=-\frac{1}{96\pi}|\underline{x}-\underline{x}'|^3 \qquad (10.1.4b)$$

Operating (10.1.4b), which is regular everywhere, by Δ yields

$$\Delta G_b(\underline{x},\underline{x}')=-\frac{1}{8\pi}|\underline{x}-\underline{x}'| \qquad (10.1.5)$$

which together with (3.2.6) reproduces (10.1.4a). Analogous to (3.2.8), there are also homogenous solutions to (10.1.4) and these are the corresponding ones in (3.2.8a) together with $|\underline{x}|=r$ from (3.2.6b) and an $|\underline{x}|^4$ term. The solution to (10.1.3) is

$$\Phi_b(\underline{x})=\Phi_{bc}(\underline{x})+\frac{d_b}{r}+d_{b0}+d_{b1}r+d_{b2}r^2+d_{b4}r^4 \qquad (10.1.6a)$$

$$\Phi_{bc}(\underline{x})=\frac{g_s^6}{192\pi}\int d^3\underline{x}'|\underline{x}-\underline{x}'|^3\,\mathrm{Re}\left(\chi_{\{\dot{b}\dot{h}\}}^{\;\;f}(\underline{x}')\psi^{*\{\dot{b}\dot{h}\}}_{\;\;f}(\underline{x}')\right) \qquad (10.1.6b)$$

Here, the d_b's are five integration constants to the sixth order potential equation (10.1.3a) without the source term on its right side. This is the case for free baryons, for which the right side of (10.1.3a) and (10.1.6b) drop out via (10.3.12) below. These constants originate in the relative space $\underline{x}$ and are independent of baryon flavor which are characterized by the internal function $\eta^{p}{}_{r}$ in the internal space z_I, z_{II} as in (9.3.5b). They are analogous to the integration constants d_m and d_{m0} for meson case in (3.2.8) which are independent of flavor, as was mentioned below (4.4.1).

In the $|\underline{x}|\to 0$ limit, we have

$$\Phi_b\left(|\underline{x}|\to 0\right)=d_b/r \tag{10.1.7a}$$

Aymtotically,

$$\Phi_b\left(|\underline{x}|\to \infty\right)=\alpha_c^3 r^3, \qquad \alpha_c^3=\frac{g_s^6}{192\pi}\int d^3\underline{x}\,\mathrm{Re}\left(\chi_{\{\dot{b}\dot{h}\}}^{\;\;f}\left(\underline{x}\right)\psi^{*\{\dot{b}\dot{h}\}}{}_{f}\left(\underline{x}\right)\right) \tag{10.1.7b}$$

Both of these potentials are independent of the angles in the relative space. By a reason analogous to that leading to (3.2.20),

$$d_{b4}=0 \tag{10.1.8}$$

is set so that the leading confining term at large r is the nonlinear Φ_{bc} of (10.1.6b). Such a potential gives rise to a harmonic type of confinement (10.2.6) below, which is close to the form (10.2.7a) below obtained from the linearized confining potential $d_{b2}r^2$. It is in agreement with the rather successful harmonic oscillator model [F3], which does not call for a stronger confinement signified by $d_{b4}r^4$ in (10.1.6a).

Φ_{bc} of (10.1.6b) is the nonlinear confining potential for baryons analogous to Φ_c of (3.2.8b). $d_{b1}r+d_{b2}r^2$ in (10.1.6a) is the linearized confining potential for baryons which takes over confinement when Φ_{bc} vanishes. It has no counterpart in Φ_m of (3.2.8a), noting (3.2.20).

10.2. Doublet Wave Equations and Confinement

10.2.1. Doublet Wave Equations in Relative Space

Insert (10.1.1) into (10.0.1) and put $\underline{K}=0$. The quartet wave functions of (9.2.8) drop out and we obtain the rest frame doublet baryon wave equations for $\underline{2}_s$ in (9.2.7) and (9.3.7b),

$$\begin{aligned}\left(i\delta^{a\dot{b}}E_{0d}/2+\underline{\sigma}^{a\dot{b}}\underline{\partial}\right)\left(E_{0d}^2/4+\Delta\right)\chi_{0\dot{b}}\left(\underline{x}\right)&=i\left(M_b^3+\Phi_{bd}\left(\underline{x}\right)\right)\psi_0^a\left(\underline{x}\right)\\ \left(i\delta_{\dot{b}c}E_{0d}/2-\underline{\sigma}_{\dot{b}c}\underline{\partial}\right)\left(E_{0d}^2/4+\Delta\right)\psi_0^c\left(\underline{x}\right)&=i\left(M_b^3+\Phi_{bd}\left(\underline{x}\right)\right)\chi_{0\dot{b}}\left(\underline{x}\right)\end{aligned} \tag{10.2.1}$$

A similar set of equations is obtained from (10.0.2) for $\underline{2}_a$ in (9.2.6) and (9.3.7c) using the same exponential ansatz as in (10.1.1),

$$\left(i\delta^{a\dot{b}}E_{0d}/2-\underline{\sigma}^{a\dot{b}}\underline{\partial}\right)\left(E_{0d}^2/4+\Delta\right)\psi_{\dot{b}}(\underline{x})=-i\left(M_b^3+\Phi_{bd}(\underline{x})\right)\chi^a(\underline{x}) \quad (10.2.1\text{´a})$$

$$\left(i\delta_{\dot{b}c}E_{0d}/2-\underline{\sigma}_{\dot{b}c}\underline{\partial}\right)\left(E_{0d}^2/4+\Delta\right)\chi^c(\underline{x})=i\left(M_b^3+\Phi_{bd}(\underline{x})\right)\psi_{\dot{b}}(\underline{x}) \quad (10.2.1\text{´b})$$

By limiting the baryon wave functions on the right of (10.1.6b) to doublet wave functions, (10.1.6) becomes

$$\Phi_{bd}(\underline{x})=\Phi_{cd}(\underline{x})+\frac{d_b}{r}+d_{b0}+d_{b1}r+d_{b2}r^2 \quad (10.2.2\text{a})$$

$$\Phi_{cd}(\underline{x})=-\frac{g_s^6}{72\pi}\int d^3\underline{x}'|\underline{x}-\underline{x}'|^3\,\mathrm{Re}\left(\chi_{0a}^*(\underline{x}')\psi_0^a(\underline{x}')\right),\qquad \chi_{0a}^*=\left(\chi_{0\dot{a}}\right)^* \quad (10.2.2\text{b})$$

Here, the subscript d refers to specialization to doublets. Note that if $\underline{K}\neq 0$, the doublet and quartet wave function components couple and the full (10.1.2) and (10.1.6b) must be used. Analogous to the meson case mentioned at the end of Sec. 3.3, separation of r, ϑ and ϕ variables will then no longer be possible. The doublet wave functions (9.2.7) will remain as "large components" and the quartet wave functions (9.2.8) will enter as "small components" of the order of $|\underline{K}|$ in (10.1.2) and (10.1.6b).

10.2.2. Doublet Radial Wave Equations and Asymtotic Solutions

If $\Phi_{bd}(\underline{x})$ depends only upon r, which according to (10.1.7) is the case at least in the $|\underline{x}|\to 0$ and ∞ limits, (10.2.1) can be separated. One can recognize that (10.2.1) has a form similar to that of the Dirac equation (C10) with a central potential and $A, V_{PB}\to 0$. The forms of the solutions [B7] to such a Dirac equation can therefore take over here. For each total angular momentum value j, there are two possible orbital angular momentum values, namely, $l=j-½$ and $l=j+½$. In terms of the spinors (C11), the corresponding wave functions are

$$\psi_0^a(\underline{x})=\pm g_l(r)\sqrt{\frac{(l\pm m+½)}{2l+1}}\,Y_{l\,m\mp½}(\vartheta,\phi)+if_l(r)\sqrt{\frac{(l\mp m+3/2)}{2l+3}}\,Y_{l+1\,m\mp½}(\vartheta,\phi)$$
for $\quad j=l+½$ (10.2.3a)

$$\psi_0^a(\underline{x})=g_l(r)\sqrt{\frac{(l\mp m+½)}{2l+1}}\,Y_{l\,m\mp½}(\vartheta,\phi)\pm if_l(r)\sqrt{\frac{(l\pm m-½)}{2l-1}}\,Y_{l-1\,m\mp½}(\vartheta,\phi)$$
for $\quad j=l-½$ (10.2.3b)

where the upper signs hold for $a=1$ and the lower ones for $a=2$. $\chi_{0\dot{a}}$ is found by changing the signs of $f_l(r)$ in (10.2.3). With (10.2.3a, 3b), (10.2.1) yields eight radial equations which degenerate into two such;

$$\left[\frac{E_{0d}}{2}\left(\frac{E_{0d}^2}{4}+\Delta_l\right)+M_b^3+\Phi_{bd}\left(|\underline{x}|\to 0,\infty\right)\right]g_l(r)+\left(\frac{E_{0d}^2}{4}+\Delta_l\right)\left(\frac{\partial}{\partial r}+\frac{1+\kappa}{r}\right)f_l(r)=0 \quad (10.2.4a)$$

$$\left[-\frac{E_{0d}}{2}\left(\frac{E_{0d}^2}{4}+\Delta_{l\pm1}\right)+M_b^3+\Phi_{bd}\left(|\underline{x}|\to 0,\infty\right)\right]f_l(r)+\left(\frac{E_{0d}^2}{4}+\Delta_{l\pm1}\right)\left(\frac{\partial}{\partial r}+\frac{1-\kappa}{r}\right)g_l(r)=0 \quad (10.2.4b)$$

$$\text{for } j=l\pm\tfrac{1}{2},\ \begin{pmatrix}\text{upper}\\ \text{lower}\end{pmatrix}\text{ sign and } \kappa=\begin{pmatrix}l+1\\ -l\end{pmatrix} \quad (10.2.4c)$$

where Δ_l has been given in (3.4.5b) and

$$\Phi_{bd}\left(|\underline{x}|\to 0\right)=\frac{d_b}{r} \quad (10.2.5a)$$

$$\Phi_{cd}\left(|\underline{x}|\to\infty\right)=-\frac{g_s^6}{72\pi}r^3\int d^3\underline{x}\,\mathrm{Re}\left(\chi_{0a}^*(\underline{x})\psi_0^a(\underline{x})\right)=\alpha_d^3 r^3 \quad (10.2.5b)$$

according to (10.2.2).

Applying the same expansion (10.2.3) to (10.2.1´a), one obtains (10.2.4) with the sign of E_{0d} changed. For (10.2.1´b), however, (10.2.4) with the sign of g or f changed is found. These two sets of radial equations are incompatible with each other and (10.2.1´) has therefore no solution. This shows that in addition to that $\underline{8}_a$ vanishes in (9.3.5c, 7c), $\underline{2}_a$ in (9.3.7c) also drops out consistently.

Note that (10.2.4) holds for all r for free baryons for which the nonlinear confining potential Φ_{cd} of (10.2.2b) vanishes (see discussion below (10.3.14)). The assignments of signs of f and g in (10.2.3a) differ from the corresponding ones in [H15 (6.3)] which reads

$$(10.2.3a)\text{ with } g_l\to g_l,\qquad \chi_{0\dot{a}}=\psi_0^a\text{ with } f_l\to -f_l \quad (10.2.3c)$$

As a consequence, (10.2.4) differs from the corresponding [H15 (6.4)] in the sign of f.

The both $j=\pm\frac{1}{2}$ cases are convertible into each other. Putting $l\to l-1$ and $g\leftrightarrow if$ in (10.2.3a) converts it into (10.2.3b). Putting $l\to l-1$, $E_{0d}\to -E_{0d}$ and $g\leftrightarrow f$ in the upper case of (10.2.4) converts it into the lower case. If j is required to be the same, $l\to l-1$ is removed. In this way, only one set of radial equations needs be treated.

Of the six asymtotic solutions to (10.2.4, 5b), four are complex and one diverges at large r. The remaining one is of interest here and reads

$$g_l(r \to \infty) = -f_l(r \to \infty) = \text{constant} \times \exp\left(-\alpha_d^2 r^2/2\right) \tag{10.2.6}$$

which is also independent of the angles, analogous to (3.2.21). If $\alpha_d<0$, g_l changes its sign relative to that of f_l.

In the absence of the nonlinear confining potential $\Phi_{cd}(\underline{x})$, the baryon wave functions are still confined by the linearized confining potential. Replacing (10.2.5b) by $d_{b1}r+d_{b2}r^2$ in (10.2.2a), (10.2.6) goes over to

$$g_l(r \to \infty) = -f_l(r \to \infty) = \text{constant} \times \exp\left(-\frac{3}{5}\left|d_{b2}\right|^{1/3} r^{5/3}\right) \tag{10.2.7a}$$

$$g_l(r \to \infty) = -f_l(r \to \infty) = \text{constant} \times \exp\left(-\frac{3}{4}\left|d_{b1}\right|^{1/3} r^{4/3}\right), \quad d_{b2} = 0 \tag{10.2.7b}$$

The nonlinear potential (10.2.5b) is not too different from the linearized confining potential $d_{b2}r^2$ above so that the asymtotic solutions (10.2.6) and (10.2.7a) are also not too different. These solutions would correspond to the wave functions associated with the harmonic confinement model [F3], which has been rather successful in organizing the wealth of low energy baryon data.

Near the origin, the six independent solutions of (10.2.4) are

$$j = l + \tfrac{1}{2}: \quad f_l(r \to 0) = 0, \quad g_l(r \to 0) = c_{g+}(\lambda_+) r^{\lambda_+}, \quad \lambda_+ = l+2,\ l,\ -l-1$$
$$g_l(r \to 0) = 0, \quad f_l(r \to 0) = c_{f+}(\lambda_+) r^{\lambda_+}, \quad \lambda_+ = l+1,\ -l,\ -l-2 \tag{10.2.8a}$$

$$j = l - \tfrac{1}{2}: \quad f_l(r \to 0) = 0, \quad g_l(r \to 0) = c_{g-}(\lambda_-) r^{\lambda_-}, \quad \lambda_- = l,\ -l+1,\ -l-1$$
$$g_l(r \to 0) = 0, \quad f_l(r \to 0) = c_{f-}(\lambda_-) r^{\lambda_-}, \quad \lambda_- = l+1,\ l-1,\ -l \tag{10.2.8b}$$

where the c's are constant amplitudes.

Inserting (10.2.3) into (10.2.2b) leads to that Φ_{cd} depends upon the angles for $l>0$ and (10.2.1) with (10.2.3) is no longer separable, except at the $r\to 0$ and $r\to\infty$ limits governed by (10.2.4). The separable solutions (10.2.3) with (10.2.8) and (10.2.6) at $r\to 0$ and $r\to\infty$ are to be connected by the unknown nonseparable solutions to (10.2.1, 2). This situation is analogous to the $l>0$ case for mesons in §3.4.2 discussed at the end of §5.7.1.

In the absence of orbital excitations, $l=0$ and only $j=l+½$ of (10.2.3a) applies. Φ_{cd} of (10.2.2b) turns out to be independent of the angles for all r, just like the case for the $l=0$ mesons treated in §3.4.1. Equation (10.2.1) can now be separated. Inserting (10.2.3a) with $l=0$ into (10.2.2b) and making use of the procedure intervening (3.2.14) and (3.2.17), we obtain analogously

$$\Phi_{cd}(r) = -\frac{g_s^6}{360\pi}\left[\begin{array}{l}\int_0^r dr' r'^2\left(f_0^2(r') - g_0^2(r')\right)\left(5r^4 + 10r^2r'^2 + r'^4\right)/r + \\ + \int_r^\infty dr' r'\left(f_0^2(r') - g_0^2(r')\right)\left(5r'^4 + 10r^2r'^2 + r^4\right)\end{array}\right] \quad (10.2.9)$$

In the r=0 and ∞ limits, (10.2.9) yields

$$\Phi_{cd}(r \to 0) = -\frac{g_s^6}{72\pi}\int_0^\infty drr^5\left(f_0^2(r) - g_0^2(r)\right) \quad (10.2.10)$$

$$\Phi_{cd}(r \to \infty) = -\frac{g_s^6}{72\pi}r^3\int_0^\infty dr' r'^2\left(f_0^2(r') - g_0^2(r')\right) = \alpha_{d0}^3 r^3 \quad (10.2.11)$$

Equation (10.2.4) now goes over to

$$\left[\frac{E_{0d}^3}{8} + M_b^3 + \Phi_{bd}(r) + \frac{E_{0d}}{2}\Delta_0\right]g_0(r) + \left(\frac{E_{0d}^2}{4} + \Delta_0\right)\left(\frac{\partial}{\partial r} + \frac{2}{r}\right)f_0(r) = 0 \quad (10.2.12a)$$

$$\left[\frac{E_{0d}^3}{8} - M_b^3 - \Phi_{bd}(r) + \frac{E_{0d}}{2}\Delta_1\right]f_0(r) - \left(\frac{E_{0d}^2}{4} + \Delta_1\right)\frac{\partial}{\partial r}g_0(r) = 0 \quad (10.2.12b)$$

which holds for all r. Here, $\Phi_{bd}(r)$ is given by (10.2.2a) with $\underline{x} \to r$ together with (10.2.9). Equation (10.2.12) determines the properties of J^P=½$^+$ baryons, including their radially excited states.

10.2.3. Gound State Doublet Wave Functions at Origin

The six negative λ values in (10.2.8) will in general lead to wave functions diverging at the origin and hence are discarded. In the absence of orbital excitations, l=0, only (10.2.8a) applies and λ_+=0, 1 and 2. It is natural to associate the first one, i.e., $\lambda_+ = l$, l=0, with the ground state J^P=½$^+$ baryons, such as the nucleons. For this state, (10.2.3a, 8a) yields

$$m = \frac{1}{2} \qquad \psi_0^1(r \to 0) = \frac{1}{\sqrt{4\pi}}c_{g+}(0), \qquad \psi_0^2(r \to 0) = 0 \quad (10.2.13a)$$

$$\chi_{0\dot{a}}(r \to 0) = \psi_0^a(r \to 0)$$

$$m = -\frac{1}{2} \qquad \psi_0^1(r \to 0) = 0, \qquad \psi_0^2(r \to 0) = -\frac{1}{\sqrt{4\pi}} c_{g+}(0)$$
$$\chi_{0\dot{a}}(r \to 0) = \psi_0^a(r \to 0) \tag{10.2.13b}$$

For the other two states, (10.2.3a, 8a) with m=½ leads to

$$\lambda_+ = l + 1, \; l = 0; \qquad \psi_0^1(r \to 0) = i\frac{1}{\sqrt{4\pi}} c_{f+}(1) r \cos\vartheta$$
$$\psi_0^2(r \to 0) = i\frac{1}{\sqrt{4\pi}} c_{f+}(1) r \sin\vartheta \exp(i\phi), \quad \chi_{0\dot{a}}(r \to 0) = \psi_0^a(r \to 0) \tag{10.2.14}$$

$$\lambda_+ = l + 2, \; l = 0; \qquad \psi_0^1(r \to 0) = -\frac{1}{\sqrt{4\pi}} c_{g+}(2) r^2$$
$$\psi_0^2(r \to 0) = 0, \qquad \chi_{0\dot{a}}(r \to 0) = -\psi_0^a(r \to 0) \tag{10.2.15}$$

10.2.4. Relations to Dirac's Equation

The wave equations in relative space (10.2.1) for a doublet baryon at rest can be related to Dirac's equation in two-spinor form (B23) with V_{SB}=0. Multiplying (10.2.1) by the exponential factor on the right of (10.1.1) with $\underline{K}$=0 leads to wave functions of the form

$$\begin{pmatrix} \psi_0^1(\underline{x}) \\ \psi_0^2(\underline{x}) \end{pmatrix} \exp\left(-iE_0 X^0 + i\omega_0 x^0\right) = \begin{pmatrix} \psi_0^1(X^0, x) \\ \psi_0^2(X^0, x) \end{pmatrix}$$
$$\begin{pmatrix} \chi_{0\dot{1}}(\underline{x}) \\ \chi_{0\dot{2}}(\underline{x}) \end{pmatrix} \exp\left(-iE_0 X^0 + i\omega_0 x^0\right) = \begin{pmatrix} \chi_{0\dot{1}}(X^0, x) \\ \chi_{0\dot{2}}(X^0, x) \end{pmatrix} \tag{10.2.16}$$

which are basically (10.1.1) with $\underline{K}$=0. Let now the quark at x_{II} merge into the diquark at x_I so that the relative coordinates $x = x_{II} - x_I = 0$ in (3.1.3a). In this way, the originally doublet baryon having finite extension becomes a point particle, such as a electron. Replace the wave functions in (10.2.1) by $\chi_{0\dot{b}}(X^0, x \to 0)$ and $\psi_0^a(X^0, x \to 0)$, it is seen that the operators $\underline{\partial}$ and Δ operating in the relative space $\underline{x}$ drop out and (10.2.2a) leads in the absence of $\underline{x}$ to

$$\Phi_b(\underline{x}) = d_{b0} = \text{constant} \tag{10.2.17}$$

Equation (10.2.1), dropping the subscript d, now becomes

$$i\delta^{a\dot{b}}E_0\chi_{0\dot{b}}\left(X^0,x\to 0\right)=i\left(8\left(M_b^3+d_{b0}\right)/E_0^2\right)\psi_0^a\left(X^0,x\to 0\right)$$
$$i\delta_{\dot{b}c}E_0\psi_0^c\left(X^0,x\to 0\right)=i\left(8\left(M_b^3+d_{b0}\right)/E_0^2\right)\chi_{0\dot{b}}\left(X^0,x\to 0\right) \quad (10.2.18)$$

Since the baryon is at rest, its energy can be read from (10.2.1) with (10.2.17),

$$E_0=2\left(M_b^3+d_{b0}\right)^{1/3}=m_{eff} \quad (10.2.19)$$

where M_b has been given in (9.3.19). Making use of (C3a) and noting (10.2.16) and (10.2.19), (10.2.18) becomes

$$\partial_X^{a\dot{b}}\chi_{0\dot{b}}\left(X^0,0\right)=im_{eff}\psi_0^a\left(X^0,0\right)$$
$$\partial_{X\dot{b}c}\psi_0^c\left(X^0,0\right)=im_{eff}\chi_{0\dot{b}}\left(X^0,0\right) \quad (10.2.20)$$

The wave functions are obtainable from (10.1.1) with $\underline{x}$=0 and $\underline{K}$=0 together with (10.2.13),

$$\begin{pmatrix}\psi_0^1\left(X^0,0\right)\\ \psi_0^2\left(X^0,0\right)\end{pmatrix}=\exp\left(-iE_0X^0\right)\frac{1}{\sqrt{4\pi}}c_{g+}(0)\begin{pmatrix}1, & 0\\ 0, & -1\end{pmatrix}, \quad \chi_{0\dot{a}}\left(X^0,0\right)=\psi_0^a\left(X^0,0\right) \quad (10.2.21)$$

where the commas separate the m=½ from the m=−½ solutions. With (C11), (10.2.21) can be expressed in the conventional four component Dirac wave functions for a positive energy point baryon with $\underline{K}$=0,

$$\left(\psi_1,\psi_2,\psi_3,\psi_4\right)=\exp\left(-iE_0X^0\right)\frac{1}{\sqrt{4\pi}}c_{g+}(0)\begin{pmatrix}1, & 0, & 0, & 0\\ 0, & -1, & 0, & 0\end{pmatrix} \quad \text{for } m=\pm\tfrac{1}{2} \quad (10.2.22)$$

which apart from a phase factor agrees with (A13) and (A14) with N_D=$c_{g+}(0)/\sqrt{4}\pi$. Thereefore, m=½ and −½ in (10.2.3a) correspond to the spin up and down, respectively.

As was mentioned at the end of §10.2.1, a doublet baryon in motion ($\underline{K}$≠0) can no longer be described by the doublet wave functions (9.2.7) alone but require also the quartet wave functions (9.2.8). For small K, however, the quartet wave functions are small and of order K, just like that the vector meson wave functions are small and of order K in §3.5.3-4. If one ignores terms of order K^2, (10.0.1) with (10.1.1) reduces to (10.2.1) with a K term in the first parentheses and a $\underline{K}\partial$ term in the second. In the $\underline{x}\to 0$ limit, this modified (10.2.1) reduces to a modified form of of (10.2.18) with (10.2.19),

$$\left(i\delta^{a\dot{b}}E_0 - i\underline{K\sigma}^{a\dot{b}}\right)\chi_{0\dot{b}}(X,0) = im_{eff}\psi_0^a(X,0)$$
$$\left(i\delta_{\dot{b}c}E_0 + i\underline{K\sigma}_{\dot{b}c}\right)\psi_0^c(X,0) = im_{eff}\chi_{0\dot{b}}(X,0) \qquad (10.2.23)$$

where (3.1.4) has been used. Equation (10.2.23) can be written as

$$\partial_X^{a\dot{b}}\chi_{0\dot{b}}(X,0) = im_{eff}\psi_0^a(X,0)$$
$$\partial_{X\dot{b}c}\psi_0^c(X,0) = im_{eff}\chi_{0\dot{b}}(X,0) \qquad (10.2.24)$$

which has the same form as the Dirac equation for a free particle (B23) with V_{SB}=0. Equation (10.2.24) replaces (10.2.20) and holds for small K. With this equation, the classical energy moemtum relation (3.5.31) is eatablished to order K for the doublet baryons.

By reducing a baryon with extension to a point particle, much of the underlying physics determining the baryon properties are irretrievably lost. Nevertheless, the point particle description (10.2.24) of a slow baryon can be a good approximation if the baryon behaves like a point particle when interacting with another particle. For instance, in low energy pion-nucleon scattering, the short range strong potential of the nucleon may be considered to originate from a point source. Similarly, in low energy Mott scattering, the proton charge may be approximated by a point charge. In these cases, the quark structure of the nucleon does not enter the problem.

The point particle approximation does however not apply to weak interactions in which the different quarks, broadly speaking, interact differently with the gauge boson and the $\underline{x}\to 0$ limit leading to (10.2.24) cannot be taken. At large K, the Dirac description (10.2.24) breaks down; the full (9.3.16) needs be resorted to.

10.3. NORMALIZATION OF DOUBLET WAVE FUNCTIONS

The amplitudes of the doublet baryon wave functions in (10.2.3), hence also the amplitude constants in (10.2.6-8), are not free but, as it turns out, are approximately fixed by a normalization procedure analogous to that of Sec. 4.2 for mesons. Analogous to (10.1.3b), we define

$$\psi_0^{*\dot{a}} = \left(\psi_0^a\right)^*, \qquad \chi_{0a}^* = \left(\chi_{0\dot{a}}\right)^* \qquad (10.3.1)$$

Take the comlex conjugate of (9.3.16b) and multiply it by $\psi_{\dot{d}}^{\{bh\}}(I,II)$. Add the resulting equation to (9.3.16a) multiplied by $\chi_{\{ag\}}^{*\dot{e}}(I,II)$ and rearrange the indices to obtain

$$\chi_{\{ag\}\dot{e}}^*\partial_I^{a\dot{b}}\partial_{II}^{f\dot{e}}\partial_I^{g\dot{h}}\chi_{\{\dot{b}\dot{h}\}f} + \psi^{\{bh\}\dot{d}}\partial_{I\dot{c}b}\partial_{II\dot{d}e}\partial_{I\dot{k}h}\psi^{*\{\dot{c}\dot{k}\}e} = 0 \qquad (10.3.2a)$$

Take the complex conjugate of (9.3.16a) and multiply it by $\chi^{e}_{\{\dot{a}\dot{g}\}}(I, II)$. Add the resulting equation to (9.3.16b) multiplied by $\psi_{d}^{*\{\dot{b}\dot{h}\}}(I, II)$ and rearrange the indices to obtain

$$\chi_{\{\dot{a}\dot{g}\}e}\partial_{I}^{b\dot{a}}\partial_{II}^{e\dot{f}}\partial_{I}^{h\dot{g}}\chi^{*}_{\{bh\}\dot{f}} + \psi^{*\{\dot{b}\dot{h}\}d}\partial_{I\dot{b}c}\partial_{II\dot{e}d}\partial_{I\dot{h}k}\psi^{\{ck\}\dot{e}} = 0 \qquad (10.3.2b)$$

which is the complex conjugate of (10.3.2a).

If (9.3.16) is replaced by a Dirac equation in analogous spinor form, the corresponding sum of (10.3.2a) and (10.3.2.b) becomes the continuity equation for the probability current density associated with the Dirac particle and leads to a conserved total probability. This total probability is conventially set to 1 and this fixes the amplitude of the Dirac wave function. However, the probability interpretation is not unique; the conserved total probability can for instance be any constant without affecting the predictive power of the Dirac equation. The situation is entirely analogous to that for a Schrödinger and Klein- Gordon particle considered in Sec. 4.1. This arbitrariness in wave function amplitude is largely removed for doublet baryons below.

Analogous to (4.2.2), (10.3.2a) is put in the form

$$C_b = \partial_{I}^{a\dot{b}}\chi^{*}_{\{ag\}\dot{e}}\partial_{II}^{f\dot{e}}\partial_{I}^{g\dot{h}}\chi_{\{\dot{b}\dot{h}\}f} + \partial_{I\dot{c}b}\psi^{\{bh\}\dot{d}}\partial_{II\dot{d}e}\partial_{I\dot{k}h}\psi^{*\{\dot{c}\dot{k}\}e} = \qquad (10.3.3a)$$

$$= D_b = \left(\partial_{I}^{a\dot{b}}\chi^{*}_{\{ag\}\dot{e}}\right)\left(\partial_{II}^{f\dot{e}}\partial_{I}^{g\dot{h}}\chi_{\{\dot{b}\dot{h}\}f}\right) + \left(\partial_{I\dot{c}b}\psi^{\{bh\}\dot{d}}\right)\left(\partial_{II\dot{d}e}\partial_{I\dot{k}h}\psi^{*\{\dot{c}\dot{k}\}e}\right) \qquad (10.3.3b)$$

which is Lorentz invariant by virtue of its spinor form. Therefore,

$$\frac{1}{2}\int d^4x_I d^4x_{Ii}\left(C_b + C_b^*\right) = \frac{1}{2}\int d^4x_I d^4x_{Ii}\left(D_b + D_b^*\right)$$

$$= \text{real Lorentz invariant with dimension of inverse of four volume} \quad \text{Alt}\, a \qquad (10.3.4a)$$

$$= \text{real Lorentz invariant scalar} \quad \text{Alt}\, b \qquad (10.3.4b)$$

Unlike the corresponding (4.2.3) for mesons, Alt a of (10.3.4a) is not dimensionless. This is due to the merging of two quark coordinates $x_{III} \to x_I$ in (9.2.12) so that the four volume $\int d^4x_{III}$, which would have been applied to (10.3.4), drops out. This choice is an extension of the point particle consideration that led to (4.1.2) and the two quark treatment that led to (4.2.8); both amplitudes are independent of the particle energy. Alt b of (10.3.4b) however is consistent with (4.2.3a) and (7.1.12a) which are both Lorentz scalars.

Analogous to Sec. 4.2 for mesons, we specialize (10.3.4) to the doublets (9.2.7) at rest ($\underline{K}$=0). In this case, the quartet wave functions (9.2.8) can be omitted. Formally, this can be accomplished by letting $g \to e$ and $h \to d$ in (10.3.3). Noting (9.2.7), (10.3.3) turns into

$$\frac{1}{2}C_b \to C_{bd} = \partial_I^{a\dot{b}}\chi_{0a}^{*}\partial_{II}^{f\dot{e}}\partial_I^{e\dot{h}}\chi_{\{\dot{b}\dot{h}\}f} + \partial_{I\dot{c}b}\psi_0^{b}\partial_{II\dot{d}e}\partial_{I\dot{k}d}\psi^{*\{\dot{c}\dot{k}\}e}$$
$$= D_{bd} = \left(\partial_I^{a\dot{b}}\chi_{0a}^{*}\right)\left[\partial_{II}^{f\dot{e}}\partial_I^{e\dot{h}}\chi_{\{\dot{b}\dot{h}\}f}\right] + \left(\partial_{I\dot{c}b}\psi_0^{b}\right)\left[\partial_{II\dot{d}e}\partial_{I\dot{k}d}\psi^{*\{\dot{c}\dot{k}\}e}\right] \leftarrow \frac{1}{2}D_b \quad (10.3.5)$$

This expression can also be obtained by multiplying (10.0.1a) by χ^{*}_{0a} and adding the resulting equation to the complex conjugate of (10.0.1b) multiplied by ψ^{b}_{0}.

Inserting (9.2.7) with the dependence of (10.1.1) into the brackets of (10.3.5) yields

$$\partial_{II}^{f\dot{e}}\partial_I^{e\dot{h}}\chi_{\{\dot{b}\dot{h}\}f}(I,II) = -2\left(E_{0d}^2/4+\Delta\right)\chi_{0\dot{b}}(X,x) \quad (10.3.6a)$$

$$\partial_{I\dot{k}d}\partial_{II\dot{d}e}\psi^{*\{\dot{c}\dot{k}\}e}(I,II) = -2\left(E_{0d}^2/4+\Delta\right)\psi_0^{*\dot{c}}(X,x) \quad (10.3.6b)$$

Inserting (10.3.6) into (10.3.5) and applying (3.1.4, 10a) leads to

$$C_{bd} = \left(\frac{1}{2}-\frac{\omega_0}{E_{0d}}\right)\frac{\partial}{\partial X^0}\left[\chi_{0b}^{*}(\underline{x})\left(\frac{E_{0d}^2}{4}+\Delta\right)\chi_{0\dot{b}}(\underline{x}) + \psi_0^{b}(\underline{x})\left(\frac{E_{0d}^2}{4}+\Delta\right)\psi_0^{*\dot{b}}(\underline{x})\right]$$
$$+\underline{\sigma}_{\dot{b}a}\underline{\partial}\left[\psi_0^{a}(\underline{x})\left(\frac{E_{0d}^2}{4}+\Delta\right)\psi_0^{*\dot{b}}(\underline{x})\right] - \underline{\sigma}^{a\dot{b}}\underline{\partial}\left[\chi_{0a}^{*}(\underline{x})\left(\frac{E_{0d}^2}{4}+\Delta\right)\chi_{0\dot{b}}(\underline{x})\right]$$
$$= D_{bd} = -\left[\left(\frac{i}{2}E_{0d}\delta^{a\dot{b}} + \underline{\sigma}^{a\dot{b}}\underline{\partial}\right)\chi_{0a}^{*}(\underline{x})\right]\left(\frac{E_{0d}^2}{4}+\Delta\right)\chi_{0\dot{b}}(\underline{x})$$
$$-\left[\left(\frac{i}{2}E_{0d}\delta_{\dot{b}a} - \underline{\sigma}_{\dot{b}a}\underline{\partial}\right)\psi_0^{a}(\underline{x})\right]\left(\frac{E_{0d}^2}{4}+\Delta\right)\psi_0^{*\dot{b}}(\underline{x}) \quad (10.3.7)$$

We further limit ourselves to doublet baryons without orbital excitation, l=0. Equation (10.2.3a) ff with l=0 yields

$$m=\frac{1}{2},\quad \psi_0^1(\underline{x}) = \frac{1}{\sqrt{4\pi}}\left(g_0(r)+if_0(r)\cos\vartheta\right),\qquad \chi_{0\dot{1}}(\underline{x}) = \left(\psi_0^1(\underline{x})\right)^{*}$$
$$\psi_0^2(\underline{x}) = \frac{1}{\sqrt{4\pi}}if_0(r)\sin\vartheta\exp(i\phi),\qquad \chi_{0\dot{2}}(\underline{x}) = -\psi_0^2(\underline{x}) \quad (10.3.8a)$$

$$m=-\frac{1}{2},\quad \psi_0^1(\underline{x}) = \frac{1}{\sqrt{4\pi}}if_0(r)\sin\vartheta\exp(-i\phi),\qquad \chi_{0\dot{1}}(\underline{x}) = -\psi_0^1(\underline{x})$$
$$\psi_0^2(\underline{x}) = \frac{1}{\sqrt{4\pi}}\left(-g_0(r)+if_0(r)\cos\vartheta\right),\qquad \chi_{0\dot{2}}(\underline{x}) = \left(\psi_0^2(\underline{x})\right)^{*} \quad (10.3.8b)$$

Similarly, (10.2.3c) ff with l=0 yields

$$m=\frac{1}{2},\quad \psi_0^1(\underline{x})=\frac{1}{\sqrt{4\pi}}\left(-g_0(r)+if_0(r)\cos\vartheta\right),\quad \chi_{0\dot{1}}(\underline{x})=-\left(\psi_0^1(\underline{x})\right)^*$$
$$\psi_0^2(\underline{x})=\frac{1}{\sqrt{4\pi}}if_0(r)\sin\vartheta\exp(i\phi),\qquad \chi_{0\dot{2}}(\underline{x})=\psi_0^2(\underline{x}) \tag{10.3.8c}$$

$$m=-\frac{1}{2},\quad \psi_0^1(\underline{x})=\frac{1}{\sqrt{4\pi}}if_0(r)\sin\vartheta\exp(-i\phi),\quad \chi_{0\dot{1}}(\underline{x})=\psi_0^1(\underline{x})$$
$$\psi_0^2(\underline{x})=\frac{1}{\sqrt{4\pi}}\left(g_0(r)+if_0(r)\cos\vartheta\right),\quad \chi_{0\dot{2}}(\underline{x})=-\left(\psi_0^2(\underline{x})\right)^* \tag{10.3.8d}$$

The analog of (4.2.4, 5) is obtained by inserting (10.3.8) into (10.3.7) and integrating over $\underline{x}$;

$$\int d^3\underline{x}\left(C_{bd}+C_{bd}^*\right)=\frac{\partial}{\partial X^0}N_{dr}=\int d^3\underline{x}\left(D_{bd}+D_{bd}^*\right)=0 \tag{10.3.9a}$$

$$N_{dr}=2\int drr^2\left[\frac{E_{0d}^2}{4}\left(g_0^2(r)+f_0^2(r)\right)-\left(\underline{\partial}g_0(r)\right)^2-\left(\underline{\partial}f_0(r)\right)^2-\frac{2}{r^2}f_0^2(r)\right] \tag{10.3.9b}$$

In arriving at these relations f_0 and g_0 must vanish at large r, as in (10.2.6) or (10.2.7). With (10.3.9a, 5) and (3.1.3c), (10.3.4) becomes

$$\int dX^0\frac{\partial}{\partial X^0}\int dx^0\Omega N_{dr}=0,\qquad \Omega=\int d^3\underline{X} \tag{10.3.10}$$

which is of dimension $1/l_t l_s^3$ for (10.3.4a) and scalar for (10.3.4b). l_t and l_s are unit length associated with the time and space components, respectively, of a position four vector.

For (10.3.4a), ΩN_{dr} is a conserved quantity of dimension $1/l_t^2 l_s^3$. $1/l_t$ is of dimension energy and the only observable energy available for the rest frame doublet baryon under concideration is its mass E_{0d}. l_s is of dimension length but there is no natural length scale associated with doublet baryons. Therefore, no unique, natural normalization can be achieved here for Alt *a* of (10.3.4a).

For Alt *b* of (10.3.4b), ΩN_{dr} is the baryon equivalent to N_{mJ} of (4.2.4). Thus, analogous to (4.2.6) for mesons, the identification

$$\Omega\left|N_{dr}\right|=\frac{E_{0d}}{2}\frac{\Omega}{\Omega_{cb}} \tag{10.3.11}$$

is made. Ω_{cb} is similarly interpreted as the finite volume in laboratory space $\underline{X}$ occupied by the doublet baryon wave functions which acquire finite amplitudes due to interaction with other hadrons by a similar mechanism described in Sec. 4.5. For a free doublet baryon, i.e., a baryon far away from other particles, the analog of (4.2.7) is

$$\Omega_{cb} \to \Omega \to \infty \tag{10.3.12}$$

Combining (10.3.11) and (10.3.9b) leads to the normalization condition corresponding to (4.2.8),

$$\frac{|N_{dr}|2}{E_{0d}^2} = \int drr^2 \left| g_0^2(r) + f_0^2(r) - \frac{4}{E_{0d}^2}\left[(\partial g_0(r))^2 + (\partial f_0(r))^2 + \frac{2}{r^2} f_0^2(r) \right] \right| = \frac{1}{E_{0d}\Omega_{cb}} \tag{10.3.13}$$

which leads to

$$(f_0(r),\ g_0(r)) = \frac{a_g}{\sqrt{\Omega_{cb}}}(f_{00}(r),\ g_{00}(r)), \quad g_{00}(0) = 1, \quad a_g^2 = \frac{E_{0d}}{2|N_{dr0}|} \tag{10.3.14}$$

$$N_{dr0} = N_{dr} \text{ in } (10.3.9\text{b}) \text{ with } f_0(r), g_0(r) \text{ replaced by } f_{00}(r), g_{00}(r)$$

Here, $f_{00}(r)$ and $g_{00}(r)$ are dimensionless; examples will be given in Figure 11.1 below. It is possible to choose a different normalization by multiplying the right of (10.3.13) by a constant κ_b. In this case, a_g^2 will be multiplied by κ_b. Note that $f_0(r)$ and $g_0(r)$ do not vanish if Ω_{cb} is finite and the confinement potential in (10.2.2a) is nonlinear. The baryon can be interacting with another particle that gives rise to the nonlinear confining potential accoording to the same mechanism for mesons given in Sec. 4.5. Or it is in radially excited state, by the same line of reasoning given in Sec. 4.6 for excited mesons.

For freely moving baryons, the reasoning of §4.3.1 analogously holds and the baryon wave equation (9.3.16) must be linear so that the superposition principle holds and baryon wave packets can be constructed. This requires that Φ_b in (9.3.16) or (10.1.6) is independent of the baryon wave functions or that their amplitudes in (10.1.6b), hence also (10.1.7b), vanish. By (10.3.14), (10.3.12) is recovered and the baryon wave functions now spread over the entire laboratory space $\underline{X}$. However, these baryon wave functions are still confined in relative space $\underline{x}$ by the linearized confining potential in (10.1.6a). This confining mechanism is absent in the potential (3.2.8a) with (3.2.20) for mesons. Thus, the baryons are more strongly and permanently confined than do the mesons.

The normalization (10.3.13) differs from (4.2.8) for mesons in that derivatives of the baryon wave functions appear in (10.3.13). This is due to that each of the baryon wave equations in (9.3.16) is one order higher than that of the corresponding meson wave equations in (2.3.22). The dimension of f_0 and g_0 is by (10.3.13) (length)$^{-2.5}$ which also differs from that of ψ for mesons in (4.3.2) which by (4.7.3) is (length)$^{-3}$.

10.4. DOUBLET RADIAL EQUATIONS OF REDUCED ORDER

The masses E_{0d} of l=0, doublet baryons, including their radially excited states, are determined by the effectively sixth order (10.2.12) and (10.2.2a) with Φ_{cd}=0, subjected to the boundary condition (10.2.7). The solutions near r=0 have been given by (10.2.13-15) and are developable in integer powers of r. On the other hand, the asymtotic solutions (10.2.7) can only be expressed in powers of $r^{1/3}$. Therefore, an analytic solution in the intermediate r region that bridges these two types of solutions at r=0 and ∞ seems not to be easily found. Although numerical integration can be carried out, it is hampered by the large number of unknown integration constants, d_b, d_{b0}, d_{b1} and d_{b2} in (10.2.2a).

An analytic, but highly approximative solution can however be found by introducing the following two rather drastic simplifications. Firstly, it is assumed that d_b is large so that d_b/r dominates in (10.2.2a). The nonlinear confining potential Φ_{cd} can be dropped according to Sec. 10.6 below. Secondly, it is assumed that Δ_l and Δ_{l+1}, which correspond conventionally to "kinetic energies" associated with the quarks, are small next to the mass term $E_{0d}^2/4$ so that they can be dropped in (10.2.4) with $j=l+½$. This reduces the order of (10.2.4) from six to two and greatly simplifies the treatment. However, this simplification comes at the expense of limiting ourselves to heavy baryons, in which the quarks move "nonrelativistically" in conventional terms. This is akin to the heavy meson approximation of (3.5.9a). We obtain

$$\left[-\frac{E_{0d}^3}{8}-M_b^3-d_{b0}-\frac{d_b}{r}\right]g_l(r)-\frac{E_{0d}^2}{4}\left(\frac{\partial}{\partial r}+\frac{l+2}{r}\right)f_l(r)=0 \tag{10.4.1a}$$

$$\left[-\frac{E_{0d}^3}{8}+M_b^3+d_{b0}+\frac{d_b}{r}\right]f_l(r)+\frac{E_{0d}^2}{4}\left(\frac{\partial}{\partial r}-\frac{l}{r}\right)g_l(r)=0 \tag{10.4.1b}$$

These equations are nearly the same as the radial equations for a Dirac particle of mass m_D moving in Coulomb potential instead of the V's in (C10). With

$$\psi^a, \chi_{\dot{b}} \propto \exp\left(-iE_D X^0\right)$$

and (10.2.3a), we obtain [B7],

$$\left(E_D-m_D+\frac{e^2}{r}\right)g_{Dl}(r)+\left(\frac{\partial}{\partial r}+\frac{l+2}{r}\right)f_{Dl}(r)=0 \tag{10.4.2a}$$

$$\left(E_D+m_D+\frac{e^2}{r}\right)f_{Dl}(r)-\left(\frac{\partial}{\partial r}-\frac{l}{r}\right)g_{Dl}(r)=0 \tag{10.4.2b}$$

The subscript D refers to this Dirac particle. Further, l is the orbital angular momentum related to the total angular momentum j_D by

$$j_D = l + \frac{1}{2} \tag{10.4.2c}$$

This equation yields the well-known result

$$E_D = \frac{m_D}{\sqrt{1 + e^4/(n_r + s + 1)}}, \qquad s + 1 = \sqrt{(l+1)^2 - e^4} \tag{10.4.3}$$

where n_r is the radial quantum number. Equation (10.4.1) is of the form of (10.4.2) with e^2/r changing its sign in (10.4.2b). This change corresponds to replacing the Coulomb potential, which is the time component of a form vector, by a scalar potential. The treatment [B7] of (10.4.2) can be largely taken over for the case with changed sign to yield

$$E_D = \pm m_D \sqrt{1 - e^4/(n_r + s + 1)}, \qquad s + 1 = \sqrt{(l+1)^2 + e^4} \tag{10.4.4}$$

By identifying the constants in (10.4.1) and (10.4.2), (10.4.4) with the lower sign is converted to

$$E_{0d} = 2\left(M_b^3 + d_{b0}\right)^{1/3}\left(1 - 16d_b^2 / E_{0d}^4 (n_r + s + 1)^2\right)^{1/6} \tag{10.4.5a}$$

$$s + 1 = \sqrt{(l+1)^2 + 16d_b^2 / E_{0d}^4} \tag{10.4.5b}$$

In the absence of radial excitations, n_r=0 and

$$E_{0d} = 2\left(M_b^3 + d_{b0}\right)^{1/3}\left(1 + 16d_b^2 / E_{0d}^4 (l+1)^2\right)^{-1/6} \tag{10.4.6}$$

This approximate formula holds better for the ground state or l=0 for which the wave function is large at $r \approx 0$ where d_b/r is large;

$$E_{0d} = 2\left(M_b^3 + d_{b0}\right)^{1/3}\left(1 + 16d_b^2 / E_{0d}^4\right)^{-1/6} \tag{10.4.7}$$

The wave functions associated with this state is obtained from (10.4.1),

$$f_0(r) = A_f r^{s_0} \exp(-\alpha_0 r) \tag{10.4.8a}$$

$$\alpha_0 = \frac{E_{0d}}{2}\sqrt{64\frac{\left(M_b^3 + d_{b0}\right)^2}{E_{0d}^6} - 1} \tag{10.4.8b}$$

$$s_0 = \sqrt{1+16d_b^2/E_{0d}^4} - 1 \tag{10.4.8c}$$

$$g_0(r) = -\sqrt{\frac{M_b^3 + d_{b0} + E_{0d}^3/8}{M_b^3 + d_{b0} - E_{0d}^3/8}} f_0(r) \tag{10.4.8d}$$

The amplitude constant A_f can be fixed by a relation similar to (10.3.14) with $\omega_0=0$ together with (10.3.13). In passing, we note that $s_0>0$ so that (10.2.3a) is convergent at the origin, contrary to the ground state wave functions in the Coulomb case in (10.4.2) which are known to diverge at $r=0$ [B7] (see also second of (10.4.3)).

According to the discussion above (10.4.1), (10.4.5-8) is a kind of "nonrelativistic" approximation which was however invalidated by the numerical results to be mentioned near the end of Sec. 11.1 and Sec. 11.4 below. Further, the reduction of the order of (10.2.4) and (10.2.12) by four corresponds to keeping solutions associated with $\lambda_+=l$, $-l-1$ in (10.2.8a) only and removing solutions having the other four λ_+ values. Therefore, (10.4.7) cannot be used.

10.5. Quartet Wave Equations and Confinement

10.5.1. Quartet Wave Equations in Relative Space

Remove the doublet components described by (10.2.1) from the rest frame baryons equations (10.1.2, 6). We are left with the quartet baryon wave equations

$$\begin{aligned}
&\left(i\delta^{a\dot{b}}E_{0q}/2+\underline{\sigma}^{a\dot{b}}\underline{\partial}\right)\left(i\delta^{g\dot{h}}E_{0q}/2+\underline{\sigma}^{g\dot{h}}\underline{\partial}\right)\left(i\delta_{\dot{e}f}E_{0q}/2+\underline{\sigma}_{\dot{e}f}\underline{\partial}\right)\chi_{\{\dot{b}\dot{h}}{}^{f\}}(\underline{x})\\
&=-i\left(M_b^3+\Phi_{bq}(\underline{x})\right)\psi^{\{ag}{}_{\dot{e}\}}(\underline{x})\\
&\left(i\delta_{\dot{b}c}E_{0q}/2-\underline{\sigma}_{\dot{b}c}\underline{\partial}\right)\left(i\delta_{\dot{h}k}E_{0q}/2-\underline{\sigma}_{\dot{h}k}\underline{\partial}\right)\left(i\delta^{d\dot{e}}E_{0q}/2-\underline{\sigma}^{d\dot{e}}\underline{\partial}\right)\psi^{\{ck}{}_{\dot{e}\}}(\underline{x})\\
&=-i\left(M_b^3+\Phi_{bq}(\underline{x})\right)\chi_{\{\dot{b}\dot{h}}{}^{d\}}(\underline{x})
\end{aligned} \tag{10.5.1}$$

$$\Phi_{bq}(\underline{x}) = \Phi_{cq}(\underline{x}) + \frac{d_b}{r} + d_{b0} + d_{b1}r + d_{b2}r^2 \tag{10.5.2a}$$

$$\Phi_{cq}(\underline{x}) = \frac{g_s^6}{192\pi}\int d^3\underline{x}'\left|\underline{x}-\underline{x}'\right|^3 \mathrm{Re}\left(\chi_{\{\dot{b}\dot{h}}{}^{f\}}(\underline{x}')\psi^{*\{\dot{b}\dot{h}}{}_{f\}}(\underline{x}')\right) \tag{10.5.2b}$$

Here, the χ's and ψ's are totally symmetric in their indices, the subscript q refers to specialization to quartets and (10.1.8) has been consulted. In terms of (9.2.8), (10.5.2b) becomes

$$\Phi_{cq}(\underline{x}) = \frac{g_s^6}{192\pi}\int d^3\underline{x}'|\underline{x}-\underline{x}'|^3 \sum_{\mu=-3/2}^{\mu=3/2} \mathrm{Re}\left(\chi_\mu(\underline{x}')\psi_\mu^*(\underline{x}')\right) \quad (10.5.3)$$

Note that if $\underline{K}\neq 0$, the quartet and doublet wave function components in (9.2.7, 8) couple and the full (9.3.16) with (10.1.1, 6b) must be used. Analogous to the doublet case at the end of §10.2.1, the quartet wave functions (9.2.8) will remain as "large components" and the doublet wave function (9.2.7) will enter as "small components" of the order $|\underline{K}|$ in (10.1.2, 6b).

10.5.2. Quartet Radial Wave Equations for $j \geq 3/2$

The totally symmetric wave functions in (10.5.1, 2) have been grouped into two quartets according to (9.2.8). There are four values of the orbital angular momentum l that can combine with the spin angular momentum s=3/2 to produce a total angular momentum j, namely,

$$j = l+\frac{3}{2},\ l+\frac{1}{2},\ l-\frac{1}{2},\ \text{and } l-\frac{3}{2} \quad (10.5.4)$$

This result as well as $j=\pm½$ in (10.2.3) are special cases of a more general theorem on the eigenvalues $j(j+1)$ of the sum of the squares of two commuting angular momentum operators, here the orbital angular momentum operator $\underline{L}=\underline{r}\times\underline{\partial}$ and the spin operators $\underline{S}=(S^1, S^2, S^3)$ [B6]

$$S^1 = \begin{pmatrix} 0 & \frac{1}{2\sqrt{3}} & 0 & 0 \\ \frac{1}{2\sqrt{3}} & 0 & 1 & 0 \\ 0 & 1 & 0 & \frac{1}{2\sqrt{3}} \\ 0 & 0 & \frac{1}{2\sqrt{3}} & 0 \end{pmatrix},\quad S^2 = i\begin{pmatrix} 0 & \frac{-1}{2\sqrt{3}} & 0 & 0 \\ \frac{1}{2\sqrt{3}} & 0 & -1 & 0 \\ 0 & 1 & 0 & \frac{-1}{2\sqrt{3}} \\ 0 & 0 & \frac{1}{2\sqrt{3}} & 0 \end{pmatrix},\quad S^3 = \begin{pmatrix} \frac{3}{2} & 0 & 0 & 0 \\ 0 & \frac{1}{2} & 0 & 0 \\ 0 & 0 & \frac{-1}{2} & 0 \\ 0 & 0 & 0 & \frac{-3}{2} \end{pmatrix} \quad (10.5.5a)$$

$$\underline{S}^2 = s(s+1), \qquad s = 3/2 \quad (10.5.5b)$$

$$\underline{J} = \underline{L} + \underline{S}, \qquad \underline{J}^2 = j(j+1) \quad (10.5.5c)$$

For the doublet case, $\underline{S}$ are simply the Pauli matrices.

Analogous to the expansion of the doublet wave functions in (10.2.3), each of the four combinations in (10.5.4) can be expanded into spherical harmonics. Consider at first $j=l+3/2$ and let m be the eigenvalue of the third component of $\underline{J}$. The expansion is

$$j = l+\frac{3}{2},\quad \chi_\mu(\underline{x}) = \sum_{\nu=-3/2}^{\nu=3/2} C\left(j+\nu, \frac{3}{2}, j; m-\mu, \mu\right) Y_{j+\nu, m-\mu}(\vartheta,\phi)\, g_{j+\nu}(r)$$

(10.5.6a)

$$\psi_\mu(\underline{x}) = \sum_{\nu=-3/2}^{\nu=3/2} (-)^{\nu+1/2} C\left(j+\nu, \frac{3}{2}, j; m-\mu, \mu\right) Y_{j+\nu, m-\mu}(\vartheta, \phi) g_{j+\nu}(r) \qquad (10.5.6b)$$

where $\mu=\pm 3/2$ and $\pm 1/2$. The C's are Clebsch-Gordan coefficients and have for instance been tabulated by Condon and Shortly [C2]. In the notation of Rose [R4], these are

$$g_{j+3/2} = if_j, \quad C\left(j+\frac{3}{2}, \frac{3}{2}, j; m\mp\frac{3}{2}, \pm\frac{3}{2}\right) = \mp\sqrt{\frac{(j\mp m+1)(j\mp m+2)(j\mp m+3)}{(2j+3)(2j+2)(2j+4)}}$$

$$C\left(j+\frac{3}{2}, \frac{3}{2}, j; m\mp\frac{1}{2}, \pm\frac{1}{2}\right) = \pm\sqrt{\frac{3(j+m+1)(j-m+1)(j\mp m+2)}{(2j+3)(2j+2)(2j+4)}}$$

$$g_{j+1/2} = g_j, \quad C\left(j+\frac{1}{2}, \frac{3}{2}, j; m\mp\frac{3}{2}, \pm\frac{3}{2}\right) = \sqrt{\frac{3(j\pm m)(j\mp m+1)(j\mp m+2)}{2j(2j+2)(2j+3)}}$$

$$C\left(j+\frac{1}{2}, \frac{3}{2}, j; m\mp\frac{1}{2}, \pm\frac{1}{2}\right) = -(j\pm 3m)\sqrt{\frac{j\mp m+1}{2j(2j+2)(2j+3)}}$$

$$g_{j-1/2} = ih_j, \quad C\left(j-\frac{1}{2}, \frac{3}{2}, j; m\mp\frac{3}{2}, \pm\frac{3}{2}\right) = \mp\sqrt{\frac{3(j\pm m-1)(j\pm m)(j\mp m+1)}{(2j-1)2j(2j+2)}}$$

$$C\left(j-\frac{1}{2}, \frac{3}{2}, j; m\mp\frac{1}{2}, \pm\frac{1}{2}\right) = \mp(j\mp 3m+1)\sqrt{\frac{j\pm m}{(2j-1)2j(2j+2)}}$$

$$g_{j-3/2} = k_j, \quad C\left(j-\frac{3}{2}, \frac{3}{2}, j; m\mp\frac{3}{2}, \pm\frac{3}{2}\right) = \sqrt{\frac{(j\pm m-2)(j\pm m-1)(j\pm m)}{(2j-2)(2j-1)2j}}$$

$$C\left(j-\frac{3}{2}, \frac{3}{2}, j; m\mp\frac{1}{2}, \pm\frac{1}{2}\right) = \sqrt{\frac{3(j+m)(j-m)(j\pm m-1)}{(2j-2)(2j-1)2j}} \qquad (10.5.7)$$

Here, $g_{j+\nu}$ has been renamed. The spherical harmonics and radial wave functions contained in (10.5.6) are

$$j = l+\frac{3}{2}, \qquad Y_l k_j, \qquad Y_{l+1} h_j, \qquad Y_{l+2} g_j, \quad \text{and} \quad Y_{l+3} f_j \qquad (10.5.8)$$

Inserting (10.5.6) into (10.5.3) leads to that Φ_{cq} depends upon the angles, including the $l=0$ case, so that (10.5.1) is not separable. In the $|\underline{x}|\to 0$ and ∞ limits, (10.5.2) becomes independent of the angles;

$$\Phi_{bq}\left(\left|\underline{x}\right|\to 0\right)=\frac{d_b}{r}$$
$$\Phi_{cq}\left(\left|\underline{x}\right|\to\infty\right)=\frac{g_s^6}{192\pi}r^3\int d^3\underline{x}\sum_{\mu=-3/2}^{\mu=3/2}\mathrm{Re}\left(\chi_\mu(\underline{x})\psi_\mu^*(\underline{x})\right)=\alpha_q^3 r^3 \tag{10.5.9}$$

In these limits, (10.5.1) is separable. Insert (9.2.8) and (10.5.3) into each of the eight component equations of (10.5.1) and make use of (10.5.7). Each of the four spherical harmonics in (10.5.6) is associated with a radial equation. After considerable algebra, it is shown that all the eight component equations in (10.5.1) yield the same set of radial equations

$$\left[\frac{E_{0q}^3}{8}-M_b^3-\Phi_{bq}\left(\left|\underline{x}\right|\to 0\text{ or }\infty\right)+\frac{E_{0q}}{2}\frac{3}{2j+2}\Delta_{j+3/2}\right]f_j(r)$$
$$-\frac{E_{0q}}{2}\frac{\sqrt{3(2j-1)(2j+3)}}{2j+2}\Delta_{(j-1/2)-}h_j(r)-\sqrt{\frac{(2j-1)(2j+3)}{4j(j+1)}}\Delta_{(j-1/2)-}\left(\frac{\partial}{\partial r}-\frac{j-3/2}{r}\right)k_j(r)$$
$$+\left[\frac{E_{0q}^2}{4}\sqrt{\frac{3j}{j+1}}+\sqrt{\frac{3}{4j(j+1)}}\Delta_{j+3/2}\right]\left(\frac{\partial}{\partial r}-\frac{j+1/2}{r}\right)g_j(r)=0 \tag{10.510a}$$

$$\left[-\frac{E_{0q}^3}{8}-M_b^3-\Phi_{bq}\left(\left|\underline{x}\right|\to 0\text{ or }\infty\right)+\frac{E_{0q}}{2}\frac{4j-3}{2j}\Delta_{j+1/2}\right]g_j(r)$$
$$+\frac{E_{0q}}{2}\frac{\sqrt{3(2j-1)(2j+3)}}{2j}\Delta_{(j-3/2)-}k_j(r)+\left[\frac{E_{0q}^2}{4}\sqrt{\frac{3j}{j+1}}+\sqrt{\frac{3}{4j(j+1)}}\Delta_{j+1/2}\right]\left(\frac{\partial}{\partial r}+\frac{j+5/2}{r}\right)f_j(r)$$
$$+\sqrt{\frac{(2j-1)(2j+3)}{j(j+1)}}\left[\frac{E_{0q}^2}{4}-\frac{1}{2}\Delta_{j+1/2}\right]\left(\frac{\partial}{\partial r}-\frac{j-1/2}{r}\right)h_j(r)=0 \tag{10.5.10b}$$

$$\left[\frac{E_{0q}^3}{8}-M_b^3-\Phi_{bq}\left(\left|\underline{x}\right|\to 0\text{ or }\infty\right)-\frac{E_{0q}}{2}\frac{4j+7}{2j+2}\Delta_{j-1/2}\right]h_j(r)$$
$$-\frac{E_{0q}}{2}\frac{\sqrt{3(2j-1)(2j+3)}}{2j+2}\Delta_{(j+3/2)+}f_j(r)+\left[\frac{E_{0q}^2}{4}\sqrt{\frac{3(j+1)}{j}}-\sqrt{\frac{3}{4j(j+1)}}\Delta_{j-1/2}\right]\left(\frac{\partial}{\partial r}-\frac{j-3/2}{r}\right)k_j(r)$$
$$+\sqrt{\frac{(2j-1)(2j+3)}{j(j+1)}}\left[\frac{E_{0q}^2}{4}-\frac{1}{2}\Delta_{j-1/2}\right]\left(\frac{\partial}{\partial r}+\frac{j+3/2}{r}\right)g_j(r)=0 \tag{10.5.10c}$$

$$\left[-\frac{E_{0q}^3}{8}-M_b^3-\Phi_{bq}\left(\left|\underline{x}\right|\to 0 \text{ or } \infty\right)+\frac{E_{0q}}{2}\frac{3}{2j}\Delta_{j-3/2}\right]k_j(r)$$
$$+\frac{E_{0q}}{2}\frac{\sqrt{3(2j-1)(2j+3)}}{2j}\Delta_{(j+1/2)+}g_j(r)-\sqrt{\frac{(2j-1)(2j+3)}{4j(j+1)}}\Delta_{(j+1/2)+}\left(\frac{\partial}{\partial r}+\frac{j+5/2}{r}\right)f_j(r)$$
$$+\left[\frac{E_{0q}^2}{4}\sqrt{\frac{3(j+1)}{j}}-\sqrt{\frac{3}{4j(j+1)}}\Delta_{j-3/2}\right]\left(\frac{\partial}{\partial r}+\frac{j+1/2}{r}\right)h_j(r)=0 \qquad (10.5.10d)$$

where the Δ operators have been defined in (3.4.5b). For free quartet baryons, the nonlinear confinement potential Φ_{cq} (10.5.3) must vanish according to the reasoning given near the end of Sec. 10.3. $\Phi_{bq}(\underline{x})$ in (10.5.2a) now depends only upon r and (10.5.10) holds for all r.

Equation (10.5.10) is a system of four third order equations and has 12 solutions and holds generally for $j=l+3/2$, $l\geq 0$. The asymtotic solutions are of the form of (10.2.6) with $\alpha_d\to\alpha_q$ or (10.2.7). This can be verified by putting f_j, g_j, h_j, and k_j to such asymtotic forms and inserting them into (10.5.10).

Near the origin, we make the expansion

$$\begin{pmatrix} f \\ g \\ h \\ k \end{pmatrix} = r^{\lambda}\begin{pmatrix} f_0+f_1r+\ldots \\ g_0+g_1r+\ldots \\ h_0+h_1r+\ldots \\ k_0+k_1r+\ldots \end{pmatrix} \qquad (10.5.11)$$

Inserting (10.5.11) into (10.5.10) leads to the following 12 roots of the indicial equation;

$$f_0=h_0=0, \qquad \lambda=j+\frac{5}{2},\ j+\frac{1}{2},\ j+\frac{1}{2},\ j-\frac{3}{2},\ -j+\frac{1}{2},\ -j-\frac{3}{2}$$
$$g_0=k_0=0, \qquad \lambda=j+\frac{3}{2},\ j-\frac{1}{2},\ -j+\frac{3}{2},\ -j-\frac{1}{2},\ -j-\frac{1}{2},\ -j-\frac{5}{2}$$
$$(10.5.12)$$

Analogous to (3.4.7), the six λ values with $-j$ will in general lead to diverging solutions at $r=0$ and are discarded. Among the remaining ones, the $\lambda=j+1/2$ and $j+3/2$ cases turn out to yield indeterminate k_0/g_0 ratios and are left out for eventual future investigation. Therefore, only the following three solutions remain;

$$\lambda=j-\frac{3}{2}, \qquad f_0=h_0=g_0=0, \qquad \chi_\mu, \psi_\mu \propto k_0 r^{j-3/2}Y_{j-3/2} \qquad (10.5.13a)$$

$$\lambda = j - \frac{1}{2}, \qquad f_0 = k_0 = g_0 = 0, \qquad \chi_\mu, \psi_\mu \propto h_0 r^{j-1/2} Y_{j-1/2} \tag{10.5.13b}$$

$$\lambda = j + \frac{5}{2}, \qquad f_0 = h_0 = 0, \qquad k_0 = -\left(\frac{j}{2}+1\right)\sqrt{\frac{(2j-1)(2j+3)}{3}}\, g_0$$
$$\chi_\mu, \psi_\mu \propto r^{j+5/2}\left(g_0 Y_{j+1/2} + k_0 Y_{j-3/2}\right) \tag{10.5.13c}$$

$$j = l + \frac{3}{2} \geq \frac{3}{2} \tag{10.5.13d}$$

The form of (10.5.13) is similar to that of the corresponding solutions (3.4.8, 6, 4) for triplet, $l>0$ mesons.

10.5.3. Quartet Radial Wave Equations for $j = 1/2$

Equation (10.5.10) holds for $j>½$ and contains four radial wave functions. It corresponds to (3.4.5) for the triplet, $l>0$ meson case, which contains three radial wave functions. Two of these three functions drop out for the $l=0$ case in §3.4.3. Similarly, the number of radial functions in (10.5.10) will also be reduced for the $j=½$ case.

As is seen in (10.5.4), the $j=½$ case can only be obtained from the last three combinations there but not from the $j=l+3/2$ case treated in §10.5.2. This situation is reminiscent of the doublet case discussed below (10.2.5) where $j=l-½$ wave functions (10.2.3b) can be obtained from the $j=l+½$ wave functions (10.2.3a) by some symbol changes. In an analogous manner, the quartet wave functions for the last three combinations in (10.5.4) can be obtained from (10.5.6) with $j\to j-1$, $j-2$ and $j-3$. The spherical harmonics and radial wave functions associated with these cases are by analogy to (10.5.8),

$$j = l + \frac{1}{2}, \qquad Y_{l-1}k_{j-1}, \quad Y_l h_{j-1}, \quad Y_{l+1} g_{j-1}, \quad \text{and} \quad Y_{l+2} f_{j-1} \tag{10.5.14a}$$

$$j = l - \frac{1}{2}, \qquad Y_{l-2}k_{j-2}, \quad Y_{l-1} h_{j-2}, \quad Y_{l} g_{j-2}, \quad \text{and} \quad Y_{l+1} f_{j-2} \tag{10.5.14b}$$

$$j = l - \frac{3}{2}, \qquad Y_{l-3}k_{j-3}, \quad Y_{l-2} h_{j-3}, \quad Y_{l-1} g_{j-3}, \quad \text{and} \quad Y_{l} f_{j-3} \tag{10.5.14c}$$

It is obvious that the radial wave equations associated with these cases are of the same form as (10.5.10) with $j\to j-1$, $j-2$ and $j-3$, respectively.

Consider the $l=0$ case for (10.5.10) which now holds for $j=3/2$. Conversion of this case to the $l=0$ case for (10.5.14a) starts by letting $j=3/2\to j=1/2$. According to (10.5.6, 7), (10.5.10d) is associated with Y_{l-1}, which however does not exist for l=0. Therefore, (10.5.10d) drops out. The associated radial function $k_{½}$ in (10.5.10) and $k_{-½}$ in (10.5.14a) become meaningless and can thus be put to zero. Making use of this result and inserting j=½ into (10.5.10c) further leads to $h_{½}$=0. In the remaining two equations, (10.5.10a, 10b), therefore, only the $f_{½}$ and $g_{½}$ terms remain. This is further insured by observing that the coefficients of the $h_{½}$ and $k_{½}$ terms in (10.5.10a, 10b) vanish for j=½.

For this case, the sum in (10.5.3) turns out to be independent of the angles and proportional to $f_{½}^2(r)-g_{½}^2(r)$. The angular integrations in (10.5.3) can now be carried out, as in (10.2.9), to yield

$$\Phi_{cq}(r)=\frac{g_s^6}{1920\pi}\left[\begin{array}{l}\int_0^r dr'r'^2\left(f_{\frac{1}{2}}^2(r')-g_{\frac{1}{2}}^2(r')\right)\left(5r^4+10r^2r'^2+r'^4\right)/r\\+\int_r^\infty dr'r'\left(f_{\frac{1}{2}}^2(r')-g_{\frac{1}{2}}^2(r')\right)\left(5r'^4+10r^2r'^2+r^4\right)\end{array}\right] \qquad (10.5.15)$$

Equation (10.5.10) now reduces to

$$j=\frac{1}{2},\qquad \left[\frac{E_{0q}^3}{8}+M_b^3+\Phi_{bq}(r)+\frac{E_{0q}}{2}\Delta_1\right]g_{\frac{1}{2}}(r)-\left(\frac{E_{0q}^2}{4}+\Delta_1\right)\left(\frac{\partial}{\partial r}+\frac{3}{r}\right)f_{\frac{1}{2}}(r)=0$$
$$\left[\frac{E_{0q}^3}{8}-M_b^3-\Phi_{bq}(r)+\frac{E_{0q}}{2}\Delta_2\right]f_{\frac{1}{2}}(r)+\left(\frac{E_{0q}^2}{4}+\Delta_2\right)\left(\frac{\partial}{\partial r}-\frac{1}{r}\right)g_{\frac{1}{2}}(r)=0 \qquad (10.5.16)$$

which holds for all r because (10.5.15), hence (10.5.2a), depends only upon r and separation according to (10.5.6) is possible. Equations (10.5.15) and (10.5.16) are nearly the same as the corresponding doublet equations (10.2.9) and (10.2.12). In fact, (10.5.15) is formally –3/16 times (10.2.9). Equations (10.5.16) and (10.2.12) are formally obtained from (10.2.4) with upper sign by putting l=1 and 0, respectively. The relations between these quartet and doublet cases are entirely analogous to those between the singlet, l=0 and triplet, l=0 cases of §3.4.1 and §3.4.3, respectively.

10.5.4. Quartet, j = 1/2 Equations of Reduced Order

Due to the complexity of (10.5.10), even for l=0, we shall limit ourselves to the j=½ case (10.5.16). The closeness of this case to the doublet, l=0 case considered near the end of Sec. 10.2 enables us to take over the results worked out for the latter case.

In the $r\to0$ and ∞ limits, (10.5.15) yields

$$\Phi_{cq}(r\to 0)=\frac{g_s^6}{384\pi}\int_0^\infty dr\, r^5\left(f_{\frac{1}{2}}^2(r)-g_{\frac{1}{2}}^2(r)\right) \qquad (10.5.17a)$$

$$\Phi_{cq}(r \to \infty) = \frac{g_s^6}{384\pi} r^3 \int_0^\infty dr' r'^2 \left(f_{1/2}^2(r') - g_{1/2}^2(r') \right) = \alpha_q^3 r^3 \tag{10.5.17b}$$

analogous to (10.2.10, 11). Asymtotic solutions to (10.5.16) are the same as those mentioned above (10.5.11) for (10.5.10). Near the origin, (10.5.13) no longer holds but the solution is given by (10.2.8a) with l=1,

$$g_{1/2}(r \to 0) = c_{gq}(\lambda_q) r^{\lambda_q}, \quad f_{1/2}(r \to 0) = 0, \quad \lambda_q = 3,\ 1,\ -2 \tag{10.5.18a}$$

$$f_{1/2}(r \to 0) = c_{fq}(\lambda_q) r^{\lambda_q}, \quad g_{1/2}(r \to 0) = 0, \quad \lambda_q = 2,\ -1,\ -3 \tag{10.5.18b}$$

Radial wave functions converging at r=0 are associated with λ_q=3 and 1 in (10.5.18a) and λ_q=2 in (10.5.18b).

As to the solution of (10.5.16), the discussion given in the beginning of Sec. 10.4 applies. The same type of assumptions that led to (10.4.1) are made. The order of (10.5.16) is analogously reduced by four to yield

$$\begin{aligned}
&\left[-\frac{E_{0q}^3}{8} - M_b^3 - d_{b0} - \frac{d_b}{r} \right] g_{1/2}(r) + \frac{E_{0q}^2}{4} \left(\frac{\partial}{\partial r} + \frac{3}{r} \right) f_{1/2}(r) = 0 \\
&\left[-\frac{E_{0q}^3}{8} + M_b^3 + d_{b0} + \frac{d_b}{r} \right] f_{1/2}(r) - \frac{E_{0d}^2}{4} \left(\frac{\partial}{\partial r} - \frac{1}{r} \right) g_{1/2}(r) = 0
\end{aligned} \tag{10.5.19}$$

These equations are basically (10.4.1) with l=1. The approximate mass formula (10.4.6), which is more accurate for heavy baryons, becomes

$$E_{0q} = 2\left(M_b^3 + d_{b0}\right)^{1/3} \left(1 + 4d_b^2 / E_{0q}^4\right)^{-1/6} \tag{10.5.20}$$

Similar to the consideration given at the end of Sec. 10.4, this relation has no application.

10.6. LINEAR BARYON RADIAL EQUATIONS

For ground state doublet baryons, such as the nucleons, Λ, etc. the wave equations must be linear to accomodate the superposition principle and building of wave packets, just like the case for pseudoscalar and vector mesons of Sec. 4.3. This implies that the nonlinear confining potential (10.2.2b) vanishes and (10.3.13) applies. Unlike the mesons considered in Sec. 4.3, which are not confined, these baryon wave functions are still confined by the linearized confining potential given by the last two terms in (10.2.2a). Therefore, separation of variables according to (10.2.3) is possible and the linearized doublet radial equations (10.2.4) with Φ_{bd}

given by (10.2.2a) and Φ_{cd}=0 hold. Since excited states are described by the same equations and are therefore similarly confined, there is no need to introduce nonlinear confinement for these states. This is in contrast to the excited mesons of Sec. 5.5 and 5.7, which in general are no longer confined by the d_m/r term in (3.4.1-3, 5a); the nonlinear confining potential (3.2.8b) has to step in to confine the meson wave functions. Therefore, (10.2.4) and (10.2.2a) with Φ_{cd}=0 are assumed to hold for both the ground state and radially excited doublet baryons. From a practical point of view, linear, radial equations are much easier to deal with than nonlinear and nonseparable equations.

The linearized (10.2.4) and (10.2.2a) with Φ_{cd}=0 are a special case of the analogously linearized (9.3.16) and (9.2.13b), to which we have to turn if the doublet baryon is in motion, discussed at the end of §10.5.1. From the linearized (9.3.16), the quartet radial equations (10.5.10) are derived and are hence also linear. This circumstance is analogous to that for pseudoscalar and vector mesons discussed in the beginning of Sec. 4.4.

Even these linearized radial equations, (10.2.4) and (10.2.2a) with Φ_{cd}=0 for doublet baryons and (10.5.10, 16) and (10.5.2a) with Φ_{cq}=0 for quartet baryons, are not easily handled. As was pointed out at the beginning of Sec.10.4, the integer power series (10.2.8) and (10.5.13) near r=0 are not easily connected to the fractional power series solutions of the form (10.2.7) for large r. To facilitate their solution, these linearized equations are transformed into a first order system of equations below, analogous to the conversion of (3.4.5) to (3.4.12).

10.7. FIRST ORDER SYSTEMS AND RECURSION FORMULA

According to the last section, Φ_{bc} is dropped in (10.1.6a), which together with (10.1.8) is inserted into (10.2.4) to yield

$$\left(\left(\frac{\partial}{\partial r}+\frac{L+2}{r}\right)\left(\frac{\partial}{\partial r}-\frac{L}{r}\right)+\frac{E_{bL}^2}{4}\right)\left(\frac{\partial}{\partial r}+\frac{L+2}{r}\right)f_{bL}(r)$$
$$+\left(\frac{E_{bL}}{2}\left(\frac{\partial}{\partial r}+\frac{L+2}{r}\right)\left(\frac{\partial}{\partial r}-\frac{L}{r}\right)+\frac{E_{bL}^3}{8}+M_b^3+d_{b0}+\frac{d_b}{r}+d_{b1}r+d_{b2}r^2\right)g_{bL}(r)=0$$
$$\left(-\frac{E_{bL}}{2}\left(\frac{\partial}{\partial r}+\frac{L+3}{r}\right)\left(\frac{\partial}{\partial r}-\frac{L+1}{r}\right)-\frac{E_{bL}^3}{8}+M_b^3+d_{b0}+\frac{d_b}{r}+d_{b1}r+d_{b2}r^2\right)f_{bL}(r)$$
$$+\left(\left(\frac{\partial}{\partial r}+\frac{L+3}{r}\right)\left(\frac{\partial}{\partial r}-\frac{L+1}{r}\right)+\frac{E_{bL}^2}{4}\right)\left(\frac{\partial}{\partial r}-\frac{L}{r}\right)g_{bL}(r)=0 \qquad (10.7.1)$$

where $L=l$ in (10.2.4) for $j=l+1/2$. $f_{bL}(r)$ and $g_{bL}(r)$ are eigenfunctions regular at r=0 and vanishing at $r=\infty$ together with the corresponding discrete, real and positive eigenvalues E_{bL}, if they exist, are sought. For $L=l=0$, $f_{b0}(r)$ and $g_{b0}(r)$ are the same as $f_0(r)$ and $g_0(r)$ in (10.2.12).

Equation (10.7.1) is now converted to a standard form of a first order equation system for which formal solutions near the singularities are known [C4 Ch 4, 5]. Introduce the analog of (3.4.10),

$$w_1(r) = -f_{bL}(r), \quad w_2(r) = \left(\frac{\partial}{\partial r} + \frac{L+2}{r}\right) w_1(r), \quad w_3(r) = \left(\frac{\partial}{\partial r} - \frac{L}{r}\right) w_2(r)$$
$$w_4(r) = g_{bL}(r), \quad w_5(r) = \left(\frac{\partial}{\partial r} - \frac{L}{r}\right) w_4(r), \quad w_6(r) = \left(\frac{\partial}{\partial r} - \frac{L+1}{r}\right) w_5(r)$$

(10.7.2)

We obtain

$$w_k{}'(r) = \left(\frac{A_{kn}}{r} + B_{kn} + C_{kn}\left(r, r^2\right)\right) w_n(r) \tag{10.7.3a}$$

$$A_{kn} = \begin{pmatrix} -L-2 & & & & & \\ & L & & & & \\ & & -L-2 & d_b & \frac{E_{bL}(2L+3)}{2} & \\ & & & L & & \\ & & & & L+1 & \\ d_b & & & & & -L-3 \end{pmatrix} \tag{10.7.3b}$$

$$B_{kn} = \begin{pmatrix} & 1 & & & & \\ & & 1 & & & \\ & -\frac{E_{bL}^2}{4} & & \frac{E_{bL}^3}{8} + M_b^3 + d_{b0} & & \frac{E_{bL}}{2} \\ & & & & 1 & \\ & & & & & 1 \\ -\frac{E_{bL}^3}{8} + M_b^3 + d_{b0} & & -\frac{E_{bL}}{2} & & -\frac{E_{bL}^2}{4} & \end{pmatrix}$$

(10.7.3c)

$$C_{kn}=\begin{pmatrix} 0 & 0 & 0 & 0 & 0 & 0 \\ 0 & & & & & \\ 0 & & & d_{b1}r+d_{b2}r^2 & & \\ 0 & & & & & \\ 0 & & & & & \\ d_{b1}r+d_{b2}r^2 & & & & & \end{pmatrix} \quad (10.7.3d)$$

The conversion (10.7.2) is similarly chosen such that the singularities at r=0 and ∞ do not get stronger in (10.7.3). The indicial equation of (10.7.3) is

$$\det\left|\lambda\delta_{kn}-A_{kn}\right|=(\lambda+L+3)(\lambda+L+2)^2(\lambda-L)^2(\lambda-L-1)=0 \quad (10.7.4)$$

λ is related to λ_+ in (10.2.8a) and to λ_q in (10.5.18), L=0 refers to l=0 in (10.2.8a) and L=1 to l=1 in (10.2.8a) but applied to (10.5.18). By means of (10.7.3b) and (10.7.2), it can be seen that the roots of (10.7.4) reproduce those given by (10.2.8a) and for L=1, also those of (10.5.18).

More conventionlly [C4 §4.5], (10.7.2, 3) are replaced by

$$w_1(r)=-f_{bL}(r), \qquad w_4(r)=g_{bL}(r)$$

$$\frac{\partial w_1}{\partial r}=\frac{w_2}{r}, \qquad \frac{\partial w_2}{\partial r}=\frac{w_3}{r}, \qquad \frac{\partial w_4}{\partial r}=\frac{w_5}{r}, \qquad \frac{\partial w_5}{\partial r}=\frac{w_6}{r}$$

$$\frac{\partial w_3}{\partial r}=w_1\left[L(L+1)(L+2)\frac{1}{r}-\frac{E_{bL}^2}{4}(L+2)r\right]+w_2\left[\left(L^2+2L+2\right)\frac{1}{r}-\frac{E_{bL}^2}{4}r\right]-w_3\frac{L+1}{r}$$
$$+w_4\left[d_b r+\left(\frac{E_{bL}^3}{8}+M_b^3+d_{b0}\right)r^2+d_{b1}r^3+d_{b2}r^4-\frac{E_{bL}}{2}L(L+1)\right]+(w_5+w_6)\frac{E_{bL}}{2}$$

$$\frac{\partial w_6}{\partial r}=w_1\left[d_b r+\left(-\frac{E_{bL}^3}{8}+M_b^3+d_{b0}\right)r^2+d_{b1}r^3+d_{b2}r^4+\frac{E_{bL}}{2}(L+1)(L+2)\right]-(w_2+w_3)\frac{E_{bL}}{2}$$
$$+w_4\left[-L(L+1)(L+2)\frac{1}{r}+\frac{E_{bL}^2}{4}Lr\right]+w_5\left[\left(L^2+2L+2\right)\frac{1}{r}-\frac{E_{bL}^2}{4}r\right]+w_6\frac{L+1}{r}$$

(10.7.5)

which yields an indicial equation equivalent to (10.2.8).

A more direct approach is to start from (10.7.1) and write

$$f_{bL}(r) = -\sum_{\nu=0}^{\infty} a_\nu r^{\lambda+\nu}, \qquad g_{bL}(r) = \sum_{\nu=0}^{\infty} b_\nu r^{\lambda+\nu} \tag{10.7.6}$$

Insertion into (!0.7.1) leads to the recursion formula

$$a_\nu = -\frac{1}{(\lambda+\nu+L+2)(\lambda+\nu-L-1)}\frac{E_{bL}^2}{4}a_{\nu-2} + \frac{1}{\lambda+\nu+L+2}\frac{E_{bL}}{2}b_{\nu-1}$$
$$+\frac{1}{(\lambda+\nu+L+2)(\lambda+\nu+L)(\lambda+\nu-L-1)}\left[d_b b_{\nu-2} + \left(M_b^3 + d_{b0} + \frac{E_{bL}^3}{8}\right)b_{\nu-3} + d_{b1}b_{\nu-4} + d_{b2}b_{\nu-5}\right] \tag{10.7.7a}$$

$$b_\nu = -\frac{1}{(\lambda+\nu+L+1)(\lambda+\nu-L)}\frac{E_{bL}^2}{4}b_{\nu-2} - \frac{1}{\lambda+\nu-L}\frac{E_{bL}}{2}a_{\nu-1}$$
$$+\frac{1}{(\lambda+\nu+L+1)(\lambda+\nu-L)(\lambda+\nu-L-2)}\left[d_b a_{\nu-2} + \left(M_b^3 + d_{b0} - \frac{E_{bL}^3}{8}\right)a_{\nu-3} + d_{b1}a_{\nu-4} + d_{b2}a_{\nu-5}\right] \tag{10.7.7b}$$

Chapter 11

BARYON WAVE FUNCTIONS AND CLASSIFICATION

Although the equations of motion for mesons [H5] and for baryons [H15] were both published in the early 1990's, subsequent work took almost wholly in the meson sector. This due to that the meson wave equations can be decomposed into second order differential equations (3.2.3-4) which have analytical solutions, providing the zeroth order background upon which higher order perturbational problems can be built and treated. Much success and new insights about mesons have been achieved, as are seen in Chapters 6-8.

On the other hand, the two coupled third order equations for baryon doublets (10.2.1) cannot be decoupled and reduced to lower order equations. The are eventually reduced to the six first order equations (10.7.5) which have to be solved numerically. Further, the interquark potential (10.2.2a) contains four integration constants that enter (10.7.5). Very lengthy work has been spent in finding these constants and solving (10.7.5) by computer. This is the reason that the results to be presented below come so many years later.

11.1. CONFINED DOUBLET BARYON WAVE FUNCTIONS

11.1.1. Numerical Solutions of Radial Wave Equations

For the doublet baryons of Sec. 10.2, L=0 in (10.7.5) and M_b there is given by (9.3.19) in which the quark masses m_ν are consistently taken to be the same as those for the mesons obtained in Table 5.2. There are also four unknown d_b constants which are independent of flavor, i. e., baryon species, as was pointed out below (10.1.6). The procedure adopted is to insert the known E_b and M_b of a given baryon into (10.7.5) and the vary the four d_b's over suitable ranges such that the solutions satisfy the boundary condition at $r\to\infty$ or converge there.

Near the origin $r\to 0$, the radial wave functions can be expanded in power series in r according to (10.7.6-7). At large r, (10.2.7a) shows that the solutions can be approximated by power series in $r^{1/3}$. Therefore the series of (10.7.6-7) must be inifinite and (10.7.5) has to be solved numerically.

Consider (10.7.5) and expand the w's around the regular singular point r=0,

$$w_\alpha(r) \rightarrow w_{(\lambda+)\alpha}(r) = \sum_{\beta=0}^{\infty} r^{\lambda+} w_{(\lambda+)\alpha\beta} r^{\beta}, \qquad \alpha = 1, 2, \ldots.6 \qquad (11.1.1)$$

where λ_+ are given by (10.2.8a). The three cases with possible $\lambda_+ < 0$ are excluded because they lead to diverging $w_\alpha(r=0)$. The remaining cases, apart from the normalization considerations of (10.3.13, 14), yield for orbital quantum number $l=0$

$$\lambda_+ = 0, \qquad w_{(0)40} = 1, \qquad w_{(0)10} = 0 \qquad (11.1.2a)$$

$$\lambda_+ = 1, \qquad w_{(1)40} = 0, \qquad w_{(1)10} = w_{(1)} \qquad (11.1.2b)$$

$$\lambda_+ = 2, \qquad w_{(2)40} = w_{(2)}, \qquad w_{(2)10} = 0 \qquad (11.1.2c)$$

where the amplitude in (11.1.2a) has been normalized to unity. $w_{(1)}$ and $w_{(2)}$ are amplitudes for the remaining two solutions. The general solution near $r=0$ reads

$$w_\alpha(r) = w_{(0)\alpha}(r) + w_{(1)\alpha}(r) + w_{(2)\alpha}(r) \qquad (11.1.3)$$

Substituting this into (10.7.5), which is now solved as an initial value problem with initial conditions (11.1.2).

The six unknown parameters d_b, d_{b0}, d_{b1}, d_{b2}, $w_{(1)}$, and $w_{(2)}$ are adjusted so that the six boundary conditions $w_\alpha(r\rightarrow\infty)\rightarrow 0$ for a given M_b^3 and the associated $E_{b0} = E_{0d}$. The calculations are done via a Fortran program employing Runge-Kutta integration subroutine on computers of the Dept. of Information Technology at Uppsala University. In the paramater range of interest, it was found that integration of (10.7.5) and summation of the series (10.7.6-7) give the nearly the same results for $r \leq 5$-8 Gev^{-1}. Beyond this value, (10.7.6-7) become unreliable due to accumulated computational errors. Similarly, (10.7.5) yields diverging solutions at about $r > 7$-12 Gev^{-1}. This is because that (10.2.7a) without the minus sign in the exponent is also a solution which tends to overshadow (10.2.7a) due to accumulated computer errors and takes over at large r.

This procedure has been applied to the neutron, Σ^0 and Ξ^0 with masses 0.9397, 1.1926 and 1.3148 Gev, respectively, for huge numbers of combinations of the six parameters. Only a very narrow range of values for these parameters leads to converging $w_\alpha(r)$.

Near the origin $r=0$ and at large r, the potential (10.2.2a) is dominated by the d_b/r and $d_{b2}r^2$, respectively, and the flavor dependent quark mass term M_b^3 in (10.7.5) can be dropped. The solutions $f_{b0}(r)$ and $g_{b0}(r)$ in these r regions are independent of the baryon flavor. Thus, d_b and d_{b2} are flavor independent constants on par with the corresponding meson sector's d_m and d_{m0} which are independent of flavor according to Sec. 4.4. The small r region is very small and the large r region determines the asymtotic behavior of $f_{b0}(r)$ and $g_{b0}(r)$.

Therefore, d_{b2}, which determines the "confinement strength" via (10.2.7a), is used to lable a set of the six unknown parameters and chosen first. Extensive computer calculations showed that the remaining five constants are uniquely fixed if the solutions g_{b0} and f_{b0} are to converge at $r\rightarrow\infty$. Some examples of the results are given in Table 11.1.

Table 11.1. Values of the four d_b constants in (10.7.5) with Gev as basic unit, the spread σ_{db1}, σ_{db0} and σ_{db} from the mean values of the four d_b costants and the sum spread σ_s according to (11.1.4) below, and $w_{(1)}$ and $w_{(2)}$ in (11.1.2) for some converging solutions of (10.7.5)

	d_{b2}	d_{b1}	d_{b0}	d_b	σ_s	$w_{(1)}$	$w_{(2)}$
Neutron	–0.140	1.0439	–3.0	0.695		0.016	0.0288
Σ^0	–0.1411	1.336	–4.06	1.085		0.0415	0.0691
Ξ^0	–0.141	1.111	–3.568	0.924		0.0019	0.0851
σ_d		10.7%	12.2%	17.7%	40.6%		
Neutron	–0.1641	1.334	–3.719	1.013		0.0668	0.045
Σ^0	–0.1641	1.279	–3.706	0.990		0.0353	0.0756
Ξ^0	–0.1641	1.220	–3.691	0.9177		0.0523	0.0721
σ_d		3.6%	0.3%	4.2%	8.1%		
Neutron	–0.3202	2.272	–4.922	1.024		0.1827	0.0674
Σ^0	–0.3202	2.167	–4.783	1.032		0.0586	0.0662
Ξ^0	–0.3202	2.142	–4.899	1.083		0.1191	0.1061
σ_d		2.6%	1.2%	2.5%	6.3%		
Neutron	–0.4462	2.968	–5.749	1.035		0.247	0.0859
Σ^0	–0. 4462	2.831	–5.541	1.066		0.1624	0.090
Ξ^0	–0. 4462	2.768	–5.516	1.033		0.121	0.0974
σ_d		2.9%	1.9%	1.4%	6.2%		
Neutron	–0.575	3.658	–6.55	1.064		0.2943	0.102
Σ^0	–0.575	3.471	–6.224	1.073		0.203	0.0997
Ξ^0	–0. 575	3.383	–6.147	1.029		0.195	0.1164
σ_d		2.2%	2.8%	1.8%	6.8%		
Neutron	–0.975	5.572	–8.328	0.8173		0.4278	0.1576
Σ^0	–0.975	5.319	–7.895	0.9048		0.316	0.1298
Ξ^0	–0.975	5.090	–7.441	0.6859		0.2902	0.1391
σ_d		3.7%	4.6%	11.2%	19.5%		

The wave functions for d_{b2}=–0.1641 and –0.4462 are plotted in Figure 11.1 below.

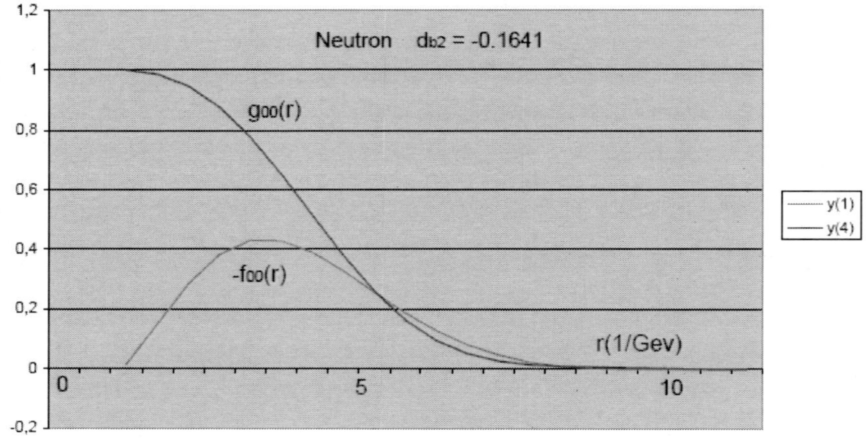
Neutron db2 = -0.1641
g00(r)
-f00(r)
r(1/Gev)
y(1)
y(4)
1,2
1
0,8
0,6
0,4
0,2
0
-0,2
0
5
10

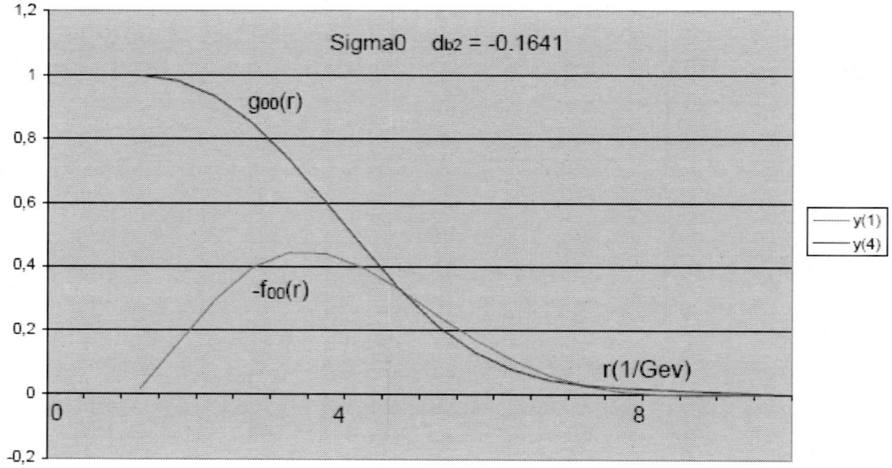
Sigma0 db2 = -0.1641
g00(r)
-f00(r)
r(1/Gev)
y(1)
y(4)
1,2
1
0,8
0,6
0,4
0,2
0
-0,2
0
4
8

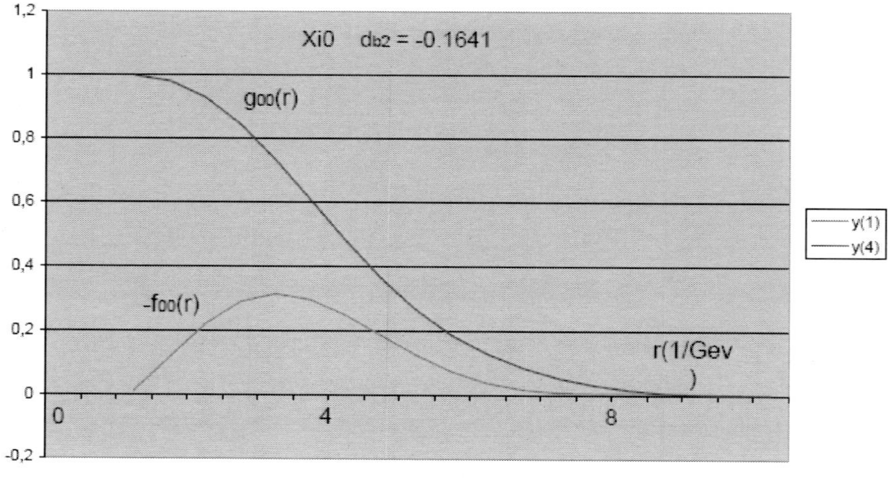
Xi0 db2 = -0.1641
g00(r)
-f00(r)
r(1/Gev
)
y(1)
y(4)
1,2
1
0,8
0,6
0,4
0,2
0
-0,2
0
4
8

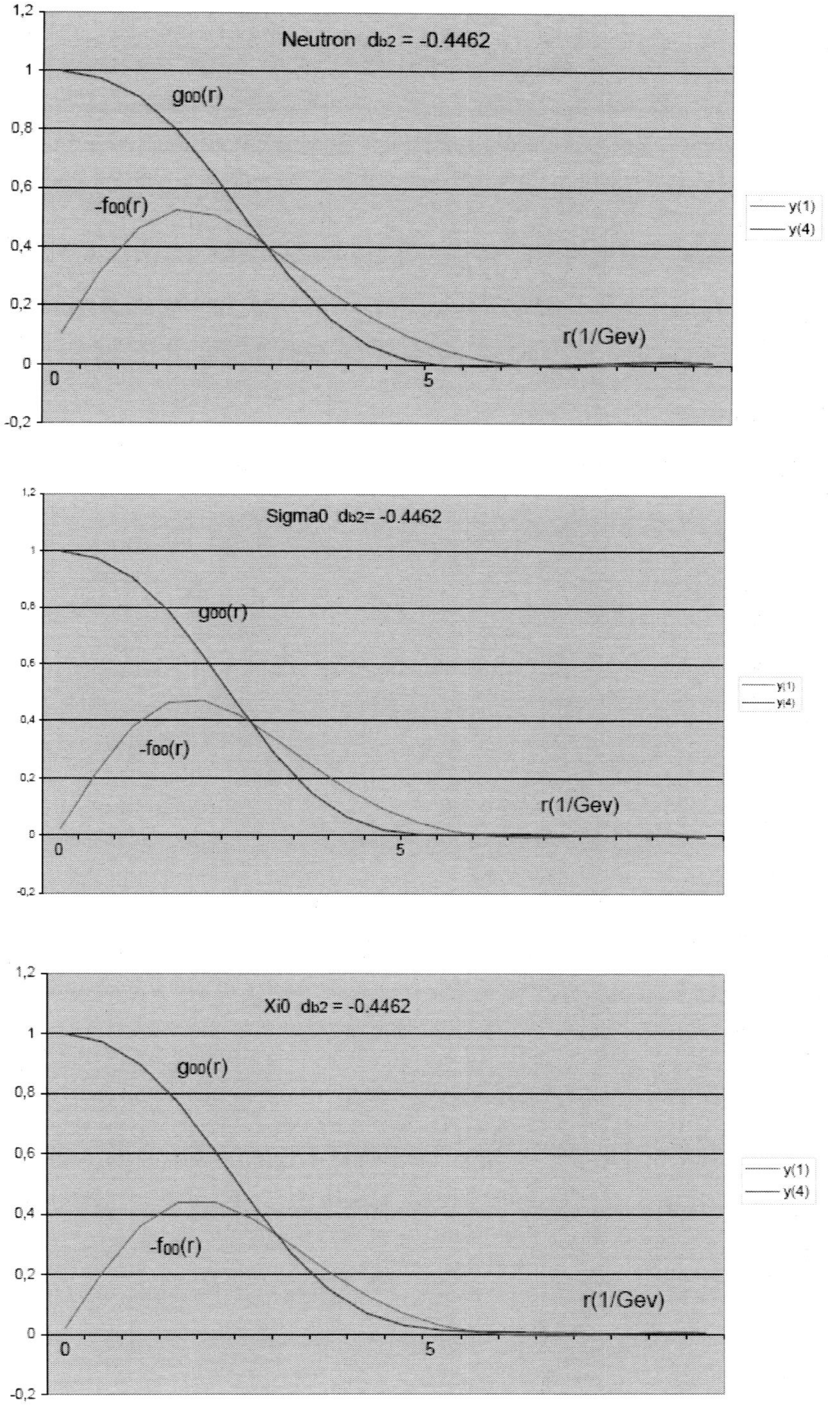

Figure 11.1. Baryon radial wave functions $f_{b0}(r)$ and $g_{b0}(r)$ in (10.7.5-7) normalized according to (10.3.14) for the $d_{b2} = -0.1641$ and -0.4462 cases in Table 11.1. r is the quark-diquark distance.

Convergent solutions have been found for continuous ranges of d_{b2} values largely in the region –0.1 to –1.4, although some values outside this region seem also to lead to convergence and some values inside this region, for instance from –0.23 to –0.27, do not. It may be noted that it is to some degree arbitrary to regard a set of solutions to be convergent or not. Due to accumulation of computer errors at large r, all solutions eventually diverge for sufficiently large r. Convergence is regarded as good if $f_{b0}(r)$ and $g_{b0}(r)$ are nearly zero over a "sufficiently" large range of r when r is large(see Figure 11.1).

The mean spread σ_s in Table 11.1 is defined as follows. Let

$$\sigma_{db} = \frac{1}{\bar{d}_b}\sqrt{\sum\nolimits_{\text{baroyn species}}\left(d_b - \bar{d}_b\right)^2/3}, \qquad \bar{d}_b = \text{average value of } d_b \tag{11.1.4a}$$

Repeat this step for d_{b0} and d_{b1} and define

$$\sigma_s = \sigma_{db} + \sigma_{db0} + \sigma_{db1} \tag{11.1.4b}$$

which is a measure of how much the d_b, d_{b0} and d_{b1} values deviate from the averages $\bar{d}_b$, $\bar{d}_{b0}$ and $\bar{d}_{b1}$.

If a set of d_{b2}, d_{b1}, d_{b0}, and d_b values that lead to convergent solutions using the quark masses obtained from meson spectra in Table 5.2 and the known masses for all three baryons in Table 11.1, a solution to the present baryon spectra problem has been found. Table 11.1 shows that this is not the case but some sets possess values that are rather close to each other for the three baryons and may therefore be regarded as approximate solutions to the baryon spectra problem here. Possible nature of these approximations are consisdered in the next paragraph.

The set that yields d_{b2}, d_{b1}, d_{b0}, and d_b values which are closest to each other for the three baryons in Table 11.1 is the d_{b2}=–0.4462 case with a minimum spread σ_{db}=1.4% for d_b. The mean spread σ_s=6.2% is also a minimum. This set is therefore associated with an approximate solution to the baryon spectra problem here.

However, Table 11.1 shows that this minimum is very shallow one and sets of d_{b2}, d_{b1}, d_{b0}, and d_b values with d_{b2} values in the range around –0.32 to –0.45 may also be qualified to yield approximate solutions. This range is supported by the approximative agreement of the calculated and experimental values of the half life and A and B asymmetry coefficients of free neutron decay for the d_{b2} = –0.3202 case in §12.3.6 below.

The flavor independence of d_b mentioned above Table 11.1 is largely confirmed in this table by the small σ_{db} there. In addition, this value is independent of the value of the labling d_{b2} constant. It corresponds to the meson sector's d_m in (5.2.3) is close to $\bar{d}_b$ for d_{b2} values in the range around d_{b2} = –0.32 to –0.45 in Table 11.1 and is

$$d_b \approx 1.04 \text{ Gev}^2 \tag{11.1.5}$$

The above results have been obtained using (10.2.8a) or j=l+1/2. If (10.2.8b) or j=l–1/2, which corresponds to the assignments of [P1] in Table 11.2 below were chosen, the d_{b0} values for the three baryons in Table 11.1 would deviate from each other so much that d_{b0} can no longer be considered as an approximately flavor independent constant.

The "reduced order" baryon spectra (10.4.7) was obtained assuming nonrelativistic qaurks, i. e., $E_{0d}^2/4 >> \Delta_0$, Δ_1 mentioned above (10.4.1). However, it can be estimated from Figure 11.1 that the opposite holds. Therefore, (10.4.7), hence also (10.5.20), cannot be used. The quark and diquark in the baryons are highly relativistic just like the quarks in the mesons are, mentioned at the end of Sec. 5.3.

11.1.2. On Functions Nonseparable in Relative Space *x* and Internal Space *z*

Consider a two electron system. When the both electrons are far apart, it is described by two independent Dirac wave functions $\psi_\alpha(x_I)$ and $\psi_\beta(x_{II})$ totaling eight wave function components. When they are closer to each other the product wave fucntions $\psi_\alpha(x_I)\psi_\beta(x_{II})$ must be generalized to the nonseparable $\psi_{\alpha\beta}(x_I,x_{II})$ having 16 components governed by the Bethe-Salpeter equation which includes the interaction between the both electrons. The eight extra wave function components are associated with this interaction and are small perturbations when the both electrons are not too close to each other. This examples illustrates the conjecture below.

The approximate solutions in §11.1.1 are based upon the construction that the total baryon wave functions are separable in x and z in (9.3.6) or, more specifically, in relative space wave function $\psi_0{}^a(\underline{x})$ and internal space functions $\eta_r{}^p(z_I,z_{II})$ in (9.3.7b). According to the epistemological considerations in Sec. 5.4 of [H16 or Appendix G], these both spaces are "hidden", on par with each other, and at the so-called "level logic2" and can be combined to form a larger manifold (x,z_I,z_{II}). In this case the product form $\psi_0{}^a(\underline{x})\eta_r{}^p(z_I,z_{II})$ in (9.3.7b) needs be generalized to the nonseparable form $\Psi_0{}^a{}_r{}^p(x,z_I,z_{II})$. A conjecture is now made that the mass operator $m_{3op}(z_I,z_{II})$ of (9.3.14), which contains $\partial/\partial z$ via (2.3.13), is also generalized to a nonseparable, as yet unknown form $m_{3op}(z_I,z_{II},x)$, which may also contain $\partial/\partial x$, analogous to the generalization of the masses operator $m_{2op}(z_I,z_{II})$ to $m_{2op}(z_I,z_{II},x)$ in Sec.5.4 of [Appendix G]. This generalization may for instance be such that when $r=|\underline{x}|$ falls outside some region of values, $m_{3op}(z_I,z_{II},x)$ degenerates back to $m_{3op}(z_I,z_{II})$.

The product wave fucntion $\psi_0{}^a(\underline{x})\eta_r{}^p(z_I,z_{II})$ in (9.3.7b) has 10 components, two from a=1, 2 and eight from p=r=1, 2, 3(less the singlet). The generalized, nonseparable $\Psi_0{}^a{}_r{}^p(x,z_I,z_{II})$ has 2×8=16 wave fucntion components, six more than 10 components. These six extra components are associated with this presently unknown dependence of $m_{3op}(z_I,z_{II},x)$ upon x. They may be the cause of the approximative nature of the results in Table 11.1.

Actually, the generalization from separable to nonseparable forms can already be formally introduced at the quark level, because the quark space-time x and internal space z are also "hidden" variables at "level logic2" in Sec. 5.4 of [H16 or Appendix G]. In this case, the following generalization is to be introduced into (2.3.11),

$$\chi_{\dot{b}}(X)\xi^p(z) \to \Xi_{\dot{b}}{}^p(X,z), \qquad \psi^a(X)\xi^p(z) \to \Psi^{ap}(X,z)$$
$$m_{1op}(z) \to m_{1op}(z,X) \tag{11.1.6}$$

Again, the form of $m_{1op}(z_I,X)$ is unknown.

11.2. Classification Scheme for Doublet Baryons and Ground State Baryon Wave Functions

An attempt is made to classify most of the known doublet baryons according to (10.2.8), much like the classification of the triplet, l>0 mesons in §5.7.2. The starting point is [P1] with its quark model assignments in terms of flavor-spin SU(6) basis and "bands" referring to unspecified excitations. These assignments are replaced by those in (10.2.8) containing the six linearly independent solutions of (10.2.1-4) near the origin and are shown in Table 11.2. The angular momenta and parity assignments of [P1] have also been changed due to considerations mentioned in relation to Table 11.3 below.

Table 11.2. Classification of doublet baryons given in [P1] according to (10.2.8) with parity $P=(-)^{l+1}$. The subscripts to $\lambda_\pm$ refer to the multiplet $l_{2l,\ 2j}$ and *a, b* or *c* refers to a specific $\lambda_\pm$ value in (10.2.8) or to a linear combination of solutions associated with different $\lambda_\pm$ values. To avoid divergence at r=0, $\lambda_\pm \geq 0$. Assignments given in [P1] are shown in parentheses and correspond to (10.2.8b) or $j=l-1/2$. They are replaced by the neighboring ones corresponding to (10.2.8a) or $j=l+1/2$.

$l_{2l,\ 2j}$	j^P	Octet members	Singlets	$j=l\pm½$	$\lambda_\pm$(10.2.8)
$(P_{11})S_{11}$	$(1/2^+)1/2^-$	$N(939)$, $\Lambda(1116)$, $\Sigma(1193)$, $\Xi(1318)$		1/2=(1−1/2)0+1/2	λ_{+S11a}
$(P_{11})S_{11}$	$(1/2^+)1/2^-$	$N(1440)$, $\Lambda(1600)$, $\Sigma(1660)$, $\Xi(?)$		1/2=(1−1/2)0+1/2	λ_{+S11b}
$(P_{11})S_{11}$	$(1/2^+)1/2^-$	$N(1710)$, $\Lambda(1810)$, $\Sigma(1880)$, $\Xi(?)$	$\Lambda(?)$	1/2=(1−1/2)0+1/2	λ_{+S11c}
$(S_{11})P_{11}$	$(1/2^-)1/2^+$	$N(1535)$, $\Lambda(1670)$, $\Sigma(1620)$, $\Xi(?)$	$\Lambda(1405)$	1/2=(0+1/2)1−1/2	λ_{-P11a}
$(S_{11})P_{11}$	$(1/2^-)1/2^+$	$N(1650)$, $\Lambda(1800)$, $\Sigma(1750)$, $\Xi(?)$		1/2=(0+1/2)1−1/2	λ_{-P11b}
$(P_{13})D_{13}$	$(3/2^+)3/2^-$	$N(1720)$, $\Lambda(1890)$, $\Sigma(?)$, $\Xi(?)$		3/2=(1+1/2)2−1/2	λ_{-D13a}
$(D_{13})P_{13}$	$(3/2^-)3/2^+$	$N(1520)$, $\Lambda(1690)$, $\Sigma(1670)$, $\Xi(1820)$	$\Lambda(1520)$	3/2=(2−1/2)1+1/2	λ_{+P13a}
$(D_{13})P_{13}$	$(3/2^-)3/2^+$	$N(1700)$, $\Lambda(?)$, $\Sigma(?)$, $\Xi(?)$		3/2=(2−1/2)1+1/2	λ_{+P13b}
$(F_{15})D_{15}$	$(5/2^+)5/2^-$	$N(1680)$, $\Lambda(1820)$, $\Sigma(1915)$, $\Xi(2030)$		5/2=(3−1/2)2+1/2	λ_{+D15a}
$(D_{15})F_{15}$	$(5/2^-)5/2^+$	$N(1675)$, $\Lambda(1830)$, $\Sigma(1775)$, $\Xi(?)$		5/2=(2+1/2)3−1/2	λ_{+F15a}
$(G_{17})F_{17}$	$(7/2^-)7/2^+$	$N(2190)$, $\Lambda(?)$, $\Sigma(?)$, $\Xi(?)$	$\Lambda(2100)$	7/2=(4−1/2)3+1/2	λ_{+F17a}
$(H_{19})G_{19}$	$(9/2^+)9/2^-$	$N(2220)$, $\Lambda(2350)$, $\Sigma(?)$, $\Xi(?)$		9/2=(5−1/2)4+1/2	λ_{+G19a}
$(G_{19})H_{19}$	$(9/2^-)9/2^+$	$N(2250)$, $\Lambda(?)$, $\Sigma(?)$, $\Xi(?)$		9/2=(4+1/2)5−1/2	λ_{-H19a}

11.3. Classification Scheme for Quartet Baryons

For the quartet baryons, there are four possible values for each j in (10.5.4), rather than two in the doublet case (10.2.3). For each j and l pair, there are three radial modes (10.5.13), analogous to the radial modes associated with the $\lambda_{\pm}=l+\ldots$ roots in (10.2.8) for doublets. These three radial modes are characterized by the power λ in (10.5.11) which has the values

$$\lambda = \lambda_l,\ \lambda_{l+1},\ \lambda_{l+4} \tag{11.3.1}$$

For j=3/2, (10.5.13d, 13a) yields

$$j = l + \frac{3}{2}, \qquad \lambda_l = j - \frac{3}{2} = l \tag{11.3.2a}$$

By analogy to (10.5.14), one obtains

$$j = l + \frac{1}{2}, \qquad \lambda_l = l - 1 \tag{11.3.2b}$$

$$j = l - \frac{1}{2}, \qquad \lambda_l = l - 2 \tag{11.3.2c}$$

$$j = l - \frac{3}{2}, \qquad \lambda_l = l - 3 \tag{11.3.2d}$$

An attempt to classify the decuplet baryons proceeds along the same line applied to doublet baryons in Sec. 11.2. The results are given in Table 11.3.

Table 11.3. Tentative classification of the decuplet baryons including those given in [P1] according to (11.3.1, 2). Notations, restrictions as well as the assignments of the λ values are analogous to those in Sec. 11.2.

$l_{2I,\,2j}$	j^P	Decuplet members	j in (11.3.1, 2)		λ in (11.3.1)
P_{31}	$1/2^+$	$\Delta(1910)$, $\Sigma(?)$, $\Xi(?)$, $\Omega(?)$	$j=l-½$	$l=1$	λ_{P31a}
S_{31}	$1/2^-$	$\Delta(1620)$, $\Sigma(?)$, $\Xi(?)$, $\Omega(?)$	$j=l+½$	$l=0$	λ_{S31a}
P_{33}	$3/2^+$	$\Delta(1232)$, $\Sigma(1385)$, $\Xi(1530)$, $\Omega(1672)$	$j=l+½$	$l=1$	λ_{P33a}
P_{33}	$3/2^+$	$\Delta(1600)$, $\Sigma(?)$, $\Xi(?)$, $\Omega(?)$	$j=l+½$	$l=1$	λ_{P33b}
D_{33}	$3/2^-$	$\Delta(1700)$, $\Sigma(?)$, $\Xi(?)$, $\Omega(?)$	$j=l-½$	$l=2$	λ_{D33a}
F_{35}	$5/2^+$	$\Delta(1905)$, $\Sigma(?)$, $\Xi(?)$, $\Omega(?)$	$j=l-½$	$l=3$	λ_{F35a}
D_{35}	$5/2^-$	$\Delta(1930)$, $\Sigma(?)$, $\Xi(?)$, $\Omega(?)$	$j=l+½$	$l=2$	λ_{D35a}
F_{37}	$7/2^+$	$\Delta(1950)$, $\Sigma(2030)$, $\Xi(?)$, $\Omega(?)$	$j=l+½$	$l=3$	λ_{F37a}
$H_{3,11}$	$11/2^+$	$\Delta(2420)$, $\Sigma(?)$, $\Xi(?)$, $\Omega(?)$	$j=l+½$	$l=5$	$\lambda_{H3,11a}$

Any assignment of the λ's are far more uncertain than those in the doublet case. A numerical, confined solution of the effectively 12th order radial equations (10.5.10) is far more complex to obtain than that for the effectively 6th order equations (10.2.4) for the doublet baryons.

Chapter 12

NEUTRON DECAY AND POSSIBLE NONCONSERVATION OF ANGULAR MOMENTUM

This chapter has been adapted from [H19].

12.1. BACKGROUND

The present theory of nuclear β-decay is based upon the electroweak part of the standard model [P1]. The origin of this part is the four fermion point interaction Lagrangian density

$$L_F = -C_V\left(\overline{\psi}_p\gamma_\mu\psi_n\right)\left(\overline{\psi}_e\gamma_\mu\psi_\nu\right) \tag{12.1.1}$$

for neutron decay

$$n \to p + e^- + \overline{\nu} \tag{12.1.2}$$

first proposed by Fermi in 1934. Here C_V is a constant. Subsequently, (12.1.1) has been generalized to the β-interaction Hamiltonian currently in use [K3 (13.9)],

$$H_W = \frac{1}{\sqrt{2}}\sum_i \int d^3\underline{X}\left(\overline{\psi}_p O_i\psi_n\right)\left(C_i\overline{\psi}_e O_i\psi_\nu + C_i'\overline{\psi}_e O_i\gamma_5\psi_\nu\right) + h.c. \tag{12.1.3}$$

Here, i refers to the scalar, vector, tensor, axial vector, and pseudoscalar interactions between the nucleon and lepton currents. The O_i's contain γ_μ and γ_5 and the C's are generally complex.

Based upon (12.1.3), lepton kinematics and conventional conservation laws, including angular momentum conservation, Jackson et al [2] derived in 1957 a number of decay rate formulae. The most frequently used one is

$$dW = \text{constant} \times d^3\underline{K}_e d^3\underline{K}_\nu \delta\left(E_e + E_\nu + E_p - E_n\right)$$

$$\times \xi \left[1 + a\frac{\underline{K}_e \underline{K}_\nu}{E_e E_\nu} + \frac{\langle \underline{J}_n \rangle}{J_n} \left(A\frac{\underline{K}_e}{E_e} + B\frac{\underline{K}_\nu}{E_\nu} + D\frac{\underline{K}_e \times \underline{K}_\nu}{E_e E_\nu} + R\frac{\underline{K}_e \times \underline{\sigma}_e}{E_e} \right) + \ldots \right] \quad (12.1.4)$$

in which the final spins have been summed over for a given initial neutron polarization $\langle \underline{J}_n \rangle$. Here, the K's denote momenta, E energy and $\underline{\sigma}_e$ the electron polarization. The constants a, A, B, D, R, and ξ depend upon the C's and the nucleon current consisting of a vector and an axial vector part in the so-called V-A theory. Experiments on neutron decay have since then largely been devoted to determine these constants in this nearly 50 year old (12.1.4) and related formulae [L6, S8-11 and references therein]. These have, however, yielded little physical insight into the decay mechanism.

The standard Hamiltonian (12.1.3) is a phenomenological model, not derivable from any first principles' theory. It treats nucleon as a point particle and hence ignores its quark structure. There are in principle 20 real C constants in (12.1.3), leaving the theory with little predictive power. This hints at a superfluousness of (12.1.3), which is not invariant under SU(2) gauge transformations. Such an invariance would give rise to an intermediate vector boson W which couples to a left-handed $\bar{\nu} e$ pair.

Therefore, despite its noble origin and general acceptance, this model (12.1.3) does not differ in principle from the large number of recent phenomenological models, constructed for different and narrow application angles, present in a vast body of literature.

Angular momentum conservation has not been established experimentally in free neutron decay. Most of the experiments make use of nuclei and the conclusion is that angular momentum is conserved. But a nucleus poses a highly complex, unsolved many body problem and experiments with it cannot lead to any firm conclusion on angular momentum conservation in free neutron decay.

In this chapter, free neutron decay will be treated using the equations of motion for doublet baryon and its development in Chapters 9-11, incorporating vector gauge fields of §7.1.2 and new tensor gauge fields. The results obtained, not reachable from (12.1.4), include a prediction of the half life of the neutron in approximate agreement with data and a rather accurate prediction of the A or B asymmetry coefficient.

The scalar strong interaction hadron theory consists of a meson sector and a baryon sector. The former is achored on many data points in Chapters 5-8. The two contact points with data just mentioned lend some credibility to this theory also in the baryon sector, hence to this theory as a whole.

12.2. Introduction of Vector and Tensor Gauge Fields

12.2.1. Action-Like Integrals for Doublet Baryons

The starting point is the equation of motion for doublet baryons (10.0.1). Multiply (10.0.1a) from the left by χ_{0a}* and (10.0.1b) by $-\psi_0^{\dot{b}*}$, subtract the resulting expressions from their complex conjugates and integrate over x_I and x_{II} to obtain

$$S'_\chi = \frac{1}{2i}\int dx_I^4 dx_{II}^4 \left(\chi_{0a}^* \partial_{II}^{f\dot{e}} \partial_I^{e\dot{h}} \partial_I^{a\dot{b}} \tfrac{1}{2}\chi_{\{\dot{b}\dot{h}\}f} + i\left(M_b^3 + \Phi_b\right)\chi_{0a}^* \psi_0^a - c.c.\right) \tag{12.2.1}$$

$$S'_\psi = \frac{1}{2i}\int dx_I^4 dx_{II}^4 \left(-\psi_0^{*\dot{b}} \partial_{II\,\dot{e}h} \partial_{I\,\dot{h}k} \partial_{I\,\dot{b}c} \tfrac{1}{2}\psi^{\{ck\}\dot{e}} + i\left(M_b^3 + \Phi_b\right)\psi_0^{*\dot{b}} \chi_{0\dot{b}} - c.c.\right) \tag{12.2.2}$$

where * is an extra sign denoting complex conjugate. The positions of ∂_I and ∂_{II} have been changed so that the summations over the e and h indices conform to matrix multiplication convention and that the both ∂_I's appear next to each other to reflect that they operate on the diquark part indicated by the braces.

These integrals will not be varied with respect to χ_{0a}* and $-\psi_0^{\dot{b}*}$ in an attempt to reproduce (10.0.1); handling of the last *c.c.* term and the necessary boundary conditions will require efforts beyond the scope of this chapter. Nor is such a variation necessary for the present purposes. If the solutions to (10.0.1) is inserted into (12.2.1-2), we obtain

$$S'_\chi \to S'_{\chi 0} = 0, \qquad S'_\psi \to S'_{\psi 0} = 0 \tag{12.2.3}$$

12.2.2. NonMinimal Substitution and Tensor Gauge Field

The minimal substitution of (7.1.4) led to the introduction of the vector gauge boson field W_I which is naturally associated with the vector part V of the V-A theory mentioned beneath (12.1.4). The axial vector part A is an asymmetrical part of a tensor which can be introduced by the following nonminimal substitution,

$$\begin{aligned} &\partial_I^{e\dot{h}} \partial_I^{a\dot{b}} \chi_{\{\dot{h}\dot{b}\}f} \to \partial_I^{e\dot{h}} \partial_I^{a\dot{b}} \chi_{\{\dot{h}\dot{b}\}f} \\ &+\frac{i}{4} g\left[W^{e\dot{h}}(X)\partial_I^{a\dot{b}} + W^{a\dot{b}}(X)\partial_I^{e\dot{h}} + \frac{1}{2}T^{e\dot{h}a\dot{b}}(X) + \frac{i}{4} g W^{e\dot{h}}(X) W^{a\dot{b}}(X)\right]\chi_{\{\dot{b}\dot{h}\}f} \end{aligned} \tag{12.2.4}$$

$$\begin{aligned} &\partial_{I\dot{h}k}\partial_{I\dot{b}c}\psi^{\{kc\}\dot{e}} \to \partial_{I\dot{h}k}\partial_{I\dot{b}c}\psi^{\{kc\}\dot{e}} \\ &+\frac{i}{4} g\left[W_{\dot{h}k}(X)\partial_{I\dot{b}c} + W_{\dot{b}c}(X)\partial_{I\dot{h}k} + \frac{1}{2}T_{\dot{h}k\dot{b}c}(X) + \frac{i}{4} g W_{\dot{h}k}(X) W_{\dot{b}c}(X)\right]\psi^{\{ck\}\dot{e}} \end{aligned} \tag{12.2.5}$$

Here, the subscript I in W_I of (7.1.5) has been dropped. The right side of (12.2.4) can readily be shown to be invariant under the U(1) gauge transformations

$$\begin{aligned} &W^{a\dot{b}}(X) \to W^{a\dot{b}}(X) - \partial_X^{a\dot{b}}\phi_s(X), \quad T^{e\dot{h}a\dot{b}}(X) \to T^{e\dot{h}a\dot{b}}(X) - \partial_X^{e\dot{h}}\partial_X^{a\dot{b}}\phi_s(X) \\ &\chi_{\{\dot{h}\dot{b}\}f} \to \chi_{\{\dot{h}\dot{b}\}f} \exp\left(\tfrac{i}{2} g\phi_s(X)\right) \end{aligned} \tag{12.2.6}$$

where $\phi_s(X)$ is a local phase and (7.1.4) has been consulted. The right side of (12.2.5) transform analogously.

12.2.3. SU(3) Tensor Gauge Fields and Gauge Invariance

These expressions are now generalized to include SU(3) gauge fields analogous to (2.4.2, 3) and (7.1.4). Limiting ourselves to baryon doublets in (9.3.7b), (2.2.1-2) with (12.2.4-5) are generalized, with the sign of the tensor term changed in (12.2.5), to

$$S_\chi = \frac{1}{2i}\int dx_I^4 dx_{II}^4 \chi^*_{tp\,0a} \times$$
$$\left[\begin{array}{l}\left\{\partial_{II}^{f\ddot{e}}\delta_{ps} + \frac{i}{4}g(\lambda_l)_{ps}W_l^{f\ddot{e}}\right\} \\ \times\left\{\partial_I^{e\dot{h}}\partial_I^{a\dot{b}}\delta_{sq}\delta_{qr} + \frac{i}{4}g\left[\begin{array}{l}\delta_{sq}(\lambda_l)_{qr}W_l^{e\dot{h}}\partial_I^{a\dot{b}} + (\lambda_l)_{sq}W_l^{a\dot{b}}\delta_{qr}\partial_I^{e\dot{h}} + \frac{1}{2}(\lambda_l)_{sq}(\lambda_{l'})_{qr}T_{ll'}^{e\dot{h}a\dot{b}} \\ +\frac{i}{4}g\{(\lambda_l)_{sq}W_l^{e\dot{h}}\}\{(\lambda_{l'})_{qr}W_{l'}^{a\dot{b}}\}\end{array}\right]\right\}\chi_{rt\{\dot{b}\dot{h}\}f} \\ +i\left(M_b^3 + \Phi_b\right)(-2)\psi^a_{pt} - c.c.\end{array}\right]$$

(12.2.7)

$$S_\psi = \frac{1}{2i}\int dx_I^4 dx_{II}^4 (-)\psi^{*\dot{b}}_{tp\,0} \times$$
$$\left[\begin{array}{l}\left\{\partial_{II\,\dot{e}h}\delta_{ps} + \frac{i}{4}g(\lambda_l)_{ps}W_{l\,\dot{e}h}\right\} \\ \times\left\{\partial_{I\,\dot{h}k}\partial_{I\,\dot{b}c}\delta_{sq}\delta_{qr} + \frac{i}{4}g\left[\begin{array}{l}\delta_{sq}(\lambda_l)_{qr}W_{l\,\dot{h}k}\partial_{I\,\dot{b}c} + (\lambda_l)_{sq}W_{l\,\dot{b}c}\delta_{qr}\partial_{I\,\dot{h}k} - \frac{1}{2}(\lambda_l)_{sq}(\lambda_{l'})_{qr}T_{ll'\,\dot{h}k\dot{b}c} \\ +\frac{i}{4}g\{(\lambda_l)_{sq}W_{l\,\dot{h}k}\}\{(\lambda_{l'})_{qr}W_{l'\dot{b}c}\}\end{array}\right]\right\}\psi_{rt}^{\{ck\}\dot{e}} \\ +i\left(M_b^3 + \Phi_b\right)2\chi_{pt\,0\dot{b}} - c.c.\end{array}\right]$$

(12.2.8)

Equation (12.2.7) is invariant with respect to the SU(3) gauge transformations (7.1.7), with the obvious replacement of the two meson indices by the three baryon indices associated with χ in (12.2.7), together with a generalization of the second of (12.2.6),

$$(\lambda_l)_{sq}(\lambda_{l'})_{qr}T_{ll'}^{e\dot{h}a\dot{b}}(X)\chi_{rt\{\dot{b}\dot{h}\}f} \to \left((\lambda_l)_{sq}(\lambda_{l'})_{qr}T_{ll'}^{e\dot{h}a\dot{b}}(X) + \frac{2i}{g}\left(\partial_X^{e\dot{h}}\partial_X^{a\dot{b}}U_{3sq}(X)\right)U_{3qr}^{-1}(X)\right)\chi_{rt\{\dot{b}\dot{h}\}f}$$

(12.2.9)

Analogously, the same invariance also holds for (12.2.8) with (12.2.9) replaced by

$$(\lambda_l)_{sq}(\lambda_{l'})_{qr}T_{ll'\dot{h}k\dot{b}c}(X)\psi_{rt}^{\{ck\}\dot{e}} \rightarrow \left((\lambda_l)_{sq}(\lambda_{l'})_{qr}T_{ll'\dot{h}k\dot{b}c}(X)-\frac{2i}{g}\left(\partial_{X\dot{h}k}\partial_{X\dot{b}c}U_{3sq}(X)\right)U_{3qr}^{-1}(X)\right)\psi_{rt}^{\{ck\}\dot{e}}$$

(12.2.10)

For application to neutron decay, only the SU(2) part of (7.1.4-5) is needed and l and l' run from 1 to 3. Apart from the flavor indices l and l', the tensor $T_{ll'}^{eha\dot{b}}$ has 16 components, 10 symmetrical and 6 asymmetrical, which in its turn is grouped into a vector E (electric field in electromagnetism) and an axial vector H (magnetic field), which is assigned to the axial vector A mentioned above (12.2.4). Identification of the tensor components corresponding to H has been given in [L1 §4-5] which are found by means of the invariant aymmetrical operator ε_{kl} of (B15). These are

$$\begin{aligned} &T_{ll'}^{\dot{h}\dot{b}} = T_{ll'}^{eha\dot{b}}\varepsilon_{ea} = \frac{1}{2}\left(T_{ll'}^{\dot{h}\dot{b}21} - T_{ll'}^{\dot{h}\dot{b}12}\right) \\ &T_{ll'}^{\{\dot{1}\dot{2}\}} = T_{ll'}^{\{\dot{2}\dot{1}\}} = 2iH_3, \qquad T_{ll'}^{\dot{1}\dot{1}} = -2i(H_1 + iH_2), \qquad T_{ll'}^{\dot{2}\dot{2}} = 2i(H_1 - iH_2) \end{aligned} \qquad (12.2.11)$$

The remaining 13 components of $T_{ll'}^{eha\dot{b}}$ do not enter here and are put to zero.

12.3. First Order Relations

The gauge boson and tensor field in (12.2.7-8) are decay products of the neutron whose wave functions now acquire a weak time dependence. Following (6.4.1) or (7.3.1), let the nucleon wave functions in (12.2.7-8) take the form

$$\begin{aligned} &\chi_{\{\dot{b}\dot{h}\}f}(X,x) = \left(a_{op} + a_{op}^{(1)}(X^0)\right)\chi_{\{\dot{b}\dot{h}\}f}(\underline{x})\exp\left(-iEX^0 + i\underline{K}\,\underline{X}\right) \\ &= \left(1 + \frac{a_{op}^{(1)}(X^0)}{a_{op}}\right)\chi_{0\{\dot{b}\dot{h}\}f}(X,x) = \chi_{0\{\dot{b}\dot{h}\}f}(\underline{x})\exp\left(-iEX^0 + i\underline{K}\,\underline{X}\right) + \chi_{1\{\dot{b}\dot{h}\}f}(X,x) \\ &\chi \rightarrow \psi \end{aligned} \qquad (12.3.1)$$

where E is the energy, $\underline{K}$ the momentum, a_{op}=1 and $a_{op}^{(1)}(X^0)$ is a first order quantity varying slowly with time. Both can be elevated to operators in quantized case, as are described beneath (6.4.2). Ordering of these small quantities $a_{op}^{(1)}(X^0)$, g, W_l etc is the same as that in (7.3.2). The subscripts *0*, *1* denote zeroth order and first order quantities, respectively.

Following the rudimentary quantization procedure of (6.4.12-15), let the initial and final states be denoted by

$$|i\rangle = |n(\underline{K}_n = 0)\rangle, \qquad \langle f| = \langle p(\underline{K}_p), W(E_W, \underline{K}_W), T(E_W, \underline{K}_W)| \tag{12.3.2}$$

respectively. n, p, W, T denote the neutron, proton, gauge boson, and tensor gauge field, respectively. Further,

$$\langle 0|i\rangle = \langle f|0\rangle = \langle f|i\rangle = 0, \qquad \langle 0|0\rangle = \langle i|i\rangle = \langle f|f\rangle = 1 \tag{12.3.3}$$

Let a_{op} in (12.3.1) and its hermetian conjugate $a_{op}{}^{+}$ be elevated to annihilation and creation operators analogous to that mentioned below (6.4.2). Insert (12.3.1) into (12.2.7) and sandwich the resulting expression between <f| and |i>. There are two types of first order terms: i) those containing $a_{op}{}^{(1)}(X^0\to\infty)$ and ii) those linear in gW and gT.

Carrrying out integration over the time X^0 and employing (3.1.4) and (3.5.7), the type i) terms read

$$S_{fi} \int d^3\underline{X} \int d^4x \left\{ -i\chi^*_{0\dot{a}}(\underline{x}) \left(\frac{E_n^2}{4} + \Delta \right) \chi_{0\dot{a}}(\underline{x}) \right\}, \quad S_{fi} = \langle f| a^+_{op} a^{(1)}_{op} (X^0 \to \infty) |i\rangle \tag{12.3.4}$$

in which the $c.c.$ term in (12.2.7) contributes equally. S_{fi} is the decay amplitude. Here, (9.2.7) has been used putting the quartet functions in (9.2.8) to zero which leads to

$$\chi_{\{\dot{1}\dot{1}\}1} = \tfrac{4}{3}\chi_{\dot{1}}, \quad \chi_{\{\dot{1}\dot{2}\}2} = \tfrac{2}{3}\chi_{\dot{1}}, \quad \chi_{\{\dot{2}\dot{1}\}1} = \tfrac{2}{3}\chi_{\dot{2}}, \quad \chi_{\{\dot{2}\dot{2}\}2} = \tfrac{4}{3}\chi_{\dot{2}}, \quad \chi_{\{\dot{1}\dot{1}\}2} = \chi_{\{\dot{2}\dot{2}\}1} = 0$$
$$\psi^{\{11\}\dot{1}} = -\tfrac{4}{3}\psi^1, \quad \psi^{\{12\}\dot{2}} = -\tfrac{2}{3}\psi^1, \quad \psi^{\{21\}\dot{1}} = -\tfrac{2}{3}\psi^2, \quad \psi^{\{22\}\dot{2}} = -\tfrac{4}{3}\psi^2, \quad \psi^{\{11\}\dot{2}} = \psi^{\{22\}\dot{1}} = 0 \tag{12.3.5}$$

For the evaluation of the type ii) terms, the final state <f| in (12.3.2) will contain the gauge fields,

$$\left(W_1^{a\dot{b}}(X) - iW_2^{a\dot{b}}(X)\right)/\sqrt{2} = W^{-a\dot{b}}(X) = W^{-0}\delta^{a\dot{b}} - \underline{\sigma}^{a\dot{b}}\underline{W}^- = w^{a\dot{b}} \exp\left(iE_W X^0 - i\underline{K}_W \underline{X}\right)$$
$$= \left(w_0\delta^{a\dot{b}} - \underline{\sigma}^{a\dot{b}}\underline{w}\right)\exp\left(iE_W X^0 - i\underline{K}_W \underline{X}\right) \tag{12.3.6}$$

$$\left(T_{13}^{e\dot{h}a\dot{b}} - T_{31}^{e\dot{h}a\dot{b}} + iT_{32}^{e\dot{h}a\dot{b}} - iT_{23}^{e\dot{h}a\dot{b}}\right)/\sqrt{2} = T^{e\dot{h}a\dot{b}}(X) = t^{e\dot{h}a\dot{b}} \exp\left(iE_W X^0 - i\underline{K}_W \underline{X}\right) \tag{12.3.7}$$

where (7.1.4-5) limited to its SU(2) part has been consulted. The initial and final nucleon states in (12.3.2) will have a laboratory space time dependence of the form given in (12.3.1) with subscripts p for proton and n for neutron attached to the variables there. Here, use has been made of (9.3.7b) and of (9.3.18a) which gives tp=31 for proton and rt=23 for neutron in (12.2.7). After summing over the flavor indices t, p, s, q, and r and carrying out the integration over X, the type ii) terms become

$$\frac{g}{4\sqrt{2}}(2\pi)^4\delta\left(\underline{K}_p+\underline{K}_W\right)\delta\left(E_p+E_W-E_n\right)\times$$

$$\int d^4x\left\{w_0\left[\chi^*_{p0a}(\underline{x})\left(-2\left(\frac{E_n^2}{4}+\Delta\right)-E_n^2\right)\chi_{n0\dot{a}}(\underline{x})+c.c.\right]-i\left[\chi^*_{p0a}(\underline{x})\partial^{f\dot{e}}_{II}t^{\dot{h}bea}\chi_{n0\{\dot{b}\dot{h}\}f}(\underline{x})+c.c.\right]\right\}$$

(12.3.8)

where (12.3.5) has been used and E_W and K_W terms have been neglected because they are small relative to $|\partial_{I,II}\,\chi|/|\chi|$. With (12.2.11), (12.3.7) and (B5), the 16 t's in (12.3.8) reduce similarly to three for an axial vetor:

$$t^{\dot{h}\dot{b}}=t^{e\dot{h}a\dot{b}}\varepsilon_{ea}=\frac{1}{2}\left(t^{\dot{h}\dot{b}21}-t^{\dot{h}\dot{b}12}\right),\qquad t^{ea}=t^{e\dot{h}a\dot{b}}\varepsilon_{\dot{h}\dot{b}}=\frac{1}{2}\left(t^{ea\dot{2}\dot{1}}-t^{ea\dot{1}\dot{2}}\right)$$

$$t^{\dot{1}\dot{1}}=t^{22}=t_{\dot{2}\dot{2}}=t_{11},\qquad t^{\dot{2}\dot{2}}=t^{11}=t_{\dot{1}\dot{1}}=t_{22},\qquad t^{12}=t^{21}=-t^{\{\dot{1}\dot{2}\}}=t_{\{\dot{1}\dot{2}\}}=-t_{\{12\}}$$

$$T^{\dot{h}\dot{b}}(X)=t^{\dot{h}\dot{b}}\exp\left(iE_WX^0-i\underline{K}_W\underline{X}\right),\qquad T^{ea}(X)=t^{ea}\exp\left(iE_WX^0-i\underline{K}_W\underline{X}\right)$$

(12.3.9)

12.4. Decay Amplitude

The decay amplitude $S_{fi\chi}$ for the χ function is found by putting (12.3.4) to the negative of (12.3.8). Letting $\Omega=\int d^3\underline{X}$, the result is

$$S_{fi\chi}=-\frac{ig}{4\sqrt{2}}(2\pi)^4\delta\left(\underline{K}_p+\underline{K}_W\right)\delta\left(E_p+E_W-E_n\right)\times$$

$$\frac{\int d^3\underline{x}\left\{w_0\left[\chi^*_{p0a}(\underline{x})\left(-2\left(\frac{E_n^2}{4}+\Delta\right)-E_n^2\right)\chi_{n0\dot{a}}(\underline{x})+c.c.\right]-i\left[\chi^*_{p0a}(\underline{x})\partial^{f\dot{e}}_{II}t^{\dot{h}bea}\chi_{n0\{\dot{b}\dot{h}\}f}(\underline{x})+c.c.\right]\right\}}{\Omega\int d^3\underline{x}\left\{\chi^*_{0a}(\underline{x})\left(\frac{E_n^2}{4}+\Delta\right)\chi_{0\dot{a}}(\underline{x})\right\}}$$

(12.4.1)

In an analogous fashion, The decay amplitude $S_{fi\psi}$ for the ψ function is obtained from (12.2.8) and reads

$$S_{fi\psi}=-\frac{ig}{4\sqrt{2}}(2\pi)^4\delta\left(\underline{K}_p+\underline{K}_W\right)\delta\left(E_p+E_W-E_n\right)\times$$

(12.4.2 continued)

$$\frac{\int d^3\underline{x}\left\{w_0\left[\psi^{*\dot{a}}_{p0}(\underline{x})\left(-2\left(\frac{E_n^2}{4}+\Delta\right)-E_n^2\right)\psi^{a}_{n0}(\underline{x})+c.c.\right]+i\left[\psi^{*\dot{a}}_{p0}(\underline{x})\partial_{II\dot{e}d}t_{\dot{a}ahb}\psi^{\{bh\}\dot{e}}_{n0}(\underline{x})+c.c.\right]\right\}}{\Omega\int d^3\underline{x}\left\{\psi^{*\dot{a}}_{0}(\underline{x})\left(\frac{E_n^2}{4}+\Delta\right)\psi^{a}_{0}(\underline{x})\right\}}$$

(12.4.2)

The starred wave functions in nominators of (12.4.1-2) represent final states or proton, irrespective the nucleon lables; the *c.c.* terms will turn out to contribute equally and can be dropped together with an overall factor 2 multiplying the right of (12.4.1-2). The wave functions in denoiminators of (12.4.1-2) are those of the initial neutron, as <f| has been included in S_{fi} of (12.3.4).

The proton may have a different m or spin value in (10.3.8) relative to that pertaining to the initial neutron in (12.4.1-2). There are four combinations which are denoted by

$$\begin{array}{lcccc} \text{neutron} & m=\frac{1}{2} & \frac{1}{2} & -\frac{1}{2} & -\frac{1}{2} \\ \text{proton} & m=\frac{1}{2} & -\frac{1}{2} & \frac{1}{2} & -\frac{1}{2} \\ \text{notation} & F+ & GT+ & GT- & F- \end{array} \qquad (12.4.3)$$

As there are two equally correct solutions (10.3.8a-8b) and (10.3.8c-8d), (12.3.4) and (12.3.8) are to be averaged over these two equally probable solutions. The averaging will turn out to not affect (12.3.4) so that it can be carried out for $S_{fi\chi}$ and $S_{fi\psi}$ of (12.4.1-2) directly to obtain $S_{fi\chi av}$ and $S_{fi\psi av}$. It removes terms of containing $f_0(r)g_0(r)$ or their derivatives that will appear in (12.4.1-2) after application of (10.3.8a-8b) or (10.3.8c-8d) and (12.4.3). Such terms will for instance differentiate between neutron spin up and spin down decay rates contray to the measured A asymmetry coefficient [P1]. Inserting (10.3.8a-8b) and (10.3.8c-8d) into (12.4.1) and (12.4.2) for the four combinations of (12.4.3), making use of (3.1.4) and (3.5.7) for ∂_{II}, summing over the spinor indices, employing (12.3.9), and integrating over the angles ϑ and ϕ in the relative space, one finds that the both averaged decay amplitudes $S_{fi\chi av}$ and $S_{fi\psi av}$ are the same, as may be expected from the symmetry between the χ part (12.2.7) and the ψ part (12.2.8);

$$S_{fi\chi av}=S_{fi\psi av}=\begin{pmatrix} S_{fiF+} \\ S_{fiGT+} \\ S_{fiGT-} \\ S_{fiF-}\end{pmatrix}=-i\frac{g}{2\sqrt{2}}\frac{(2\pi)^4}{\Omega}\delta\left(E_p+E_W-E_n\right)\delta\left(\underline{K}_p+\underline{K}_W\right)\frac{1}{N_{dr}}$$

(12.4.4 continued)

$$\times\left[-2w_0\left(N_{dr}+E_n^2\left(I_{g0}+I_{f0}\right)\right)\begin{pmatrix}1\\0\\0\\1\end{pmatrix}-i\frac{2}{3}E_n\left(I_{g0}-\frac{I_{f0}}{3}\right)\begin{pmatrix}t^{\{12\}}\\-t^{22}\\t^{11}\\-t^{\{12\}}\end{pmatrix}\right] \tag{12.4.4}$$

$$I_{g0}=\int_0^\infty drr^2g_0^2(r),\quad I_{f0}=\int_0^\infty drr^2f_0^2(r),\quad I_{g00}=\int_0^\infty drr^2g_{00}^2(r),\quad I_{f00}=\int_0^\infty drr^2f_{00}^2(r) \tag{12.4.5}$$

where N_{dr} is given by (10.3.9b) and I_{g00} and I_{f00}. are defined (10.3.14).

12.5. Expressions for Vector and Tensor Gauge Fields

12.5.1. Mass Generation of *W* Boson Via Virtual π^0

In the analogous meson case, the pion beta decay $\pi^+\rightarrow\pi^0e^+\nu_e$ of §7.4.5, the gauge boson W_I decays into a pair of leptons, as is seen in (7.4.27). The mass of the W_I boson comes from the $(gW_I)^2$ terms in the meson action integral S_{m3} (7.1.8) via (7.4.4) and (7.4.6b) and is generated by the integral of the |pion wave functions|^2 over the relative space x. It can be seen from (7.4.3a, 4) that the energy of pion does not enter and can be zero. In this case, the pion is virtual.

In the corresponding action-like integrals S_χ and S_ψ of (12.2.7-8), there is no such $(gW_I)^2$ mass term, only terms of the form $(gW_I)(g\partial_{I,II}W_I)$. Therefore, the gauge boson W_I from neutron decay cannot decay into a pair of leptons via the integrals (12.2.7-8) for neutrons.

The interpretation of this situation is that gauge boson W_I in the neutron case here decays into a pair of leptons via a virtual π^0 action as has been considered in §7.6.2 3) and employed earlier for muon decay in §7.6.3 4). In the pion beta decay, the W_I is positively charged and decays into a lepton pair via (7.4.10). In neutron decay, the W_I or W is negatively charged and W_I^- in (7.4.5) is replaced by W_I^+ so that (7.4.5) and (7.4.6a), neglecting the first three terms there and noting (7.4.10), are modified to

$$\frac{M_W^2}{2}W_I^{-b\dot{a}}=\frac{M_W^2}{2}\begin{pmatrix}W_0-W_3 & -W_1+iW_2\\-W_1-iW_2 & W_0+W_3\end{pmatrix}=\delta S_L/\delta W_I^{+a\dot{b}}$$
$$=\frac{g}{2\sqrt{2}}\psi_{L\dot{b}}^{(+)}\psi_{\nu La}^{(-)}=\frac{g}{2\sqrt{2}}\begin{pmatrix}\psi_{L\dot{1}}^{(+)}\psi_{\nu L1}^{(-)} & \psi_{L\dot{1}}^{(+)}\psi_{\nu L2}^{(-)}\\ \psi_{L\dot{2}}^{(+)}\psi_{\nu L1}^{(-)} & \psi_{L\dot{2}}^{(+)}\psi_{\nu L2}^{(-)}\end{pmatrix} \tag{12.5.1}$$

where the gauge boson mass is given by (7.4.6b) and (7.4.9).

12.5.2. Variation of the Total Action for Neutron Decay

Having found an expression for w_0 in (12.4.4) from (12.5.1) via (12.3.6), expressions for the other unknowns t^{ab} there will be obtained in this and the next paragraph. The vector gauge boson fields (12.5.1) was found by varying the total action (7.1.1), after removing its last term, with respect to W^- in §7.4.2. For neutron decay, the meson action S_{m3} in (7.1.1) is to be replaced by the corresponding baryons actions (12.2.7-8), $S_\chi \pm S_\psi$, and the vector gauge boson action S_{GB} be replaced by a gauge boson-tensor field action S_{GBT}. Equation (7.1.1) is here replaced by

$$S_{Tn} = S_{GBT} + \kappa_b \left(S_{\chi av} - S_{\psi av}\right) + S_L + S_{Lm} \tag{12.5.2}$$

Here, the lepton actions $S_L + S_{Lm}$ are the same as those in (7.1.16-17). κ_b is a dimensionless proportional constant that, for instance, signifies that the action-like baryon integrals S_χ and S_ψ differ basically from the meson action S_{m3} in (7.1.1) in that the normalization of the wave function amplitudes are different. $S_{\chi av}(S_{\psi av})$ denotes that $S_\chi(S_\psi)$ has been averaged over the both equally probable solutions (10.3.8a-8b) and (10.3.8c-8d) for the baryon wave functions that enter it, just like that $S_{fi\chi}$ and $S_{fi\psi}$ of (12.4.1-2) have been averaged to $S_{fi\chi av}$ and $S_{fi\psi av}$ in (12.4.4).

The − sign in (12.5.2) is chosen because (12.5.2) will lead to an expression for t^{ab} in (12.4.4); if this sign is replaced by +, the needed (12.5.3) below will vanish. Further, this − sign will remove terms linear in gW, as is implied by $S_{fi\chi av} - S_{fi\psi av} = 0$ from (12.4.4). This will cause $S_{\chi av} - S_{\psi av}$ in (12.5.2) to possess $(gW)^2$ terms, apart from the t^{ab} terms, to that order and render it to have an extremum when solutions near the correct ones are inserted into $S_{\chi av} - S_{\psi av}$.

Unlike S_{GB} in (7.1.2) given by (7.1.2), S_{GBT} in (12.5.2) is unknown. If the tensor gauge field is to have an equation of motion like that for the gauge boson (7.4.6), S_{GBT} is expected to contain terms of the form $(\partial_X W)(\partial_X T)$. Physical existence and interpretation of the tensor gauge field are not known and will be left to eventual future work. For the present purpose, it is sufficient to obtain an expression for t^{ab} from (12.5.2) for use in (12.4.4).

Follow the steps of §7.4.2 and vary (12.5.2) with respect to $W_I^{+\dot{a}b}$. The unknown S_{GBT} is expected to give rise to energy-momentum terms of the form $(\partial_X^2 T)$ corresponding to the first terms on the left of (7.4.6a), which have been neglected because they are much smaller than the gauge boson mass term that follows them in (7.4.6a). Similarly, the obtained $(\partial_X^2 T)$ terms can also be dropped on the same ground so the exact but unknown form of S_{GBT} is of no concern here. Variation of the $S_L + S_{Lm}$ terms in (12.5.2) with respect to $W_I^{+\dot{a}b}$ is analogous to that given by (7.4.5) and leads to (12.5.1).

12.5.3. Expressions for Tensor and Vector Gauge Fields

Variation of $S_{\chi av} - S_{\psi av}$ in (12.5.2) with respect to $W_I^{+\dot{a}b}$ is limited to terms of order g^2. When evaluating the average $S_{\chi av}(S_{\psi av})$ using (10.3.8a-8b) and (10.3.8c-8d), it is practically sufficient to use one of them, for instance (10.3.8a-8b) for $S_\chi(S_\psi)$ of (12.2.7-8), and drop the $f_0(r)g_0(r)$ terms, as was mentioned beneath (12.4.3).

In the evaluation of (12.2.7-8), use is made of (12.3.1), (12.3.5), (3.1.4), (3.5.7). When summing over the flavor indices t, p, s, q, and r, tp=32 and rt=23 for neutron and (12.3.6-7) are employed. Carrying out the angular integrations, one finds

$$\frac{\delta\kappa_b\left(S_{\chi av}-S_{\psi av}\right)}{\delta W^{+a\dot{b}}}=U_{\dot{b}a}=\begin{pmatrix}U_0+U_3 & U_1-iU_2\\ U_1+iU_2 & U_0-U_3\end{pmatrix} \tag{12.5.3}$$

$$U_0=\frac{\kappa_b g^2}{96}E_n\tau_0\left(I_{g0}-\frac{1}{3}I_{f0}\right)7W_3^-(X)\begin{pmatrix}1\\-1\end{pmatrix}=-\frac{g}{4\sqrt{2}}\left(\psi_{L\dot{1}}^{(+)}\psi_{\nu L1}^{(-)}+\psi_{L\dot{2}}^{(+)}\psi_{\nu L2}^{(-)}\right) \tag{12.5.4}$$

$$U_3=-\frac{\kappa_b g^2}{96}E_n\tau_0\left(I_{g0}-\frac{1}{3}I_{f0}\right)3W_0^-\begin{pmatrix}1\\-1\end{pmatrix}-i\frac{\kappa_b g^2}{24}\tau_0\left(I_{g0}+I_{f0}\right)T^{12}(X)\begin{pmatrix}1\\1\end{pmatrix}=-\frac{g}{4\sqrt{2}}\left(\psi_{L\dot{1}}^{(+)}\psi_{\nu L1}^{(-)}-\psi_{L\dot{2}}^{(+)}\psi_{\nu L2}^{(-)}\right) \tag{12.5.5}$$

$$U_1-iU_2=-i\frac{\kappa_b g^2}{24}\tau_0\left(-T^{11}\right)(X)\begin{pmatrix}2I_{f0}/3\\ I_{g0}+I_{f0}/3\end{pmatrix}=-\frac{g}{4\sqrt{2}}\psi_{L\dot{1}}^{(+)}\psi_{\nu L2}^{(-)} \tag{12.5.6}$$

$$U_1+iU_2=-i\frac{\kappa_b g^2}{24}\tau_0 T^{22}(X)\begin{pmatrix}I_{g0}+I_{f0}/3\\ 2I_{f0}/3\end{pmatrix}=-\frac{g}{4\sqrt{2}}\psi_{L\dot{2}}^{(+)}\psi_{\nu L1}^{(-)},\qquad \tau_0=\int dx^0 \tag{12.5.7}$$

where the upper and lower rows in the U's refer to neutron spin up m=1/2 and spin down m=−1/2, respectively, in (10.3.8). Further, (12.5.3) has been equated to the negative of (12.5.1) as is prescribed in (12.5.2). Note that U_0 in (12.5.4) does not contain W_0, which however enters (12.5.1) and stems from the virtual π^0 action in (7.4.4).

Comparing the X dependence of W's and T's and the ψ's in (12.5.4-7) via (12.3.6, 9) and (7.4.19) and replacing L and νL there by e and ν, respectively, we find

$$E_W=E_\nu+E_e^{(+)},\qquad \underline{K}_W=\underline{K}_e^{(+)}-\underline{K}_\nu \tag{12.5.8}$$

Removing the X dependence in (12.5.4-7) and observing (7.4.6b) and (12.3.6, 9), we obtain

$$0=\frac{1}{M_W^2\Omega}\frac{g}{4\sqrt{2\Omega_\nu\Omega_e}}\left(u_{L\dot{1}}^{(+)}u_{\nu L1}^{(-)}+u_{L\dot{2}}^{(+)}u_{\nu L2}^{(-)}\right)+\frac{7}{96}E_n\left(I_{g0}-I_{f0}/3\right)\kappa_b w_3\begin{pmatrix}1\\-1\end{pmatrix} \tag{12.5.9}$$

$$t^{\{12\}} = -i\frac{24}{M_W^2\Omega(I_{g0}+I_{f0})}\frac{g}{4\sqrt{2\Omega_\nu\Omega_e}}\left(u_{L\dot{1}}^{(+)}u_{\nu L1}^{(-)} - u_{L\dot{2}}^{(+)}u_{\nu L2}^{(-)}\right) + i\frac{3}{4}E_n\frac{(I_{g0}-I_{f0}/3)}{(I_{g0}+I_{f0})}\kappa_b w_0\begin{pmatrix}1\\-1\end{pmatrix} \tag{12.5.10a}$$

$$-t^{11}\kappa_b\begin{pmatrix}2I_{f0}/3\\ I_{g0}+I_{f0}/3\end{pmatrix} = -i\frac{24}{M_W^2\Omega}\frac{g}{4\sqrt{2\Omega_\nu\Omega_e}}u_{L\dot{1}}^{(+)}u_{\nu L2}^{(-)}$$
$$t^{22}\kappa_b\begin{pmatrix}I_{g0}+I_{f0}/3\\ 2I_{f0}/3\end{pmatrix} = -i\frac{24}{M_W^2\Omega}\frac{g}{4\sqrt{2\Omega_\nu\Omega_e}}u_{L\dot{2}}^{(+)}u_{\nu L1}^{(-)} \tag{12.5.10b}$$

The w's in (12.3.6) are similarly found from (12.5.1) and read

$$w_0 = \frac{1}{M_W^2}\frac{g}{2\sqrt{2\Omega_\nu\Omega_e}}\left(u_{L\dot{1}}^{(+)}u_{\nu L1}^{(-)} + u_{L\dot{2}}^{(+)}u_{\nu L2}^{(-)}\right),\quad w_3 = \frac{1}{M_W^2}\frac{g}{2\sqrt{2\Omega_\nu\Omega_e}}\left(u_{L\dot{1}}^{(+)}u_{\nu L1}^{(-)} - u_{L\dot{2}}^{(+)}u_{\nu L2}^{(-)}\right) \tag{12.5.11}$$

12.5.4. Decay Amplitude as Function of Lepton Wave Functions

Equations (12.5.9-11) can now be inserted into (12.4.4). Here, it is noted that the neutron spin is up or m=1/2 for the upper two amplitudes corresponding to the upper case in (12.5.9-10) and down or m=−1/2 for the lower two amplitudes corresponding to the lower case in (12.5.9-10). Thus, the w_0 in the third component of a triplet (12.5.5, 10a) become a singlet to be combined to the w_0 in (12.5.11). Analogously, the singlet (12.5.9), when inserted into (12.4.4) becomes the third component of a triplet to be combined with (12.5.10a). Effectively, the $7W_3$ in (12.5.4) and $3W_0$ in (12.5.5) switch place after insertion into (12.4.4). These two terms are not the dominating ones but will lead to corrections given by the two c's(< 0.4 or so) in (12.6.16) below. The resulting decay amplitude, noting (12.5.8) and (12.4.5), reads

$$S_{fi} = \begin{pmatrix}S_{fiF+}\\ S_{fiGT+}\\ S_{fiGT-}\\ S_{fiF-}\end{pmatrix} = i\frac{g^2}{8}\frac{1}{M_W^2}\frac{(2\pi)^4}{\Omega\sqrt{\Omega_\nu\Omega_e}}\delta\left(E_p+E_\nu+E_e^{(+)}-E_n\right)\delta\left(\underline{K}_p+\underline{K}_e^{(+)}-\underline{K}_\nu\right)\frac{1}{N_{dr}}$$
$$\times\begin{bmatrix}b_{F+0} & 0 & 0 & b_{F+3}\\ 0 & b_{GT+p} & 0 & 0\\ 0 & 0 & b_{GT-m} & 0\\ b_{F-0} & 0 & 0 & b_{F-3}\end{bmatrix}\begin{pmatrix}\left(u_{L\dot{1}}^{(+)}u_{\nu L1}^{(-)} + u_{L\dot{2}}^{(+)}u_{\nu L2}^{(-)}\right)\\ u_{L\dot{2}}^{(+)}u_{\nu L1}^{(-)}\\ u_{L\dot{1}}^{(+)}u_{\nu L2}^{(-)}\\ \left(u_{L\dot{1}}^{(+)}u_{\nu L1}^{(-)} - u_{L\dot{2}}^{(+)}u_{\nu L2}^{(-)}\right)\end{pmatrix} \tag{12.5.12}$$

$$b_{F+0} = b_{F-0} = -2\left(N_{dr} + E_n^2\left(I_{g0} + I_{f0}\right)\right)\frac{1}{1 + c_{F0}} \tag{12.5.13}$$

$$b_{F+3} = -b_{F-3} = -\frac{8E_n}{\kappa_b}\frac{\left(I_{g0} - I_{f0}/3\right)}{a_g^2\left(I_{g00} + I_{f00}\right)}\left(1 + c_{F3}\right) \tag{12.5.14}$$

$$b_{GT+p} = b_{GT-m} = b_{GT} = \frac{16E_n}{\kappa_b}\frac{\left(I_{g0} - I_{f0}/3\right)}{a_g^2\left(I_{g0} + I_{f0}/3\right)} \tag{12.5.15}$$

$$c_{F0} = \frac{1}{16}E_n\kappa_b a_g^2\left(I_{g00} - I_{f00}/3\right), \qquad c_{F3} = \frac{7}{48}E_n a_g^2\left(I_{g00} - I_{f00}/3\right) \tag{12.5.16}$$

Because κ_b in (12.5.2) can be incorporated into the wave functions χ and ψ, it can be absorbed into the normalization condition (10.3.13) by chosing a different normalization constant or equivalently a different normalized amplitude ag given by (10.3.14).

12.6. Decay Rate and Asymmetry Coefficients *A* and *B*

12.6.1. Decay Rate

As in (7.5.1), the decay rate is

$$\Gamma = \sum\nolimits_{\text{final states}} \left|S_{fi}\right|^2 / T_d \tag{12.6.1}$$

where T_d is a long decay time. The subscript "final states" refers to the four final lepton spin states and all possible momenta of the proton, electron and antineutrino, like that in (7.5.2). The decay rates are

$$\begin{pmatrix} \Gamma_{F\pm} \\ \Gamma_{GT\pm} \end{pmatrix} = \frac{\Omega\Omega_e\Omega_\nu}{(2\pi)^9}\int d^3\underline{K}_p \int d^3\underline{K}_e \int d^3\underline{K}_\nu \frac{1}{T_d}\sum\nolimits_{e\,\text{spins}}\sum\nolimits_{\nu\,\text{spins}} \begin{pmatrix} \left|S_{fiF\pm}\right|^2 \\ \left|S_{fiGT\pm}\right|^2 \end{pmatrix} \tag{12.6.2}$$

The square of S_{fi} contains squares of the δ functions in (12.5.12), which are "linearized" by (7.5.6) type of formula. Following the common approach, integration over the recoil momenta $\underline{K}_p$ in (12.6.2) is carried out first. By (12.5.8), this gives $\underline{K}_p$=$\underline{K}_\nu$–$\underline{K}_e$, where the superscript (+) has been dropped. Introduce

$$\underline{K}_\nu = K_\nu\left(k_{\nu 1},\, k_{\nu 2},\, k_{\nu 3}\right), \quad k_{\nu\pm} = k_{\nu 1} \pm ik_{\nu 2} = \sin\vartheta_n \exp\left(\pm i\phi_\nu\right), \quad k_{\nu 3} = \cos\vartheta_\nu$$
$$K_{e\pm} = K_{e1} \pm iK_{e2} = K_e \sin\vartheta_e \exp\left(\pm i\phi_e\right), \quad K_{e3} = K_e\cos\vartheta_e \tag{12.6.3}$$

These and (12.5.12) are inserted into the decay rate expression (12.6.2) to produce

$$\begin{pmatrix}\Gamma_{F\pm}\\ \Gamma_{GT\pm}\end{pmatrix}=\frac{g^4}{8192\pi^5 M_W^4 N_{dr}^2}\int dK_e K_e^2\int dK_\nu K_\nu^2\delta\left(E_p+E_e+E_\nu-E_n\right)\left(1+\frac{m_e}{E_e}\right)\begin{pmatrix}I_{F\pm}\\ I_{GT\pm}\end{pmatrix} \tag{12.6.4}$$

$$\begin{pmatrix}I_{F\pm}\\ I_{GT\pm}\end{pmatrix}=\int d\vartheta_e\sin\vartheta_e d\phi_e\int d\vartheta_\nu\sin\vartheta_\nu d\phi_\nu\sum\nolimits_{e\text{spins}}\sum\nolimits_{\nu\text{spins}}\begin{pmatrix}i_{F\pm}\\ i_{GT\pm}\end{pmatrix} \tag{12.6.5}$$

$$\sum\nolimits_{e\text{spins}}\sum\nolimits_{\nu\text{spins}}\left(i_{F\pm}\right)=i_{F\pm\uparrow\uparrow}+i_{F\pm\uparrow\downarrow}+i_{F\pm\downarrow\uparrow}+i_{F\pm\downarrow\downarrow} \tag{12.6.6}$$

$$\sum\nolimits_{e\text{spins}}\sum\nolimits_{\nu\text{spins}}\left(i_{GT\pm}\right)=i_{GT\pm\uparrow\uparrow}+i_{GT\pm\uparrow\downarrow}+i_{GT\pm\downarrow\uparrow}+i_{GT\pm\downarrow\downarrow} \tag{12.6.7}$$

where the both arrows denote the spin directions, separated by the commas in (7.4.19), of the electron and the antineutrino and

$$i_{F\pm\uparrow\uparrow}=\left|\left(-b_{F\pm0}+b_{F\pm3}\right)\left(1-\frac{K_{e3}}{E_e+m_e}\right)k_{\nu-}+\left(b_{F\pm0}+b_{F\pm3}\right)\frac{K_{e-}}{E_e+m_e}\left(1-k_{\nu3}\right)\right|^2 \tag{12.6.8a}$$

$$i_{F\pm\uparrow\downarrow}=\left|\left(b_{F\pm0}-b_{F\pm3}\right)\left(1-\frac{K_{e3}}{E_e+m_e}\right)\left(1+k_{\nu3}\right)+\left(-b_{F\pm0}-b_{F\pm3}\right)\frac{K_{e-}}{E_e+m_e}k_{\nu+}\right|^2 \tag{12.6.8b}$$

$$i_{F\pm\downarrow\uparrow}=\left|\left(-b_{F\pm0}-b_{F\pm3}\right)\left(1+\frac{K_{e3}}{E_e+m_e}\right)\left(1-k_{\nu3}\right)+\left(b_{F\pm0}-b_{F\pm3}\right)\frac{K_{e+}}{E_e+m_e}k_{\nu-}\right|^2 \tag{12.6.8c}$$

$$i_{F\pm\downarrow\downarrow}=\left|\left(b_{F\pm0}+b_{F\pm3}\right)\left(1+\frac{K_{e3}}{E_e+m_e}\right)k_{\nu+}+\left(-b_{F\pm0}+b_{F\pm3}\right)\frac{K_{e+}}{E_e+m_e}\left(1+k_{\nu3}\right)\right|^2 \tag{12.6.8d}$$

$$i_{GT+\uparrow\uparrow}=\left|\left(b_{GT+p}\right)\left(1-\frac{K_{e3}}{E_e+m_e}\right)\left(1-k_{\nu3}\right)\right|^2,\quad i_{GT-\uparrow\uparrow}=\left|\left(-b_{GT-m}\right)\frac{K_{e-}}{E_e+m_e}k_{\nu-}\right|^2 \tag{12.6.9a}$$

$$i_{GT+\uparrow\downarrow}=\left|\left(-b_{GT+p}\right)\left(1-\frac{K_{e3}}{E_e+m_e}\right)k_{\nu+}\right|^2,\quad i_{GT-\uparrow\downarrow}=\left|\left(b_{GT-m}\right)\frac{K_{e-}}{E_e+m_e}\left(1+k_{\nu3}\right)\right|^2 \tag{12.6.9b}$$

$$i_{GT+\downarrow\uparrow} = \left|\left(-b_{GT+p}\right)\frac{K_{e+}}{E_e+m_e}\left(1-k_{\nu 3}\right)\right|^2, \quad i_{GT-\downarrow\uparrow} = \left|\left(b_{GT-m}\right)\left(1+\frac{K_{e3}}{E_e+m_e}\right)k_{\nu-}\right|^2 \tag{12.6.9c}$$

$$i_{GT+\downarrow\downarrow} = \left|\left(b_{GT+p}\right)\frac{K_{e+}}{E_e+m_e}k_{\nu+}\right|^2, \quad i_{GT-\downarrow\downarrow} = \left|\left(-b_{GT-m}\right)\left(1+\frac{K_{e3}}{E_e+m_e}\right)\left(1+k_{\nu 3}\right)\right|^2 \tag{12.6.9d}$$

Let

$$\Delta_m = E_n - E_p \approx m_n - m_p = 1.2933 \text{ Mev} \tag{12.6.10}$$

Equation (12.6.5) can now be evaluated using (12.6.8-9), and (12.6.3). Carrying out the angular integrations, it is found that the cross terms in (12.6.8) drop out and one finds

$$\begin{pmatrix} I_{F\pm} \\ I_{GT\pm} \end{pmatrix} = 4\pi^2 \begin{pmatrix} 2b_{F\pm 0}^2 + 2b_{F\pm 3}^2 \\ b_{GT}^2 \end{pmatrix} \frac{16E_e}{E_e+m_e} \tag{12.6.11}$$

which is independent of the antineutrino energy $E_\nu = K_\nu$. Carrying out the K_ν integration in (12.6.4) using (12.6.10), one gets

$$\int dK_\nu K_\nu^2 \delta\left(E_p + E_e + E_\nu - E_n\right)\begin{pmatrix} I_{F\pm} \\ I_{GT\pm} \end{pmatrix} = \left(\Delta - E_e\right)^2 \begin{pmatrix} I_{F\pm} \\ I_{GT\pm} \end{pmatrix} \tag{12.6.12}$$

where $0 \le K_\nu \le \Delta - E_e$. Inserting (12.6.11-12) into (12.6.4), changing the variable $dK_e K_e$ to $dE_e E_e$ and noting (12.5.13-15) leads to

$$\begin{pmatrix} \Gamma_{F\pm} \\ \Gamma_{GT\pm} \end{pmatrix} = \begin{pmatrix} \Gamma_F \\ \Gamma_{GT} \end{pmatrix} = \frac{1}{128\pi^3}\frac{g^4}{M_W^4}\frac{1}{N_{dr}^2}\int dE_e P_F\left(E_e\right)\begin{pmatrix} 2b_{F\pm 0}^2 + 2b_{F\pm 3}^2 \\ b_{GT}^2 \end{pmatrix} \tag{12.6.13}$$

$$P_F\left(E_e\right) = \sqrt{E_e^2 - m_e^2}\, E_e \left(\Delta - E_e\right)^2 \tag{12.6.14}$$

where $m_e \le E_e \le \Delta + m_e$ and $P_F(E_e)$ is the conventional Fermi electron energy spectrum.

The half life of the neutron to be compared to the known τ_{exp}=885.7 sec. is

$$\tau_{th}\left(\log 2\right) = \frac{h}{2\pi}\frac{1}{\left(\Gamma_F + \Gamma_{GT}\right)}\left(\log 2\right) \tag{12.6.15}$$

where h is the Planck constant.

12.6.2. Asymmetry Coefficients *A* and *B*

The asymmetry coefficients *A* and *B* are obtained from (12.6.5), just like (12.6.11), but without carrying out integration over θ_ν and θ_e, respectively. One finds,

$$\begin{pmatrix} I_{F+} + I_{GT+} \\ I_{F-} + I_{GT-} \end{pmatrix} = \int d\vartheta_e \sin\vartheta_e 4\pi^2 \begin{pmatrix} b_{FGT+} \\ b_{FGT-} \end{pmatrix} \frac{8E_e}{E_e + m_e} \left(1 + A_\pm \frac{K_e}{E_e} \cos\vartheta_e \right)$$

$$b_{FGT} = b_{FGT+} = 2b_{F+0}^2 + 2b_{F+3}^2 + b_{GT+p}^2 = b_{FGT-} = 2b_{F-0}^2 + 2b_{F-3}^2 + b_{GT-m}^2$$

$$A_+ = \left(4b_{F+0}b_{F+3} - b_{GT+p}^2\right)/b_{FGT+} = -A_- = -\left(4b_{F-0}b_{F-3} + b_{GT-m}^2\right)/b_{FGT-} \qquad (12.6.16)$$

The antineutrino moves in the opposite direction relative to that of the neutrino so that $\underline{K}_\nu$ is to be replaced by $-\underline{K}_\nu$ in (12.5.8), hence also (12.5.12). With this replacement, both $\underline{K}_e$ and $\underline{K}_\nu$ are now on equal footing in these both expressions and (12.6.5) can be written in the form

$$\begin{pmatrix} I_{F+} + I_{GT+} \\ I_{F-} + I_{GT-} \end{pmatrix} = \int d\vartheta_\nu \sin\vartheta_\nu 4\pi^2 \begin{pmatrix} b_{FGT+} \\ b_{FGT-} \end{pmatrix} \frac{8E_e}{E_e + m_e} \left(1 + B_\pm \cos\vartheta_\nu \right)$$

$$B_+ = \left(4b_{F+0}b_{F+3} + b_{GT}^2\right)/b_{FGT} = -B_- = -\left(4b_{F-0}b_{F-3} - b_{GT}^2\right)/b_{FGT} \qquad (12.6.17)$$

12.6.3. Comparison with Data

The expressions in (12.6.15-17) have been evaluated using (12.5.13-16), (12.4.5) and the normalized radial wave functions $g_{00}(r)$ and $f_{00}(r)$ for the neutron associated with some of the confinement cases in Table 11.1. The results are summarized in Table 12.1 below.

These results have been derived starting from (10.0.1), as was mentioned in §12.2.1, and using (12.3.5) which stems from putting the quartet wave functions (9.2.8) entering (10.0.1) to zero.

If the symmetric quark postulate [H15] mentioned beneath (9.3.7) is used, (10.0.1) is replaced by (10.0.6) or equivalently (10.0.1) together with (10.0.7). The expressions (12.6.15-17) remain valid if c_{F0} and c_{F3} in (12.5.16) are replaced by c'_{F0} and c'_{F3}, respectively

$$c'_{F0} = \frac{1}{2} c_{F0}, \qquad c'_{F3} = \frac{9}{14} c_{F3} \qquad (12.6.18)$$

The corresponding results are similarly summarized inside parentheses in Table 12.1.

Table 12.1. Values of the calculated decay rate $\tau_{th}(log2)$, *A* and *B* asymmetry coefficients given by (12.6.15-17) and Γ_{GT}/Γ_F by (12.6.13) are presented for a number of confinement strengths d_{b2} given in Table 11.1. The normalization constant κ_b in (12.5.13-16) is chosen such that in one case the *A* coefficient agrees with data [P1] and in the other case the *B* coefficient agrees with data. The integrals appearing in (12.5.13-16), given by (12.4.5), depend upon the approximate, normalized radial wave functions for the neutron $g_{00}(r)$ and $f_{00}(r)$ obtained in Sec. 11.2 and shown in Figure 12.1 below for the d_{b2}=−0.3202 case. The corresponding results stemming from symmetric quark postulate beneath (9.3.7) using (12.6.18) are given inside parentheses. $\Gamma_A/\Gamma_V = \left(2b_{F+3}^2 + b_{GT+p}^2\right)/2b_{F+0}^2$ ≈4.8–5.5 for all these cases.

PDF data [P1] A_+=−0.1173 B_+=0.9807 τ_{exp}=885.7 sec

d_{12}	κ_b	A	B	Γ_{GT}/Γ_F	$\tau_{th}(log2)$ sec.
−0.1621	3.436	0.0595	0.9807	0.854	73.5
	2.245	−0.1174	0.9984	1.26	38.0
	(2.928	0.0130	0.9808	0.938	56.1)
	(2.203	−0.1172	1.0000	1.27	36.6)
−0.1641	3.472	0.0000	0.9806	0.962	91.1
	2.639	−0.1172	0.9996	1.26	59.9
	(3.049	−0.0381	0.9807	1.04	73.0)
	(2.571	−0.1171	0.9965	1.26	56.7)
−0.1670	3.413	−0.0580	0.9807	1.08	109.2
	2.982	−0.1172	0.9937	1.25	89.16
	(3.081	−0.0883	0.9807	1.15	91.60)
	(2.897	−0.1172	0.9879	1.24	83.73)
−0.3202	9.585	−0.1173	0.9781	1.21	1036
	9.374	−0.1265	0.9807	1.24	1001
	(8.815	−0.1173	0.9732	1.20	872)
	(8.347	−0.1422	0.9807	1.28	805)
−0.3622	12.41	−0.1173	0.9679	1.19	1956
	11.25	−0.1578	0.9807	1.32	1684
	(11.30	−0.1173	0.9631	1.18	1614)
	(10.05	−0.1708	0.9807	1.36	1359)
−0.4042	17.40	−0.1173	0.9571	1.16	4345
	14.66	−0.1861	0.9807	1.40	3349
	(15.55	−0.1174	0.9526	1.15	3455)
	(13.03	−0.1969	0.9807	1.43	2671)

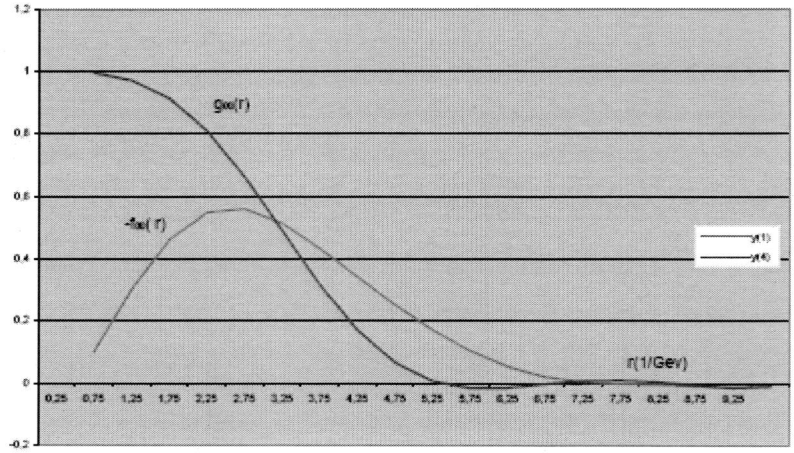

Figure 12.1. Neutron radial wave functions $f_{b0}(r)$ and $g_{b0}(r)$ in (10.7.5-7), same as $f_0(r)$ and $g_0(r)$ in (10.3.14) and normalized there to become $f_{00}(r)$ and $g_{00}(r)$, for the d_{b2}=−0.3202 case in Table 11.1. r is the quark-diquark distance.

Sec. 11.1 shows that the approximate solutions to the baryon spectra problem have confinement strength constant d_{b2} values in the range around −0.32 to −0.45. Table 12.1 shows that neutron wave functions associated with the d_{b2}=−0.3202 case leads to half life $\tau_{th}(log2)$ which are about 15%(12%) off from data [P1]. The associated A and B asymmetry coefficients for this d_{b2}=−0.3202 case are also rather close to data. These are obtained by adjusting the only free parameter or chosen normalization constant κ_b such that the calculated A coefficient coincides with data. The so predicted B coefficients deviate only 0.3%(0.8%) from data [P1], close to experimental error. If κ_b is adjusted such that the calculated B coefficient coincides with data, the so predicted A coefficient deviates from data [P1] by 8%(21%).

For $d_{b2}\leq-0.3622$, the predicted half life time are too long and the both the predicted A and B coefficients are farther way from data. For $d_{b2}>-0.3202$, it was pointed out beneath Figure 11.1 that no satisfactory convergent solutions were found in the range $-0.23>d_{b2}>-0.27$.

For $d_{b2}\geq-0.1670$, the predicted half life times are too short and the both the predicted A and B coefficients are also farther way from data, particularly for the A coefficient.

Thus, agreement of predictions of the half life $\tau_{th}(log2)$ and A or B asymmetry coefficients with data [P1] is best for d_{b2}=−0.3202 and deteriorates considerably for larger or smaller d_{b2} values. This supports the results of Sec. 11.1 that confinement strength constant d_{b2} lies in the range −0.32 to −0.45.

In conclusion, the present treatment leads to two approximate predictions, bearing in mind that they are based upon the approximate results of §11.1.1. Possible source of the approximations there has been conjectured in §11.1.2. Firstly, the predicted half life time for the chosen confinement strength d_{b2}=−0.3202 is consistent with the approximate solution to the baryon spectra problem given in §11.1.1. Secondly, this approximate solution also leads to B coefficient in agreement with data [P1].

12.6.4. Detachment of Weak and Electromagnetic Couplings

Γ_{GT}/Γ_F and Γ_A/Γ_V in Table 12.1 have not been measured. Γ_A/Γ_V is the ratio between the decay rates stemming from the axial vector or tensor part and the vector part in the decay amplitude (12.4.4). Both these ratios are > 1 and shows that the axial vector or tensor part of the amplitude is greater than that of the vector part. They behave qualitatively in a similar way as does the conventional $(g_A/g_V)^2$=1.611 [P1], which lies between Γ_{GT}/Γ_F and Γ_A/Γ_V but cannot be related to them.

In the present theory, there is only one weak coupling constant g in (12.2.4-6) identified or associated with g_V in the literature [P1]. Gauge transformations in (12.2.4-6) does not allow that the axial vector or tensor field is associated with a different coupling constant g_A. That $\Gamma_A > \Gamma_V$ is due to that the normalization type of constant κ_b in (12.5.2), the only free parameter in these calculations, is connected to Γ_A but not to Γ_V and κ_b is rather large in Table 12.1.

Even this g or g_V will drop out in the decay rate. Equation (12.6.13) shows that the magnitude of neutron decay rate is proportional to $(g/M_W)^4$. Now, M_W^2 itself is proportional to g^2 according to (7.4.6b) so that

$$\frac{g^4}{M_W^4} = 32G_F^2 = \left(\frac{\Omega}{\int dx^0}\right)^2 \tag{12.6.19}$$

is independent of g. Here, G_F is the Fermi constant of (7.4.29). Since $g=e/\sin\vartheta_W$ where ϑ_W is the Weinberg angle and $-e$ the electron charge, (12.6.19) is independent of e.

Because ϑ_W is not a basic constant in the present theory but can be derived as in (7.2.3, 12), the more genuine weak coupling (12.6.19) is detached from the much stronger electromagnetic coupling characterized by e, just like such a detachment found in the meson case in §7.5.3. *Nature is too economical* to deal out two fundamental constnts g and e that are so close to each *other*. Instead, the strength of weak interactions is characterized by the dimensionless constant $F_W^2=8.684\times10^{-14}$ in (7.5.22a) which is much smaller than the corresponding strength of electromagnetic interactions $\alpha=1/137$.

On the contrary, based upon an anlysis of some vector meson decay rates, the strong coupling α_s and electromagnetic coupling $\alpha=1/137$ are unified into one single "electrostrong" coupling via the hypothesis [H13 (9.2)] or (8.3.11), $\alpha_s^4=\alpha$ or $\alpha_s=0.2923$.

That the Fermi constant G_F in (12.6.19) is a ratio between a large volume and a long relative time indicates that the weak interaction is related to the large scale, long time or low energy aspects of physics.

12.7. Possible Nonconservation of Angular Momentum

12.7.1. Theoretical Background of Possible $\underline{J}$ Nonconservation

The angular momentum $\underline{J}$ of an observable, spin ½ point particle in conventional form

$$\underline{J} = -i\underline{X} \otimes \frac{\partial}{\partial \underline{X}} + i\begin{pmatrix} \underline{\sigma} & \\ & \underline{\sigma} \end{pmatrix} \tag{12.7.1}$$

is a constant of motion, hence a conserved quantity. Therefore, the total angular momentum of an ensemble of such particles is also conserved.

However, a baryon is neither a point particle nor a detachable ensemble of such particles. Therefore, conservation of $\underline{J}$ in (12.7.1) cannot be applied to a baryon without reservation. $\underline{J}$ conservation also does not apply to the quarks that constitute the baryon because a quark is not observable in the sense implied by (12.7.1). For a slowly moving doublet baryon, however, §10.2.4 shows that (10.2.1) can be reverted to a Dirac equation when the diquark coordinate x_I merges into the quark coordinate x_{II}. This merger reduces the present nonlocal description (9.3.11) to a local one so that (12.7.1) is applicable and $\underline{J}$ is conserved.

By reducing a baryon with extension into a point particle, a great amount of underlying physics leading to various observable baryon phenomena is irretrievably lost. However, the point particle approximation of a baryon can be valid in certain low energy interactions. For instance, the strong charge of a baryon may be considered to be concentrated at a point source in pion-nucleon scattering. Analogously, in Rutherford scattering, the proton can be regarded as having a point charge. In these cases, (12.7.1) can be applied and $\underline{J}$ is conserved.

In weak interactions, however, the gauge boson interacts differently with the differently flavored quarks, broadly speaking. Therefore, reduction of the baryon to a point particle cannot be made. This is evident in the introduction of a gauge fields in (7.1.4), where $\partial^{a\dot{b}}$ operating in the relative space of the diquark and quark cannot be neglected in the ensuing calculations. Thus, the angular momentum $\underline{J}$ of (12.7.1) is not applicable to the baryon wave functions in (9.3.10), which depend upon both the laboratory coordinates X and the relative space coordinates x. Therefore, $\underline{J}$ needs not be conserved in neutron decay. This is supported by noting the following.

In the four possile spin combinations for the nucleons in (12.4.3), the total spin of the lepton pair is fixed, being zero for the Fermi decays and unity for the Gamow-Teller decays. However, (12.6.6-9) show that this total spin can also be unity for the Fermi decays and zero for the Gamow-Teller decays in violation of angular momentum conservation. Thus, a qualitative prediction is that angular momentum is not conserved in free neutron decay and hence in weak interactions in general.

12.7.2. Experimental Tests of $\underline{J}$ Conservation Involving Nucleons

$\underline{J}$ in (12.7.1) is conserved in muon decay which involves four spin ½ point particles.

As was mentioned in Sec. 12.1, (12.1.4) makes use of conservation of angular momentum. When combined with Coulomb correction functions, its predictions are relatively consistent with nuclear β-decay and free neutron decay data. Therefore, there is a prevalent view that angular momentum is conserved in such decays. Arguments aginst this view has been given in Sec. 12.1. The conclusion given at the end of §12.7.1 thus invalidates (12.1.4) by Jackson et al and hence also the interpretations of the experimental results that ensue from it.

Already in their classical paper of 1956, Lee and Yang [L7] remarked that in weak interaction experiments up to that time, the baryon number, electric charge, energy and momentum are conserved. Conservation of angular momnetum $\underline{J}$ and parity P as well as invariances under charge conjugation C and time reversal T had however not been established. Nonconservation of P and C was soon discovered and P violation has been extensively measured [S10 and references therein]. A small violation of T has also been detected and subjected to many experimental investigations [L6, S9 and references therein].

In contrast, no experiment dedicated to test conservation of $\underline{J}$ in nuclear β-decay or free neutron decay has been performed to my knowledge. In fact, no experiment exists that directly distinguishes Fermi from Gamow-Teller transitions in free neutron decay without making use of (12.1.3).

Therefore, specific tests on $\underline{J}$ conservation in such decays seem to be called for. Such a test is however not strictly a test of the present theory which holds for free neutron decay only. As was indicated below (12.1.4), the unknown effects of internucleon interactions intervene the theory and eventual experimental results. To this end, decays of free neutrons are needed and will give results of more fundamental importance. That this has not been done is due to the great technical difficulty of such an experiment.

In view of the wealth of raw data available on coincidental experiments with free, polarized neutrons, it may be possible to obtain some indication as to whether $\underline{J}$ is conserved in free neutron decay. No analysis along this line has been carried out to my knowledge. Important clues may be obtained by reviewing existing data.

12.8. Epilogue, QCD and High Energy End

The equations of motion for ground state baryons (9.3.10) and for mesons (2.4.3) in the present scalar strong interaction hadron theory differ fundamentally from QCD [P1]. The interquark force is of scalar nature rather than the colored gauge field force of QCD. Also, no quark wave function is present in the meson or baryon wave equations while quark fields are explicitly present in the QCD action. Another difference is that the present theory remains largely on the quantum mechanical level while QCD has been quantized. However, we may remind ourselves that most of our low energy knowledge of atoms are obtained at the quantum mechanical level.

The approximate agreements between the scalar strong interaction hadron theory predictions and a relatively wide area of low energy mesonic data have been achieved by analytical means. Such a contact area has not been obtained starting from QCD, which in addition requires extensive numerical computations (lattice QCD).

Chapters 3 and 10 further show that the bulk of hadron physics takes place in the relative space among the quarks. These physics come out of solutions to partial differential equations and cannot be adequately represented by parameters and structure functions introduced into local phenomenological models widely used at present. The relative time among the quarks, absent in these models, turns out to be a chief contributor to the new physics found in the present theory.

Turning to the high energy region, QCD appears to be rather successful in accounting for data. On the other hand, the scalar strong interaction hadron theory has not been quantized presently and can therefore not be applied to that region. If the present theory is correct, an eventual quantized version of it must also be able to parallel or exceed QCD's predictive power at higher energies. Such a possibility is made more feasible by the following observations.

The internal degrees of freedom in form of the three internal coordinates z_I, z_{II} and z_{III} of Sec. 9.3 can play some of the roles of the three colors in QCD, excluding the color confinement mechanism. At high energies, the interquark potential Φ_m and the quark mass term M_m^2 in the meson wave equation (2.3.22) can be neglected over sizeable interquarks distances. In this region, (2.3.22) shows that the nonseparable meson wave functions can revert to separable free, massless quark wave functions according to the reverse of (2.2.1). A similar reversion of the baryon wave functions in (9.3.10) to massless quark wave functions also takes place at high energies. Quarks now behave as if they were free in a fairly large region and this reminds one of the phenomenon of "asymtotic freedom" in QCD.

Returning to the low energy end, many of the present problems are computational. The wave functions of moving mesons in Sec. 3.5 need be found and the classical energy-momentum relation (3.5.31) needs be established for large meson momenta $\underline{K}$. The physical meaning of ∞/∞ of §7.5.4 needs be understood. The baryon sector has been only touched upon. The formulation of meson-meson (four quarks), meson-baryon (three quarks and a diquark) and baryon-baryon (two quarks and two diquarks) interaction problems should be of great interest to nuclear physics.

APPENDICES

Appendices A-C serve to introduce notations and present well-known results in forms that provide starting points of the present theory. For proof and further details, the reader is referred to the literature cited. Appendices D-F are reprints on this subject from the proceedings of the 9th, 7th and 6th international conferences on hadron spectroscopy. Appendices G is a recent paper on the epistemological and historical background underlying the present theory. Appendices H gives a brief account on how I, an outsider to the high energy physics community, has been received.

A. NOTATION AND DIRAC'S EQUATION

Natural units are used; $h/2\pi = c = 1$. The unit of mass, energy, momentum, and the inverses of length and of time generally used is 10^9 electron volts or Gev. The fine structure constant is $\alpha = e^2/4\pi = 1/137.04$, where $-e$ is the electron charge. The action S is dimensionless. The notations of Schweber et al [S1] will be used. The space time coordinate is denoted by the contravariant vector

$$X^\mu = (X^0, \underline{X}) = (X^0, X^k) = (X^0, X^1, X^2, X^3) \qquad \text{(A1a)}$$

The metric tensor $g^{\mu\nu}=g_{\mu\nu}$ has the nonvanishing elements $g^{00} = g_{00} = 1$ and $g^{kl} = g_{kl} = -1$ for $k = l = 1, 2$ and 3. The associated covariant space time vector is given by

$$X_\mu = (X^0, -\underline{X}) = (X^0, -X^k) = g_{\mu\nu} X^\nu \qquad \text{(A1b)}$$

Repeated Greek indices are summed over from 0 to 3. The corresponding gradients are

$$\partial_{X\mu} = (\partial/\partial X^0, \partial/\partial \underline{X}), \qquad \partial^\mu_X = (\partial/\partial X^0, -\partial/\partial \underline{X}) \qquad \text{(A2)}$$

The Dirac equation for a spin ½ particle reads

$$(-i\gamma^\mu \partial_{X\mu} + m)\psi(X) = I_B \qquad \text{(A3)}$$

where ψ is a column vector (ψ_ν) with $\nu = 1, 2, 3,$ and 4 and is called a bispinor or four-spinor in the terminology of Appendix B, m is the particle mass and

$$\gamma^0 = \begin{pmatrix} I & 0 \\ 0 & -I \end{pmatrix}, \qquad \gamma^k = \begin{pmatrix} 0 & \sigma^k \\ -\sigma^k & 0 \end{pmatrix}$$

$$\sigma^1 = \begin{pmatrix} 0 & 1 \\ 1 & 0 \end{pmatrix}, \quad \sigma^2 = \begin{pmatrix} 0 & -i \\ i & 0 \end{pmatrix}, \quad \sigma^3 = \begin{pmatrix} 1 & 0 \\ 0 & -1 \end{pmatrix}, \quad I = \sigma^0 = \begin{pmatrix} 1 & 0 \\ 0 & 1 \end{pmatrix} \tag{A4}$$

I_B denotes the interaction term. If the particle is free or interacts with a pseudoscalar potential V_{PB}, or with a scalar potential V_{SB}, or with an electromagnetic potential A_μ, then

$$I_B = 0 \tag{A5a}$$

$$I_B = \gamma_5 V_{PB} \psi \tag{A5b}$$

$$I_B = V_{SB} \psi \tag{A5c}$$

$$I_B = i q_r \gamma^\mu A_\mu \psi \tag{A5d}$$

$$\gamma_5 = -i \begin{pmatrix} 0 & I \\ I & 0 \end{pmatrix} \tag{A6}$$

where q_r is the charge of the ψ field. If these potentials are generated by another spin ½ particle B with wave function $\psi_B = (\psi_{B\nu})$, then

$$\Box V_{PB} = g_s^2 \overline{\psi}_B \gamma_5 \psi_B \tag{A7a}$$

$$\Box V_{SB} = g_s^2 \overline{\psi}_B \psi_B \tag{A7b}$$

$$\overline{\psi} = \psi^+ \gamma^0, \qquad \psi^+ = \left(\psi_1^*, \psi_2^*, \psi_3^*, \psi_4^*\right) \tag{A7c}$$

$$\Box = \partial_{X\mu} \partial_X^\mu = \left(\partial / \partial X^0\right)^2 - \left(\partial / \partial \underline{X}\right)^2 \tag{A8}$$

where g_s^2 is the interaction coupling constant between the both particles and * denotes complex conjugation.

Equations (A3), (A5) and (A7) are invariant under the homogeneous Lorentz transformations [S1]

$$X^{\mu\prime} = L^\mu_{\ \nu} X^\nu \tag{A9}$$

where $L^{\mu}{}_{\nu}$ is required to conserve

$$s_X^2 = X_\mu{}' X^{\mu}{}' = X_\mu X^\mu = \left(X^0\right)^2 - \left(X^1\right)^2 - \left(X^2\right)^2 - \left(X^3\right)^2 \tag{A10}$$

This leads to

$$L_\nu^{\mu} L^{\nu}_{\lambda} = \delta^{\mu}{}_{\lambda} = \begin{pmatrix} 1 & \text{for } \mu = \lambda \\ 0 & \text{for } \mu \neq \lambda \end{pmatrix} \tag{A11}$$

Equation (A9) is the self representation of the Lorentz group and X^{μ} spans the smallest nontrivial tensorial representation space. The product of n X^{μ}'s spans the 4^n dimensional tensorial representation space for the Lorentz group.

If (A9) is restricted to represent the proper Lorentz group L_p, then

$$L^0{}_0 > 0, \quad \det\left|L^{\mu}{}_{\nu}\right| = +1 \tag{A12}$$

L_p is characterized by six real parameters, three angles of rotations and boosts in three directions.

For a free particle (A3)-(A5a) have the solution

$$\left(\psi_\nu(X)\right) = \left(u_\nu\right) \exp\left(-iK_\mu X^\mu\right) \tag{A13a}$$

$$K_\mu = \left(E, -\underline{K}\right) \tag{A13b}$$

$$E = \pm\sqrt{m^2 + \underline{K}^2} \tag{A13c}$$

where E denotes the energy of the particle and $\underline{K}$ its momentum.

$$(u_\nu) = \begin{pmatrix} u_1 \\ u_2 \\ u_3 \\ u_4 \end{pmatrix} = N_D \begin{pmatrix} 1 & , & 0 \\ 0 & , & 1 \\ K^3/(E+m) & , & \left(K^1 - iK^2\right)/(E+m) \\ \left(K^1 + iK^2\right)/(E+m), & & -K^3/(E+m) \end{pmatrix} \quad \text{positive energy in (A13c)}$$

$$(u_\nu) = \begin{pmatrix} u_1 \\ u_2 \\ u_3 \\ u_4 \end{pmatrix} = N_D \begin{pmatrix} \left(-K^1 + iK^2\right)/(E+m), & & K^3/(E+m) \\ K^3/(E+m) & , & \left(K^1 + iK^2\right)/(E+m) \\ 0 & , & -1 \\ 1 & , & 0 \end{pmatrix} \quad \text{negative energy in (A13c)}$$

$$N_D = \sqrt{\frac{E+m}{2E\Omega}} \qquad (A14)$$

are the four solutions for (u_ν) with the commas separating the spin up and spin down solutions. Ω is a large normalization volume. In the nonrelativistic limit $|\underline{K}|<<E$, ψ_1 and ψ_2 are called the large components and ψ_3 and ψ_4 the small components.

In the extreme relativistic limit, $|\underline{K}|\approx E$ and m can be dropped. The four component (A3) with (A5a) decomposes into two decoupled two component equations

$$\gamma^\mu \partial_{X\mu}\psi_L = 0, \qquad \gamma^\mu \partial_{X\mu}\psi_R = 0 \qquad (A15)$$

where ψ_L and ψ_R are left and right handed, respectively, and are given by

$$\psi_R = \tfrac{1}{2}(1+i\gamma_5)\psi = \frac{1}{2}(\psi_1+\psi_3)\begin{pmatrix}1\\0\\1\\0\end{pmatrix} + \frac{1}{2}(\psi_2+\psi_4)\begin{pmatrix}0\\1\\0\\1\end{pmatrix} \qquad (A16a)$$

$$\psi_L = \tfrac{1}{2}(1-i\gamma_5)\psi = \frac{1}{2}(\psi_1-\psi_3)\begin{pmatrix}1\\0\\-1\\0\end{pmatrix} + \frac{1}{2}(\psi_2-\psi_4)\begin{pmatrix}0\\1\\0\\-1\end{pmatrix} \qquad (A16b)$$

B. SPINORS AND LORENTZ INVARIANCE

The material in this appendix comes largely from [W1, L1, C1, R1]. That the Dirac equation (A3) with (A5a) is invariant under proper Lorentz transformations [D1, S1] suggests that there are other representations of the Lorentz group than the vectorial representation of (A9) and the tensorial representations of dimension 4^n mentioned below (A11). Such representations were first uncovered by van der Waerden in 1929 [W1] and are called spinorial representations. Basis vectors generating such representations are called spinors, a word coined by Ehrenfest [W1] shortly after the proposal of the Dirac equation in 1928.

Let two complex variables ψ_a, $a = 1, 2$ be defined in a two dimensional complex space and transform according to

$$\psi_a' = \psi_b s^b{}_a \qquad (B1)$$

$$\det\left|s^b{}_a\right| = +1 \qquad (B2)$$

ψ_a has two components and is called a two-spinor, to be distinguished from the four component bispinor or four-spinor to which the Dirac wave function ψ in (A3) belongs. More precisely, it is called an elementary spinor or a covariant spinor of first rank but will often be called a spinor for short. Dummy spinor indices *a, b,...* are summed over *1* and *2*. Let χ_c be another spinor transforming in the same way as ψ_a does in (B1). Making use of (B2), we find

$$\psi_1'\chi_2' - \psi_2'\chi_1' = \psi_1\chi_2 - \psi_2\chi_1 = \text{invariant} \tag{B3}$$

An analog of the antisymmetric ε tensors in tensor analysis in spinor analysis is the raising operator ε^{ab} with

$$\varepsilon^{12} = 1, \qquad \varepsilon^{21} = -1, \qquad \varepsilon^{11} = \varepsilon^{22} = 0 \tag{B4}$$

A contravariant spinor is defined as

$$\psi^a = \varepsilon^{ab}\psi_b \tag{B5a}$$

so that

$$\psi^1 = \psi_2, \quad \psi^2 = -\psi_1 \tag{B5b}$$

Conversely, the lowering operator is defined as

$$\varepsilon_{ba} = \varepsilon^{ab}, \qquad \varepsilon_{ab}\varepsilon^{ab} = -1 \quad \text{(no summation)} \tag{B6}$$

It converts a contravariant spinor to a covariant one;

$$\psi_b = \varepsilon_{ba}\psi^a \tag{B7}$$

which also yields (B5b). The invariant (B3) now takes the form of the scalar product

$$\psi_a'\chi^{a\prime} = -\psi^{a\prime}\chi_a' = \psi_a\chi^a = -\psi^a\chi_a = \text{invariant} \tag{B8}$$

Raising the lower index and lowering the upper repeated index leads to a sign change.

By analogy to (B1),

$$\psi^{a\prime} = S^a{}_b\psi^b \tag{B9}$$

Forming the scalar product of (B1) and (B9) and noting (B8) leads to

$$\psi_a{}'\psi^{a\prime} = \psi_a s^a{}_b S^b{}_c \psi^c = \psi_a \psi^a \tag{B10}$$

whence

$$s^a{}_b S^b{}_c = \delta^a{}_c \tag{B11}$$

which relates the transformations (B1) and (B9). Here, the contraction operator $\delta^a{}_c=1$ for $a=c$ and 0 otherwise. Taking the determinant of (B10) and noting (B2), we obtain

$$\det\left|S^b{}_c\right| = +1 \tag{B12}$$

The complex conjugates of the spinors ψ_a and ψ^a are respectively denoted by $\psi_{\dot a}$ and $\psi^{\dot a}$ which are called dotted spinors. Since the spinor ψ_a and the transformation matrix $s^b{}_a$ are already complex, the forms of (B1)–(B12) hold also for the dotted spinors, which however transform genuinely differently;

$$\psi_{\dot a}{}' = \psi_{\dot b} s^{\dot b}{}_{\dot a}, \qquad \psi^{\dot a\prime} = S^{\dot a}{}_{\dot b} \psi^{\dot b} \tag{B13}$$

$$\det\left|s^{\dot b}{}_{\dot a}\right| = \det\left|S^{\dot b}{}_{\dot a}\right| = 1, \qquad s^{\dot a}{}_{\dot b} S^{\dot b}{}_{\dot c} = \delta^{\dot a}{}_{\dot c} \tag{B14}$$

$$\varepsilon^{\dot 1\dot 2} = 1,\quad \varepsilon^{\dot 2\dot 1} = -1,\quad \varepsilon^{\dot 1\dot 1} = \varepsilon^{\dot 2\dot 2} = 0,\quad \varepsilon_{\dot b\dot a} = \varepsilon^{\dot a\dot b},\quad \varepsilon_{\dot a\dot b}\varepsilon^{\dot a\dot b} = -1 \ \ \text{(no summation)} \tag{B15}$$

$$\psi^{\dot a} = \varepsilon^{\dot a\dot b}\psi_{\dot b}, \qquad \psi_{\dot a} = \varepsilon_{\dot a\dot b}\psi^{\dot b}, \qquad \psi_{\dot 2} = \psi^{\dot 1}, \qquad \psi_{\dot 1} = -\psi^{\dot 2} \tag{B16}$$

$$\psi_{\dot a}{}'\chi^{\dot a\prime} = -\psi^{\dot a\prime}\chi_{\dot a}{}' = \psi_{\dot a}\chi^{\dot a} = -\psi^{\dot a}\chi_{\dot a} = \text{invariant} \tag{B17}$$

The last equation and (B8) provide the basic invariants in spinor analysis. The ε and δ matrices or spinors of second rank in (B4), (B11), (B14), and (B15) are invariant under the transformations (B1), (B9) and (B13), applied to each of the spinor indices.

The set of unimodular matrices $s^a{}_b$ in (B1) and (B2) forms a group and is a representation of the SL(2,C) group (Special(≈ unimodular or (B2)) Linear group of Complex dimension 2). So do also $S^a{}_b$, $s^a{}_b$, $S^{\dot a}{}_{\dot b}$, $s^{\dot a}{}_{\dot b}$ as can be inferred from (B11), (B13) and (B14). The group SL(2,C) is characterized by six real parameters; one relation (B2) holds between four complex numbers in $s^a{}_b$, just like the proper Lorentz group L_p in Appendix A is. Van der Waerden [W1] has pointed out that L_p is homomorphic to SL(2,C). The unimodularity condition (B2) has its origin in the conservation of the "length" s_X in (A10).

These representations of SL(2,C) contain genuine spinorial representations of the proper Lorentz group, in addition to the tensorial representations mentioned in Appendix A. They have the dimensionality $(2j+1)(2j'+1)$. The elements of the corresponding representation space are spinors (undotted) of order $2j$ and dotted spinors of order $2j'$ belonging to the Minkowski space [R1].

Representations with $j+j'=$ integer coincide with the tensorial representations of L_p in Appendix A. They are single valued, i.e., corresponding to any element of L_p there is only one transformation in a representation space of given dimensionality.

Representations with $j+j'=$ half integer are genuine spinoral representations of L_p not covered by tensorial representations. They are double valued, i.e., corresponding to any proper Lorentz transformation, there is a matrix representation as well as its negative. Therefore, any equation which is covariant under SL(2,C) transformations will be also covariant under L_p transformations.

Some results of spinor calculus are quoted below. The basic invariants of spinor calculus have been given by (B8) and (B17). Putting $\psi=\chi$ there, we find that the "value" of the spinor of first rank vanishes;

$$\psi^a\psi_a = -\psi_a\psi^a = 0, \qquad \psi^{\dot{a}}\psi_{\dot{a}} = -\psi_{\dot{a}}\psi^{\dot{a}} = 0 \tag{B18a}$$

Indeed, the "value" of any spinor of odd rank vanishes,

$$\psi^{\dot{a}bc}\psi_{\dot{a}bc} = -\psi_{\dot{a}bc}\psi^{\dot{a}bc} = 0 \tag{B18b}$$

Higher rank spinors can be formed by multiplying together spinors of first rank. For example,

$$w^{ab} = \psi^a\chi^b, \qquad w^{a\dot{b}}_{\ \ c} = \psi^a\chi_c\chi^{\dot{b}} \tag{B19}$$

The spinor

$$w_{ab....\dot{c}}^{\ \ \ \ \ \ \ \ d......\dot{e}} \tag{B20a}$$

is generally defined as a quantity transforming under SL(2,C) as the product of first rank spinors

$$\psi_a\chi_b.....\xi_{\dot{c}}\varphi^d......\eta^{\dot{e}} \tag{B20b}$$

While the positions of the spinors in (B20b) can be rearranged, the positions of any pair of undotted or dotted upper or lower indices in (B20a) cannot be interchanged. There exists, however, a subset of this spinor for which such an interchange is allowed. This subset is called a symmetric spinor. An interchange between a dotted index and an undotted index is however allowed, because the obey different transformation rules. The complex conjugate of a spinor of any rank is obtained by changing all dotted indices to undotted ones and vice versa.

The rank of a spinor can be lowered by means of the two types of invariant spinors ε and δ in (B4), (B6), (B11), (B14) and (B15). For example,

$$\varepsilon^{\dot{a}\dot{b}} w^{\dot{c}}_{e\dot{a}\dot{b}} = u^{\dot{c}}_{e}, \qquad \delta^{a}{}_{d} w^{dc}_{ba\dot{e}} = v^{c}_{b\dot{e}} \tag{B21}$$

Such contractions are often applied in the reduction of a higher dimensional representation space to lower dimensional invariant subspaces.

Any equation that can be written in tensor form is covariant under L_p. For instance, the Maxwell equation

$$\partial F^{\mu\nu} / \partial X^{\nu} = J^{\mu} \tag{B22}$$

is manifestly covariant under L_p, where $F^{\mu\nu}$ denotes the electromagnetic field tensor and J^{μ} the source current. Similarly, any equation that can be written in spinor form is manifestly covariant under SL(2,C) and hence also covariant under L_p. This will be shown explicitly for the following example.

Consider Dirac's equation in van der Waerden's spinor form (C10) below, put V_{PB} and q_r there to 0 for simplicity and use (B17). One can write

$$\partial^{a\dot{b}}_{X} \chi_{\dot{b}}(X) = -\partial^{a}_{X\dot{b}} \delta^{\dot{b}}{}_{\dot{c}} \chi^{\dot{c}}(X) = i(m + V_{SB}(X))\psi^{a}(X) \tag{B23a}$$

$$\partial_{Xb\dot{c}} \psi^{c}(X) = \partial_{Xb\dot{c}} \delta^{c}{}_{d} \psi^{d}(X) = i(m + V_{SB}(X))\chi_{\dot{b}}(X) \tag{B23b}$$

Perform proper Lorentz transformation L_p similar to (A9)

$$X^{\mu} \to X^{\mu\prime} = L^{\mu}_{p\nu} X^{\nu} \tag{B24}$$

The homomorphism between L_p and SL(2,C) enables a determination, up to a sign, of the corresponding genuine spinorial representations (B1) and (B9),

$$\psi_{a}'(X') = \psi_{b}(X) s^{b}{}_{a}, \qquad \psi^{a\prime}(X') = S^{a}{}_{b} \psi^{b}(X) \tag{B25}$$

and their complex conjugates (B13). Operating (B23a) by $S^{d}{}_{a}$ and (B23b) by $S^{\dot{a}}{}_{\dot{e}} \varepsilon^{\dot{e}\dot{b}} S^{a}{}_{e} \varepsilon^{eb}$ and replacing the δ spinors by (B11) and (B14) leads to

$$-S^{d}{}_{a} \partial^{a}_{X\dot{b}} s^{\dot{b}}{}_{\dot{e}} S^{\dot{e}}{}_{\dot{c}} \chi^{\dot{c}}(X) = i\left(m + V_{SB}\left(\left(L^{\mu}{}_{\nu}\right)^{-1} X'\right)\right) S^{d}{}_{a} \psi^{a}(X) \tag{B26a}$$

$$S^{\dot{a}}{}_{\dot{e}} \partial^{\dot{e}}_{Xc} s^{c}{}_{b} S^{b}{}_{d} \psi^{d}(X) = i\left(m + V_{SB}\left(\left(L^{\mu}{}_{\nu}\right)^{-1} X'\right)\right) S^{\dot{a}}{}_{\dot{e}} \chi^{\dot{e}}(X) \tag{B26b}$$

Apply now (B1), (B9) and (B13) to (B26) and make the identifications

$$S^{d}{}_{a}\partial^{a}_{X\dot{b}}s^{\dot{b}}{}_{\dot{e}} = \partial'^{d}_{\dot{e}}, \qquad S^{\dot{a}}{}_{\dot{e}}\partial^{\dot{e}}_{Xc}s^{c}{}_{b} = \partial'^{\dot{a}}_{b} \tag{B27}$$

where the prime denotes differentiation with respect to $X^{\mu}{}'$. After another application of (B17) and (B16), we obtain

$$\partial'^{d\dot{e}}\chi'_{\dot{e}}(X') = i(m + V'_{SB}(X'))\psi'^{d}(X') \tag{B28a}$$

$$\partial'_{\dot{a}b}\psi'^{b}(X') = i(m + V'_{SB}(X'))\chi'_{\dot{a}}(X') \tag{B28b}$$

$$V'_{SB}(X') = V_{SB}\left((L^{\mu}{}_{\nu})^{-1}X'\right) = V_{SB}(X) \tag{B29}$$

Equation (B28) with (B29) has the same form as (B23) and covariance under SL(2,C) is demonstrated.

Under space inversion, invariance of the Dirac equation (A3) with (A5c) and (A7b) requires that the Dirac wave function transforms like [J1]

$$\psi'(X'^{0}, \underline{X'}) = \gamma^{0}\psi(X^{0}, \underline{X}), \qquad X'^{0} = X^{0}, \qquad \underline{X'} = -X \tag{B30}$$

Applying this relation to (C11) below leads to that

$$\chi_{\dot{a}}(X^{0}, \underline{X}) \leftrightarrow \psi^{a}(X^{0}, -\underline{X}) \tag{B31}$$

under space inversion [W2].

C. Formulae Relating Spinors and Tensors

The material in this appendix is again largely gathered from [W1, L1, R1]. The spinor ψ_a is specified by two complex or four real quantities. Therefore, the product spinor $\psi_a\psi_{\dot{b}}$ is also determined by four real quantities. This spinor transforms as a mixed spinor of second rank $X_{\dot{b}a}$. Let

$$X_{\dot{b}a} = \begin{pmatrix} X^{0} + X^{3} & X^{1} - iX^{2} \\ X^{1} + iX^{2} & X^{0} - X^{3} \end{pmatrix} = \delta_{\dot{b}a}X^{0} + \underline{\sigma}_{\dot{b}a}\,\underline{X}$$

$$X^{b\dot{a}} = \begin{pmatrix} X^{0} - X^{3} & -X^{1} + iX^{2} \\ -X^{1} - iX^{2} & X^{0} + X^{3} \end{pmatrix} = \delta^{b\dot{a}}X^{0} - \underline{\sigma}^{b\dot{a}}\,\underline{X} \tag{C1a}$$

where the second relation follows from the first one. Equation (C1a) also contains only four real quantities, i. e., X^{μ} with μ=0-3. It further yields

$$\tfrac{1}{2} X_{\dot{b}a} X^{a\dot{b}} = \left(X^{0}\right)^{2} - \left(\underline{X}\right)^{2} \tag{C1b}$$

which is a Lorentz invariant. Therefore, (X^0, $\underline{X}$) is a vector, here assigned to the space-time vector. With this association, the spinor ψ_a above has the dimension (length)^{½}. $\delta_{\dot{b}a}$ and $\delta^{a\dot{b}}$ are defined as 1 for a=b and 0 otherwise. They are introduced for convenience and do not transform as spinors.

The Pauli matrices

$$\underline{\sigma}_{\dot{b}a} = \underline{\sigma}^{b\dot{a}} = \left[\begin{pmatrix} 0 & 1 \\ 1 & 0 \end{pmatrix}, \begin{pmatrix} 0 & -i \\ i & 0 \end{pmatrix}, \begin{pmatrix} 1 & 0 \\ 0 & -1 \end{pmatrix}\right] \tag{C2}$$

are classified as spin-tensors here. Since the positions of the undotted and dotted indices can be interchanged, the following convention will be followed in this book; *the first index is dotted (undotted) if it is a covariant (contravariant) index and refers to the rows of the matrices. The second index refers to the columns.* Corresponding to (C1), define

$$\partial_{X\dot{b}a} = -\delta_{\dot{b}a}\partial_{X0} + \underline{\sigma}_{\dot{b}a}\partial_{\underline{X}}, \qquad \partial_X^{b\dot{a}} = -\delta^{b\dot{a}}\partial_{X0} - \underline{\sigma}^{b\dot{a}}\partial_{\underline{X}} \tag{C3a}$$

$$\partial_{X0} = \partial/\partial X^{0}, \qquad \partial_{\underline{X}} = \partial/\partial \underline{X} \tag{C3b}$$

which are mixed gradient spinors of second rank.

Extending (C2) to

$$\sigma_{\mu\dot{b}a} = \sigma^{\mu b\dot{a}} = \left(\delta_{\dot{b}a}, \underline{\sigma}_{\dot{b}a}\right) = \left(\delta^{b\dot{a}}, \underline{\sigma}^{b\dot{a}}\right) \tag{C4}$$

the relations between the vector components (A1) and (A2) and the spinor components (C1a) and (C3a) are

$$\begin{aligned} &X_{\dot{b}a} = \sigma_{\mu\dot{b}a}X^{\mu}, \quad X^{b\dot{a}} = \sigma^{\mu b\dot{a}}X_{\mu}, \quad X^{\nu} = \tfrac{1}{2}\sigma^{\nu b\dot{a}}X_{\dot{a}b}, \quad X_{\nu} = \tfrac{1}{2}\sigma_{\nu\dot{b}a}X^{a\dot{b}} \\ &\partial_{X\dot{b}a} = -\sigma_{\mu\dot{b}a}\partial_X^{\mu}, \quad \partial_X^{b\dot{a}} = -\sigma^{\mu b\dot{a}}\partial_{X\mu}, \quad \partial_X^{\nu} = -\tfrac{1}{2}\sigma^{\nu b\dot{a}}\partial_{X\dot{a}b}, \quad \partial_{X\nu} = -\tfrac{1}{2}\sigma_{\nu\dot{b}a}\partial_X^{a\dot{b}} \end{aligned} \tag{C5}$$

Further,

$$\tfrac{1}{2}\partial_{X\dot{b}a}\partial_X^{a\dot{b}} = \square, \qquad \tfrac{1}{2}\partial_{X\dot{b}a}y^{a\dot{b}} = \partial_{X\mu}y^{\mu} \tag{C6}$$

Application of the first of (C3a) to the second of (C1a) with *X* replaced by *y* yields

$$\partial_{X\dot{b}a} y^{a\dot{c}} = \left(-\delta_{\dot{b}a}\partial_{X0} + \underline{\sigma}_{\dot{b}a}\partial_{\underline{X}}\right)\left(\delta^{a\dot{c}} y^0 - \underline{\sigma}^{a\dot{c}} \underline{y}\right)$$
$$= -\delta_{\dot{b}}{}^{\dot{c}}\partial_{X0} y^0 + \underline{\sigma}_{\dot{b}}{}^{\dot{c}}\left(\partial_{\underline{X}} y^0 + \partial_{X0}\underline{y}\right) - \delta_{\dot{b}}{}^{\dot{c}}\partial_{\underline{X}}\underline{y} - i\underline{\sigma}_{\dot{b}}{}^{\dot{c}}\left(\partial_{\underline{X}} \times \underline{y}\right) \tag{C7}$$

where the Dirac formula [D1]

$$\left(\underline{\sigma}\partial_{\underline{X}}\right)\left(\underline{\sigma}\,\underline{y}\right) = \partial_{\underline{X}}\underline{y} + i\underline{\sigma}\left(\partial_{\underline{X}} \times \underline{y}\right) \tag{C8}$$

has been employed. Here,

$$\delta_{\dot{b}a}\delta^{a\dot{c}} = \delta_{\dot{b}}{}^{\dot{c}}\begin{pmatrix} 1 & \text{for } b = c \\ 0 & \text{otherwise} \end{pmatrix}, \qquad \underline{\sigma}_{\dot{b}}{}^{\dot{c}} = \underline{\sigma}_{\dot{b}a}\delta^{a\dot{c}} = \delta_{\dot{b}a}\underline{\sigma}^{a\dot{c}} \tag{C9}$$

where *b* is the row number. $\delta^a{}_c$ and $\underline{\sigma}^a{}_c$ can be analogously constructed with *a* being the row number. These δ and $\underline{\sigma}$ and the corresponding ones with dotted indices have been introduced for convenience of notation. They do not transform as spinors, apart from the coincidence of the δs with those in (B11) and (B14).

The Dirac equation (A3) and (A5b-d) can now be put in the spinor form

$$\left(\partial_X^{a\dot{b}} + iq_r A^{a\dot{b}}\right)\chi_{\dot{b}} = i\left(m + V_{SB} - V_{PB}\right)\psi^a \tag{C10a}$$

$$\left(\partial_{X\dot{b}c} + iq_r A_{\dot{b}c}\right)\psi^c = i\left(m + V_{SB} + V_{PB}\right)\chi_{\dot{b}} \tag{C10b}$$

where (C4) and (C5) have been used. Here,

$$\chi_{\dot{1}} = \psi_1 + \psi_3, \qquad \chi_{\dot{2}} = \psi_2 + \psi_4, \qquad \psi^1 = \psi_1 - \psi_3, \qquad \psi^2 = \psi_2 - \psi_4 \tag{C11}$$

Inserting this expression into (A7) yields

$$\Box V_{PB} = \frac{i}{2} g_s^2\left(\psi_B^b \chi_{Bb} - \psi_B^{\dot{b}} \chi_{B\dot{b}}\right) \tag{C12a}$$

$$\Box V_{SB} = \frac{1}{2} g_s^2\left(\psi_B^b \chi_{Bb} + \psi_B^{\dot{b}} \chi_{B\dot{b}}\right) \tag{C12b}$$

Proper Lorentz invariance of (C10) with V_{PB}=0 as well as its invariance under space inversion have been demonstrated at the end of Appendix B. Using the same method, it is readily seen that the full (C10) and (C12) are also covariant under these transformations.

Putting m, V_{SB}, V_{PB} and q_r to 0 in (C10), it becomes

$$\partial_X^{a\dot{b}} \chi_{\dot{b}} = 0 \qquad (\chi_{\dot{b}} \quad \text{right handed}) \tag{C13a}$$

$$\partial_{X\dot{b}c} \psi^c = 0 \qquad (\psi^c \quad \text{left handed}) \tag{C13b}$$

which is the spinor form of the bispinor equations (A15) and (A16). The first one is sometimes called the Weyl equation and describes neutrinos.

D. Scalar Strong Interaction Hadron Theory -An Alternative to QCD

(Reprinted from *Hadron '01* [H1])

F. C. Hoh

Dragarbrunnsg. 9B, 75332 Uppsala and associated with Department of Radiation Sciences, Uppsala University, Box 535, SE-751 21 Uppsala, Sweden

Abstract

Reasons for not accepting QCD as the correct strong interaction theory given at Hadron '95 and Hadron '97 are expanded and made more precise. The scalar strong interaction hadron theory, recently presented in a book, provides an alternative to QCD. The motivation for and basic steps leading to this theory are outlined.

INTRODUCTION

On the theoretical side of this meeting, the contributions consist largely of the latest versions of phnomenological models, mostly on hadron spectra. These are but a tiny increment to the large body of such models already existing in the literature, or as some put it, on the "market". In his summary talk, Ted Barnes hinted at this situation by putting forth the formula: If there are n theorists, there are n+1 theories. Actually, the number n+1 is too small and may be replaced by αn, where $\alpha > 1$. This is due to the fact that the same theorist can have different models for different applications. Moreover, there may be several models for the same application area as he updates and improves his earlier models.

There are hundreds of such kinds of models with different assumptions, parameters, form factors, structure functions,... While these models work well within the relatively narrow application angle they are designed for, they provide little understanding of hadronic phenomena in any coherent fashion. This aspect together with the large number of such models makes it very difficult to extract from them eventual guidance to a correct hadron theory. In this respect, the above-mentioned models seem to differ from the models that preceeded quantum mechanics. The atomic models of Bohr and Sommerfeld and the Rydberg formula played impotant roles in the development and verification of the Schrödinger theory.

The great freedom in constructing these models stems from the lack of a working first principles' hadron theory. The current strong interaction theory, quantum-chromodynamics(QCD), is nonperturbative at low energies and has not been able to provide useful predictions ever since it was proposed nearly 30 years ago. By contrast, 30 years after the appearance of the Maxwell equations, nearly all basic electromagnetic phenomena were explained. Similarly, 30 years after the publication of the Schrödinger-Dirac equations, most of the atomic phenomena were understood.

Furthermore, Dirac's book appeared already seven years after he published his equation and Schiff's text came out 23 years after the appearance of the Schrödinger equation. Still, there is no book covering QCD.

These circumstances indicate that QCD is not the correct strong interaction theory. Reasons for taking this standpoint have been given in Hadron '95 and Hadron '97[1]. This contribution is a further and more precise development of these references.

QUARKS, FERMIONS, QCD, AND MODELS

QCD as a hadron theory does not include two basic properties of quarks, namely, *property Q1*: *A quark cannot be observed alone* and *property Q2*: *A quark is always accompanied by an antiquark or two other quarks or a diquark.*

Prior to the quark hypothesis of 1963-64, a fermion is defined to be a lepton or a point particle having higher odd half-integer spin. Since the latter has not been observed, it is dropped for the present considerations. A fermion has two basic proper-ties, namely, *property F1*: *A fermion can exist freely and is described by Dirac's wave functions* and *property F2*: *In weakly interacting aggregates, fermions obey Pauli's exclusion principle.*

After the quark hypothesis, quarks were soon embraced into the fermion family. This inclusion is present in QCD and in chiral symmetry models for hadrons because at the times of the proposals of these theories, the nonobservation of free quarks or *property Q1* was not firmly established. Only afterwards was this *property Q1* accepted as an experimental fact.

Accordingly, a quark is not a fermion in the sense of *property F1*. Consequently, there is no reason that *property F2* should hold for quarks. With these observations, the color parts of QCD and chiral symmetry models are no longer needed. Further, quark wave functions similar to those of *property F1* appear in QCD and chiral symmetry models. According to quantum mechanics, these wave functions can be used to form expectation values of dynamical variables such as energy, momentum and position. But such formations would contradict *property Q1*. Therefore, these theories cannot be correct. Also, the concept of spontaneous symmetry breaking in chiral symmetry models is borrowed from solid state physics. But there is no reason why such a mechanism would apply in the totally different case of hadrons.

On par with the chiral symmetry models are the constituent quark models and the Bethe-Salpeter(BS) equation approach[2]. Although no quark wave function appears in the BS approach explicitly, they are present implicitly. This can be seen in the positronium case, which is described by a BS equation with 16 wave function components. The positronium can however disintegrate into a free electron and a free positron and the BS equation decomposes into two Dirac equations describing these free particles. But a meson cannot disintegrate into a free quark and a free antiquark. This difference prevents that the same BS formalism can be applied to a meson. Because of *properties Q1* and *Q2,* quarks in a meson have less degrees of freedom as do the constituents of positronium. Hence the number of meson wave function components should be less than 16(see eq. (3) ff below). Similarly, the number of baryon wave function components should be less than 64(see last §).

The constituent quark models are more phenomenological in nature and have been the most successful ones. The interaction potential among the quarks are usually assumed to be of vector type in order to conform to QCD. In this connection, the spin-orbit terms in the potential have caused quite some controversy[3]. More importantly, confinement potential must be assumed to fit data.

SCALAR STRONG INTERACTION HADRON THEORY

In view of the above considerations, alternatives to QCD are obviously called for. I am as surprised as I was at Hadron '95[1] at that no voice on this issue has been heard also at this meeting. One such alternative was proposed in 1993[4]. Since then, it has been developed in 10 papers which have recently been synthesized and amplified into a book[5].

The approach of this theory is to construct sets of differential equations for hadrons. These sets are then taken as basic postulates and are solved for various applications, just like that we solve the Maxwell and the Schrödinger-Dirac equations for different purposes and get results. They stand or fall with their ability to account for data.

To begin with, it is observed that the conventional form of the Dirac equation involving 4×4 γ matrices is not Lorentz invariant by inspection. van der Warden[6] has transformed this equation into a set of manifestly Lorentz invariant two-spinor equations. Consider the meson case. Quark A at x_I and antiquark B at x_{II} interacting with each other via a massless scalar potential V_S are each described by a set of such two-spinor equations. For quark A, it reads

$$\begin{aligned} &\partial_I^{a\dot{b}}\chi_{A\dot{b}}(x_I) - iV_{SB}(x_I)\psi_A^a(x_I) = im_A\psi_A^a(x_I) \\ &\partial_{I\dot{b}c}\psi_A^c(x_I) - iV_{SB}(x_I)\chi_{A\dot{b}}(x_I) = im_A\chi_{A\dot{b}}(x_I) \\ &\Box_I V_{SB}(x_I) = \tfrac{1}{2}g_s^2\left(\psi_B^c(x_I)\chi_{Bc}(x_I) + c.c.\right) \end{aligned} \tag{1}$$

Here, a, b, c = 1, 2 and $\chi_{\dot{a}} = (\chi_a)^*$, $g_s^2/4\pi$ is the strong quark-quark coupling, I refers to x_I, and m_A is the mass of quark A. The both sets are multiplied together and the basic assumption of the present theory is the generalization of the product wave functions into nonseparable quantities according to

$$\chi_{A\dot{b}}(x_I)\chi_{Ba}(x_{II}) \rightarrow \chi_{\dot{b}a}(x_I, x_{II}),, \qquad \psi_A^c(x_I)\psi_B^{\dot{a}}(x_{II}) \rightarrow \psi^{c\dot{a}}(x_I, x_{II}) \tag{2a}$$

$$\chi_{A\dot{b}}(x_I)\psi_{B\dot{e}}(x_{II}) \rightarrow \chi_{\dot{b}\dot{e}}(x_I, x_{II}), \qquad \psi_A^c(x_I)\chi_B^f(x_{II}) \rightarrow \psi^{cf}(x_I, x_{II}) \tag{2b}$$

$$V_{SA}(x_{II})V_{SB}(x_I) \rightarrow \Phi_m(x_I, x_{II}) \tag{2c}$$

where Φ_m is the interquark potential. The right sides of (2a) contain mixed spinors of second rank and are assigned to represent mesons. The right sides of (2b) contain unmixed spinors and transform like diquarks. In the so-gneralized equations, these last quantities and unpaired quarks are dropped to conform to *properties Q1* and *Q2*. The resulting meson wave equations read

$$\partial_I^{a\dot{b}}\partial_{II}^{f\dot{e}}\chi_{\dot{b}f}(x_I,x_{II})-\left(M_m^2-\Phi_m(x_I,x_{II})\right)\psi^{a\dot{e}}(x_I,x_{II})=0$$
$$\partial_{I\dot{b}c}\partial_{II\dot{e}d}\psi^{c\dot{e}}(x_I,x_{II})-\left(M_m^2-\Phi_m(x_I,x_{II})\right)\chi_{\dot{b}d}(x_I,x_{II})=0$$
$$\Box_I\Box_{II}\Phi_m(x_I,x_{II})=-\tfrac{1}{2}g_s^4\,\mathrm{Re}\,\psi^{b\dot{a}}(x_I,x_{II})\chi^*_{\dot{a}b}(x_I,x_{II}) \qquad (3)$$

M_m turns out to be an eigenvalue equal to the average quark mass[5, §2.3.5]. It is obtained after having generalized the quark mass in (1) into a mass operator, similar to that employed in the Gell Mann-Okubo formula.

We note that there is no quark wave function in (3) so that *property Q1* is present. The two indices of the meson wave functions χ and ψ on the right sides of (2a) and in (3) show that a quark and an antiquark appear together in agreement with *property Q2*. On the other hand, *properties F1* and *F2* are absent. There are eight meson wave function components, half the number 16 for the corresponding BS amplitudes. Equations (3) consist of two coupled second order equations containing an interaction potential term that depends nonlinearly upon the meson wave functions.

Equations (3) provide the starting point from which all two quark meson phenomena are to be accounted for, similar to that the Schrödinger-Dirac equations are the basis for understanding all atomic phenomena. So far, (3) has been rather successful when compared to QCD.

A set of baryon wave equations has been analogously constructed. In ground state, a baryon consists of a quark and a diquark and is described by 12 wave function components, far less than 64 in the BS case. However, contact with data has hitherto been hampered by difficulties in solving the differential equations, which are much more complicated than those for the mesons. The present theory has so far only been applied to low energy data. It has not been worked out for application to high energy hadronic phenomena.

REFERENCES

[1] Hoh, F. C., "QCD versus Spinor Strong Interaction Theory" in *Hadron '95 The 6th International Conference on Hadron Spectroscopy,* edited by M. C. Birse et al., World Scientific, 1995, pp. 368-370 and "On the Foundation of QCD and an Overview of the Spinor Strong Interaction Theory" in *Hadron Spectroscopy 7th International Conference,* edited by S-U. Chung et al., AIP Conference Proceedings 432, New York, 1997, pp. 181-184.

[2] Metsch, B., these proceedings.

[3] Barnes, T., these proceedings.

[4] Hoh, F. C., *Int. J. Theor. Phys.* **32** 1111-1133 (1993).

[5] Hoh, F. C., *Two-Spinor Hadron Theory, http://www4.tsl.uu.se/~ho*h 2001.

[6] van der Waerden, B., *Göttinger Nachrichten* 100-109 (1929).

E. On the Foundation of QCD and an Overview of the Spinor Strong Interaction Theory

(Reprinted from *Hadron '97* [H1])

F. C. Hoh

Dragarbrunnsg. 9B, 75332 Uppsala, Sweden

Abstract

Based upon its predictive power as well as its theoretical foundation, it is reasoned that QCD is not the correct strong interaction theory for hadrons. The meson and baryon wave equations of the spinor strong interaction theory are presented. Contact of this theory with data is good for a number of basic issues at low energies. At high energies, this theory possesses properties similar to those of QCD.

INTRODUCTION

The material presented at this meeting has been rich. It reinforces and makes clearer our picture and the status of hadron theory that were present in the Hadron '95 conference two yers ago in Manchester. This contribution can be considered as an updated and expanded version of a similar one to Hadron '95(1).

There are two distinct but incomplete approaches to account for data, dedicated largely to mesons so far, i.e., phenomenology and quantum chromo-dynamics(QCD).

In phenomenological models(2), some essential features of the Bethe-Salpeter equation and QCD are collected and certain inter-quark potentials are assumed. The models are mostly semi-relativistic and are separately constructed for application to specific low energy regions, where each of them is highly successful in accounting for data.

These models are rather independent of each other, pay no attention to gauge invariance considerations, cannot be derived from any first principles' formulation, such as QCD, and hence lack a self-consistent physical basis. We may at most hope that some of them may eventually constitute certain limits of a correct strong interaction theory, in the manner that the Rydberg formula preceeded the Schroedinger theory.

QCD is a Lorentz and gauge invariant theory. It has been shown to work well in the high energy region in which perturbational calculations can be carried out. In fact, QCD has its origin in that region. For low energy phenomena, however, the QCD equations are largely non-perturbative and must be approximated(3) or solved numerically on a lattice(4). In the former case, QCD's relatively weak predictive power has deteriorated. For instance, $f_0(1510)$ and $\xi(2230)$ were strong glueball candidates in Hadron '95(5) two years ago but are no longer so at this meeting(3,6). Confinement, on the other hand, has not been proved theoretically.

It has been conjectured that these approximations may be too coarse and the issues need be resolved by lattice calculations. These have been carried out to some extent with much effort. They have however led to little physical understanding(4). The predictions can be made to fit a relatively small number of data. On the whole, they are not convincingly supported by data or have too large error margins or require greater computing capacity.

The present status of hadron theory is therefore highly unsatisfactory.

ON THE FOUNDATION OF QCD

An important underpinning of QCD is provided by the three jet data from JADE experiments in 1979. One of the jets differed from the other two, could be associated with a spin one structure and was assigned to a jet spear-headed by a gluon. However, this underspinning is removed if the gluon is replaced by a diquark, which has a spin one part and is also not observable. This possibility has not been disproved to my knowledge.

On principle grounds, one may raise two debatable points, which also hold for some common phenomenological models. In the first place, why nature prefers a vector(gluon) force among the quarks and bypasses the more fundamental scalar and pseudoscalar forces? Second, why are the quarks described by Dirac's equation like any other observable fermion(a Newtonian particle in the spinless and low speed limit) is, in spite of the fact that they are not observable?

Conception and Description of a Particle

Our conception of a particle is in the simplest case a Newtonian particle, i.e., a point mass at some space-time point moving with a certain veocity, which depends upon this mass and the force acting on it. This is nearly automatically formed by our daily experiences and is so ancient and so ingrained in our minds that we are almost unconscious of it. This conception has been greatly dented by the rise of quantum mechanics, which attaches a wave property to the particle and renders its position and speed diffuse.

Nevertheless, this Newtonian conception remains with us intuitively and was taken over to apply to quarks, without paying much attention to whether quarks are observable or not. For some physicists, this may be construed as a slight oversight. Nature, however, is incapable of distinguishing an unintentional violation of her laws from an intentional one. She is relentless on such basic issues. If one particle is observable and another is not, they are to be described differently.

In QCD, this difference is provided by the hypothesis of colored quarks. I think that this hypothesis is a superficial one and serves only to "paper" over the non-observation of quarks. Further, quarks do not have to obey the spin-statistics theorem so that no color is needed in this regard. Beneath this "paper", QCD provides as a detailed description of quarks as that of an observable fermion. Such a detailed description is superfluous from the viewpoint of nature since it cannot be verified anyway. Nature is economical and concise; if a description can be dispensed with consistenly, it will be.

Two Approaches to Construct a Physical Theory

During the first part of this century, two different approaches to the construction of basic physical theories were much debated on. In the first and older one, advocated by Eddington and Einstein, one starts from basic and known physical concepts and general physical principles or postulates and builds up a mathematically self-consistent theory. This approach is largely based upon physical reasoning and is the common one. Thermodynamics, statistical mechanics and the theory of relativity were constructed in this way. The description of quarks in QCD and the Bethe-Salpeter equation also follow this approach in essence.

The second and newer one is associated with logical positivism, propounded by R. Carnap and others of the Vienna school in the 1920's. It downplays the significance of physical concepts and is more pragmatic. In practice, any self-consistent mathematical construction that leads to predictions that systematically account for data with sufficient coverage is a good theory. New mathematical entities entering the construction are subsequently associated with some physical concepts(words) which we may comprehend.

This approach is to a considerable degree guided by mathematic. It is in line with the Kantian thesis that "the thing-itself is unknowable". It is employed more seldom and only when we delve into regions where existing concepts fail and we can only rely on data and self-consistent mathematical representation. Older physical concepts are dropped if required by mathematical consistency. The Maxwell, Schroedinger and Dirac equations were constructed along this line.

Comparison to Earlier Theories

The mathematical formulation of the known basic disciplines, classical gravitation, classical electromagnetism, quantum mechanics, and quantum electrodynamics all began with differential equations. Integral formulations, i.e., Lagrangians, are later constructed from these equations. In QCD, however, one starts with a Langrangian; the differential equations derived from it are not useful since they contain open color.

Furthermore, a criterion of a good fundamental theory is its high degree of mathematical elegance. QCD does not appear to fulfill this criterion; its aesthetic level is much lower than those of the basic disciplines above.

On ground of its predictive power as well as its theoretical foundation according to the above, I think that QCD is not the correct strong interaction theory for hadrons and will in time be replaced by a correct such.

OVERVIEW OF THE SPINOR STRONG INTERACTION THEORY

It is against earlier versions of such a background that the spinor strong interaction theory was proposed a few years ago(7) and subsequently developed(8). Its construction follows the second approach above(logical positivism) and makes full use of the invisibility of quarks.

This theory is based upon manipulations of van der Waerden's two component spinors which are right- and left-handed and hence suitable to represent relativistic fermions such as quarks and neutrinos. Dirac's equations in this spinor form for quarks interacting with each other via pseudoscalar and scalar forces are used to construct hadron wave equations. Afterwards, the quark and diquark wave functions are discarded so that the resulting hadron wave equations are much simpler than the QCD and the Bethe-Salpeter equations. These meson and baryon wave equations are Lorentz and gauge invariant and read

$$
\begin{aligned}
&\partial_I^{a\dot{b}} \chi_{\dot{b}}^{f}(x_I, x_{II}) \partial_{II\, f\dot{e}} = \left(\Phi_P(x_I, x_{II}) - M_m^2\right) \psi_{\dot{e}}^{a}(x_I, x_{II}) \\
&\partial_{I\, \dot{c}b} \psi_{\dot{e}}^{b}(x_I, x_{II}) \partial_{II}^{\dot{e}d} = \left(\Phi_P(x_I, x_{II}) - M_m^2\right) \chi_{\dot{c}}^{d}(x_I, x_{II}) \\
&\Box_I \Box_{II} \Phi_P(x_I, x_{II}) = \tfrac{1}{2} g_A^2 g_B^2 \left[\psi_{\dot{b}}^{a}(x_{II}, x_I) \left(\chi_{\dot{a}}^{b}(x_{II}, x_I)\right)^*\right]
\end{aligned}
\qquad (1)
$$

$$
\begin{aligned}
&\partial_I^{a\dot{b}} \partial_I^{g\dot{h}} \chi_{\{\dot{b}\dot{h}\}}^{f}(I\ II) \partial_{II\, f\dot{e}} = -i\left(M_b^3 + \Phi_S(I\ II)\right) \psi^{\{ag\}}{}_{\dot{e}}(I\ II) \\
&\partial_{I\, \dot{b}c} \partial_{I\, \dot{h}k} \psi^{\{ck\}}{}_{\dot{e}}(I\ II) \partial_{II}^{\dot{e}d} = -i\left(M_b^3 + \Phi_S(I\ II)\right) \chi_{\{\dot{b}\dot{h}\}}^{d}(I\ II) \\
&\Box_I \Box_I \Box_{II} \Phi_S(I\ II) = \frac{1}{4} g_q^6 \left(\chi_{\{\dot{b}\dot{h}\}}^{f}(I\ II) \left(\psi^{\{bh\}}{}_{\dot{f}}(I\ II)\right)^* + c.c.\right)
\end{aligned}
\qquad (2)
$$

respectively. Here, ψ and χ are the hadron wave functions, $x_{\mathrm{I}} = \mathrm{I}$ and $x_{\mathrm{II}} = \mathrm{II}$ the quark coordinates(I for diquark in equation (2)), M the average quark mass, and g's strong pseudoscalar(meson) and scalar(baryon) coupling constants which can however be absorbed into ψ and χ. Two of the quarks in baryon is taken to form a diquark with respect to the strong force. In this way, the three body problem of the baryon is reduced to a much simpler two body one.

The method by which equations (1) and (2) are constructed is heuristic and intuitive. Justification of these equations lies ultimately in their ability to account for data.

Equations (1) and (2) naturally lead to confinement and a new hadron classification scheme. The ground state hadron spectra fit well into this scheme, which leaves no unobserved states. Scalar mesons, much discussed in the literature, during Hadron '95(5) and at this conference, do not exist according to equation (1) because the quark are not confined. Equation (1) also largely successfully addresses the following basic problems: the ground state meson spectra, the U(1) problem, the absence of Higgs bosons, the Cabbibo and Weinberg angles, the conserved vector current hypothesis, and radiative decays of some open flavor vector mesons.

Internal functions $\xi(z)$ dependent upon internal coordinates $z=z_{\mathrm{I}}$, z_{II} have been introduced into equations (1) and (2) via the substitutions ψ, χ->$\psi\xi(z)$, $\chi\xi(z)$. These functions contain flavor dependence and are needed to distinguish some members of different SU(3) multiplets. The M's are eigenvalues of the corresponding mass operator operating on $\xi(z)$. For electromagnetic interactions, $z=z_{\mathrm{I}}$, z_{II}, z_{III}, has been introduced into equation (2) and may play the role of color in QCD. In the high energy region, equations (1) and (2) possess asymtotic freedom type of property and may therefore not contradict QCD results. No work has as yet been done in that region.

The spinor strong interaction theory is by no means free from problems. A basic problem has been the determination of integration constants which accompany these higher order equations (1) and (2) and the associated Lagrangians.

REFERENCES

[1] Hoh, F. C., "QCD versus Spinor Strong Interaction Theory" in *Hadron '95, The 6th International Conference on Hadron Spectroscopy,* 1995, pp. 368-370.

[2] Barnes, T., Godfrey, S., Iachello, F., Metsch. B., these proceedings.

[3] Swanson, E. S., these proceedings.

[4] Michael, C., these proceedings.

[5] Pennington, M. R., "Hadron '95-Summary: Part I" in *Hadron '95, The 6th International Conference on Hadron Spectroscopy,* 1995, pp. 3-18.

[6] Shen, X., these proceedings.

[7] Hoh, F. C., *Int. J. Theor. Phys.* **32**, 1111-1133 (1993), **33**, 2325-2349 (1994).

[8] Hoh F. C., *Int. J. Mod. Phys.* **A9**, 365-381 (1994), *J. Phys.* **G22**, 85-98 (1996), *Int. J. Theor. Phys.* **33**, 2351-2363 (1994), **36**, 509-531 (1997), "Radiative Decay of Vector Meson V->P+Gamma in the Spinor Strong Interaction Theory", "The Spinor Strong Interaction Theory and Strong Decay of Heavy Vector Meson V->PP", "The Weinberg Angle and Pion Beta Decay in the Spinor Strong Interaction Theory", submitted.

F. QCD Versus Spinor Strong Interaction Theory

(Reprinted from *Hadron '95* [H2])

F. C. Hoh

Dragarbrunnsg. 9B, Uppsala, Sweden

QCD is critically viewed in regard to predictive power, comparison to other basic disciplines and historical development. Reasons are given that it cannot be a good strong interaction theory. A brief introduction to the recently developed spinor strong interaction theory is presented. The large blocks of data successfully covered by it are pointed out.

1. Quantum Chromodynamics

As an outsider to the high energy physics community, I was struck, even surprised, by the passivity and uncritical attitude toward QCD at this meeting. There are staunch supporters of QCD. There are those who go along and say that QCD is a good strong interaction theory but has not yet reached the stage of making low energy predictions. But I heard no one actively voicing that QCD may be basically wrong and that alternative strong interaction theories ought to be sought.

This attitude is present in face of the fact that QCD, two decades after its proposal, can practically make no predictions on basic hadronic phenomena at low energies. Since QCD cannot be proven to be confining, there are no basic equations for hadrons. Therefore, hadron spectra and decay and other low energy phenomena cannot be predicted on a firm theoretical basis. Predictions in these areas are phenomenological and semi-relativistic and hence unreliable. If QCD is to be adhered to more strictly, complicated computer calculations are required to obtain even the simplest meson spectra approximately and sparsely. A prediction characteristic to QCD, the glueballs, has on the other hand not been confirmed by data. At higher energies, the non-Abelian gauge properties of QCD lead to asymtotic freedom and some success has been met.

The predictive power of QCD as a basic theory for strong interactions is orders of magnitude below those of equivalent basic disciplines, such as QED, C(lassical)ED, classical gravitation, and to some extent the standard model. At low energies and for simple configurations, the basic equations of these disciplines reduce to very simple forms that account for bulk of basic data with natural ease and elegance. Only at high energies or for complicated geometries do they exhibit calculational difficulties and their validity becomes questionable.

It has sometimes been argued that QCD differs from the above disciplines in that the coupling constant is large. However, this should not be an obstacle in treating stationary states such as meson spectra. For example, if the fine structure constant is allowed to be unity, the Schroedinger equation for a hydrogen atom still yields the same form of spectra although with different values. Difficulties arise first when higher order effects are considered. Another example is fluid mechanics, although it is not a basic discipline in the above sense. The governing equations are highly nonlinear but are still capable of producing stationary solutions to simple geometries rather easily.

That QCD, contrary to the other basic disciplines, claims some success at high energies but fails at low energies is possibly related to its development during the early 1970's, when existence of free quarks was not firmly rules out. Color was originally introduced to appease the spin statistics theorem applied to the delta++ baryon consisiting of three u quarks. Later it was thought to offer a means for confinement as well. Around that time, a renormalizable non-Abelian gauge theory proved to be highly successful in the standard model. Impressed and supported by the great success of QED as a renormalizable gauge theory, QCD was composed to incorporate the above features for quarks. Not much attention was paid to low energy phenomena such as hadron spectra.

In the two decades that follow, non-existence of free quarks has been well established. The need for color drops out since the spin statistics theorem which holds for observable particles does not have to hold for non-observable quarks. Further, quark wave functions satisfying Dirac type of equations no longer need to appear in the theory. I believe that many physicists will agree that nature does not "play games" by offering us the elegant Dirac equation that describes observable particles, such as leptons, and at the same time also describe non-observable particles like quarks, irrespective color. In a sense, some basic motivations underlying QCD has been removed.

2. Spinor Strong Interaction Theory

The spinor strong interaction theory[1-4] has been developed in the above background. It follows the development of the above-mentioned disciplines, except for QCD, and emphasizes the low energy end. The theory is so named because it is based upon van der Waerden's two component spinors. These are either left or right handed and hence suitable for description of quarks and leptons which are essentially relativistic. The commonly used Dirac four component bispinors are more suited for approaching the non-relativistic limit, but lead to cumbersome transportation of some gamma matrices when describing quarks and leptons.

Quarks obeying Dirac's equations are assumed to act upon each other via massless scalar forces. For quark-antiquark interaction, the force is pseudoscalar. The quark wave functions are generalized to include internal functions and the quark masses and charges to operators operating upon these internal functions. The so-generalized Dirac equations and associated massless interaction potential equations in spinor form are multiplied into the corresponding ones for the other quark or quarks. The resulting product wave functions and operators are subsequently generalized to non-separable quantities[1].

In the resulting hadron equations, the remaining quark wave functions are put to zero on account of that they have not been observed; they have been used as "scaffolding" in constructing the hadron equations and removed afterwards. Baryon is assumed to consist of a quark-diquark pair structure and the 64 components of its Bethe-Salpeter amplitude are reduced to 12. The usual 16 components of the meson Bethe-Salpeter amplitude are reduced to 8. The differential equations for the interaction potentials are of fourth and sixth orders which lead to linear and harmonic confinement for meson and baryons, respectively.

The so-obtained meson equations[1] have proved to be successful in classifying the known two quark mesons[3]. The ground state scalar and axial vector mesons are not confined and hence do not exist. The two quark pseudoscalar isosinglets such as eta/b are forbidden by U(1) gauge invariance considerations. For mesons without angular excitation, the basic meson equations can be reduced, without approximation, to radial equations like that for the hydrogen atom yielding a simple mass formula. Using a number of meson masses to fix the quark masses and as input, the remaining masses of unmixed mesons are mostly correctly predicted.

Lorentz and SU(2)XU(1) gauge invariances of the meson equations have been established[1,2]. It is shown that the kaon doublet plays the role of the Higgs doublet in generating the masses of the gauge bosons, thus rendering the Higgs bosons unnecessary. The large body of rather successful results of leptonic interactions predicted by the standard model can thus be carried over to hold in the spinor strong interaction theory[4]. For semi-leptonic interactions, the standard model is far less successful. In the spinor strong interaction theory, however, no unknown Cabbibo angle is needed to account for the ratio of the kaon and pion rates of decay to muons. The equivalent Cabbibo angle is correctly predicted.

At high energies, the quark mass and interaction potential terms in the meson and baryon equations can be neglected. The non-separable hadron wave functions become now separable into nearly free quark wave functions while retaining non-Abelian gauge invariance in the meson case. This is similar to the asymtotic freedom of QCD.

3. SUMMARY

A comparison of the predictive power of the over two decade old QCD to that of the newly developed spinor strong interaction theory so far show that the latter power is much greater and more natural than the former at the low energy end. At high energies, both theories lead to that quarks are nearly free but no work has yet been done employing the latter theory.

REFERENCES

[1] F. C. Hoh, *Int. J. Theor. Phys.* **32**, 1111 (1993), **33**, 2325, 2351 (1994).

[2] F. C. Hoh, *Int. J. Mod. Phys.* **A9**, 365 (1994).

[3] F. C. Hoh, "Meson Classification and Spectra in the Spinor Strong Interaction Theory", *J. Phys.* G, to appear.

[4] F. C. Hoh, "Meson Lepton Interaction in the Spinor Strong Interaction Theory", submitted.

G. Epistemological and Historical Implications or Elementary Particle Theories

(Reprint of [H16])

F. C. Hoh

Abstract

The epistemological credo of Einstein is further developed and specified in greater detail for practical applications. The results are applied to quasi-historical creation processes of established physical theories and to current theories for elementary particles. The outcome of these applications is then considered from epistemological viewpoints. Reasons underlying the basic difficulties of the current main stream particle theory, the standard model, are given. The steps in the creation of an alternative approach, the scalar strong interaction hadron theory, are delineated.

1. Introduction

The current main stream elementary particle theory is the standard model(SM) [1, 2] which includes quantum chromodynamics(QCD) and the electro-weak interaction model(EWM). Despite the claim that QCD's predictions agree with many accurate experiments [2], QCD cannot account for low energy hadronic data, such as those given by the Particle Data Group(PDG) [1]. In spite of the claim that EWM has been "wonderfully successful" [2], the Higgs boson, on which EWM hinges, has not been seen.

In view of these difficulties, an alternative to low energy QCD and EWM, the scalar strong interaction hadron theory(SSI) has been proposed [3-5]. This theory naturally accounts for many basic low energy hadronic phenomena that cannot be accounted for by QCD and EWM. The approach of SSI however drastically differs from that of SM and appears to be unfamiliar to most physicists.

To clarify this situation, therefore, EWM, QCD and SSI will be examined from two basic points of view. One of them is provided by epistemology or theory of knowledge, which in earlier centuries has been the main branch of philosophy but is presently much less dominant. Another one is given by the history of the creation of the established physical theories.

The epistemologies of interest are sketched in Appendix A. The teachings of these are generally accepted and are, is short, the common knowledge that theory needs be tested against experiment. But how this is practiced varies over a wide range of rigor. Allowing epistemology to bear upon the creation of physical theories will sharpen this practice. As the early Wittgenstein expressed: "The object of philosophy is the logical clarification of thoughts. Philosophy is not a theory but an activity. A philosophical work consists essentially of elucidations. The result of philosophy is not a number of philosophical propositions, but to make propositions clear".

In Sec. 2, Einstein's epistemological belief is further developed and specified in greater detail for practical applications. The results are then applied to creation processes of established physical theories from historical points of view in Sec. 3. In Sec. 4, the same procedure is carried out for EWM, QCD and SSI. The results of these studies are then discussed from epistemological viewpoints in Sec. 5. Sec. 6 gives the main conclusion of the present investigations. For reference, the epistemologies of interst are sketched in Appendix A. Appendix B presents difficulties of QCD. In Appendix C, the steps leading to SSI are given making use of Sec. 2-3.

2. Further Development of Einstein's Exposition

For the present investigations, the epistemologies mentioned in Appendix A need be specified in greater detail. Closest to the present application is Einstein's exposition in §A3, which will be the starting point. To obtain practical rules for application, the ideas of §A3 are extended and the results are represented in Fig. 2.1 below. His statements Ea) in §A3, which is equivalent to LPb) in §A2, is expanded to *domain Data* and Eb) in §A3, which is equivalent to LPa) in §A2, to *domain Logic*. His *connection* between Ea) and Eb) is expanded to an intervening *domain Physics*. These domains are set up in order to avoid possible mix-up of mathematical and physical concepts.

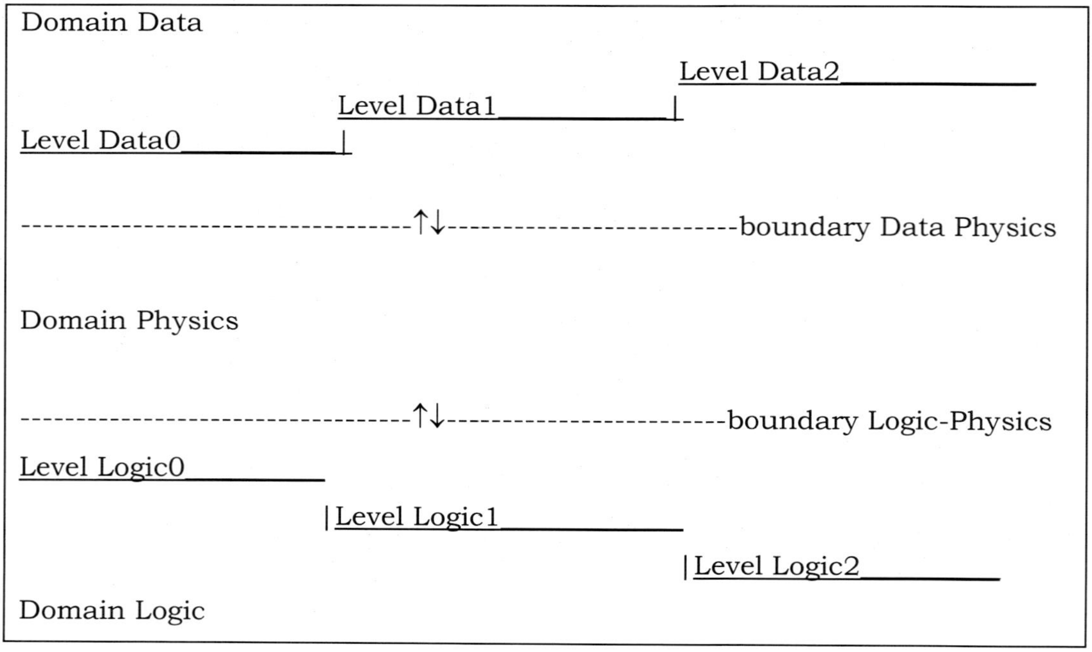

Figure. 2.1. It shows the three domains separated by two boundaries . The ↑↓ signs indicate that the contents in domains Data and Logic can be moved across the boundaries Data-Physics and Logic-Physics, respectively, into domain Physics and be brought back from it. Roughly speaking, the vertical distance between two levels indicates the distance between measurement and basic theory. The horisontal axis is some measure of time since renaissance. The left, middle and right parts of this figure are further considered in §5.2, 5.3 and 5.4 below.

2.1. Domain Data

It consists of several levels Data0, Data1, Data2,... The highest level contains raw experimental data, which by a series of treatments, which can for instance include Monte-Carlo calculations, are converted to quantities on lower levels and finally to the lowest level Data0. This level contains data in form of quantities that can be identified with corresponding quantities on level Logic0 in §2.2 below when brought into domain Physics. Algebraical combinations of such quantities are also on this level. Examples of the content of level Data0 are coordinates of a particle, temperature of a substance, magnetic field, electric current density, mass and decay time of a hadron, and in general data given by PDG [1].

2.2. Domain Logic

It contains only logic and pure mathematics. Since Newton, such logical or *mathematical structures* have been differential equations for all known and established basic physical disciplines. For the sake of clarity, all symbols in domain Logic are attached by a prime ´ which denotes that these symbols are mathematical quantities. The definitions, statements, propositions, and equations need only be self-consistent and obey conventions of logic, but are otherwise free from any restraint. They are exact but have no physical meaning and belong to LPa) in §A2 and Eb) in §A3 below. This domain presently consists of three levels, Logic0, Logic1 and Logic2.

Level Logic0 contains mathematical structures(differential equations) whose variables, or some quantities formed algebraically from them, when brought into domain Physics, can all be associated with and tested against the corresponding data quantities from level Data0. Examples are Newton's equations, thermodynamical relations and Ampére's law which becomes on this level

$$\nabla' \times \underline{B}'(\underline{x}') = \underline{j}'(\underline{x}') \tag{2.1}$$

This mathematical structure states that a vector field $\underline{j}'$(not electric current density here) is defined as the curl of another vector field $\underline{B}'$ in a flat three dimensional space $\underline{x}'$.

Level Logic1 is more abstract. The mathematical structures(differential equations) therein will now be the source of the physics. Here, some *hidden dependent variables* which, after crossing the boundary Logic-Physics, have no corresponding quantities from level Data0 to identify with and compare to. Examples of such variables are the Schrödinger wave function

ψ' and the metric tensor $g'_{\mu\nu}$ in general relativity. By mathematical manipulations the content on this level can be converted to that on level Logic0.

Level Logic2 is still more abstract. The mathematical structures (differential equations) therein will now be the source of the physics. However, in addition to hidden dependent variables, there are also some *hidden independent variables* and functions of them which, after crossing the boundary Logic-Physics, have no corresponding quantites from level Data0 to identify with and compare to. Examples of such variables are the relative space time between two quarks and coordinates in flavor space in SSI(see Appendix C). By mathematical manipulations, the content on level Logic2 can be converted to that on level Logic1 and, if necessary, subsequently to level Logic0.

2.3. Domain Physics

The contents or data on level Data0 are temporarily be brought across the boundary Data-Physics into domain Physics in Figure 2.1. Simultaneously, a *mathematical structure* on level Logic0 is also temporarily carried across the boundary Logic-Physics and thereby loses the primes ´ attached to the symbols therein and becomes a *theory* in domain Physics. The mathematical symbols in this theory are now intuitively assigned to various physical quantities. The theory has now physical meaning but is no longer exact. It is then worked out and tested against the corresponding data quantities in this domain. If the test is successful, the theory can be, but does not have to be, "right"(see §A4). If there is disagreement or conflict, the theory in domain Physics and hence also the corresponding mathematical structure in domain Logic need be changed and more experiments may be called for. The contents in domain Physics are then brought across the boundaries Data-Physics and Logic-Physics back to levels Data0 and Logic0, respectively.

It is also possible to bring the mathematical structures on levels Logic1 and Logic2 directly arosss the boundary Logic-Physics into domain Physics to becomes physical theories, which can then be reduced to theories corresponding to those on level Logic0. This reduction was done in domain Logic in §2.2. These both ways of reduction are equivalent but the latter is simpler conceptually because one has only to follow the rules of logic and can leave physics aside in the meantime.

2.4. Development of Physical Theories

The history of the development of physical theories shows that the creation of a new theory can be considered to proceed in the following stages in terms of activities in the three domains of §2.1-3.

Stage I: *Recent data* This stage takes place in domain Data in which some new or recent data become available.

Stage II: *Conflict* This stage takes place in domain Physics. The *Recent data* are brought across the boundary Data-Physics into domain Physics. Simultaneously, the associated *existing mathematical strutcure* in domain Logic, preferably reduced to level Logic0, is also carried over the boundary Logic-Physics into domain Physics to become the *existing theory*, which is then tested against the *Recent data*. A *Conflict* between the existing theory and the recent data arises. The existing theory is then returned into domain Logic to resume its role as the existing mathematical strutcure.

Stage III: *Leap* This stage takes place in domain Logic. The *Recent data* in domain Physics in Stage II are taken across the boundary Logic-Physics to become a *recent mathematical requirement* in domain Logic. The *Conflict* in domain Physics becomes an *Inconsistency* in domain Logic between this recent mathematical requirement and the existing mathematical strutcure. A *Leap* or a change is then introduced into the existing mathematical strutcure to obtain a modified or *new mathematical strutcure* consistent with logical rules. This new mathematical strutcure may now be consistent with the recent mathematical requirement.

Stage IV: *Hypothesis-Theory* The *Leap* made in domain Logic is brought across the boundary Logic-Physics to become a physical *Hypothesis* in domain Physics. The *new mathematical structure* becomes in this domain a *new theory* that may now agree with the *Recent data*.

If however there is still conflict, the above cycle or part of it is repeated. New experiments and new theories can be needed. The development of physical theories may be considered to consist of the chain D-P, L-P, L-P, D-P, D-L-P, and so on, where D-L-P stands for "activities in domain Data followed by activities in domain Logic followed by activities in domain Physics".

The leap made in Stage III takes place in domain Logic and has no physical meaning. It is therefore completely arbitray, as long as logic is maintained, and can be done in an infinite number of ways. Such a leap or a guess cannot be derived in any way. The choice of the leap obvioully takes into account the *Recent data* and is such that the new theory will hopefully remove the *Conflict* in Stage II or agree with new data. If there is no data that can guide the choice, the simplest form of the new mathematical structure may be chosen, as is indicated in Ed) ff of §A3 below.

The choice of new experiments is also in principle arbitrary but is largely aimed at testing the new theories. New technologies can also lead to new or more accurate experiments and thereby produce new or more accurate data. After these experiments, the content on level Data0 will be new *Recent data* which, upon crossing the boundary Data-Physics, tests the theory again.

3. Historical Survey of Physical Theories

3.1. Criteria in the Formation of New Physical Theories

The steps in §2.4 are now applied to the creation processes of established physical theories in §3.2 below. The result of this application shows that in the creation processes the following criteria CR1 and CR2 are satisfied.

CR1: Between Stages II and III,

a) the only new quantities crossing the boundary Logic-Physics into domain Logic are the *Recent data*, which become recent mathematical requirements to be satisfied by performing suitable leaps in the existing mathematical structure. In particular,

b) no physical quantities or concepts foreign to the existing theory in form of any physical hypothesis made in domain Physics make this crossing to become foreign mathematical symbols, relations and requirements that need be compatible with the existing mathematical structure in domain Logic.

CR2: The new theory satisfy the requirements of

a) pragmatsim in §A1,

b) the original verifiability principle of logical postivism LPa) and LPb) in §A2 and

c) Einstein's criteria Ec) and Ed) in §A3.

3.2. Historical Cases

In this subsection, the creation processes of the physical theories below are sketched in terms of the stages in §2.4. The criterion CR2 is considered to be satisfied in each of the following well-established cases. The criterion CR1 and the levels in domain Logic will be considered separately below.

1) Copernican universe ~1543

Recent data: Astronomical data were acquired for astrological purposes.

Conflict: Increasing number of geocentric models conflicted with each other and disagreed with data. Astronomers disagreed as to whether Venus and Mecury were inside or outside the orbit of the sun. The civil calendar had fallen seriously out of alignment with the sun's positions.

Leap-Hypothesis: The existing mathematical or logical structures are those pertaining to the various geocentrical models. The recent mathematical or logical requirements originate in *Recent data*; CR1 is satisfied. The leap-hypothesis consists of new assignments of the roles of the heavenly bodies and the replacement of geocentrical models by the heliocentric model so that the inconsistencies corresponding to the *Conflict* are removed. The new logical structure remains on level Logic0 because the quantities involved all have their counterparts on level Data0.

2) Newtonian Mechanics ~1687

Recent data: They were observations by Kepler and others.

Conflict: There was no existing theory in form of differential equations; hence there was no conflict.

Leap-Hypothesis: The definitions and assumptions, to which the mathematical structure of Newton's equations together with the inverse square law of gravity can be reduced to, are taken as the leaps. The new mathematical structures are differential equations and are on level Logic0 because the quantities involved therein, when moved across the boundary Logic-Physics into domain Physics, have all counterparts on level Data0. Since there was no conflict, CR1 does not apply.

3) Thermodynamics and Statistical Mechanics 1700's and late 1800's

Recent data: They were work on heat engines for the first topic. For the last topic, chemical and mechanical experiments brought down the caloric theory of heat so that the atomistic view of matter could regain ground.

Conflict: There were no viable existing theories in form of differential equations in these areas; hence there was no conflict.

Leap-Hypothesis: In thermodynamics, it is the impossibility to create perpetuum mobile. Although the concept of caloric was used, it did not cross the boundary Logic-Physics into domain Logic to become a recent mathmatical requirement; CR1 is not violated in practice. The quantities in the equations of thermodynamics are obviously on level Logic0. For statistical mechanics, it is the atomic hypothesis and mechanical nature of heat. The quantities in the associated differential expressions can all be measured and are hence on level Logic0. Since there was no conflict, CR1 does not apply.

4) Electromagnetism ~1873

Recent data: They were data summarized by the laws of Coulomb, Ampère and Faraday.

Conflict: Ampère's law disagreed with conservation of charge.

Leap-Hypothesis: The existing mathematical strutcures in domain Logic are the above three laws in form of differential equations(with prime ´ attached to the symbols), which may be considered to have been moved from domain Data via domain Physics into domain Logic. CR1 is therefore satisfied. Among these existing mathematical strutcures, Maxwell discovered an inconsistency corresponding to the *Conflict*. The hypothesis corresponding to the leap consists of adding a displacement current(derivative of the electric field with respect to time) to the Ampère law (2.1), whereby the inconsistency is removed. The quantitites in Maxwell's equations are all measurable and the new mathematical structure is therefore on level Logic0.

Later developments showed that the electric and magnetic fields can be written as derivatives of a vector potential A_μ which in domain Logic satisfies

$$\left(\nabla'^2 - \frac{1}{c^2}\frac{\partial^2}{\partial t'^2}\right)A'_\mu\left(\underline{x}',t'\right) = j'_\mu\left(\underline{x}',t'\right) \tag{3.1}$$

where j_μ denotes the electric current density and c the speed of light. These equations are simpler but are of second order while the Maxwell equations are of first order. A_μ can however not be observed directly and hence do not have any counterpart on level Data0 and (3.1) is therefore assigned to level Logic1. The mathematical manipulation that converts A'_μ to the fields on level Logic0 is derivation.

5) Quanta 1900, 1905

Recent data: The black body radiation spectrum was accurately measured and fitted empirically to a formula, the Planck law.

Conflict: This spectrum could not be calculated from clasical physics at that time, which could also not be modified to obtain this law.

Leap-Hypothesis: The established thermodynamics and Boltzmann's statistical mechanics can be brought across the boundary Logic-Physics to become the existing mathematical structures in domain Logic. Apart from that, only the Planck law is brought across the same boundary to become the recent mathematical requirement; CR1 is satisfied. The inconsistency between them corresponding to the *Conflict* is eliminated by introducing the leap $E'=h\nu'$ into the existing mathematical structures. The variables in Planck's law are all measurables and the new mathematical structures are therefore on level Logic0.

The physical meaning of $E=h\nu$ came in1905, when Einstein noticed that the entropies of low intensity monochromatic radiation and of an ideal gas varied in the same way with the volume and inferred that light behaves also like material particle.

6) Special relativity 1905

Recent data: Accurate measurements of the velocity of light c showed that it was the same in different inertial systems.

Conflict: c = constant was inconsistent with absolute time in Newtonian mechanics.

Leap-Hypothesis: The existing mathematical structure is that pertaining to Newtonian mechanics and Galileo transformations. Only c = constant is taken across the boundary Logic-Physics to become the recent mathematical requirement; CR1 is satisfied. The hypothesis corresponding to the leap consists of removal of the above absolute time concept and allow time to be different in different inertial systems. The inconsistency associated with the *Conflict* is now removed in the new mathematical structure pertaining to Lorentz transformations. This structure remains on level Logic0.

7) General relativity 1915

Recent data: They were accurate measurements showing that inertial and gravitational masses were equivalent.

Conflict: Lorentz transformations could not accomodate this equivalence.

Leap-Hypothesis: The existing mathematical structure is that pertaining to special relativity. The recent mathematical requirement originates in the equivalence of inertial and gravitational masses; CR1 is satisfied. The first leap consists of replacing Lorentz transformations, in which $g_{\mu\nu}$ are constants, by general nonlinear transformations $g_{\mu\nu}(x_\sigma)$ and thereby the inconsistency due to the *Conflict* is eliminatd. This is followed by the leap in setting the Ricci tensor $R_{\lambda\nu}(g_{\mu\nu}(x_\sigma)$ and its derivatives up to second order) = 0, which is the simplest possible choice, in accordance with CR2c)Ed) ff in §A3, because there was no guidance from data. g_{00} $-1(g_{00}{}^{-1})$ is identified as the Newtonian gravitational potential, when it is weak, and hence does not have counterpart on level Data0; only the derivative of the gravitational potential is measurable and can appear on level Data0. The other components in $g_{\mu\nu}(x_\sigma)$, identified as the generalized gravitational potentials can analogously not be observed directly. The new mathematical structure and the leap are therefore on level Logic1. The mathematical manipulation that converts $g'_{\mu\nu}(x'_\sigma)$ to fields on level Logic0 is derivation, analogous to the conversion of A'_μ to electric and magnetic fields.

8) Quantum Mechanics 1923-26

Recent data: Compton(1923) confirmed that light wave behaves also like material particle.

Conflict: de Broglie(1924) pointed out that this would be inconsistent with that material particle were not considered as wave.

Leap-Hypothesis: Debye said that a wave equation is needed. Schrödinger followed this up and considered the table below, which refers to domain Physics [6]. Boxes 2 and 3 can be taken across the boundary Logic-Physics into domain Logic to beome the existing mathematical structures. The content of *Recent data* is similarly brought across the same boundary to become the recent logical requirement; CR1 is obeyed. This requirement is fulfilled by the leap that creates the logical counterpart of box 1, which can be derived from box 2. Now an inconsistency corresponding to the *Conflict* arises and is removed by the leap creating box 4. Because the contents of box 1 and box 3 have the same form(Hamilton 1834), the form of box 4 is therefore the same as that of box 2. Since the material particle in classical mechanics is a scalar, let its wave function be a scalar function ψ. Replacing A_μ in box 2 by ψ and make use of p in box 3, the Schrödinger equation is obtained in box 4. These last steps take place in domain Physics but can also be considered to have taken place in domain Logic, analogous to the both alternatives mentioned at the end of §2.3. The Schrödinger equation is on level Logic1 because ψ' replaces A_μ' in (3.1).

	Particle Description	Wave Description
Light	*Box 1:* Eikonal equation of geo-metrical optics in a medium with refractive index n	*Box 2:* Wave equation (3.1) with no prime, j_μ=0, light speed c replaced by $u=c/n$, and frequency ν
Material Particle	*Box 3:* Hamilton Jacobian equation with energy E, potential V, mass m, and mementum $p=(2m(E-V))^{1/2}$	*Box 4:* An analogous wave equation with wave length $\lambda=u/\nu=h/p$; the Schrödinger equation

The manipulation that converts ψ' to quantities on level Logic0 is the formation of expectation values.

9) Relativistic Quantum Mechanics 1928

Recent data: These were summarized by Schrödinger's equation and special relativity.

Conflict: Schrödinger's equation was inconsistent with special relativity.

Leap-Hypothesis: The Schrödinger equation is taken across the boundary Logic-Physics to become the existing mathematical structure. The same is done to Lorentz transformations which become the recent mathematical requirement; CR1 is satisfied. The inconsistency derived from the *Conflict* is eliminated by the leap that requires that the Schrödinger equation be modified to a new mathematical structure linear in momenta in order to conform to special relativity. The new mathematical structure is that pertaining to Dirac's equation (C1.2) and is similarly on level Logic1. The extra ψ' components are on the same level as ψ' in the previous case. The manipulation that converts ψ' to quantities on level Logic0 is the same as that in case 8).

10) Parity nonconservation 1956

Recent data: The $\tau(K_L)$ and $\vartheta(K_S)$ mesons were found to be the same meson but their decay products had different parities.

Conflict: These results were inconsistent with the view at that time that parity was always conserved.

Leap-Hypothesis: The existing mathematical structure is that pertaining to Dirac's equation or more generally quantum electrodynamics(QED) which conserves parity. The recent mathematical requirement is derived from *Recent data* so that CR1 is satisfied. It corresponds to nonconservation of parity. This inconsistency corresponding to *Conflict* is eliminated by the leap that removes the requirement of parity conservation in the existing mathematical structure when applied to weak interactions. The new mathematical structure later evolved into the Glashow-Salam-Weinberg(GSW) [7] model for electro-weak interactions. It is on level Logic1, similar to the previous case.

In summary, the criterion CR1, where applicable, is satisfied by all the above classical cases. The mathematical structures for classical physics in cases 1-6 are om level Logic0. Those for general relativity and quantum physics in cases 7-10 are on level Logic1.

4. EWM, QCD AND SSI

The steps applied to the above ten cases are now also applied to the current main stream particle theories EWM and QCD and also to SSI.

4.1. EWM 1960's

Recent data: They were experimental results that follow §3.2.10) and the experimental verification of the $W^{\pm}$ and Z gauge bosons.

Conflict: The then existing GSW model did not allow finite masses of these gauge bosons.

Leap-Hypothesis: The existing mathematical structure is that pertaining to the GSW model. The recent mathematical requirement is derived from *Recent data* and CR1 is satisfied in this respect. According to the *Conflict*, it is inconsistent with the existing mathematical structure due to the finite masses of the gauge bosons. In §3.2, such inconsistencies have been overcome by leaps performed in the existing mathematical structure in domain Logic. In EWM, however, the hypotheses that the symmetry of vacuum is spontaneusly broken and vacuum consists of Higgs condensate are made in domain Physics. These are then brought across the boundary Logic-Physics to become additional recent mathematical requirements. Because these hypotheses contain concepts absent in the existing GSW structure and in *Recent data*, CR1b) is *violated.* Moreover, Higgs boson is not present on level Data0 because it has not been seen. Consequently, it also violates CR2b)LPb) and CR2c)Ec). Further, quark wave functions are put on par with lepton wave functions in the GSM model, contrary to the observation that "no free quark exists" in §4.3 below. Thus, it can be shown that the creation of the GSM model itself violates CR1. The new mathematical structure involved is on level Logic1, just like that for the case §3.2.10).

4.2. QCD Early 1970's

Recent data: Experiments on hadrons in the 1950's and 1960's led to the quark(fractionally charged spin ½ point particle) hypothesis. Nucleons were found to contain point-like objects, called partons, which could not be shown to be quarks. Although no free quark had been seen, its absence was not fully established experimentally at that time. Therefore, the interpretation was that "quarks are confined" which implied that quarks existed but were somehow confined(see §B3 below). It differs from that "no free quark exists" in §4.3 below. There were also a large amount of data on hadrons.

Conflict: The then existing QED could not account for quark confinement and other hadronic data.

Leap-Hypothesis: The existing mathematical structure is that pertaining to QED. However, the content of *Recent data* about quarks is not taken directly across the boundary Logic-Physics into domain Logic to become recent mathematical requirements, as in the cases of §3.2. Instead, this content is modified into the hypotheses that quarks come in 3 colors and interact with each other via gauge fields having 8 colors in domain Physics. These are then brought across the boundary Logic-Physics to become recent mathematical requirements and relations as are seen in (B1) below. Because these physical quantities and concepts are absent in the existing QED structure and in *Recent data*, CR1b) is *violated.* Further, these colored objects are not present on level Data0 because they cannot be observed. Therefore, they also violate CR2b)LPb) and CR2c)Ec). QCD also violates CR2a), CR2b)LPb) and CR2c)Ec) because QCD cannot account for low energy hadronic data in PDG [1]. The new mathematical structure involved is formally on level Logic1 but there are no straightforward mathematical manipulations that can convert this structure to level Logic0.

Additional difficuties of QCD are given in Appendix B.

4.3. SSI 1990's

Recent data: Data showed convincingly that "no free quark and no free diquark exist". This differed from the interpretation that "quarks are confined" in §4.2 above. There were also a huge amount of data on hadrons.

Conflict: The existing QED, in form of Bethe-Salpeter(BS) equation, could not account for confinement and other hadronic data.

Leap-Hypothesis: The existing mathematical structure is that pertaining to the BS equation The content of *Recent data* regarding quark and diquark, when taken across the boundary Logic-Physics into domain Logic, becomes the recent mathematical requirement that wave functions to be associated with quarks and diquarks cannot be present in the new mathematical structure. The construction of SSI is given in Appendix C and §C5 shows that CR1 is satisfied. The inconsistency related to confinement in *Conflict* is removed in [3]. The new mathematical structure involved is on level Logic2.

Summarizingly, the criteria CR1 are CR2 are satisfied by SSI but not by EWM and QCD. The mathematical structures of EWM and QCD are fomally on level Logic1. For SSI, it is on level Logic2.

5. Interaction of Particle Theories with Epistemology

5.1. Hypotheses on Vacuum

Vaccum can carry gravitational and electromagnetic energies and waves and can yield particle-antiparticle pairs. In the late 1940's, vacuum was found to be slightly polarizable by electromagnetic fields and such effects are exactly calculated in QED. The virtual particles in QED are however *not new*, but have observable counterparts.

Comtemporary hadron physicists believe that vaccum can also carry other types of *new, not observed* media. In EWM, vacuum is assumed to be a Higgs condensate. For color confinement in QCD, vacuum is assumed to be a perfect color dia-electric. These virtual media have however no observable counterparts and these two vacua hence differ from the QED vacuum in this respect.

Attempts to resort to vacuum as a means to explain physical phenomena are not new. Over a century ago, the "vacuum" at that time was thought to be filled by a universal ether in which light propagates and this view was widely accepted by physicists. It persisted in spite of the fact that Maxwell's theory of light does not require any ether but was finally brought down by special relativity.

Over two centuries ago, the then "vacuum" was analogously thought to contain a hypothetical weightless fluid known as caloric representing heat. This caloric hypothesis helped Carnot to arrive at his discoveries in thermodynamics and was widely accepted in the 1700's. But by the mid-1800's, many kinds of experiments showed that heat is a form of mechanical energy and the caloric concept had to be abandoned.

These earlier "vacua" however did not interfere with the mathematical formalisms of Huygens and Carnot. The difficulties were comceptual. The "vacua" in EWM and QCD also differ from these earlier ones in that they are specific and characterize the mathematical formalisms.

Nevertheless, this is some resemblance between these two "vacua" and the two earlier ones. The caloric concept held for more than a century. The history of ether was even longer. Judging from these time scales, the concepts of vacuum as Higgs condensate and color dia-electric may continue to prevail for some time to come. However in a few years, when the Large Hadron Collider will deliver data, the existence of Higgs boson will face a decisive test.

5.2. Level Logic0

As was mentioned at the end of Sec. 3, the mathematical structures(differential equations) for physical theories up to about 1910 are on level Logic0. As was defined in §2.2, level Logic0 can be "seen" form level Data0 via domain Physics. The left part of Figure 2.1 concerns these levels. The concepts are therefore all familiar and "Gedanken" or thought experiments, so forcefully employed by Einstein and Bohr, could be performed. The results of algebraical manipulations on this level can likewise be "understood" in terms of known concepts.

5.3. Level Logic1 and Its Implications

The mathematical structures for general relativity and quantum physics are on level Logic1. Because level Logic1 is "hidden" from direct view from level Data0 across domain Physics, new degrees of freedom become possible. In quantum mechanics, such a freedom is expressed in form of the existence of an arbitrary local phase associated with a wave function which is a hidden dependent variable. This arbitrary phase in its turn necessitates a U(1) gauge field, identified as the electromagnetic field.

In addition, purely quantum mechanical effects having no counterparts on level Data0 before quantum mechanics, such as exchange effects, tunnelling, Bose-Einstein condensation, etc, emerge. These new physics comes from mathematical operations on level Logic1 and cannot be "understood", like we "understand" Newtonian mechanics, because the associated concepts are unrelated to the then known concepts on level Data0(see §5.5 below). Words like "tunneling"(osmosis or infiltration could as well be used instead) are names assigned to certain mathematical results, confirmed by data, so as to give us some feeling what they are like in terms of familiar concepts.

Einstein's standpoint, that quantum mechanics offers no useful point of departure for future development, was based upon the hypothesis of "spatial separability", a result of "Gedanken" experiment, applied to Schrödinger's ψ [8]. This viewpoint is untenable here; "Gedanken" experiments make use of familiar concepts on level Logic0 and cannot be reliably initiated from level Logic1. This hypothesis was introduced in domain Physics and hence violates CR1b).

The purely quantum mechanical results can only be accepted as new concepts that one 'gets used to" them after some time.

5.4. Level Logic2 and Its Implications

This section concerns the right part of Figure 2.1 which commences in the 1990's. Level Logic2 is still farther away and more "hidden" from direct view from level Data0, as was predicted in Ed) ff of §A3. Therefore, still more freedoms are allowed on this level, as is evidenced by choices of the forms of quark-antiquark interaction, the mass operators and the meson internal or flavor functions in Appendix C. Here, the example set in §3.2.7) and the "inner perfection" criterion Ed) of §A3 come to aid.

The hidden independent variables on this level allow for new physics not obtainable on level Logic1. These include the relative space $\underline{x}$ and relative time x^0 between quarks(see (C5.1) below) which give rise to confinement and finite $W\pm$ and Z boson masses(without Higgs boson), respectively, in SSI considered in Appendix C.

The internal or flavor coordinates z_I, z_{II} in (C5.2-3) below can be considered to be on par with the "hidden" relative space coordinate x mentioned above. Thus, the mass operator $m_{2op}(z_I,z_{II})$ in (C5.3) may for instance be generalized to $m_{2op}(z_I,z_{II},x)$ such that Lorentz invariance is preserved. The internal or flavor space $z_Iz_{II}^*$, when real, and the relative space $\underline{x}$ between quarks are now coupled and this may lead to additional new physics. This may be related to that the masses of mesons with closed flavors($z_Iz_{II}^*$ real) differ somewhat from the predictions in Table 5.5 of [5].

5.5. Physics Comes Out of Mathematics

The results obtained from levels Logic1 and Logic2 cannot be "understood" at first. This is due to that we can only "understand" by referring back to familiar concepts. However, the physics that arises from mathematics on these two levels are associated with new, unknown concepts which can therefore not be "understood". A theorist's task is to provide a concise "account", not "explanation", of data, here using partial differential equations and their solutions. Whether the new concepts in the account can be "explained" or "understood" or not is of no concern in the beginning. Eventual "understanding" may come later when people get used to such new concepts because they work.

CR1 of §3.1 does not allow physical quantities or concepts outside the existing theory and *Recent data* in domain Physics to be "put into" domain Logic. From the new mathematical structure obtained in domain Logic, however, comes new physical concepts when this structure is brought into domain Physics to become a new theory. As V. Weisskopf has pointed out: physics comes out of mathematics.

Examples are time dilatation and Lorentz contraction on level Logic0, red shift, curved space time, quantum mechanical exchange effects and tunneling, and Bose-Einstein condensation on level Logic1, and confinement and mass generation for the $W^{\pm}$ and Z gauge bosns(without Higgs boson) in SSI on level Logic2. Apart from those on level Logic0, these new concepts cannot be obtained by manipulations of the old concepts, such as "Gedanken" experiments or constructions of phenomenological models because these can only contain known, hence not new, concepts. The Bohr-Sommerfeld models and potential models for hadron spectra are examples.

The *Conflict* in Stage II of §2.4 is not resolved in that stage in domain Physics. It is resolved in Stage III, in which the *Conflict* turns into an *Incocsistency* that is eliminated by the *Leap*. This leap is the *turning point* in the stages of development in §2.4 and is of purely logical nature or some general principle because Stage III takes place in domain Logic.

In the cases of §4.1-2, however, physical quantities and concepts(Higgs bosons, colored objects) foreign to the existing theories are created in Stage II in domain Physics in form of hypotheses which are then "forced" into domain Logic in Stage III. This violates of CR1 of §3.1. These concepts have their origins in analogies with existing physical concepts in solid state physics and are basically old. In fact, any physical hypothesis created in this stage can invariably only contain known, hence old, physical concepts because these are the only ones we have at our disposal.

These specific, second guesses of nature fix beforehand the directions of subsequent developments in domain Logic in Stage III and thereby deprive nature of the possibility to take its own course through mathematical developments from a more "loose" and general new mathematical structure, observing CR1, as in the cases of §3.2. CR1 means that physical hypotheses cannot "move" in the direction from domain Physics in Stage II into domain Logic in Stage III to become leaps or additional recent mathematical requirements. If it is not heeded, the new mathematical structure will be forced to live with forms of old concepts that may thwart the emergence of genuinely new physics. This is also the case for phenomenological models which are therefore stop-gap theories having narrow, specific application angles.

The direction of "motion" is the opposite. It is the *Leap,* which is logical and not physical, made in Stage III in domain Logic that "moves" into domain Physics in Stage IV to be suitably interpreted as a physical *Hypothesis.*

5.6. Impact of Quark Physics on Our Conception of the Universe

Up to a century ago, people's daily experiences and concepts formed from them originated in the Newtonian world. Time was absolute, space Euclidean and physics continuous and deterministic. The parameter region in which this Newtonian conception of the universe holds were subsequently expanded by new experiments and this conception had to be modified in the new parameter regions.

Thus, at high speeds, the above space and time become relative and interconnected to become space-time according to special relativity. Considering the large masses of the heavenly bodies, the above space-time become connected to mass to become space-time-mass or "curved space-time" in accordance to general relativity. In the atomic region, the continuous and deterministic conceptions of the universe become discrete and statistical described by quantum mechanics.

The emergence of quarks and the theories to account for them again introduce new conceptions in the subnuclear region of parameters. The new feature is the "hidden relative space-time x" and "hidden complex flavor space z"(see §C5 and §C1 below) or shortly "hidden space" necessitated by that quarks are not observable individually. Such "hidden spaces", or synonymously "hyperspaces", are not limited to SSI here but have to be present in any realistic hadron theory containing quarks at different space-times so that the first line of (C5.1) below holds. There is at least one set of "hidden spaces" or "hyperspaces" associated with each hadron. Thus, *our conception of the universe now includes infinitely many such "hyperspaces", in which essential physics takes place.*

Conventionally, hadrons are conceived to have finite sizes in laboratory space. However, such sizes are inferred from experiments involving collisions and do not necessarily apply to a hadron not under interaction. Such a hadron is described by the wave equations (C5.2) or (C6.1) and (C5.1) applies. After having solved these eqautions and integrated over the "hyperspaces", the wave functions reduced to the exponential form in (C5.1) which describes a free point particle in laboratory space X^{μ}. This puts the hadron on equal footing with the lepton; the difference is that the hadron, but not the lepton, is accompanied by "hidden relative spaces".

Thus, *the four dimensional space time X^{μ} we live in contains only vacuum and point particles and is "empty"*. This conception is not new but has for instance been arrived at long ago by buddists in akin form.

Classical physical theories were largely constructed from observed data. Thus, Maxwell set out to put Faraday's experimental results in mathematical form which became part of his theory and Kepler's data played role in the formation of Newton's theory. The newer ones, such as those mentioned in §3.2-7), §3.2-8) and §4.3, were no longer constructed in that way but were "guessed mathematically", based upon earlier mathematical forms, hints from data, intuition, and the criteria of §A3Ec)-Ed) below.

The basic language of nature may be considered to be solutions to partial differential equations. New concepts derived from these equations represent the basic features of that part of nature pertaining to these equations and have no counterparts in the Newtonian world. This is why ordinary people, including some philosophers not familiar with the creation process of newer physical theories, find the new physics difficult to "understand".

6. CONCLUSIONS

Past experiences show that in the creation of a new physical theory, one does not put in physics, which invariably refer to known concepts, into new theories but relies on logic or some basic principles in the creation of the new mathematical structure. New physics and concepts then comes out of mathematics development from this structure.

These experiences in form of CR1 were not heeded in the creation processes leading to EWM and QCD, as was mentioned in §5.5. CR2 is also not fulfilled. Appendix B indicates further that difficulties of QCD are too fundamental to be overcome by modifications. From epistemological as well as historical viewpoints, the Higgs related part of EWM and low energy QCD therefore appear to be stop-gap theories to be replaced by a more correct theory, for which SSI may be a candidate.

APPENDIX A. RELEVANT EPISTEMOLOGIES

Ever since renaissance, scientists and philosophers have influenced each other. Copernicus and Galilei influenced Descartes, who in his turn influenced Newton who influenced Kant. Hume and Mach influenced Einstein who influenced the Vienna Circle. Many of the philosophers had scientific background or started off as scientists. On the whole, science had more impact upon epistemology than vice versa.

Four epistesmologies of interest to the present investigations are sketched below for reference.

A1. Pragmatism [9]

"Pragmatism is a school of philosophy founded by Peirce(1877)... It is based on the principle that the usefulness, workability, and practicality of ideas and proposals are the criteria of their merit. Ideas borrow their meanings from their consequences and their truths from their verification.

Peirce's pragmatism is primarily a theory of meaning that emerged from his first-hand reflections on his own scientific work, in which the experimentalist understands a proposition as meaning that, if a prescribed experiment is performed, a stated experience will result. The method has two different uses: (1) It is a way of showing that when disputes permit no resolution, the difficulties are due to misuses of language, to subtle conceptual confusions. (2) The method may be employed for clarification. Consider what effects, that might conceiveably have practical bearings, we conceive the object of our conception to have. Then our conception of these effects is the whole of our conception of the object."

Peirce' also introduced the term *retroduction*, which means the forming and accepting on probation of a hypothesis to explain surprising facts. This was once his main theme of pragmatism.

A2. Logical Positivism [9]

Logical Positivism comes out of the Vienna Circle(1923-38) founded by Schlick. Einstein(special relativity) had significant impact on it. Its members paid much attention, firstly, to the form of scientific theories, in the belief that the logical structure of any particular scientific theory could be specified quite apart from its content. Second, they formulated a "verifiability principle" or criterion of meaning, a claim that the meaningfulness of a proposition is grounded in experience and observation. In its negative form, the principle said that no statement could both be a statement about the world and have no method of verification attached to it. In other words, all *meaningful* discourse consists either of

LPa) *the formal sentences of logic and mathematics* or

LPb) *the factual propositions of the special sciences.* Any assertion that claims to be factual has meaning only if it is possible to say how it might be verified. Metaphysical assertions, coming under neither of the two classes, are meaningless.

Because this principle is by itself not verifiable, they could only recommmend its use. During 1930-60, verifiability was replaced by a more tolerant version expressed in terms of testability or confirmability. The logical positivists continued to reformulate their criteria of factual meaningfulness. There are different versions; all of them are more lenient then the stringent original formulation.

A3. Einstein's Epistemological Credo [8]

Einstein saw on the one side

Ea) *the toality of sense*-experiences and

Eb) *the totality of concepts and propositions* on the other. The relations between the concepts and propositions among themselves and each other are of a logical nature and follow firmly laid down rules.

He wrote further: "The concepts and propositions get "meaning," viz., "content," only through their *connection* with sense-experiences. The connection of the latter with the former is purely intuitive, not itself of a logical nature. The degree of certainty with which this connection, viz., intuitive combination, can be undertaken, and nothing else, differentiates empty phantasy from scientific "truth". The system of concepts is a creation of man together with the rules of syntax, which constitute the structure of the conceptual systems. Although the conceptual systems are logically entirely arbitrary, they are bound by the aim to permit the most nearly possible certain (intuitive) and complete coordination with the totality of sense-experiences; secondly they aim at greatest possible sparsity of their logically independent elements (basic concepts and axioms), i.e., undefined concepts and underived (postulated) propositions... A proposition is correct if, within a logical system, it is deduced according to the accepted logical rules. A system has truth-content according to the certainty and completeness of its coordination-possibility to the totality of experience. A correct proposition borrows its "truth" from the truth-content of the system to which it belongs. Judgement of the theory is based upon the both conventions:

Ec) "external confirmation". The theory must not contradict empirical facts evenif its application can be quite delicate. For it is often, perhaps even always, possible to adhere to a general theoretical foundation by securing the adaptation of the theory to the facts by means of artificial additional assumptions...

Ed) "inner perfection". This is characterized by the "naturalness" or "logical simplicity" of the premises of the basic concepts and of the relations between these which are taken as a basis.

Among theories of equally "simple" foundation that one is to be taken as superior which most sharply delimits the qualities of systems in the abstract i.e., contains the most definite claims... We prize a theory more highly if, from the logical standpoint, it is not the result of an arbitrary choice among theories which, among themselves, are of equal value and analogously constructed... *in the choice of theories in the future will have to play an all the greater role the more the basic concepts and axioms distance themselves from what is directly observable, so that the confrontation of the implications of theory by the facts becomes constantly more difficult and more drawn out...* A theory is the more impressive in the greater the simplicity of its premises is, the more different to kinds of things it relates, and the more extended is its area of applicability."

A4. Feynman's Summary

Today's physicists follow the above epistemologies to varying degrees without thinking about them. These teachings have been summarized, among others, in Feynman's lectures:

"One guesses a theory and computes its consequences. If the results disagree with experiment, this theory is wrong. If the results agree with experiment, it does not mean that this theory is right."

APPENDIX B. ON THE FOUNDATION OF QCD

Apart from the remarks in §4.2 that the creation of QCD does not satisfy CR1 and at least part of CR2, the following points show that the foundation of QCD is far from being firm.

B1. QCD at High Energies

That QCD is allegedly successful at high energies does not mean that it is right(§A4). Further, this allegation is not stringent. An often cited agreement with data comes from deep inelastic scattering experiments but the test is on the level of structure function, which contains assumptions so that the derived results are no longer firmly anchored in first principles.

The circumstance may be compared to that classial mechanics agrees with data at high angular momenta but breaks down at small angular momenta and has to be replaced by quantum mechanics. Somewhat similarly, QCD holds perturbatively at high energies but becomes nonperturbative around 1 Gev and cannot produce predictions without resorting to assumptions or phenomenological models. In the low energy region, QCD analogously needs be replaced by another theory; SSI may be a candidate. Such a theory, at higher energies, has to go over to a form that will yield results compatible to those of perturbative QCD; there may be an equivalent "correspondence principle" analogous to that taking quantum mechanics to classical mechanics.

B2. Experimental Evidences

An important underpinning of QCD is provided by the three jet data from the JADE experiments in 1979. One of the jets differed from the other two, could be associated with a spin one structure and was assigned to a jet spear-headed by a gluon. However, this underspinning is removed if the gluon is replaced by a diquark, which has a spin one and is also not observable. The diquark interpretation is consistent with the successful quark-diquark claasification of baryon spectra.

Further, the predicted glueball has not been observed.

The so-called quark-gluon plasma experiments start with nucleons whose mutual interaction is scalar. In the plasma state, however, the interaction among quarks takes place via colored vector gauge fields. It is not clear how these two widely different types of interactions can "convert" into each other.

B3. QCD Action

In the notation of [1], the QCD Lagrangian reads

$$L_{QCD} = -\frac{1}{4} F_{\mu\nu}^{(a)} F^{\mu\nu(a)} + i \sum_q \overline{\psi}_q^i \gamma^\mu \left(D_\mu\right)_{ij} \psi_q^j - \sum_q m_q \overline{\psi}_q^i \psi_{qi} \tag{B1}$$

$$F_{\mu\nu}^{(a)} = \partial_\mu A_\nu^a - \partial_\nu A_\mu^a - g_s f_{abc} A_\mu^b A_\nu^c \tag{B2}$$

$$\left(D_\mu\right)_{ij} = \delta_{ij}\partial_\mu + ig_s \sum_a \frac{\lambda^a_{i,j}}{2} A^a_\mu \tag{B3}$$

where the $\psi_q^i(x)$ are the 4-component Dirac spinors associated with each quark field of (3) color i and flavor q, and $A_\mu{}^a(x)$ are the (8) Yang-Mills (gluon) fields.

As was mentioned in §2.2, the starting points of all established physical disciplines are all differential equations. These can subsequently be converted into action integrals. QCD, however, also breaks this rule in that it starts with an action integral with Lagrangian density (B1). Equations of motion derived from the QCD action involve quantities on level Logic0 such as colored electric and magnetic fields that have no corresponding quantities on level Data0 and can hence not be measured.

That (B1) contains quark wave fucntions $\psi_q^i(x)$ reflects the view of *Recent data* in §4.2 that quarks exist but are somehow confined. Here, quarks are treated as if they can be observed, contrary to experience, because one can in principle form expectation values, on level Logic0, of dynamic variables using these $\psi_q^i(x)$ on level Logic1. Confinement may be regarded as a "papering over" via the attachment of three colors indices i to each $\psi_q(x)$. This is different from the *Recent data* for SSI in §4.3, where the interpretation "no free quark exists" is adopted. This is reflected in that the hadron wave equations (C5.2), (C4.5) and (C6.1) below contain no quark wave functions.

Practically, there are also no simple ways to obtain hadron wave functions from (B1-3) so that CR2a) is not fulfilled.

B4. Applicabiblity of "Asymtotic Freedom" [10]

In QCD, phenomenological confinement is obtained if the vacuum is a perfect color dia-electric. This is allegedly supported by "asymtotic freedom". But this result was derived from an extension of QED(U(1), Abelian), which provides a tiny screening effect, to quarks belonging to an SU(n) multiplet(non-Abelian), which led to anti-screening. Now the charged leptons are directly observable point particles interacting via a U(1) gauge field in QED. The derivation cannot be simply taken over to apply to the not directly observable quarks. From this viewpoint, this "asymtotic freedom" provides no support to confinement.

APPENDIX C. SSI(SCALAR STRONG INTERACTION HADRON THEORY)

The background leading to SSI [3, 4] has been given in [5]. Here, it is summarized taking Sec. 2 and §3.1 into consideration. Unless specified to be otherwise, the symbols and formulae below are taken to refer to hypothetical cases and are within domain Logic but with the primes ´ dropped for simplicity.

The physics mentioned below serve to aid in the choices of leaps which are in principle arbitray and hence infinte in number on level Logic2, so that when the completed new mathmatical structure is taken across the boundary Logic-Physics into domain Physics in

Figure 2.1, it becomes a new theory that can be tested favorably against data from level Data0.

C1. Internal or Flavor Space

The Dirac equation in classical form reads

$$\gamma^{\mu} p_{\mu} + m = 0 \tag{C1.1}$$

where p_μ is the 4-momentum and m the mass of a particle. When going over to quantum mechanics, the momenta p_μ become operators operating on a wave function so that (C1.1) becomes the Dirac equation

$$\left(-i\gamma^{\mu}\partial_{X\mu} + m\right)\psi(X) = 0 \tag{C1.2}$$

Now, both p_μ and m are measurable quantities and should be treated on equal footing. Therefore, m should also be generalized into an operator m_{op} operating on some function ξ. This conception originated in the early 1960's and hadron masses were regarded as expectation values of mass operators. This led to the successful classification of hadron masses by the Gell-Mann-Okubo formula [11]. Referring to quarks, (C1.2) is thus generalized to

$$\left(-i\gamma^{\mu}\partial_{X\mu} + m_{op}\right)\psi(X)\xi^{p}\left(z^{q}\right) = 0 \tag{C1.3}$$

where the superscripts p and q designates quark flavor and z^q represents a complex internal or flavor space of dimension equal to the number of quark flavors under consideration. To begin with, let p, $q = 1, 2, 3$ which refer to the u, d, and s quarks, respectively. These z^q coordinates are not observable and hence are hidden independent variables on level Logic2. They constitute the basis vectors generating the first fundamental representation of the group of global SU(3) transformations. $\xi^p(z)$ transforms as z^p, but its form, as well as that of m_{op}, are arbitrary provided that

$$m_{op}\xi^{p}\left(z^{q}\right) = m_{p}\xi^{p}\left(z^{q}\right) \tag{C1.4}$$

where m_p is the mass of the quark with flavor p. Thus, (C1.2) can be replaced by (C1.3) and (C1.4).

Following the thesis Ed) of §A3, employed in §3.2-7), the simplest forms of the quantities in the eigenvalue equation (C1.4) are

$$\xi^{p}(z) = z^{p}, \qquad m_{op} = \sum_{q} m_{q}\left(z^{q}\frac{\partial}{\partial z^{q}} + c.c\right) \tag{C1.5}$$

where m_{op} will be called a mass counting operator. These assumptions are analogous to the leap in the general relativity case of §3.2.7) and are in accordance with Ed) ff in §A3; the simplest forms are chosen in the absence of physical requirements.

C2. van der Waerden's Equations

Since the advance of special relativity, the basics equations of classical mechanics and electromagnetism have been written in invariant(tensor) forms which garantee their Lorentz invariance. Dirac's equation (C1.2) is however not in such a form. This has remedied by van der Waerden [12] who rewrote this equation in the form

$$\partial_X^{a\dot{b}}\chi_{\dot{b}}(X)=im\psi^a(X), \qquad \partial_{Xb\dot{c}}\psi^{\dot{c}}(X)=im\chi_{\dot{b}}(X) \tag{C2.1}$$

where a and b are spinor indices running from 1 to 2 and ψ^a and $\chi_{\dot{b}}$ are two-spinors. (C2.1) is by its spinor form manifestly invariant under the SL(2,C) group of transformations, which include Lorentz transformations. The relations between the undotted and dotted spinors and Dirac's wave functions ψ_1, ψ_2, ψ_3, and ψ_4 in (C1.2) are

$$\begin{aligned}&\text{Right handed}: && \chi_{\dot{1}}=\psi_1+\psi_3, && \chi_{\dot{2}}=\psi_2+\psi_4\\ &\text{Left handed}: && \psi^1=\psi_1-\psi_3, && \psi^2=\psi_2-\psi_4\end{aligned} \tag{C2.2}$$

ψ_3 and ψ_4 are small components which vanish in the relativistic limit and (C1.2) is suitable when considering nonrelativistic problems. Quarks are however relativistic and the two-spinor form of the wave functions in (C2.2) is more natural.

Taking the complex conjugate of (C2.1) yields for the corresponding antiparticle,

$$\partial_X^{b\dot{a}}\chi_b(X)=im\psi^{\dot{a}}(X), \qquad \partial_{X\dot{c}b}\psi^{\dot{c}}(X)=im\chi_b(X) \tag{C2.3}$$

Noting (C1.4), the equivalent of (C1.3) becomes for a hypothetical quark

$$\partial_X^{a\dot{b}}\chi_{\dot{b}}(X)\xi^p(z^q)=im_{op}\psi^a(X)\xi^p(z^q), \qquad \partial_{Xb\dot{c}}\psi^{\dot{c}}(X)\xi^p(z^q)=im_{op}\chi_{\dot{b}}(X)\xi^p(z^q) \tag{C2.4}$$

$$\frac{\partial_X^{a\dot{b}}\chi_{\dot{b}}(X)}{\psi^a(X)}=i\frac{m_{op}\xi^p(z^q)}{\xi^p(z^q)}=im_p, \qquad \frac{\partial_{Xb\dot{c}}\psi^{\dot{c}}(X)}{\chi_{\dot{b}}(X)}=i\frac{m_{op}\xi^p(z^q)}{\xi^p(z^q)}=im_p \tag{C2.5}$$

The last line shows that the quark mass m_p is a separation constant between the space time coordinates X and the internal or flavor coordinates z^q.

For the corresponding antiquark, one takes the complex conjugate of (C2.4) to obtain

$$\partial_X^{b\dot{a}}\chi_b(X)\xi_p(z_q)=im_{op}\psi^{\dot{a}}(X)\xi_p(z_q),\quad \partial_{X\dot{c}b}\psi^{\dot{c}}(X)\xi_p(z_q)=im_{op}\chi_b(X)\xi_p(z_q)$$

(C2.6)

$$z_q=(z^p)^*=(z_1,z_2,z_3) \tag{C2.7}$$

C3. Starting Equations

Under *Conflict* in §4.3, the BS equation was considered as the existing theory. Consider at first the meson case. Let the quark therein be labeled *A* and be located at x_I and the antiquark therein be labeled *B* and be located at x_{II}. The BS equation for a system of this type is of the form

$$\left(i\gamma_I^\mu\partial_{I\mu}-m_A\right)\left(i\gamma_{II}^\nu\partial_{II\nu}-m_B\right)\psi_{BS}(x_I,x_{II})=\text{interaction terms},\ \partial_{I\mu}=\frac{\partial}{\partial x_I^\mu},\ \partial_{II\nu}=\frac{\partial}{\partial x_{II}^\nu}$$

(C3.1)

The left side is obtained by multiplying the left side of (C1.2) applied to quark *A* by the same applied to antiquark *B* together with the genealization

$$\psi_A(x_I)\psi_B(x_{II})\to\psi_{BS}(x_I,x_{II}) \tag{C3.2}$$

The BS equation (C3.1) has successfully applied to positronium but obviouly fails for meson in which the quark-antiquark interaction is not electromagnetic. If the electron and positron have large enough energy, they are no longer bound and can hence be free. This is contrary to the meson case in which the quark and antiquark cannot be free. Therefore, the quark-antiquark interaction must be of another kind. Further, the BS amplitude (C3.2) has 16 components, far too many for application to mesons. This suggests that the form (C1.2) leading to (C3.1) does not provide a suitable starting point. As was pointed out below (C2.2), the van der Waerden form (C2.1) is more suited to represent quarks.

As to the choice of the quark-antiquark interaction, there are in principle infinte many possibilities. In the absence of physical requirements, the simplest form, namely scalar interaction, is chosen. This is entirely analogous to the reasoning underlying (C1.5) and the leap in §3.2.7) and is in accordance with Ed) ff in §A3. Further, such an interaction is suggested by the known scalar nucleon-nucleon strong interaction.

With these two departures from the BS equation, the present starting points are (C2.1) complemented by a scalar inteaction term *V* and applied to quark A at x_I and (C2.3), after raising and lowering indices, analogously complemented and applied to antiquark B at x_{II}. Instead of (C3.1), one starts with

$$\partial_I^{a\dot{b}}\chi_{A\dot{b}}(x_I)-iV_B(x_I)\psi_A^a(x_I)=im_A\psi_A^a(x_I) \tag{C3.3a}$$

$$\partial_{I\dot{b}c}\psi_A^c(x_I) - iV_B(x_I)\chi_{A\dot{b}}(x_I) = im_A\chi_{A\dot{b}}(x_I) \tag{C3.3b}$$

$$\square_I V_B(x_I) = \tfrac{1}{2}g_s^2\left(\psi_B^b(x_I)\chi_{Bb}(x_I) + \psi_B^{\dot{b}}(x_I)\chi_{B\dot{b}}(x_I)\right) \tag{C3.3c}$$

$$\partial_{II\dot{e}f}\chi_B^f(x_{II}) - iV_A(x_{II})\psi_{B\dot{e}}(x_{II}) = im_B\psi_{B\dot{e}}(x_{II}) \tag{C3.4a}$$

$$\partial_{II}^{d\dot{e}}\psi_{B\dot{e}}(x_{II}) - iV_A(x_{II})\chi_B^d(x_{II}) = im_B\chi_B^d(x_{II}) \tag{C3.4b}$$

$$\square_{II} V_A(x_{II}) = \tfrac{1}{2}g_s^2\left(\psi_A^b(x_{II})\chi_{Ab}(x_{II}) + \psi_A^{\dot{b}}(x_{II})\chi_{A\dot{b}}(x_{II})\right) \tag{C3.4c}$$

Here, g_s^2 is the scalar strong coupling constant for quark-antiquark or quark-quark interaction. The terms in (C3.3a, 3b) and (C3.4a, 4b) are grouped such that the left sides only contain operators in space time and the right sides contain the quark masses which refer to internal space, as is seen in (C1.4). Equations (C3.3c) and (C3.4c) are written such that the potentials are on the left side and wave functions on the right.

C4. Construction of Meson Wave Equations

Analogous to the formal construction of (C3.1), the left and right sides of (C3.3) are multiplied into the left and right sides, respectively, of (C3.4). There are 3×3=9 equations, in which the operators are placed to the left of the wave functions. Similar to (C3.2), the separable product wave functions are generalized into two-quark wave functions not separable in x_I and x_{II} according to

$$\chi_{A\dot{b}}(x_I)\chi_B^f(x_{II}) \to \chi_{\dot{b}}^f(x_I, x_{II}), \qquad \psi_A^c(x_I)\psi_{B\dot{e}}(x_{II}) \to \psi_{\dot{e}}^c(x_I, x_{II}) \tag{C4.1a}$$

$$\chi_b^{*\dot{f}}(x_I, x_{II}) = \left(\chi_{\dot{b}}^f(x_I, x_{II})\right)^*, \qquad \chi \to \psi \tag{C4.1b}$$

$$\chi_{A\dot{b}}(x_I)\psi_{B\dot{e}}(x_{II}) \to \chi_{\dot{b}\dot{e}}(x_I, x_{II}) \tag{C4.2a}$$

$$\psi_A^c(x_I)\chi_B^f(x_{II}) \to \psi^{cf}(x_I, x_{II}) \tag{C4.2b}$$

$$V_A(x_{II})V_B(x_I) \to \Phi_m(x_I, x_{II}) \tag{C4.3}$$

The mixed spinor $\chi_{\dot{b}}^f$ of (C4.1a) has four components, which ccorrespond to those of a conventional 4-vector. It is therefore decomposable into a singlet and a triplet, which can be assigned to the rest frame pseudoscalar and vector meson wave functions, respectively. The same holds for $\psi_{\dot{e}}^{c}$. The complex conjugates of the mixed spinors of second rank (C4.1b) are still mixed spinors and behave in the same way under SL(2,C) transformations.

χ and ψ on the left of (C4.2) can be transposed and this leads to that the right members of (C4.2) are symmetric spinors of second rank each having three components only. They can therefore not represent the pseudoscalar meson. In view of the indistinguishability of a quark and an antiquark inside a meson in the context mentioned at the end of Sec. 2.1 of Ch. 2 in [5], ψ^{cf} and $\chi_{\dot{b}\dot{e}}$ of (C4.2) are associated with a diquark or an antidiquark from the space time and transformation points of view. ψ^{cf} and $\chi_{\dot{b}\dot{e}}$ are put to zero here in accordance with that "no free diquark exists" under *Recent data* in §4.3.

The quark wave functions ψ and χ not paired off according to (C4.1) and (C4.2) are also put to zero, in accordance with that "no free quark exists" under *Recent data* in §4.3. This differs from the interpretation "quarks are confined" in §4.2 which led to the presence of quark wave functions in (B1). Accordingly, the unpaired V_A or V_B also vanishes by virtue of (C3.3c) and (C3.4c).

Applying (C4.1-3) to the 9 product equations and noting the above null results, 6 of them drop out. The three surviving equations arise from the products of (C3.3a) with (C3.4a), (C3.3b) with (C3.4b) and (C3.3c) with (C3.4c). The last product necessitates (C4.3) because the right side is no longer separable in x_I and x_{II} after application of (C4.1-2).

The result is the following three coupled meson wave equations,

$$\partial_I^{ab}\partial_{II\dot{e}f}\chi_{\dot{b}}^{f}(x_I,x_{II})+(m_Am_B-\Phi_m(x_I,x_{II}))\psi_{\dot{e}}^{a}(x_I,x_{II})=0 \qquad \text{(C4.4a)}$$

$$\partial_{I\dot{b}c}\partial_{II}^{d\dot{e}}\psi_{\dot{e}}^{c}(x_I,x_{II})+(m_Am_B-\Phi_m(x_I,x_{II}))\chi_{\dot{b}}^{d}(x_I,x_{II})=0 \qquad \text{(C4.4b)}$$

$$\Box_I\Box_{II}\Phi_m(x_I,x_{II})=-\frac{g_s^4}{4}\left(\psi^{b\dot{a}}(x_I,x_{II})\chi_{\dot{a}b}^{*}(x_I,x_{II})+\psi^{*a\dot{b}}(x_I,x_{II})\chi_{\dot{b}a}(x_I,x_{II})\right) \qquad \text{(C4.5)}$$

If the quark-antiquark interaction is chosen to be vector(gauge) instead of scalar, similar calculations show that no self-consistent set of meson wave equations like (C4.4-5) can be formed.

C5. Hidden Independent Variables, Level Logic2 and CR1

The quark and antiquark coordinates x_I and x_{II} are not observables. They are transformed into an observable laboratory coordinate X^μ for the meson and a relative coordinate x^μ according to

$$x^\mu=x_{II}^\mu-x_I^\mu, \qquad X^\mu=(1-a_m)x_I^\mu+a_mx_{II}^\mu$$
$$\chi_{\dot{b}}^{f}(x_I,x_{II})=\chi_{\dot{b}}^{f}(X,x)\to\chi_{\dot{b}}^{f}(x)\exp(iK_\mu X^\mu), \qquad \chi\to\psi \qquad \text{(C5.1)}$$

when a_m is a constant and K_μ the 4-momentum of the meson. The arrow refers to cases separable in X and x. x^μ is also a hidden independent variable and belongs to level Logic2. It cannot be observed so that the requirement that the "no free quark exists" in §4.3 is fulfilled. If x^μ were observable, the quark and antiquark coordinates x_I and x_{II} would also be observable according to (C5.1), contrary to data.

Analogous to the transition of (C2.1) to (C2.4) and of (C2.3) to (C2.6), $m_A m_B$ in (C4.4) becomes an operator m_{2op} operating on an internal function $\xi^p{}_r$,

$$\partial_I^{a\dot{b}}\partial_{II\dot{e}f}\chi_{\dot{b}}^f(x_I,x_{II})\xi_r^p(z_I^q,z_{IIs})+(m_{2op}-\Phi_m(x_I,x_{II}))\psi_{\dot{e}}^a(x_I,x_{II})\xi_r^p(z_I^q,z_{IIs})=0 \quad \text{(C5.2a)}$$

$$\partial_{I\dot{b}c}\partial_{II}^{d\dot{e}}\psi_{\dot{e}}^c(x_I,x_{II})\xi_r^p(z_I^q,z_{IIs})+(m_{2op}-\Phi_m(x_I,x_{II}))\chi_{\dot{b}}^d(x_I,x_{II})\xi_r^p(z_I^q,z_{IIs})=0 \quad \text{(C5.2b)}$$

z_I and z_{II} are associated with the quark and antiquark, respectively, just like x_I and x_{II} do. They are hidden independent variables(§C1). Therefore, The meson wave equations (C5.2) and (C4.5) [5] are on level Logic2 of §2.2.

The choice of m_{2op} is, similar to (C1.5), of the simplest form and is

$$m_{2op}=m_{1op}m_{1op}^*, \qquad m_{1op}=m_{1op}^*=\sum_q m_q\left(z_I^q\frac{\partial}{\partial z_I^q}+z_{II}^q\frac{\partial}{\partial z_{II}^q}+c.c\right) \quad \text{(C5.3)}$$

m_{op} in (C1.5) can be replaced by m_{1op} in (C5.3) without affecting (C1.4). This simple product form is approximate and needs be refined. The simplest choice of $\xi^p{}_r(z_I{}^q,z_{IIs})$ is analogously to combine the first of (C1.5) with (C2.7). Let $\xi^p{}_r$ be the octet part, to be associated with the octet mesons, in the decomposition 3×3*=1+8, it then constitutes the basis vectors that generate the regular representation of the group of global SU(3) transformations.

The leaps here, i. e., departures from the BS equation (C3.1) in §4.3, orignate in domain Logic. The introduction of the internal space z in §C1 is necessitated by the requirement that the particle mass and momenta, being both observables, should be put on equal footing. The use of (C2.1) is equivalent to the use of (C1.2). The choice of scalar interaction in (C3.3) and (C3.4) is the simplest one, following Ed) of §A3, and also takes place in domain Logic. In this domain, this leap can be considered as choosing (one of)the simplest form of coupling between the mathematical systems (C2.1) and (C2.3). The recent mathematical requirements obtained when the relevant content in *Recent Data* in §4.3 are taken across the boundary Logic-Physics, i. e., no free quark or diquark exists, are satisfied in (C5.2) and (C4.5). Thus, CR1 is fulfilled.

An alternative viewpoint is to discard the BS equation as an existing theory for hadrons. Then there is no conflict in §4.3 and CR1 does not apply, similar to the cases §3.2.3).

C6. Baryon Wave Equations

The above procedure can analogously be applied to baryons. Here, the problem is greatly simplified for ground state baryons which have been successfully classified by considering them to consist of a quark and a diquark. Let the subscript *II* refer to the diquark, the corresponding baryon wave equations read [3]

$$\begin{aligned}
&\partial_I^{a\dot{b}}\partial_I^{g\dot{h}}\partial_{II\dot{e}f}\chi^{f}_{\{\dot{b}\dot{h}\}}(x_I,x_{II})\xi^{\{ps\}q}(z_I^r,z_{II}^t)=-i(m_{3op}+\Phi_b(x_I,x_{II}))\psi^{\{ag\}}_{\dot{e}}(x_I,x_{II})\xi^{\{ps\}q}(z_I^r,z_{II}^t)\\
&\partial_{I\dot{b}c}\partial_{I\dot{h}k}\partial_{II}^{d\dot{e}}\psi^{\{ck\}}_{\dot{e}}(x_I,x_{II})\xi^{\{ps\}q}(z_I^r,z_{II}^t)=-i(m_{3op}+\Phi_b(x_I,x_{II}))\chi^{d}_{\{\dot{b}\dot{h}\}}(x_I,x_{II})\xi^{\{ps\}q}(z_I^r,z_{II}^t)\\
&\qquad\Box_I\Box_I\Box_{II}\Phi_b(x_I,x_{II})=\frac{1}{4}g_s^6\left\{\chi^{f}_{\{\dot{b}\dot{h}\}}(x_I,x_{II})\psi^{\{\dot{b}\dot{h}\}}_{f}(x_I,x_{II})+c.c.\right\}\\
&\qquad m_{3op}=(m_{1op})^3 \qquad\qquad (C6.1)
\end{aligned}$$

The simple product form of the last line is likewise approximate and needs be refined.

C7. Origin and Status of SSI

The quark wave functions in (C3.3-4) have been used as "scaffolding" in constructing the hadron wave equations (C5.2), (C4.5) and (C6.1) and are removed afterwards. It may seem that these equations have been arrived in ad-hoc ways. This is indeed the case. However, recall that all the operations in this Appendix take place in domain Logic and are completely free from restraint apart from self-consistency. Acceptance of these equations is solely based upon the test results. These results are obtained by mathematical manipulations of these equations on level Logic2 so that they can be moved to level Logic1, level Logic0 and finally across the boundary Logic-Physics in Figure 2.1 into domain Physics, where they become new theories and are tested by contents(PDG data) from level Data0. This is equivalent to the criteria Ec) of §A3, LPb) in §A2 and §A1.

If the tests are successful, it makes no difference how ad-hoc or odd the above construction processes of these equations may seem. In this case, one can in principle discard all the steps leading to (C4.5), (C5.2) and (C6.1) and imagine that they have been dreamt up or conferred on as revelation. This is not different from that students accept the Newton and Schrödinger equations because they have been successfully tested, irrespective of the ways these equations were arrived at.

Equations (C4.5), (C5.2) and (C6.1) for mesons and baryons, if right, play the same role as the one played by Dirac's equation for atoms. The strong interaction part resides largely in functions dependent upon the hidden indedendent variables, the relative space x and z_I and z_{II}, and is at first aimed at low energy phenomena not successfully covered by QCD. Introducing local phases into the X dependent part of the wave functions leads to massive $W\pm$ and Z gauge bosons without any Higgs boson.

Confinement arises naturally from (C4.5) and the third of (C6.1) as a result of their higher order nature. The role of Higgs bosons is taken over by the kaons or pions by virtue of the presence of the relative time x^0 among the quarks. Ground state meson spectra and some meson decay problems involving weak, electromagnetic and strong interactions have been rather successfully treated. Thus, the Dalitz slope parameters in $K \rightarrow 3\pi$ have been largely correctly predicted [5]. Further applications are presently hampered by mathematical complexities.

The rest frame ground state meson wave functions have been found in closed form but not the baryon wave functions due to mathematical complications associated with higher order differential equations. Assuming the existence of such functions, the baryon magnetic moments are largely correctly predicted. Further, angular momentum is found to be not conserved in free neutron decay [5]. Recently, this assumption has been substantiated and baryon wave functions have been obtained numerically as eigensolutions to these differential equations.

SSI has however not been treated at high energies. Also, quantization of SSI has not been investigated; the hidden independent variable aspects of SSI renders that the conventional procedures in quantizing nonlocal theories, which led to difficulties, need be reaccessed.

REFERENCES

[1] Particle Data Group 2004 Phys Lett **B592**, A short exposition of QCD and EWM is given there.
[2] Wilczek F 2005 *Nature* **433**(no. 7023) 239
[3] Hoh F C 1993 *Int J Theoretical Physics* **32** 1111, 1994 ibid **33** 2325
[4] Hoh F C 1996 *J Physics* **G22** 85, 1999 *Int J Theoretical Physics* **38** 2617
[5] Hoh F C 2005 *Scalar Strong Interaction Hadron Theory*. It has been shown at http://www4.tsl.uu.se/~hoh and is now seen at http://web.telia.com/~u80001955
[6] Goldstein H *Classical Mechanics* 1950 Addison-Wesley, p307-314
[7] Glashow S 1961 *Nucl Phys* **22** 579, Weinberg S 1967 *Phys Rev Lett* **19** 1264 and Salam A 1968 in *Elementary Particle Physics, Nobel Symp*(ed Svartholm N) **8** 367
[8] "Einstein's autobiography" in *Albert Einstein-Philosopher-Scientist* 1949, Library of living Philosophers, ed. P. A. Schilpp
[9] *Encyclopedia Britannica*
[10] Gross D and Wilczek F 1973 *Phys Rev Lett* **30** 1343, Politzer H 1973 ibid **30** 1346
[11] Okubo S 1962 *Prog Theor Phys* **27** 949
[12] van der Waerden B 1929 *Göttinger Nachrichten* 100

H. "Quasi-White Book" on the Particle Physics Community

This appendix does not deal with physics itself but gives my observations and views on the particle theory community in regard to the present theory. Furture science historians may find it to be of some interest.

The reactions to the scalar strong interaction hadron theory I received at meetings and visits to institutions range from avoidance to anger. The theory has been published in a series of papers in *Int. J. Theor. Phys.* except for an early one in *Int J Modern Phys. A* and another in *J. Phys. G.* The standard journals, including the last two journals lately have rejected them. The reason for rejection given by *Phys. Rev. Dk*(*PRD*) ranged from that it is too speculative and its wish to keep its readership to that there is no reason to provide any alternative to QCD. During 1992-2005, nine papers were submitted to and rejected by *PRD* and none was sent out to referee for review. Still, the Editor-in-Chief concludes that the hearing has been fair. *Nucl. Phys. B* said: "We choose not to publish your papers". *European Phys. J. C* defended QCD, considered a paper of mine as crackpot and asked me to refrain from submitting paper on scalar strong interaction hadron theory. Los Alamos' arXiv considered my book submission to be "inappropriate".

The nonacceptance may partly be due to that to penetrate the underlying ideas, procedures and work can require much time and effort, perhaps more than what the reviewers are willing to put down. But the following may be more important.

A science journalist recently wrote that all academic areas, including particle theory, are institutionalized so that anyone acts or even thinks differently from the main stream becomes ostracized. My background is plasma physics at first. In 1992, I presented the early part of this theory to a professor in theoretical physics in Stockholm, whom I knew decades earlier. He said that particle physics is a closed society, outsiders are not let in. When I pointed out QCD's inability to account for low energy data to another professor there, he agreed but said that this is an "internal critique", implying that it is not for "outsiders" like me to interfere. This professor abandoned QCD because it could not account for the baryon magnetic moments and that the SSC(superconducting super collider) project was dropped by the US government. He went into astrophysical particles. I met his young successor, who works with neutrino models, in 2008 and showed him my last paper Appendix G above, he said that QCD is not to be tampered with too much. He further said that "it is like the church", implying, I guess, that QCD has an religious aura and the particle physicists have to tow the line in order not to be ostracized.

It seems that there is a diffuse international establishment on particle theory that tacitly and largely controls the main, relevant journals and research grants. The "official" theory is QCD and heresy is not allowed. A local professor said angrily that I threw "shit" at QCD. Another local professor said that I have been regarded as a heretic and did not allow me to give seminar on the present theory. But privately many admit that low energy QCD is wrong. A third local professor said that I, as a financially independent retiree, can afford to say this. But if a young particle physicist says this, it will not be good for his career. This and the overcrowding(article coauthored by hundered of names) have contributed to that many formerly QCD physicist have switched to astro-particle physics, cosmology, supersymmetry, string theory, where theory cannot be checked by experiment, and to fields like condensed matter, biophysics, etc. This is a *unhealthy* situation. The task of particle theorists is foremost the accounting of the data given by the Particle Data Group and we should tackle it *first*, from a first principles' theory; the scalar strong interaction hadron theory is a candidate.

A 1979 paper in *Physica A* by Weinberg extolling the virtues of phenomenological Lagrangians appeared timely to the rescue of low energy QCD which turned out to be not solvable. Since then, a huge number of phenomenological models has been built; nearly every group or sometimes person has its own model. Piles of papers get published. Already in the 1960's, Wigner said that 70-80% of *Phys. Rev.* is either wrong or irrelevant. That percentage should be much higher now. Instead of searching for other first principles' theories, I think that paper has guided the development of particle theory in a short-sighted direction and hence greatly impeded real progress.

The spectacular sucesses of fission and fusion bombs have helped to propel high energy physics into big business, turning over big money and providing steady livelihood of large groups of people. Experimentalists like the Higgs, which calls for bigger machines and hence more money. A new theory that opposes QCD and does not need Higgs is bad news; it threatens future grants and stable funding. Scientific truth can wait, monthly payments cannot.

The early scientific idealism, which led to our present day knowledge, can be swept away by "business" concerns. When I worked in plasma physics in the 1960's, I made a feasibility study of thermonuclear fusion upon a request. The efficiency was found to be way too low to be of interest. When I mentioned this to a respected colleague, he said: "Don't rock the boat". With some modification, it seems that this attitude may be taken over to particle physics, where jobs are more important for the bulk of practioners.

Centruries ago, the then young physics was surpressed by the church, who's teachings were not to be challenged. It seems that modern particle theory as frontier of physics has itself grown big and old into an institution and plays some of the roles of the old church. If some future historian happens to look into this case, he may say that "history repeats itself". Well, not exactly, inquisition is out and websites are in.

I worked as an engineer and my work on the present theory has been carried out beside it and full time since retirement. I had free hands. Einstein did his early work in his spare time, and sometimes in his work during working time, and was completely unfettered. Such freedom is nowadays greatly curtailed in academic institutions where "business" considerations have partly turned academic freedom into academic constraint with strong peer pressure. In his autobiography, Einstein wrote: "That little frail plant (meaning some promising scientific idea) which apart from stimulation, mainly stands in need of freedom, without which it dies without exception". Incidentally, a sociologist recently remarked that Einstein is an "impossibility" today. He would not get any grant; a researcher nowadays depends upon outside network that support him.

Very basic new grounds are mostly broken and spearheaded by few individuals, not by the "masses", which almost by definition are followers. Galileo described his work by: "...like an eagle soaring high above....not 100 working horses but a full blod racing horse". This is so in spite of that new theories are often rejected for good reasons because most of them turn out to be not viable. A practical example is that only a tiny fraction of new ideas are worked out to patents, only a tiny fraction, 2-3%, of patents leads to products, and only a small fraction of them becomes commercial successes. Such small fractions exists analogously in scientific theories.

Under the editorship of David Finkelstein, the *International Journal of Theoretical Physic* had a distinguished editorial board and my papers were accepted. After his retirement around 2005, the new editor has swtiched to the now common peer review system. This system may work for papers working with the main stream type of theories. It may however not work for entirely new, basic ground breaking theories because most of the peers are in the main stream and may not be capable to judge an approach completely deviating from the conventional one. Further, peer pressure will make the referee or editor to ask what may happen to him if he lets an "outsider" in? It is safer to consider my papers unsuitable, inapproapriate, outside our scope… like the standard journals do. Not the least, the funding situation in the future may be adversely affected. Scientific advancement can wait, present power is what counts, as was the church's attitude in medival times.

It will take some time for a drastically new approach to sink in. As Roosevelt observed; those who lead cannot be too far ahead of those who are led. Similarly, Schöpenhauer wrote that people listen to those whom they know and are not too much "above" them. He also noted that all truths pass through three stages; ridiculed at first, then violently counteracted and finally accepted as self-evident. Hegel divides the development of a new idea into three stages, thesis, antithesis and synthesis. The present theory is perhaps now in the second stages of these developments.

REFERENCES

[A1] ALEPH Collaboration, Abe F et al 1993 *Phys Lett* **311B** 425

[A2] Aoki S 1986 *Phys Rev Lett* **57** 3136

[A3] Adler S L 1969 *Phys Rev* **177** 2426, Bell J S and Jackiw R *Nuovo Cim* 1969 **A60** 47

[B1] Bég M A B and Ruegg H 1965 *J Math Phys* **6** 677

[B2] Blatt J M and Weisskopf V F 1979 *Theoretical Nuclear Physics* Springer, Appendix B

[B3] Bebek C J et al 1978 *Phys Rev* **D17** 1693

[B4] Bjorken J D and Drell S D 1964 *Relativistic Quantum Mechanics* McGraw-Hill

[B5] Bergström L 1990 in *Rare Decays of Light Mesons* Ed Mayer B, Frontieres, 161

[B6] Belinfante F J 1953 *Phys Rev* **92** 997

[B7] Bethe H A and Salpeter E E 1957 in *Handbook der Physik* Springer, Vol. 35 §14, §1

[C1] Corson E M 1953 *Introduction to Tensors, Spinor and Relativistc Wave Equations* Blackie and Sons, Ch. 2

[C2] Condon E U and Shortly G H1963 *Theory of Atomic Spectra* Cambridge University Press, 76

[C3] Carruthers P 1966 *Introduction to Unitary Symmetry* Interscience

[C4] Coddington E A and Levinson N 1955 *Theory of Ordinary Differential Equations* McGraw-Hill

[D1] Dirac P A M 1928 *Proc Roy Soc* **A1**17 610

[D2] Dally E B et al 1977 *Phys Rev Lett* **39** 1176, 1980 *Phys Rev Lett* **45** 232

[F1] Frank M and O'Donnell P J 1985 *Phys Lett* **159B** 174

[F2] Fierz M and Pauli W 1939 *Proc Roy Soc* **A173** 211

[F3] Feynman R P, Kislinger M and Ravendal F 1971 *Phys Rev* **D3** 2706

[G1] Goldstein H 1950 *Classical Mechanics* Addison-Wesley, Ch. 9

[G2] Gell Mann M 1962 *Phys Rev* **125** 1067

[G3] Gell Mann M 1964 *Phys Rev Lett* **8** 214

[G4] Goldberg H 1980 *Phys Rev Lett* **44** 363

[G5] Gell Mann M, Sharp D and Wagner W G 1962 *Phys Rev Lett* **8** 261

[H1] Hoh F C 2001 in *Hadron '01, The 9th Int Conference on Hadron Spectroscopy* Ed Amelin et al, AIP Conf Proc 619, 611 and 1997 in *Hadron '97, The 7th Int Conference on Hadron Spectroscopy* Ed Chung et al, AIP Conf Proc 432, 181

[H2] Hoh F C 1995 in *Hadron '95, The 6th Int Conference on Hadron Spectroscopy* Ed Birse et al, World Scientific, 368

[H3] Hurley W J 1974 *Phys Rev* **D10** 1185

[H4] Huang K and Weldon H A 1975 *Phys Rev* **D11** 257

[H5] Hoh F C 1993 *Int J Theoretical Physics* **32** 1111

[H6] Hoh F C 1996 *J Physics* **G22** 85
[H7] Harai H 1976 1980 *Phys Lett* **60B** 172
[H8] Hoh F C 1994 *Int J Modern Physics* **A9** 365
[H9] Hoh F C 1994 *Int J Theoretical Physics* **33** 2351
[H10] Hoh F C 1999 *Int J Theoretical Physics* **38** 2647
[H11] Hoh F C 1997 *Int J Theoretical Physics* **36** 509
[H12] Hoh F C 1998 *Int J Theoretical Physics* **37** 1693
[H13] Hoh F C 1999 *Int J Theoretical Physics* **38** 2617
[H14] Hoh F C 2000 *Int J Theoretical Physics* **39** 1069
[H15] Hoh F C 1994 *Int J Theoretical Physics* **33** 2325
[H16] Hoh F C 2007 *Int J Theoretical Physics* **46** 269
[H17] Hoh F C 2010 *Natural Science* **2** 398 open access at *http://www.scirp.org/journal/ns*
[H18] Hoh F C 2010 "Scalar Strong Interaction Hadron Theory(SSI)— Kaon and Pion Decay" invited and accepted chapter in "*The Large Hadron Collider and Higgs Boson Research*", Editor: Christopher J. Hong, Nova Science Publishers, scheduled for publicatiion in 2011
[H19] Hoh F C 2010 *Natural Science* **2** 929 open access at *http://www.scirp.org/journal/ns*
[J1] Jauch J M and Rohrlich F 1955 *Theory of Photons and Electrons* Addison-Wesley, 86 ff
[J2] Jackson J D 1962 *Classical Electrodynamics* Wiley, §6.3
[J3] Jackson J D, Treiman S B and Wyld Jr. H W 1957 *Phys Rev* **106** 517
[J4] Joos 1962 *Fortscht d Phys* **10** 65
[K1] Kemmer N 1938 *Proc Roy Soc* **A166** 127
[K2] Kleima D 1965 *Nucl Phys* **70** 577
[K3] Källén G 1964 *Elementary Particle Physics* Addison-Wesley
[L1] Laporte O and Uhlenbeck G E 1931 *Phys Rev* **37** 1380
[L2] Lichtenberg D B 1978 *Unitary Symmetry and Elementary Particles* Academic Press
[L3] Lichtenberg D B 1987 *Int J Modern Physics* **A2** 1669
[L4] Ludwig W and Falter C 1988 *Symmetry in Physics* Springer
[L5] Lee T D 1981 *Particle physics and an Introduction to Field Theory* Harwood Academic Publisher, Ch. 21-22
[L6] Lising L J et al 2000 *Phys Rev* **C62** 055501
[L7] Lee T D and Yang C N 1956 *Phys Rev* **104** 254
[M1] Mitra A N 1981 *Z Phys* **C8** 25, Nimai Singh N and Mitra A N 1988 *Phys Rev* **D38** 1454
[N1] Ne'eman Y 1961 *Nucl Phys* **26** 222
[O1] Okubo S 1962 *Prog Theor Phys* **27** 949
[P1] *Amsler, C. et al.* 2008 Particle Data Group, *Physics Letters* **B667**, 1
[P2] Peskin M L and Schroeder D V 1995 *An Introduction to Quantum Field Theory* Addison-Wesley
[R1] Roman P 1960 *Theory of Elementary Particles* North Holland, Ch.1
[R2] Raczka R and Fischer J 1966 *Comm Math Phys* **2** 233
[R3] van Royen R and Weisskopf V F 1967 *Nuovo Cim* **A50** 617
[R4] Rose M L 1957 *Elementary Theory of Angular Momentum* Wiley, 34
[S1] Schweber S, Bethe H A and de Hoffman F 1955 *Mesons and Fields* Harper and Row, Vol. 1 Ch. 1 §2-4
[S2] Schrödinger E 1926 *Ann d Physik* **79** 361
[S3] Salpeter E E and Bethe H A 1951 *Phys Rev* **84** 1232

[S4] Sharp R and von Baeyer H 1966 *J Math Phys* **7** 1105
[S5] Schiff L 1955 *Quantum Mechanics* McGraw-Hill
[S6] Steinberger J 1949 *Phys Rev* **76** 1180
[S7] Sakurai J J 1969 *Currents and Mesons* Univ Chicago Press
[S8] Serebrov A P et al 1998 *JETP* **86** 1074
[S9] Sromicki J et al 1996 *Phys Rev* **C53** 932
[S10] Schreckenbach K et al 1995 *Phys Lett* **B349** 427
[S11] Stratowa Chr et al 1978 *Phys Rev* **D18** 3970
[S12] Shay D 1968 *L Nuovo Cim* **53** 1039
[S13] Somalwar S V et al 1992 *Phys Rev Lett* **68** 2580
[W1] van der Waerden B 1929 *Göttinger Nachrichten* 100
[W2] van der Waerden B 1932 *Die Gruppentheoretsche Methode in der Quantenmechanik* Springer, 80 ff
[W3] Weinberg S 1964 *Phys Rev* **B133** 1318
[W4] Wick G C 1954 *Phys Rev* **96** 1124
[W5] Weinberg S 1975 *Phys Rev* **D11** 3583
[Z1] Zweig G 1964 *CERN Report* Th. 401, 412 (unpublished)

INDEX